BMA

Medical Cell Biology

THIRD EDITION

ELSEVIER
science & technology books

:• *Companion Web Site:*

http://www.books.elsevier.com/companions/9780123704580

Medical Cell Biology, **Third Edition,** by Steven R. Goodman

Resources for Professors:

- **All figures from the book available as PowerPoint slides**
- **Links to web sites carefully chosen to supplement the content of the textbook**
- **Contact the editor with questions and/or suggestions**

TOOLS FOR ALL YOUR TEACHING NEEDS
textbooks.elsevier.com

ACADEMIC PRESS

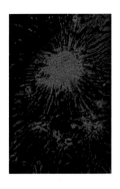

Medical Cell Biology

THIRD EDITION

Edited by Steven R. Goodman, PhD

C.L. and Amelia A. Lundell Professor of Life Sciences
The University of Texas at Dallas
Richardson, Texas
Adjunct Professor of Cell Biology
University of Texas Southwestern Medical Center
Dallas, Texas

AMSTERDAM • BOSTON • HEIDELBERG • LONDON
NEW YORK • OXFORD • PARIS • SAN DIEGO
SAN FRANCISCO • SINGAPORE • SYDNEY • TOKYO
Academic Press is an imprint of Elsevier

Academic Press is an imprint of Elsevier
30 Corporate Drive, Suite 400, Burlington, MA 01803, USA
525 B Street, Suite 1900, San Diego, California 92101-4495, USA
84 Theobald's Road, London, WC1X 8RR, UK

This book is printed on acid-free paper ∞

Library of Congress Cataloging-in-Publication Data
2007926489

British Library Cataloguing in Publication Data
A catalogue record for this book is available from the British Library.

ISBN 978-0-12-370458-0

For information on all Elsevier Academic Press publications, visit our Web site at
www.books.elsevier.com

Printed in China
09 10 11 12   9 8 7 6 5 4 3 2

I dedicate this third edition of *Medical Cell Biology* to
my family
wife, Cindy; sister, Sue; and children, Laela, Gena, Jessie, David, Christie, and Laurie
my friends
Obe, Ian, Charlie, Lynn, Santosh, Sandi, Rocky, Steve L., Stephen, Da Hsuan, and many others
my scientific heroes
Britton Chance, Aaron Ciechanover, Russell Hulse, and Alan MacDiarmid
and *my students,* past and current

Contents

Contributors

Mustapha Bahassi, PhD (Ch. 9)
Department of Cell Biology, Neurobiology and
Anatomy
University of Cincinnati Medical Center
Cincinnati, Ohio

Gail A. Breen, PhD (Ch. 4)
Department of Molecular and Cell Biology
University of Texas at Dallas
Richardson, Texas

John G. Burr, PhD (Ch. 1)
Associate Professor
Department of Molecular and Cell Biology
University of Texas at Dallas
Richardson, Texas

Santosh R. D'Mello, PhD (Ch. 10)
Professor
Department of Molecular and Cell Biology
University of Texas at Dallas
Richardson, Texas

Rockford K. Draper, PhD (Ch. 4)
Professor
Department of Molecular and Cell Biology and
Department of Chemistry
University of Texas at Dallas
Richardson, Texas

David Garrod, PhD (Ch. 6)
Professor of Developmental Biology
Faculty of Life Sciences
The University of Manchester
Manchester, England

Steven R. Goodman, PhD (Ch. 3)
Editor-in-Chief, *Experimental Biology and Medicine*
C.L. and Amelia A. Lundell Professor of Life
Sciences
Professor of Molecular and Cell Biology
University of Texas at Dallas
Richardson, Texas
Adjunct Professor of Cell Biology
University of Texas Southwestern Medical Center
Dallas, Texas

Frans A. Kuypers, PhD (Ch. 2)
Senior Scientist
Children's Hospital Oakland Research Institute
Oakland, California

Eduardo Mascareno, PhD (Ch. 8)
Department of Anatomy and Cell Biology
State University New York–Downstate
Brooklyn, New York

Stephen Shohet, MD (Clinical Cases)
Internal Medicine
San Francisco, California

M.A.Q. Siddiqui, PhD (Ch. 8)
Department of Anatomy and Cell Biology
State University New York–Downstate
Brooklyn, New York

Peter J. Stambrook, PhD (Ch. 9)
Department of Cell Biology, Neurobiology and
Anatomy
University of Cincinnati Medical Center
Cincinnati, Ohio

Michael Wagner, PhD (Ch. 8)
Department of Anatomy and Cell Biology
State University New York–Downstate
Brooklyn, New York

Danna B. Zimmer, PhD (Ch. 7)
Associate Professor of Veterinary Pathobiology
College of Veterinary Medicine & Biomedical
    Sciences
Department of Veterinary Pathobiology
Texas A&M University
College Station, Texas

Warren E. Zimmer, PhD (Ch. 3, Ch. 5)
Department of Systems Biology and Translational
    Medicine
College of Medicine
Texas A&M University, Health Science Center
College Station, Texas

Preface

The long-awaited third edition of *Medical Cell Biology* is here. It maintains the same vision as the first two editions, which is to teach cell biology in a medically relevant manner in a focused textbook of about 300 pages. We again accomplish this by focusing on human and animal cell biology, making clear the relationship of basic science to human disease. Our target audience for this textbook is health profession students (medical, osteopathic, dental, veterinary, nursing, and related disciplines) and advanced undergraduates who are future health professionals.

Although the vision remains the same, the third edition is very different from its predecessors. With the exceptions of Dr. Warren Zimmer and myself, we have an entirely new group of authors. In this edition, each chapter is written by an expert in the field, all of whom reside in different parts of the United States and England. The text, therefore, has been entirely rewritten and updated. Furthermore, this edition includes a new chapter on the important topic of cell death (Chapter 10). In addition, Chapters 2 through 10 each have two clinical vignettes that are relevant to cell biology, all of which have been beautifully written by Stephen Shohet, MD. We have stressed the importance of genomics and proteomics to our understanding of modern cell biology and medicine. We have taken a systems biology approach in several of our chapters. For example, Chapter 8, *Cell Signaling Events*, uses heart and cardiac disease to explain signaling; Chapter 9, *The Cell Cycle and Cancer*, is focused on cancer biology; and neuroscience and neurologic disorders are the platform for explaining cell death pathways in Chapter 10, *Programmed Cell Death*. All of the figures are either new or revised and are presented in full color. Academic Press has done a splendid job of helping us create an attractive and accessible textbook.

In summary, we are proud to present the third edition of *Medical Cell Biology*. The first two editions were very well received by the educational community, and we feel that the third edition is even better. We hope that lecturers will find the textbook to be an outstanding educational tool and that students will enjoy the readability of our book while they learn this fascinating material. As always, we welcome and appreciate your comments, all of which help us to make each edition better for future students.

I thank all of the authors of *Medical Cell Biology*, third edition, who have put great effort into creating a unique and beautifully crafted textbook.

STEVEN R. GOODMAN

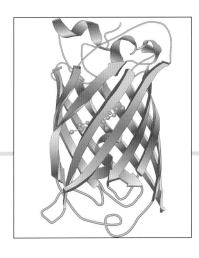

Chapter 1

Tools of the Cell Biologist

The Human Genome Project has revolutionized the study of cell biology, and it will continue to have a large impact on the practice of medicine in the decades to come. Approximately 25,000 protein-coding genes have been identified in the human genome.

Based on amino acid sequence homologies with proteins of known structure and function, some predictions can be made about the cellular roles of approximately 60% of these genes. But researchers are completely ignorant about the function of the proteins encoded by the remaining 40% of human genes because they have no identifiable sequence homologies to other proteins in the database. A major task for the future, therefore, will be to work out the functions of these thousands of novel proteins.

Because the cell is the fundamental unit of function in the organism, this project translates into searching for the functions of these newly discovered proteins in the life of a cell. The project, therefore, will be largely the task of cell biologists, using the powerful tools of modern molecular and cell biology. This chapter provides a brief review of some of these tools.

One of the first questions a cell biologist might ask in his or her search for a protein's function would be, "Where is it located in the cell?" Is it in the nucleus or the cytoplasm? Is it a surface membrane protein, or resident in one of the cytoplasmic organelles? Knowing the subcellular localization of a protein provides significant direction for further experiments designed to learn its function. (See Box 1–1 for a summary of cellular organelles and substructure.)

BOX 1–1. Organelles and Substructure of Mammalian Cells

The cell is the basic unit of life. Broadly speaking, there are two types of cells: prokaryotic and eukaryotic. Prokaryotes (eubacteria and archaea) do not have a nucleus; that is, their DNA is not enclosed in a special, subcellular compartment with a double membrane. Eukaryotic cells do have a nucleus; they are also much larger than prokaryotic cells and have numerous organelles and certain substructural elements not found in prokaryotes. The structural features of a generalized eukaryotic cell are shown in Figure 1–A.

The Nucleus

The nuclear compartment contains the **chromosomes,** the primary genetic material, as well as all the enzymes for transcribing chromosomal DNA into RNA, processing that RNA, and exporting it out to the cytoplasm; in addition, it contains all the transcription factors and chromatin remodeling factors required for regulating RNA transcription. It is surrounded by a **double membrane,** which is perforated at several thousand locations all over its surface by elaborate, protein-based pore structures (**nuclear pore**

Continued

complexes) that traverse the double membrane and regulate the entry into and exit from the nucleus of all proteins with sizes between approximately 17,000 and 60,000 daltons. Smaller molecules pass freely through the pores, whereas proteins larger than approximately 60,000 daltons are excluded. Certain large ribonucleoprotein complexes can apparently be actively deformed to permit passage through the pore. Subnuclear structures found in the nucleus include the **nucleolus** and numerous smaller structures called *Cajal bodies, gemini of coiled bodies (GEMS),* and *interchromatin granule clusters* ("speckles"). The function of the nucleolus is described in the following section; the functions of the smaller structures are less clearly understood but may include the dynamic assembly and regulation of small RNA and small nuclear ribonucleoprotein particles involved in processing and regulation of the expression of messenger RNA (mRNA) and ribosomal RNA (rRNA) molecules. The outer membrane of the nucleus is continuous with the membranes of the rough endoplasmic reticulum (ER).

The Nucleolus

The nucleolus is the most prominent substructural element observed within nuclei. Although structurally distinct, it is not surrounded by a membrane. It is the site of **transcription of the genes for rRNA** molecules, for which there are 400 genes in a diploid human cell, distributed in multiple tandem repeats on 5 different chromosomes. The rRNA gene segments on each of these five chromosomes assemble in the nucleolus for transcription, as do the various ribosomal proteins synthesized in the cytoplasm and various other proteins and ribonucleoproteins involved in processing rRNA. *Large and small ribosomes are fully assembled in the nucleolus,* and are then exported to the cytoplasm. The RNA and protein components of the enzyme **telomerase** are also assembled in the nucleolus.

Ribosomes

Ribosomes are the sites of **protein synthesis.** Eukaryotic ribosomes consist of a large (60S) and a small (40S) subunit. The large subunit consists of 3 RNA molecules (5S, 5.8S, 28S) associated with some 49 different proteins; the small subunit has a single RNA molecule (18S) and 33 proteins. In conjunction with a set of initiation factors, the 40S subunit binds first an initiator Met-transfer RNA (tRNA) molecule, and this complex then binds near the 5' end of an mRNA molecule. The large subunit is recruited, and the whole (80S) ribosome then sequentially reads the triplet codons of the mRNA, selecting the appropriate aminoacyl-tRNA molecules and ligating their associated amino acids to synthesize the protein encoded by that mRNA molecule.

Rough Endoplasmic Reticulum, the Golgi Apparatus, Transport Vesicles

The ER is a system of internal membranes that are continuous with the outer membrane of the nucleus. ER membranes nearest the nucleus are studded with ribosomes engaged in protein synthesis, and this portion of the ER is termed the **rough ER.** The ribosomes anchored to the rough ER are engaged in the *synthesis of either membrane proteins or proteins destined for secretion.* Such proteins posses a special amino-terminal signal sequence, which is recognized by a ribosome-associated particle (the "signal recognition particle"), and which then targets the ribosome with its nascent polypeptide chain to docking sites on the membrane of the rough ER. The nascent polypeptide chain is then cotranslationally extruded through a pore structure in the ER membrane and passes either partially or completely into the lumen of the rough ER. All such proteins become **glycosylated** at multiple locations along their length. Much of this glycosylation ("N-linked glycosylation") occurs on the nascent polypeptide as it passes into the lumen of the ER; the remainder ("O-linked glycosylation") occurs later, either in the lumen of the ER or in the various compartments of the Golgi apparatus. After their synthesis is complete, the proteins find their way to an adjacent region of **smooth ER** (a specialized portion of smooth ER known as **transitional ER**), from which **transport vesicles** containing the proteins bud off, and then deliver their protein cargo to the **Golgi apparatus** by fusing to form the *cis*-Golgi network. In the cis, **medial,** and **trans cisternae** of the Golgi apparatus, the oligosaccharide chains on these glycoproteins are modified in a variety of ways, and some proteins are cleaved or otherwise processed. The processed glycoproteins then leave the Golgi apparatus via vesicles that bud from the **trans-Golgi network,** for delivery to the cell surface.

Smooth Endoplasmic Reticulum

The smooth ER is a continuous extension of the rough ER, located more distally from the nucleus. Whereas the rough ER is shaped like flattened hollow pancakes in many cell types, the smooth ER is usually more tubular in structure, forming a lacelike reticulum. It is an important site of **lipid metabolism** (e.g., cholesterol biosynthesis), and, for example, in liver cells, is the site where various membrane-associated **detoxifying enzymes** (e.g., cytochrome P450 enzymes) oxidize and otherwise act to modify toxic hydrophobic molecules (e.g., phenobarbital), making them less toxic and more water soluble.

The lumen of the smooth ER also serves as an *important storage site for intracellular Ca^{2+}.* Smooth ER membranes contain ligand-regulated Ca^{2+} channels that open in response to the hormone-generated second messenger inositol 1,4,5-triphosphate (IP_3). The cytosol of all cells is virtually Ca^{2+} free under resting conditions, and the *transient appearance of Ca^{2+} in the cytosol after its release from the ER stores serves to initiate any of a number of cellular responses to extracellular signals,* depending on the cell type. The ER membrane also possesses numerous Ca^{2+} pumps that bring the transiently released Ca^{2+} back into the ER lumen. Muscle contraction is initiated by transient release of Ca^{2+} from a specialized form of smooth ER in muscle fibers, known as the **sarcoplasmic reticulum.**

Clathrin-Coated Pits, Clathrin-Coated Vesicles, Early and Late Endosomes

The receptors for certain extracellular protein ligands (low-density lipoprotein cholesterol particles, iron-bearing transferrin) are clustered in, or become recruited to, specialized dimple-like structures scattered over the surface of the cell. These dimples have an underlying hemibasket structure composed of oligomers of the protein **clathrin,** and are termed **clathrin-coated pits.** Binding of their ligands to these receptors is the first step in the process known as **receptor-mediated endocytosis,** in which polymerization of the clathrin monomers to form a spherical basket leads to the formation of an internalized, **clathrin-coated vesicle** derived from the surface membrane and its associated transmembrane receptor proteins with their ligands. Depolymerization of the clathrin coat follows, and because they have proton pumps in their membranes, the resulting uncoated vesicles begin to acidify and soon mature into structures known as **early endosomes.** The acid pH (pH 6) causes dissociation of receptor and ligand, and empty receptors are returned to the cell surface for reuse, via vesicles that bud off from the early endosome. Early endosomes become **multivesicular bodies** and continue to acidify, eventually becoming **late endosomes.** Finally, by fusing with special vesicles derived from the Golgi apparatus that contain a large variety of hydrolytic enzymes, the late endosomes mature into structures called **lysosomes** (see the following section). Alternatively, late endosomes may fuse with existing lysosomes.

Another type of "dimple" found on the cell surface is a flask-shaped structure called a *caveola* (*pl.* **caveolae**); instead of clathrin, caveolae are associated with a multipass integral membrane protein called *caveolin.* The membranes of caveolae are rich in cholesterol and sphingolipids, and are closely related to small evanescent lipid structures found in the bulk plasma membrane called **lipid rafts.** Many growth hormone receptors are concentrated in caveolae. In certain cell types, caveolae pinch off from the surface to form vesicles, and these vesicles can traverse the cell and fuse with the membrane on the opposite side of the cell, a process termed **transcytosis.**

Lysosomes

Lysosomes are membrane-enclosed, acidic (pH 5) compartments of heterogeneous size and shape that *contain more than 40 different kinds of hydrolytic enzymes,* all of which are optimally active in the acid pH of the lysosome, but have little activity at pH 7. Lysosomal hydrolases are glycoproteins that are synthesized by rough ER–associated ribosomes and are processed in the Golgi, where they are given a mannose-6-phosphate tag that targets them to lysosomes. They are capable of breaking down all the different kinds of biological macromolecules and are *responsible for the degradation of endocytosed or phagocytosed material.* Furthermore, via the process of **autophagy,** lysosomes play an essential role in the *normal turnover of all cellular macromolecules.* The amino acids, sugars, nucleotides, and so forth generated by macromolecule breakdown are transported out of the lysosome to the cytosol, for reuse. Defects in lysosomal hydrolases are responsible for a class of inherited diseases termed **lysosomal storage diseases** (e.g., Tay–Sachs disease, Gaucher's disease, Niemann–Pick disease), in which lysosomes fill with indigestible material.

Peroxisomes

Peroxisomes are small cellular organelles that play an important role in the **oxidation of cellular lipids,** especially fatty acids derived from membrane lipids. Unlike mitochondrial oxidation of fatty acids, which can produce CO_2 and adenosine triphosphate (ATP), the peroxisomal oxidation process, termed β-**oxidation,** degrades the hydrocarbon chain two carbon units at a time, yielding acetyl molecules that are transported back out to the cytosol for use in biosynthetic reactions. β-Oxidation, which is not coupled to ATP synthesis, can also occur in mammalian mitochondria, but peroxisomes are the chief site of this process in all cells, and it is only in peroxisomes that **long and very long chain fatty acids,** derived from certain membrane lipids, are oxidized. The oxidizing enzymes in peroxisomes use molecular oxygen, which is then converted to hydrogen peroxide (H_2O_2). Consequently, peroxisomes have abundant levels of the enzyme **catalase,** which uses H_2O_2 to oxidize a variety of other molecules; in this process, H_2O_2 is reduced to water. Liver peroxisomal catalase is responsible for the metabolism of a significant amount of dietary alcohol.

In addition to their important role in fatty acid oxidation, peroxisomes also have biosynthetic roles, for example, in the synthesis for certain glycerolipids. The first reactions in the synthesis of the glycerolipid **plasmalogen,** involving the synthesis of a unique ether linkage to the glycerol backbone, are catalyzed in peroxisomes, after which synthesis is completed in the cytosol. Plasmalogen makes up approximately half of the heart's phospholipids and approximately 80% to 90% of the ethanolamine phospholipid class in myelin. Defects in peroxisome function are the causes of inherited diseases such as **X-linked adrenoleukodystrophy** and **Zellweger syndrome.**

Mitochondria

Mitochondria are the *major source of ATP synthesis* in cells during aerobic respiration. They are organelles with a double membrane, approximately the size of a bacterium. In fact, they originated from symbiotic bacteria that came to reside in the cytoplasm of an ancient ancestor to today's eukaryotic cells. They retain certain bacterial features such as a circular DNA molecule and ribosomes with strikingly prokaryotic features. The mitochondrial inner membrane is highly invaginated, forming folded structures called **cristae** that protrude into the lumen (**matrix**) of the mitochondrion. The reactions of the **citric acid cycle** occur in the matrix, generating high-energy NADH and NADPH molecules, which in turn transfer their electrons to acceptor molecules located in the inner membrane; the electrons are then passed along a set of **electron carriers** to O_2, which thereby becomes reduced to H_2O. Electron transport in the inner membrane causes the accumulation of protons in the space between the inner and outer membranes, thereby producing an electrochemical potential across the inner membrane; **ATP synthase** molecules located in the inner membrane provide a channel for the return of these protons to the matrix compartment, thereby driving the synthesis of ATP, a process known as **oxidative phosphorylation.**

One of the carriers in electron transport is a molecule called **cytochrome *c*,** a small, soluble protein located in the space between the inner and outer membranes. Several years ago, it was discovered that cytochrome *c* also plays

Continued

an important role in initiating the process of programmed cell death (**apoptosis**) (see Chapter 10). In response to any of a number of circumstances, cells generate molecules (certain proapoptotic members of the Bcl-2 family of proteins) that create pores in the outer membrane of mitochondria, permitting the release of cytochrome *c* (and other apoptosis-inducing proteins) into the cytosol. Cytochrome *c* binds to a protein called Apaf-1, which in turn activates a cascade of **caspase proteases,** leading to cell death.

The Cytoskeleton

The cytoskeleton consists of three types of filamentous protein polymers, in equilibrium with a pool of subunit monomers. The three types of filaments are (in increasing order of diameter): **microfilaments, intermediate** filaments, and **microtubules.** The subunit protein of microfilaments is a small, monomeric protein called **actin;** that of microtubules is a dimeric molecule called **tubulin** (α-tubulin + β-tubulin). Intermediate filaments are heteropolymers, whose subunits vary among the various cell types in different tissues. The subunit proteins of intermediate filaments include proteins with names such as **vimentin, desmin, lamin** (lamins A, B, C), **keratin** (multiple acidic and basic keratins), **neurofilament proteins** (NF-L, NF-M, NF-H), among others.

Microtubules and microfilaments can polymerize and depolymerize dynamically in particular locations within the cell, and they also participate with various partner **motor proteins (kinesins, dyneins, myosins)** to produce cellular motility and contractility phenomena. Intermediate filaments are important in the overall structural toughness of cells and in distributing shear forces throughout one or more cells in a tissue. The nuclear lamins form a tough, resilient polymeric net around the inner surface of the nucleus.

The inherited disease **epidermolysis bullosa simplex,** whose phenotype is painful blistering in response to a light touch, is caused by a defective keratin gene.

Centrioles

Centrioles are a pair of barrel-shaped structures arranged perpendicularly to each other. The sides of each barrel are made up of nine loose, overlapping "slats"; each slat is a flat sheet of three parallel microtubules. The centriole pair is imbedded in an amorphous halo of incompletely characterized proteins. The entire unit, centriole pair plus halo, is termed the *centrosome*. The halo component of the centrosome contains multiple ring structures formed by an isoform of tubulin called γ-*tubulin*. γ-Tubulin rings nucleate the polymerization of microtubules, and most cellular

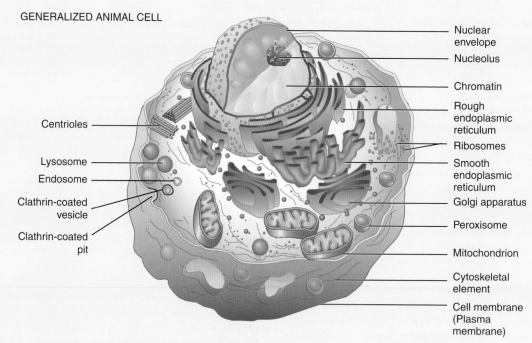

GENERALIZED ANIMAL CELL

Centrioles

Lysosome

Endosome

Clathrin-coated vesicle

Clathrin-coated pit

Nuclear envelope

Nucleolus

Chromatin

Rough endoplasmic reticulum

Ribosomes

Smooth endoplasmic reticulum

Golgi apparatus

Peroxisome

Mitochondrion

Cytoskeletal element

Cell membrane (Plasma membrane)

Figure 1–A. Structural features of animal cells. Summary of the functions of cellular organelles. **Mitochondria:** (1) Site of the Krebs (citric acid) cycle; produce ATP by oxidative phosphorylation. (2) Can release apoptosis-initiating proteins, such as cytochrome c. **Cytoskeleton:** Made up of microfilaments, intermediate filaments, and microtubules; governs cell movement and shape. **Centrioles:** Components of the microtubule organizing center. **Plasma membrane:** Consists of a lipid bilayer and associated proteins. **Nucleus:** Contains chromatin (DNA and associated proteins), gene-regulatory proteins, and enzymes for RNA synthesis and processing. **Nucleolus:** The site of ribosome RNA synthesis and ribosome assembly. **Ribosomes:** Sites of protein synthesis. **Rough ER, Golgi apparatus, and transport vesicles:** Synthesize and process membrane proteins and export proteins. **Smooth ER:** Synthesizes lipids and, in liver cells, detoxifies cells. **Lumen:** Ca++ reservoir. **Clathrin-coated pits, clathrin-coated vesicles, early and late endosomes:** Sites for uptake of extracellular proteins and associated cargo for delivery to lysosomes. **Lysosomes:** Contain digestive enzymes. **Peroxisomes:** Cause β-oxidation of certain lipids (e.g., very long chains of fatty acids). *(Modified from Freeman S., Biological Science, 1st ed., Upper Saddle River, NJ: Prentice Hall, 2002.)*

tubules originate in the centrosome, which is located close to the nucleus.

Plasma Membrane

The surface "skin" of the cell, termed the *plasma membrane,* consists of a phospholipid bilayer and associated proteins. The phospholipid bilayer is intrinsically impermeable to charged and all except the smallest hydrophilic solutes; movement across the membrane of such solutes is governed by a set of transmembrane proteins that function as channels and transporters, which function either to facilitate diffusion of certain molecules down their concentration gradients across the membrane or to actively move molecules into or out of the cell against their concentration gradient. Other membrane proteins mediate adhesion of cells to each other or to elements of the extra-

cellular matrix. Members of a third category of proteins serve as receptors for extracellular signaling molecules and initiate a cellular response to such molecules.

Cytoplasm versus Cytosol

The cytoplasm of the cell is all the material outside of the nucleus. On occasion, it is necessary to distinguish between the cytosol and the cytoplasm. The **cytosol** is defined as all the material in the cytoplasm, *excluding the contents of the various membranous organelles.* The cytosol, therefore, does include the cytoskeleton, the ribosomes, and the centrosome, together with all the other macromolecules and solutes outside the nucleus and also outside the lumen of the various cytoplasmic membranous organelles (mitochondria, ER, Golgi, transport vesicles, endosomes, and so forth).

One of the primary tools a cell biologist would use to answer the question of subcellular location would be a microscope.

MICROSCOPY: ONE OF THE EARLIEST TOOLS OF THE CELL BIOLOGIST

Microscopy, in its various forms, has historically been the primary way in which investigators have examined

the appearance and substructure of cells, and increasingly in recent decades, the location and movement of biological molecules within cells. We may speak broadly of two kinds of microscopy, **light microscopy** and **electron microscopy (EM)**, although the field of microscopy recently has been broadened by the advent of **atomic force microscopy.**

BOX 1–2. Resolution and Magnification in Microscopy

The two properties that define the usefulness of a microscope are **magnification** and **resolution.** Light microscopes use a series of glass lenses to magnify the image; electron microscopes use a series of magnets to produce the magnified image (Fig. 1–B).

However, because of the wave nature of light (and of electrons), light waves arriving at the focal point produce a magnified image in which different wave trains are either in or out of phase, amplifying or canceling each other to produce interference patterns. This phenomenon, known as **diffraction,** results in the image of straight edge appearing as a fuzzy set of parallel lines, and that of a point as a set of concentric rings (Fig. 1–C).

This fundamental limit on the clarity of an optical image, known as the limit of **resolution,** is defined as the minimum distance (**d**) between two points such that they can be resolved as two separate points. In 1873, Ernst Abbé showed that the limit of resolution for a particular light microscope is directly proportional to the **wavelength of light** used to illuminate the sample. The smaller the wavelength of light, the smaller is the value of d, that is, the better is the resolution of the magnified image. Abbé also showed that resolution is affected by two other features of the system: (1) the **light-gathering properties** of the microscope's objective lens, and (2) the **refractive index of the medium** (e.g., air or oil) between the objective lens and the sample. The light-gathering properties of the objective lens depend on its focal length, which can be

characterized by a number called the **angular aperture, α,** where α is the half angle of the cone of light entering the objective lens from a focal point in the sample (Fig. 1–D). These three parameters were quantified by Abbé in the following equation:

$$d = 0.61\lambda/n\sin\alpha,$$

where d is the resolution, λ is the wavelength of illuminating light, n is the refractive index of the medium between the objective lens and the sample, and α is the angular aperture. The denominator term in this equation (n sin α) is a property of the objective lens termed its **numerical aperture (NA)**. Because sin α has a limit approaching 1.0, and the refractive index of air is (by definition) 1.0, the best nonoil objective lenses will have numerical apertures approaching 1.0 (e.g., 0.95); because the refractive index of mineral oil is 1.52, the numerical apertures of the best oil immersion objective lenses will approach 1.5 (typically 1.4).

With a light microscope under optimal conditions, using blue light (λ approximately 400 nm [0.4 μm]), and an oil-immersion lens with a numerical aperture of 1.4, the limit of resolution will therefore be approximately 0.2 μm. This is approximately the diameter of a lysosome, and a resolution of 0.2 μm is approximately 1000-fold better than the resolution that can be attained by the unaided human eye.

Continued

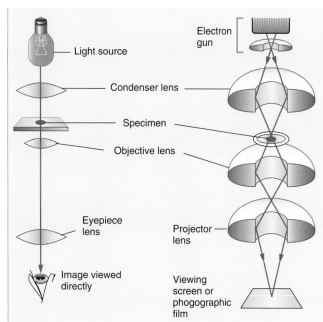

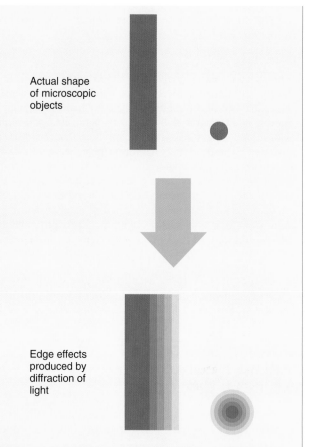

Figure 1–B. Comparison of the lens systems in a light microscope and a transmission electron microscope. In a light microscope (left), light is focused on the sample by the condenser lens. The sample image is then magnified up to 1000 times by the objective and ocular lenses. In a transmission electron microscope (right), magnets serve the functions of the condenser, objective and ocular (projection) lenses, focusing the electrons and magnifying the sample image up to 250,000 times. *(Modified from Alberts B, et al. Molecular Biology of the Cell, 4th ed. New York, NY: Garland Science, 2002.)*

Figure 1–C. Light passing through a sample is diffracted, producing edge effects. When light waves pass near the edge of a barrier, they bend and spread at oblique angles. This phenomenon is known as diffraction. Diffraction produces edge effects because of constructive and destructive interference of the diffracted light waves. These edge effects limit the resolution of the image produced by microscopic magnification. *(Modified from Alberts B, et al. Molecular Biology of the Cell, 4th ed. New York, NY: Garland Science, 2002.)*

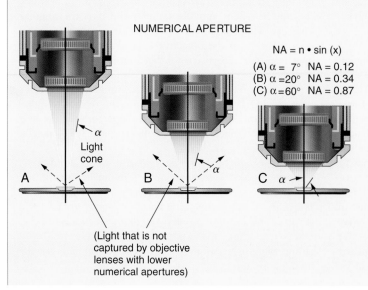

Figure 1–D. Numerical aperture (NA). The NA of a microscopic objective is defined as $n \sin \alpha$, where n is the refractive index of the medium between the sample and the objective lens (air), and α, the angular aperture, is the half angle of the cone of light entering the objective lens from a focal point in the sample. Objective lenses with increasingly high NA values (A, B, C) collect increasingly more light from the sample. *(Modified from* http://www.microscopyu.com/articles/formulas/formulasna.html.*)*

The best light microscopic images, with a resolution of 0.2 μm, can be magnified to any desired degree (i.e., photographically), but no further information will be gained. No further increase in resolution beyond 0.2 μm can be obtained with a standard light microscope, and any further magnification of the sample image would be **empty magnification,** devoid of additional information content.

In a *transmission electron microscope* with an accelerating voltage of 100,000 V, electrons are produced with wavelengths of approximately 0.0004 nm. The effective numerical aperture of an electron microscope is small (on the order of 0.02), but from Abbé's equation, one would predict a resolution on the order of 0.1 nm. Practical limitations related to sample preparation, sample thickness, contrast, and so forth result in actual resolutions on the order of 2 nm (20 Å). This is approximately *100-fold better than the resolution of light microscopy.*

The resolution of standard light microscopy is limited by the wavelength of visible light, which is comparable with the diameter of some subcellular organelles (see Box 1–2); but a variety of contemporary techniques now exist that permit light microscopic visualization of proteins and nucleic acid molecules. Chief among these new techniques are those using either organic fluorescent molecules or quantum nanocrystals ("quantum dots") to directly or indirectly "tag" individual macromolecules. Once the molecules of interest have been fluorescently tagged, their cellular location can be viewed via **fluorescence microscopy** (Fig. 1–1).

location of particular proteins. One widely used technique to fluorescently tag a protein is based on the great precision and high affinity with which an antibody molecule can bind its cognate protein antigen. This antibody-based approach has been termed **immunolabeling** (see Box 1–3 for a brief summary of the structure and function of antibodies). Because antibodies are relatively large molecules that do not cross the surface membrane of living cells, one must fix and permeabilize cells before an antibody can be used to view the location of a target protein.

Fluorescence Microscopy

In many situations, fluorescence microscopy is the first approach one might take to identify the subcellular

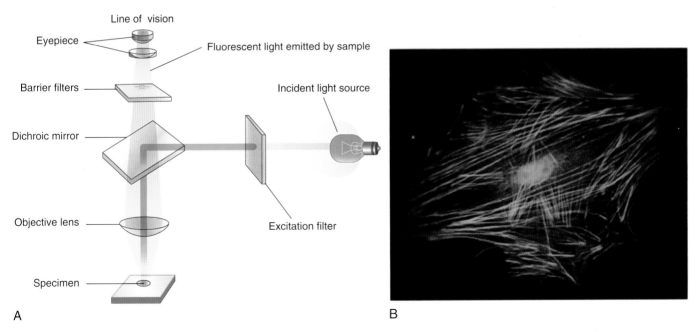

A B

Figure 1–1. **Fluorescence microscopy. A:** Optical layout of a fluorescence microscope. Incident light tuned to excite the fluorescent molecule is reflected by a dichroic mirror, and then focused on the sample; fluorescent light (longer wavelength than excitation light) emitted by the sample passes through the dichroic mirror for viewing. **B:** Immunofluorescent micrograph of a human skin fibroblast, stained with fluorescent anti-actin antibody. Cells were fixed, permeabilized, and then incubated with fluorescein-coupled antibody. Unbound antibody was washed away before viewing. *(A: Modified from Lodish H, Berk A, Zipursky SL, Matsudaira P, Baltimore D, Darnell J. Molecular Cell Biology, 4th ed. New York, NY: W.H. Freeman, 2000; B: Courtesy E. Lazerides.)*

BOX 1–3. Antibodies

Antibodies, also known as **soluble immunoglobulins,** are specialized proteins that play an important role in immunity because of their ability to bind tightly to the foreign molecules (**antigens**) expressed by pathogens that infect an individual. An antibody molecule is a Y-shaped protein, consisting of two identical **heavy chains,** plus two identical **light chains** (Fig. 1–E). The disulfide-bonded, carboxyl-terminal halves of the heavy chains (the "tail" of the antibody) are jointly called the **Fc domain;** the two arms, which bind antigens at their tips, are called the **Fab** domains.

Immunoglobulins are synthesized by a type of lymphocyte called a **B cell,** and are initially expressed as transmembrane proteins on the surface of each B cell, where they are termed **surface immunoglobulin M** (surface IgM). (A small amount of a surface immunoglobulin called **IgD** is also expressed by B cells.) Each of the millions of B cells produced by the bone marrow each day makes an immunoglobulin with a unique binding specificity. The unique binding specificity of an immunoglobulin is determined by the unique amino acid sequence (called the *variable sequence*) located at the amino-terminal end of both the heavy and the light chains of each immunoglobulin molecule.

Should a particular B cell encounter its cognate antigen, that B cell first proliferates and then differentiates into an antibody-secreting **plasma cell.** Some of the proliferating B cells differentiate early into plasma cells and secrete soluble IgM. Soluble IgM is a pentameric molecule and often has relatively weak binding affinity; sibling B cells differentiate later, after undergoing the processes of **somatic cell hypermutation** and **class switching.** During the process of somatic cell hypermutation, the DNA encoding the variable regions of the immunoglobulin chains is selectively mutated, and cells expressing mutated, higher affinity immunoglobulin are then selected. "Class switching" refers to the process whereby the gene segment

encoding an IgM-type Fc domain (Fcμ), initially expressed in all B cells, is switched out for a different gene segment, encoding a different Fc domain. Any one of three different gene segments, each encoding a different Fc domain, can be chosen to replace the Fcμ segment in the B-cell immunoglobulin gene, such that any one of three different kinds (classes) of antibody are secreted by the plasma cell after this process of class switching. These three classes of antibody are called **IgG, IgA,** and **IgE.** The class of antibody expressed depends on the identity of the pathogen causing the infection. IgE, for example, is most effective against many parasites; IgA protects against mucosal infections; and IgG is effective against many types of pathogens and is the most abundant immunoglobulin in blood. Each of these four classes of antibody (IgM, IgG, IgA, IgE) has a characteristic amino acid sequence in its Fc domain that distinguishes it from the other three classes, and each of the four Fc domains has unique effector functions that activate specific features of the immune system after binding of the antibody to its cognate antigen. The differentiation of B cells into plasma cells occurs in **secondary lymphoid tissue** such as the **lymph nodes** and the **spleen.**

The particular molecular structure on an antigen to which an antibody binds is called an **epitope.** When the antigen is a protein, the epitope typically consists of several adjacent amino acids. Injection of a foreign protein into an experimental animal typically elicits the differentiation of multiple B cells into corresponding clones of descendant plasma cells, each member of a plasma cell clonal population secreting a particular antibody that binds to just one of the multiple possible epitopes on the surface of the antigenic protein. Serum collected from an immunized animal will therefore contain a mixture of antibodies against the immunizing foreign protein, and such serum is called a **polyclonal antiserum.** This polyclonal mixture of antibodies can be purified from an antiserum and used

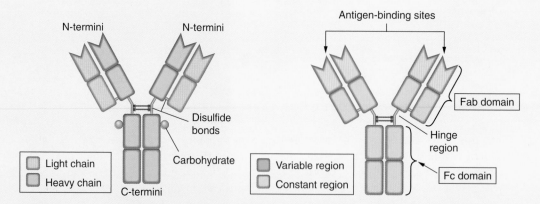

Figure 1–E. **Structure of an antibody molecule.** An antibody molecule consists of two identical heavy chains, plus two identical light chains. The disulfide-bonded, carboxyl-terminal halves of the heavy chains (the "tail" of the antibody) are jointly called the *Fc domain;* the two arms, which bind antigens at their tips, are called the *Fab domains.* Because all immunoglobulins are modified by the attachment of carbohydrate, they are examples of a type of protein termed a *glycoprotein.* The immunoglobulin shown here is an IgG molecule; class M, A, and E immunoglobulins are roughly similar; except IgM and IgE have larger Fc domains. The different immunoglobulin classes are also glycosylated at different sites. *(Modified from Parham P. The Immune System, 2nd ed. New York, NY: Garland Publishing, 2005.)*

for a variety of experimental purposes, such as Western blotting and immunofluorescence microscopy, among others.

But for many medical and diagnostic purposes, it is useful to have a preparation of pure antibodies directed against a single epitope. Such antibodies could be obtained if one had a single clone of plasma cells, able to grow indefinitely in culture and secreting a single antibody (a **monoclonal antibody**). Because plasma cells or their B-cell precursors, or both, have a limited proliferation potential, primary cultures of plasma cells have limited usefulness for the routine production of monoclonal antibodies. However, one can fuse such cells with a special line of cancerous lymphocytes called **myeloma cells.** Such myeloma cells are "immortal," that is, able to grow indefinitely in culture. The hybrid cells obtained from such a fusion, termed **hybridoma** cells, produce monoclonal antibody, like the B-cell/plasma cell parent, and yet proliferate indefinitely in culture, like the myeloma parent. In the practical application of this technique, a mouse is immunized with a particular antigen, for example, protein X; after several boosts, the animal is killed, and the mix of activated B cells and plasma cell precursors in its spleen are harvested. After fusion with myeloma cells and selection for hybridoma cells in a special selection medium (in which unfused parent cells either die or are killed), the particular hybridoma colony producing a monoclonal antibody of interest is then identified (Fig. 1–F).

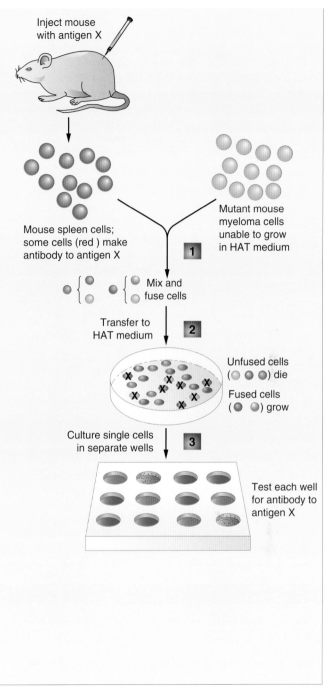

Figure 1–F. Monoclonal antibodies. *1:* Myeloma cells are fused with antibody-producing cells from the spleen of an immunized mouse. *2:* The mixture of fused hybridoma cells together with unfused parent cells are transferred to a special growth medium (HAT medium) that selectively kills the myeloma parent cells; unfused mouse spleen cells eventually die spontaneously because of their natural limited proliferation potential. Hybridoma cells are able to grow in HAT medium and have the unlimited proliferation potential of their myeloma parent. *3:* After selection in HAT medium, cells are diluted and individual clones growing in particular wells are tested for production of the desired antibody. *(Modified from Lodish H, Berk A, Matsudaira P, Kaiser CA, Krieger M, Scott MP, Zipursky SL, Darnell J. Molecular Cell Biology, 5th ed. New York, NY: W.H. Freeman, 2004.)*

In recent years, it has become possible to view the location and movement of fluorescently tagged proteins inside **living cells,** using an approach that has been broadly termed **genetic tagging.** With this approach, one uses genetic engineering to create a plasmid expressing the protein of interest, which has been fused at its amino or carboxy terminus with either a directly fluorescent tag, such as **green fluorescent protein (GFP),** or an indirect fluorescent tag, such as tetra-cysteine. Tetra-cysteine–tagged proteins when expressed in cells can bind subsequently added small, membrane-permeable fluorescent molecules such as the red or green biarsenicals FlAsH and ReAsH. The lines between immunolabeling and genetic tagging blur when one considers another type of genetic tagging, termed **epitope-tagging,** in which the recombinant protein is expressed with an antigenic amino acid sequence at one of its ends, to which commercial antibodies are readily available, such as a "myc-tag."

Let us first consider immunolabeling in more detail, and then consider genetic labeling, using the example of GFP.

Immunolabeling

Specific antibodies directed against the protein of interest, used in combination with either light microscopy or EM, are useful tools for discovering the subcellular location of the protein.

A fluorescent tag (e.g., fluorescein) can be chemically coupled to the Fc domain of antibody for use in fluorescent light microscopy. For use in transmission EM, an electron-dense tag such as the iron-rich protein ferritin or nanogold particles can be coupled to the antibody. These two techniques are referred to as immunofluorescent microscopy and immunoelectron microscopy, respectively. Figure 1–1B shows an example of the use of immunofluorescence to visualize the actin "stress fibers" in a fibroblast; an example of immunoelectron microscopy is shown later in Figure 1–4.

So how would one go about obtaining antibodies to a particular protein?

Antipeptide Antibodies

One way to obtain antibodies here would be to chemically synthesize peptides corresponding to the predicted amino acid sequence of the protein product of the gene of interest. One would then chemically couple these peptides to a carrier protein, such as serum albumin or keyhole limpet hemocyanin (commonly used), and then immunize an animal such as a rabbit with the peptide-carrier complex.

This approach has one potential problem. If dealing with one of the newly discovered human genes whose protein product is completely uncharacterized, one would not have any information about the three-dimensional structure of this protein. Consequently, one would not know whether any particular amino acid sequence chosen for immunization purposes would be exposed on the surface of the native, folded protein as found in a cell. If the selected peptide corresponded to an amino acid sequence that is buried in the interior of the folded structure, antibodies directed against it would not be able to bind the native protein in the fixed cell preparations one would be using for microscopy. It turns out that amino- or carboxyl-terminal amino acid sequences are frequently exposed on the surface of many natively folded proteins; for this reason, peptides corresponding to these terminal sequences are frequently chosen for immunization of rabbits. Also, hydrophilic sequences are generally found on the surface of folded proteins, and if one or more such sequences can be identified in the predicted amino acid sequence of the protein of interest, they too would be good candidates for immunization.

Because of the preceding considerations, antipeptide antibodies are not always successful in immunofluorescent localization experiments, where the target protein is in a native configuration. They are, however, often useful for the technique of **Western blotting** (see Box 1–4).

BOX 1–4. Standard Techniques for Protein Purification and Characterization

Proteins differ from each other in size and overall charge at a given pH (dependent on a property of a protein called its isoelectric point). Although other features of a protein can be used as a basis for purification (hydrophobicity, posttranslational modifications such as glycosylation or phosphorylation, ligand-binding properties, and so on), size and charge are the basis for several standard techniques for protein purification and characterization.

The most widely used technique for protein purification is **liquid chromatography,** in which an impure mixture of proteins containing a protein of interest (e.g., a cell extract in a buffered aqueous solvent with a defined salt concentration) is layered on top of a porous column filled with a packed suspension of fine beads with specific properties of porosity, charge, or both; the column itself is equilibrated in the same or a comparable solvent (Fig. 1–G). A "developing" solvent is then percolated through the column, carrying with it the mixture of proteins. Because of the properties of the beads, and/or the nature of the developing solvent, the various proteins pass through the column at differing rates, and the mixture is thereby resolved. Two commonly used types of beads employed resolve proteins either by size (**gel-filtration chromatography**) or by charge (**ion-exchange chromatography**). In a third approach (**affinity chromatography**), the beads can be derivatized with a molecule to which the protein of interest specifically binds; if that protein were an enzyme, for example, the beads could be coated with a substrate analog to which the enzyme tightly binds; more commonly, genetically engineered proteins have tags such as

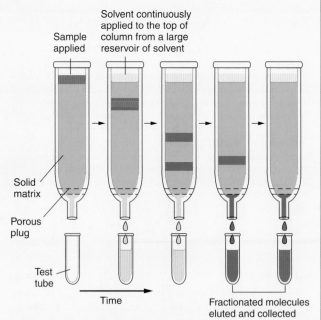

Figure 1–G. Column chromatography. A porous column of beads equilibrated in a particular solvent is prepared, and a sample containing a mixture of proteins is applied to the top of the column. The sample is then washed through the column, and the column eluate is collected in a succession of test tubes. Because of the properties of the beads in the column, proteins with different properties elute at varying rates off the column. *(Modified from Alberts B et al.,* Molecular Biology of the Cell. *New York, NY: Garland Science, 2002.)*

"6× histidine" or "glutathione S-transferase" (GST), which specifically bind to beads derivatized with Ni^{2+}-nitriloacetic acid and glutathione, respectively (Fig. 1–H). In other cases, the beads might be covered with an antibody directed against the desired protein. This is called **immunoaffinity chromatography.**

A related analytic application of these principles of protein resolution (size, charge) is the several techniques of gel electrophoresis. Electrophoresis is the movement of molecules under the influence of an electric field. A widely used analytic gel electrophoresis technique is called **sodium**

dodecyl sulfate polyacrylamide gel electrophoresis (SDS-PAGE). Polyacrylamide gels can be cast with any desired degree of porosity such that a small protein with a net charge migrates readily through the gel matrix toward an electrode, whereas a larger protein migrates more slowly through the matrix. The native charge on a protein can be the basis for its electrophoretic mobility in a gel, but it is more convenient to denature proteins with the negatively charged detergent **SDS.** SDS molecules have a hydrophobic hydrocarbon "tail," and a hydrophilic, anionic sulfate "head." The SDS molecules unfold proteins by binding via

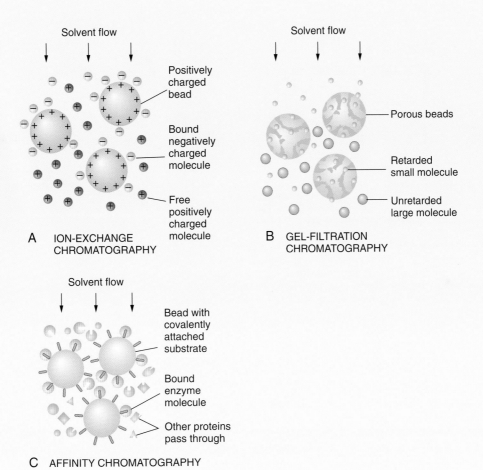

Figure 1–H. Three types of beads used for column chromatography. **A:** The beads may have a positive or a negative charge. (The positively charged beads shown in the figure might, for example, be derivatized with diethylaminoethyl groups, which are positively charged at pH 7.) Proteins that are positively charged in a pH 7 buffer will flow through the column; negatively charged proteins will be bound to the beads and can be subsequently eluted with a gradient of salt. **B:** The beads can have cavities or channels of a defined size; proteins larger than these channels will be excluded from the beads and elute in the "void volume" of the column; smaller proteins of various sizes will, to varying degrees, enter the beads and pass through them, thereby becoming delayed in their elution from the column. Such columns, therefore, resolve proteins by size. **C:** The beads can be derivatized with a molecule that specifically binds the protein of interest. In the example shown, it is a substrate (or substrate analog) for a particular enzyme; the beads could also be derivatized with an antibody to the protein of interest, in which case this would be called *immunoaffinity chromatography*. (*Modified from Alberts B et al.,* Molecular Biology of the Cell. *New York, NY: Garland Science, 2002.*)

Continued

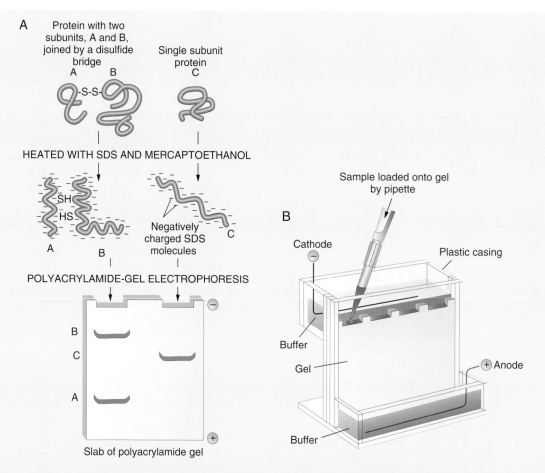

Figure 1–I. Sodium dodecyl sulfate polyacrylamide gel electrophoresis (SDS-PAGE). **A:** Proteins in the sample are heated with the negatively charged detergent SDS, which unfolds them and coats them with a uniform negative charge density; disulfide bonds (S-S) are reduced with mercaptoethanol. **B:** The sample is applied to the well of polyacrylamide gel slab, and a voltage is applied to the gel. The negatively charged detergent-protein complexes migrate to the bottom of the gel, toward the positively charged anode. Small proteins can move more readily through the pores of the gel, but larger proteins move less readily, so individual proteins are separated by size, smaller toward the bottom and larger toward the top. *(Modified from Alberts B et al.,* Molecular Biology of the Cell, *4th ed. New York, NY: Garland Science, 2002.)*

their tails at closely spaced intervals along the length of the polypeptide chain. The negative charge of the many bound SDS molecules overwhelms the intrinsic charge of a protein, and thereby gives all proteins a uniform negative charge density. SDS-denatured proteins therefore migrate as polyanions through a polyacrylamide gel *by their size* toward the positive electrode (the anode). At the end of the electrophoretic separation, smaller proteins will be found near the bottom of the gel, and larger proteins near the top (Fig. 1–I).

A useful application of SDS-PAGE is the technique of **Western blotting (immunoblotting)**. This application resolves a mixture of proteins by SDS-PAGE, and then transfers the resolved set of proteins to a special paper, such as nitrocellulose paper. Proteins adsorb strongly and nonspecifically to nitrocellulose, so that the nitrocellulose paper with the adsorbed set of proteins can subsequently be bathed in a solution containing an antibody (the 1° antibody) specific to one of the proteins in the resolved

set. The antibody will bind to the protein of interest; unbound 1° antibody is then washed away, and a second enzyme-linked antibody (the 2° antibody) that binds the 1° antibody is added. The enzyme linked to the 2° antibody (e.g., alkaline phosphatase) is able to catalyze the conversion of a substrate molecule to colored or fluorescent products, or to products that release light as a by-product of their formation. In this way, the 1° immune complex on the nitrocellulose sheet can be detected (Fig. 1–J).

A complex mixture of proteins (e.g., a whole-cell extract) will have many proteins that by chance have similar or even identical molecular weights, such that they are indistinguishable by SDS-PAGE. In this case, one can use a technique with a higher degree of resolution, namely, **two-dimensional gel electrophoresis** (Fig. 1–K). For the *first dimension* of this process, one begins by resolving the mixture of proteins based on their individual **isoelectric points**, using the method of **isoelectric focusing (IEF)**. *(The*

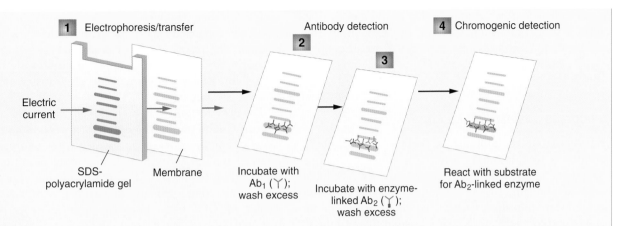

Figure 1–J. Western blotting. *1:* Proteins are resolved by sodium dodecyl sulfate polyacrylamide gel electrophoresis (SDS-PAGE). The gel with the resolved set of proteins is then placed in an apparatus that permits electrophoretic transfer of the proteins from the gel onto the surface of a special paper (e.g., nitrocellulose paper) to which proteins strongly adsorb. *2:* After transfer, the nitrocellulose sheet is incubated with an antibody (the "primary" antibody) directed against the protein of interest. (Before this incubation [not shown], the surface of nitrocellulose paper is "blocked" by incubating it with a nonreactive protein such as casein, to prevent nonspecific binding of the 1° antibody to the nitrocellulose; this casein block leaves the sample proteins still available for antibody binding.) *3:* After washing away unbound 1° antibody, an enzyme-linked 2° antibody is added, which binds the 1° antibody, and *(4)* can generate a colored product for detection. *(Modified from Lodish H et al.,* Molecular Cell Biology, *5th ed. New York, NY: W.H. Freeman, 2004.)*

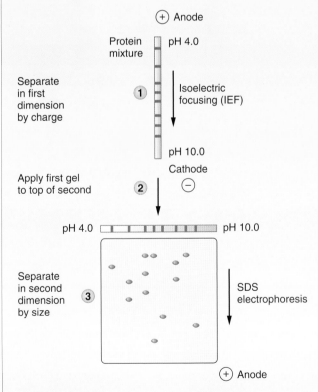

Figure 1–K. Two-dimensional gel electrophoresis. *1:* Proteins in the sample are first separated by their isoelectric points in a narrow diameter tube gel with a fixed pH gradient, by a technique called isoelectric focusing (IEF). This is the "first-dimensional" separation. *2:* The IEF gel is then soaked in SDS and laid on top of a slab SDS polyacrylamide gel for the "second-dimensional" separation of SDS-PAGE *(3)*, which resolves proteins based on their size. *(Modified from Lodish H et al.,* Molecular Cell Biology, *5th ed. New York, NY: W.H. Freeman, 2004.)*

isoelectric point of a protein can be defined as the pH at which the protein has no net charge. Many of the amino acids that comprise a protein have side chains that function as acids or bases; at a low pH, basic amino acids will be positively charged; at high pH values, acidic amino acids will be negatively charged. For every protein, there will be a pH at which the number of positively charged amino acids equals the number of negatively charged ones, such that the protein has no net charge. This is the isoelectric point of that protein.)

In the application of IEF used for two-dimensional gel electrophoresis, one first completely denatures the proteins with 8M urea. One then applies the sample to a glass tube containing a high-porosity polyacrylamide gel that has the same 8M concentration of urea, together with a mix of hundreds of small molecules (ampholytes) each with a unique isoelectric point. When a voltage is applied to the gel, the ampholytes migrate in the electric field, setting up a fixed pH gradient. The proteins in the sample migrate in the field until they reach the pH in the gradient corresponding to their isoelectric point, at which point they cease moving; that is, they become focused as a band in the gel. After all of the proteins have banded (focused) at their individual isoelectric points, the IEF gel is extruded from the tube and soaked in SDS buffer. It is then laid on top of an SDS polyacrylamide slab, and electrophoresed in the presence of SDS. This is the *second dimension* of resolution, where proteins are resolved by size. This sequential resolution of proteins, *first by charge, then by size,* produces good resolution of complex mixtures of proteins.

A convenient feature of antipeptide antibodies is that excess free peptide competes for the protein in the binding of the antibody and provides a useful control for the specificity of any antibody–protein interaction observed.

Antibodies against Full-Length Protein

The alternative to immunizing rabbits with synthetic peptides is to immunize them with either the entire protein, or a stable subdomain (e.g., the extracellular globular domain of a single-pass transmembrane protein). Immunization with the whole protein requires purification of relatively large amounts of the protein of interest (tens or hundreds of milligrams). Production of large amounts of protein (overexpression) from a cloned gene is greatly facilitated by the use of any of several plasmid or virus-based **protein-expression vectors.** Insertion of the coding sequence into an expression vector also allows creation of a "run-on" protein with a carboxyl-terminal "tag" sequence that permits subsequent rapid and efficient affinity purification. Commonly used tag sequences are "6× histidine" and "GST" tags. Such tags permit rapid and efficient affinity purification of the overexpressed protein.

Escherichia coli is often used for the expression of cloned genes, but because of different codon usage between prokaryotes and eukaryotes (and corresponding differences in the levels of the various cognate tRNA), human genes are sometimes not satisfactorily expressed in *E. coli.* Furthermore, overexpressed proteins in *E. coli* often form insoluble aggregates called **inclusion bodies,** and posttranslational modifications such as glycosylation cannot occur in bacteria. For these reasons, a human gene might preferably be expressed in a eukaryotic expression system, using either a highly inducible expression vector in yeast or the insect baculovirus *Autographa californica* in insect Sf9 cells.

Once sufficient amounts of the protein have been purified, a **polyclonal antiserum** can be obtained by immunizing rabbits; alternatively, mice can be immunized for the production of **monoclonal antibodies.**

Genetic Tagging

Green Fluorescent Protein

GFP was first identified and purified from the jellyfish *Aequorea victoria*, where it acts in conjunction with the luminescent protein aequorin to produce a green fluorescence color when the organism is excited. In brief, excitation of *Aequorea* results in the opening of membrane Ca^{2+} channels; cytosolic Ca^{2+} activates the aequorin protein and aequorin, in turn, uses the energy of ATP hydrolysis to produce blue light. By quantum mechanical resonance, blue light energy from aequorin excites adjacent molecules of GFP; these excited GFP

molecules then produce a bright green fluorescence. Thus, the organism can "glow green in the dark" when excited. The resonant energy transfer between excited aequorin and GFP is an example of a naturally occurring **fluorescence resonance energy transfer (FRET)** process (see later).

The gene for GFP has been cloned and engineered in various was to permit the optimal expression and fluorescence efficiency of GFP in a wide variety of organisms and cell types. Cloning has furthermore permitted the GFP coding sequence to be used in protein expression vectors such that a chimeric construct is expressed, consisting of GFP fused onto the amino- or carboxyl-terminal end of the protein of interest. Variant GFP proteins and related proteins from different organisms are now available that extend the range of fluorescence colors that are produced: blue (cyan) fluorescent protein (CFP), yellow fluorescent protein, and red fluorescent protein.

GFP is a β-barrel protein (its structure is shown in Fig. 1–2). Within an hour or so after synthesis and

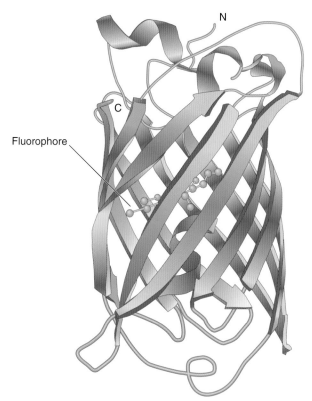

Figure 1–2. The structure of green fluorescent protein (GFP). GFP is an 11-strand β-barrel, with an α-helical segment threaded up through the interior of the barrel. The amino- and carboxyl-terminal ends of the protein are free and do not participate in forming the stable β-barrel structure. Within an hour or so after synthesis and folding, a self-catalyzed maturation process occurs in the protein, whereby side chains in the interior of the barrel react with each other and with oxygen to form a fluorophore covalently attached to the through-barrel α-helical segment, near the center of the β-barrel cavity. *(Modified from Ormö M et al., Science, 273:1392–1395, 1995.)*

folding, a self-catalyzed maturation process occurs in the protein, whereby adjacent serine, glycine, and tyrosine side chains in the interior of the barrel react with each other and with oxygen to form a fluorophore covalently attached to a through-barrel α-helical segment, near the center of the β-barrel cavity. The GFP fluorophore thus produced is excited by the absorption of blue light from the fluorescence microscope, and then decays with the release of green fluorescence.

Because the amino- and carboxyl-terminal ends of GFP are free and do not contribute to the β-barrel structure, the coding sequence for GFP can be incorporated into expression vector constructs, such that chimeric fusion proteins can be expressed with a GFP domain located at either the amino- and carboxyl-terminal ends of the protein of interest. As mentioned earlier, the great advantage of genetic tagging of proteins with fluorescent molecules such as GFP is that this technique permits one to visualize the subcellular location of the protein of interest in a living cell. Consequently, one can observe not only the location of a protein but also the path it takes to arrive at that location. For example, using a GFP-tagged human immunodeficiency virus (HIV) protein, it was discovered that after entry into cells, the HIV reverse transcription complex travels via microtubules from the periphery of the cell to the nucleus.

The FRET technique can be used to monitor the interaction of one protein with another inside a living cell. As discussed earlier in this chapter, in *Aequorea*, blue light energy from aequorin is used to excite GFP by the quantum mechanical process of resonance energy transfer. Energy transfer like this can occur only when donor and acceptor molecules are close to each other

(within 10 nm). Investigators are able to take advantage of this process to detect when or if two proteins in the cell bind each other under some circumstance. Both proteins of interest need merely be tagged with a pair of complementary (donor-acceptor) fluorescent proteins, such as CFP and GFP, and then coexpressed in the cell. CFP is excited by violet light, and then emits blue fluorescence. If the two proteins do not bind each other in the cell, only blue fluorescence will be emitted on violet light excitation; if, however, the two proteins do bind each other, resonant energy transfer from the donor CFP will be captured by the GFP-tagged partner, and green fluorescence will be detected (Fig. 1–3).

ELECTRON MICROSCOPY

There are two broad categories of EM: **transmission EM** and **scanning EM**. First, we discuss the topic of transmission EM, including the special techniques of **cryoelectron microscopy.**

Transmission Electron Microscopy

Transmission electron microscopes use electrons in a way that is analogous to the way light microscopes use visible light. The various elements in a transmission electron microscope that produce, focus, and collect electrons after their passage through the specimen are all related in function to the corresponding elements in a light microscope (see Fig. 1–B). Rather than a light source, there is an electron source, and electrons are accelerated toward the anode by a voltage differential. In an electron microscope, the electrons are focused not by optical lenses of glass, but instead by **magnets.**

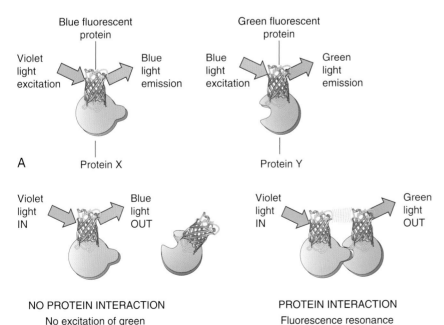

A

Violet light excitation — Blue fluorescent protein — Blue light emission

Protein X

Blue light excitation — Green fluorescent protein — Green light emission

Protein Y

B

Violet light IN — Blue light OUT

NO PROTEIN INTERACTION
No excitation of green fluorescent protein, blue light detected

C

Violet light IN — Green light OUT

PROTEIN INTERACTION
Fluorescence resonance energy transfer, green light detected

Figure 1–3. Fluorescence resonance energy transfer (FRET). **A:** The two proteins of interest are expressed in cells as fusion proteins with either blue fluorescent protein (BFP) (protein X) or GFP (protein Y). Excitation of BFP with violet light results in the emission of blue fluorescent light by BFP; excitation of GFP with blue light yields green fluorescence. **B:** If the two proteins do not bind each other inside the cell, excitation of the BFP molecule with violet light results simply in blue fluorescence. If, however, (C) the two proteins do bind each other, they will be close enough to permit resonant energy transfer between the excited BFP molecule and the GFP protein, resulting in green fluorescence after violet excitation. *(Modified from Alberts B, et al.* Molecular Biology of the Cell, *4th ed. New York, NY: Garland Science, 2002.)*

Because electrons would be scattered by air molecules, both the electron trajectory and the sample chambers must be maintained in a **vacuum.**

We are perhaps more accustomed to thinking of an electron as a particle-like object rather than as an electromagnetic wave, but of course, quantum mechanically, electrons can behave as either particles or waves. As is the case for all waves, the frequency (and hence wavelength) of an electron is a function of its energy, which in turn is a function of the accelerating voltage that drives an electron from its source in an electron microscope. Typical electron microscopes are capable of producing accelerating voltages of approximately 100,000 V, producing electrons with energies that correspond to **wavelengths of subatomic dimensions.** This would, in theory, permit subatomic resolutions! A number of factors such as lens aberrations and sample thickness, however, limit the practical resolution to much less than this. Under usual conditions with biological samples, *electron microscopic resolution is approximately 2 nm,* which is still more than 100-fold better than the resolution of a light microscope. This increased resolution, in turn, permits much larger useful magnifications, up to 250,000-fold with EM, compared with approximately 1000-fold in a light microscope with an oil-immersion lens.

Because of the high vacuum of the EM chamber, living cells cannot be viewed, and typical sample preparation involves fixation with covalent cross-linking agents such as glutaraldehyde and osmium tetroxide, followed by dehydration and embedding in plastic. Because electrons have poor penetrating power, an ultramicrotome is used to shave off extremely thin sections from the block of plastic in which the tissue is embedded. These ultrathin sections (50–100 nm in thickness) are laid on a small circular grid for viewing in the electron microscope.

Electrons would normally pass equally well through all parts of a cell, so membranes and various cellular macromolecules are given contrast by "staining" the tissue with heavy metal atoms. For example, the **osmium tetroxide** used as a fixing agent also binds to carbon-carbon double bonds in the unsaturated hydrocarbons of membrane phospholipids. Because osmium is a large, heavy atom, it deflects electrons, and osmium-stained membrane lipids appear dark in the electron microscope image. Similarly, **lead and uranium salts** differentially bind various intracellular macromolecules, thereby also staining the cell for EM.

The preceding discussion has been of how one would go about viewing the overall layout of the cell under an electron microscope, but most often we are interested in the subcellular location of a particular molecule, usually a protein. Here again a specific antibody against the protein can be brought into play, this time tagged with something electron dense; most often, this electron-scattering tag will be commercially available nanoparticles of colloidal **gold,** coated with a small antibody-binding protein, called **Protein-A** (Fig. 1–4). Gold-tagged antibodies can also be used to stain various genetically tagged proteins containing tags such as GFP, a myc tag, or any other epitope.

In some circumstances, one may wish to obtain a more three-dimensional sense of a surface feature of the cell, or of a particular object such as a macromolecular complex. Two different techniques can be used to do this with a transmission electron microscope; One is called **negative staining,** and the other is called **metal shadowing.** In the case of negative staining, the objects

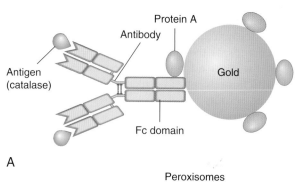

A

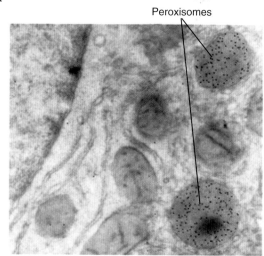

B

Figure 1–4. Protein A–coated gold particles can be used to localize antigen-antibody complexes by transmission electron microscopy (EM). A: Protein A is a bacterial protein that specifically binds the Fc domain of antibody molecules, without affecting the ability of the antibody to bind antigen (the enzyme catalase, in the example shown here); it also strongly adsorbs to the surface of colloidal gold particles. B: Anticatalase antibodies have been incubated with a slice of fixed liver tissue, where they bind catalase molecules. After washing away unbound antibodies, the sample was incubated with colloidal gold complexed with protein A. The electron-dense gold particles are thereby positioned wherever the antibody has bound catalase, and they are visible as black dots in the electron micrograph. It is apparent that catalase is located exclusively in peroxisomes. *(A: Modified from Lodish H, Berk A, Matsudaira P, Kaiser CA, Krieger M, Scott MP, Zipursky SL, Darnell J. Molecular Cell Biology, 5th ed. New York, NY: W.H. Freeman, 2004; B: From Geuze HF, et al. J Cell Biol 1981;89:653, by permission of the Rockefeller University Press.)*

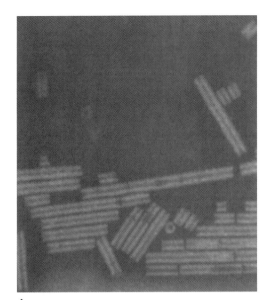

A

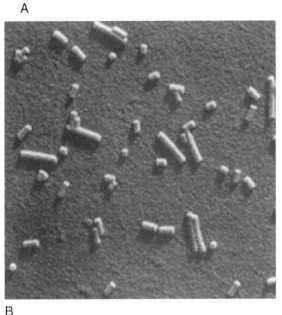

B

Figure 1–5. Electron microscopic images of negatively stained versus metal-shadowed specimens. A preparation of tobacco rattle virus was either (**A**) negatively stained with potassium phosphotungstate or (**B**) shadowed with chromium. *(Courtesy of M. K. Corbett.)*

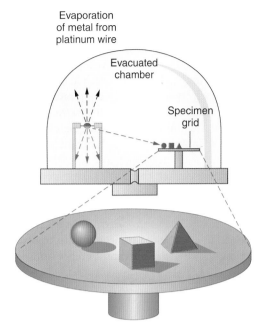

Figure 1–6. Procedure for metal shadowing. The specimen is placed in a special bell jar, which is evacuated. A metal electrode is heated, causing evaporation of metal atoms from the surface of the electrode. The evaporated metal atoms spray over the surface of the sample, thereby "shadowing" it. *(Modified from Karp G.* Cell and Molecular Biology, *3rd ed. New York: John Wiley & Sons, 2002.)*

to be viewed (e.g., virus particles) are suspended in a solution of an electron-dense material (e.g., a 5% aqueous solution of uranyl acetate), and a drop of this suspension is placed on a thin sheet of plastic, which, in turn, is placed on the EM sample grid. Excess liquid is wicked off, and when the residual liquid dries, the electron-dense stain is left in the crevices of the sample, producing images such as that shown in Figure 1–5A.

The second technique, metal shadowing, is illustrated in Figure 1–6. The chemically fixed, frozen, or dried specimen, on a clean mica sheet, is placed in an evacuated chamber, and then metal atoms, evaporated from

a heated filament located at an overhead angle to the specimen, coat one side of the elevated features on the surface of the sample, creating a **metal replica**. When subsequently viewed in the electron microscope, electrons are unable to pass through metal-coated surfaces but are transmitted through areas in the sample that were in the shadow of the object and were therefore not metal-coated. The resulting image, usually printed as the negative, is remarkably three-dimensional in appearance (see Fig. 1–5, B).

In situations that involve a frozen sample (see the following section), after metal shadowing, the entire surface of the sample can then be coated with a film of carbon. After removal of the original cellular material, the metal-carbon replica is viewed in the electron microscope. When used in conjunction with a method of sample preparation called *freeze fracture*, metal shadowing has been useful in visualizing the arrangement of proteins in cellular membranes.

Cryoelectron Microscopy

The dehydration of samples that accompanies standard fixation and embedding procedures denatures proteins and can result in distortions if one wishes to view molecular structures at high levels of magnification in the electron microscope. One solution to this difficulty is the technique of cryoelectron microscopy. Here the sample (often in suspension in a thin aqueous film on the sample stage) is rapidly frozen by plunging it into liquid propane (–42°C) or placing it against a metal block cooled by liquid helium (–273°C). Rapid freezing results in the formation of microcrystalline ice, preventing the formation of larger ice crystals that might otherwise destroy molecular structures. The frozen sample is then mounted on a special holder in the microscope, which is maintained at –160°C. In some cases, surface water is then lyophilized off ("freeze-etch") from the surface of the sample, which is then metal shadowed, producing images such as that in Figure 1–7.

In other cases, when there are many identical structures such as virus particles, computer-based averaging techniques, in combination with images from multiple planes of focus, can produce tomographic three-dimensional images with single nanometer resolution. This technique showed, for example, a previously unsuspected "tripod" structure for the HIV virus envelope spike (Fig. 1–8).

Scanning Electron Microscopy

The surfaces of metal-coated specimens can also be viewed to good advantage with another type of electron microscope, the scanning electron microscope. Unlike the case with metal shadowing in transmission EM, in this case, the entire surface of the specimen is covered with metal. The source of electrons and focusing magnets in a scanning electron microscope are like those of a traditional transmission electron microscope, except that an additional magnet is inserted in the path of the electron beam. This latter magnet is designed to sweep (scan) the focused, narrow, pencil-like electron beam in parallel lines (a raster pattern) over the surface of the specimen. Back-scattered electrons, or secondary electrons ejected from the surface of the metal-coated specimen (usually coated with gold or gold-palladium), are collected and focused to generate the scanned image. The resolving power of a scanning electron microscope is a function of the diameter of the scanning beam of electrons. Newer machines can produce extremely narrow beams with a resolution on the order of 5 nm, permitting remarkably detailed micrographic images (Fig. 1–9).

ATOMIC-FORCE MICROSCOPY

Atomic-force microscopy (AFM) was developed in the 1980s, and it has become an increasingly useful tool for cell biology. The principle of AFM is illustrated in Figure 1–10. A nanoscale cantilever/tip structure moves over the surface of the sample, and the up and down deflections of the cantilever tip are detected by a laser beam focused on its upper surface. Deflections on the order of a nanometer can be detected, producing resolutions comparable with or exceeding those of the best scanning electron microscopes.

Samples to be scanned by AFM need not be metal plated and put in a vacuum as is required for scanning EM; and a particular advantage of AFM over scanning EM is that samples immersed in aqueous buffers, or even living cells in culture medium, can be scanned by an AFM device. In this way, for example, the real-time opening and closing of nuclear pores in response to the presence or removal of Ca^{2+} (in the presence of ATP) has been demonstrated by AFM of isolated nuclear envelopes, and a novel cell-surface structure called the *fusion pore* was identified on the apical surface of living pancreatic acinar cells (Fig. 1–11).

Not all applications of AFM technology are topologic. For example, by increasing the downward force of the probe tip on the sample, nanodissections can be performed, such as taking a "biopsy" sample from a specific region of a single chromosome (Fig. 1–12)!

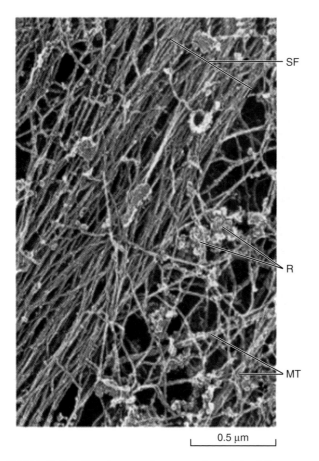

SF

R

MT

0.5 µm

Figure 1–7. **Cryoelectron microscopy of cytoskeletal filaments, obtained by deep etching.** A fibroblast was gently extracted using the nonionic detergent Triton X-100 (Sigma, St. Louis), which dissolves the surface membrane and releases soluble cytoplasmic proteins, but has no effect on the structure of cytoskeletal filaments. The extracted cell was then rapidly frozen, deep etched, and shadowed with platinum, then viewed by conventional transmission electron microscopy. MT, microtubules; R, polyribosomes; SF, actin stress fibers. *(From Heuzer JE, Kirschner M. J Cell Biol 1980;86:212, by permission of Rockefeller University Press.)*

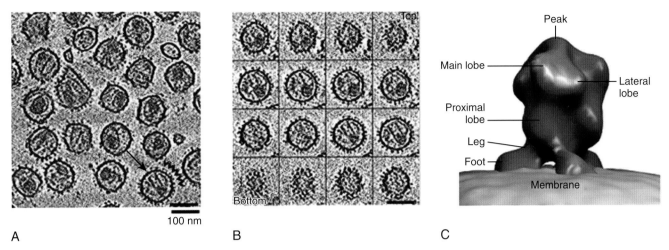

Figure 1–8. Cryoelectron microscopy and tomography of human immunodeficiency virus (HIV). Concentrated virus (HIV or SIV) in aqueous suspension was placed on a grid and rapidly frozen by plunging the grid into liquid ethane at −196°C. The frozen sample was then placed in a cryoelectron microscopy grid holder for viewing at a magnification of ×43,200. The sample holder was tilted at a succession of angles for consecutive images, from which tomograms were computed. **(A)** Sample virus field; the virus shown in this field is simian immunodeficiency virus (SIV), which has a higher density of surface spike proteins than HIV. The virus particle indicated by the *arrow* was chosen for tomographic analysis. **(B)** Computationally derived transverse sections through the selected virus particle (from top to bottom). **(C)** Tomographic structure of the virus envelope spike complex, which is a trimeric structure of viral gp120 (globular portion of the spike) and gp41 (transmembrane "foot") proteins, in the form of a twisted tripod. *(From Zhu P, et al.* Nature *2006;441:847, by permission.)*

In addition, a set of nontopologic uses of AFM technology exists that might be regarded as biophysical but which have cell biological ramifications. In these cases, the cantilever tip is used to measure interactive or deforming forces. For example, ligands or reactive molecules can be attached to the tip of the cantilever. After binding of the tip to the sample, one can measure the force required to either lift the tip or move the object to which the tip is bound. Experiments such as these yield insights into such processes as the force required to unfold modular protein domains, the strength of lectin–glycoprotein interactions, and so forth.

MORE TOOLS OF CELL BIOLOGY

In a search for the functions of novel genes demonstrated by the Genome Project, there are, of course, many other techniques in addition to microscopy that might be brought into play. The techniques of **animal cell culture, flow cytometry,** and **subcellular fractionation** are considered in the following sections.

Cell Culture

Many bacteria (auxotrophs) can be successfully grown in a medium containing merely a carbon source (e.g., sugar) and some salts. Animals (heterotrophs) have lost the ability to synthesize all their amino acids, vitamins, and lipids from scratch and require many such nutrients to be provided preformed in their diet. Mammals, for example, require 10 amino acids in their diet. Mammalian cells grown in culture require the same 10 amino acids, plus 3 others (cysteine, glutamine, tyrosine) that are normally synthesized from precursors by either gut flora or by the liver of the intact animal. By the 1960s, all the micronutrient growth requirements for mammalian cells had been worked out (amino acids, vitamins, salts, trace elements), and yet it was found that it was still necessary to supplement the growth medium with serum (typically 5–10%) to achieve cell survival and growth. Eventually, it was shown that *serum provides certain essential proteins and growth factors:* (1) **extracellular matrix proteins** such as cold-insoluble globulin (a soluble form of fibronectin), which coat the surface of the petri dish and provide a physiologic substrate for cell attachment; (2) **transferrin** (to provide iron in a physiologic form); and (3) three **polypeptide growth factors:** platelet-derived growth factor, epidermal growth factor, and insulin-like growth factor. It is now possible to provide all the required components of serum in purified form to produce a completely defined growth medium. This can be useful in certain circumstances, but for routine growth of cells, serum is still used.

Embryonic tissue is the best source of cells for growth in culture; such tissue contains a variety of cell types of both mesenchymal and epithelial origin; but one cell type quickly predominates: cells of mesenchymal origin, resembling connective tissue fibroblasts. These fibroblastic cells proliferate more rapidly than the more specialized organ epithelial cells, and hence soon outgrow their neighbors. Special procedures must be used if one wishes to study other differentiated cell types from either embryonic or adult tissue, such as liver epithelial cells, breast epithelial duct cells, and so forth, and it is

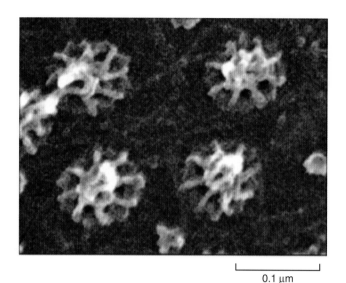

0.1 μm

A

Figure 1-9. High-resolution scanning electron micrograph of nuclear pore complexes. **A:** Purified nuclear envelopes were prepared for scanning electron microscopy. The electron micrograph shows the image of nuclear pores as viewed from the nuclear side of the pore. **B:** Current model for the structure of a nuclear pore. *(A: From Goldberg MW, Allen TD. J Cell Biol 1992;119:1429, by permission of Rockefeller University Press; B: Modified from Alberts B, et al. Molecular Biology of the Cell, 4th ed. New York, NY: Garland Science, 2002.)*

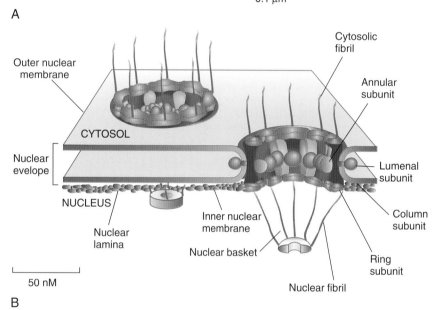

50 nM

B

often not easy to maintain the differentiated phenotype of these cells after prolonged growth in culture. Cultured fibroblasts, however, have proved useful for explorations of the fundamental details of mammalian molecular and cell biology.

To obtain cells for growth in culture, a tissue source is gently treated with a diluted solution of certain proteolytic enzymes, such as trypsin and collagenase, often in the presence of the chelating agent ethylenediamine tetra-acetic acid (EDTA). This procedure loosens the adhesions between cells and breaks up the extracellular matrix, thereby producing a suspension of individual cells. The cells are suspended in growth medium and transferred to clean glass or (more commonly) specially treated plastic petri dishes. After transfer, the cells settle to the bottom of the dish, where they attach, flatten out, and begin both moving around on the surface and proliferating. Eventually, the cells (fibroblasts) cover the bottom surface of the dish, forming a monolayer; at this

point, the growth and movement of the cells greatly slow or cease. This is known as **contact inhibition of growth.** At this point, typically 3 to 5 days after seeding, the cells can again be treated with a trypsin solution to remove them from the dish; an appropriate aliquot of the cells is then resuspended in fresh growth medium and reseeded into a new set of petri dishes. This process of cell transfer is called *trypsinizing* the cell cultures.

Cell Strains versus Established Cell Lines

Cells freshly taken from the animal initially grow well in culture, but eventually their rate of proliferation slows and stops. Depending on the animal of origin and its age, this typically occurs after anywhere from 20 to 50 cell doublings. This phenomenon is termed **cellular senescence,** and the slowing of proliferation that precedes it is termed *crisis* (Fig. 1–13). In some cases, especially with rodent cells, rare variants arise in the

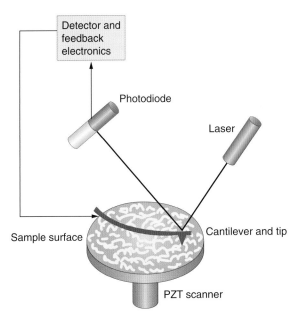

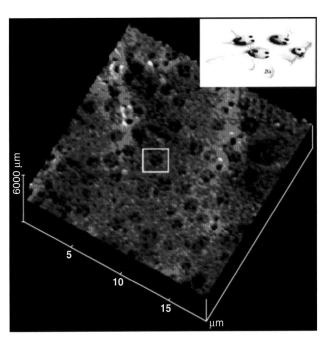

Figure 1-10. Atomic-force microscopy (AFM). In AFM, the sample is scanned by a microscale probe, consisting of a sharp tip attached to a flexible cantilever. The deflection of the probe as it moves over the sample is measured by the movement of a laser beam reflected from the top of the cantilever onto an array of photodiodes. *(Modified from the Wikipedia article "Atomic Force Microscope," http://en.wikipedia.org/wiki/Atomic_force_microscopy.)*

Figure 1-11. Atomic-force microscopy (AFM) image of fusion pores in the membrane of a living cell. The apical plasma membrane of living pancreatic acinar cells was scanned by AFM, producing this image of multiple pore structures. Pores are located in permanent pit structures (one of which is framed by the *white box*) on the apical surface membrane. **Inset:** Schematic depiction of secretory vesicle docking and fusion at a fusion pore. Fusion pores *(blue arrows)*, 100 to 180 nm wide, are present in "pits" *(yellow arrows)*. ZG, zymogen granule. *(From Hörber J, Miles M. Science 2003;302:1002, reprinted with permission from AAAS.)*

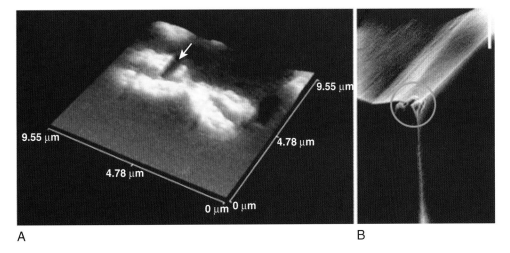

Figure 1-12. Atomic-force microscopy (AFM) "biopsy" of a human chromosome. Metaphase chromosome spreads were prepared and fixed on glass microscope slides by standard techniques. Air-dried, dehydrated chromosomes were first scanned by AFM in noncontact mode; for dissection (**A**), the probe was dragged through a previously identified location on a selected chromosome with a constant applied downward force of 17 micronewtons. (**B**) Scanning electron microscopic image of the tip of the probe used for the dissection shown in **A**; the material removed from the chromosome on the tip of the probe is *circled*. DNA in the sample could subsequently be amplified by polymerase chain reaction. *(From Fotiadis D, et al. Micron 2002;33:385 by permission of Elsevier Science, Ltd.)*

GROWTH OF MURINE CELLS IN CULTURE

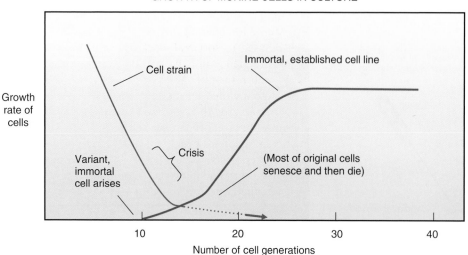

Figure 1–13. Cell strain versus established cell lines. Murine cells (e.g., mouse embryo cells) initially grow well in culture, and during this period of growth, such cells are termed a "cell strain." But the growth rate falls after several generations, and the cells enter "crisis," following which almost all cells senesce and die. Often, however, a rare variant cell will arise in the culture, capable of indefinite growth (i.e., "immortal"). The descendants of this variant cell become an "established cell line." These immortalized cells are typically aneuploid. (*Modified from Todaro GJ, Green H.* J Cell Biol *1963;17:299–313.*)

culture that have escaped the senescent restriction on cellular proliferation, and now grow indefinitely. Such cells are termed **established cell lines.** In the case of mouse embryo cells, for example, this frequently occurs, and a well-known cell line derived from mouse embryo cells in this way is called the *3T3* cell line. Cell lines are sometimes referred to as being "immortal" because of their ability to proliferate indefinitely in culture. Spontaneously arising cell lines such as mouse 3T3 cells usually have abnormalities in chromosome content and can have precancerous properties.

In the case of primary human cell cultures, this escape from crisis to form an established cell line never occurs. Human and mouse cells before crisis are referred to as **cell strains,** or sometimes, more colloquially, "primary cells." The latter term, however, is more appropriately used for cells freshly taken from the animal, before trypsinization, to produce a secondary culture.

At the heart of cellular senescence are repetitive, noncoding sequences called **telomeres,** which are found at both ends of the linear chromosomal DNA molecules of eukaryotic cells. Because of the biochemistry of DNA replication, terminal sequence information is lost each time a linear DNA molecule is replicated. The telomeric sequences of eukaryotic chromosomes protect coding DNA, because it is the "junk" telomeric DNA at the ends that shortens when chromosomal DNA is replicated. In very early embryo cells, as well as in adult germ-line cells and certain stem cells, an enzyme called **telomerase** is expressed, which maintains the length of the telomeres during cell proliferation. In most somatic cells, however, telomerase is not expressed; as a result, each time the cell replicates its DNA and divides, the telomeric DNA sequences shorten. After a certain number of cell doublings, the shortened telomeric DNA reaches a critical size limit that is recognized by the

cellular machinery responsible for activating the senescence program (i.e., the cessation of further cell proliferation).

One of the critical steps in the conversion of a normal cell into a cancer cell is reactivation of telomerase expression. Because cancer cells are therefore able to maintain telomere length, they escape senescence and are "immortal." Consequently, cancer cells, if adapted to growth in culture, grow as established cell lines. For many years there were no established lines derived from normal human cells; the one established human cell line that was available was the **HeLa** cell line. These widely used cells were derived in the 1950s from the cervical cancer tissue of a woman named Henrietta Lacks. Normal animal cells must attach and spread out to grow (the **"anchorage requirement" for growth**); but HeLa cells, like some other established lines derived from cancer cells, have lost the anchorage requirement for growth, and can be grown in suspension like bacteria or yeast cells.

Flow Cytometry

Flow cytometry is a method to count and sort individual cells based on cell size, granularity, and the intensity of one or another cell-associated fluorescent marker. The device that is most commonly used to perform the analysis is called a **fluorescence-activated cell sorter (FACS),** and the layout of a typical FACS instrument is shown in Figure 1–14. In the device, cells pass single file into sheath liquid, which in turn passes through a special vibrating nozzle that creates roughly cell-sized droplets. Most droplets contain no cell, but some droplets contain a single cell (droplets that contain no cell or aggregates of two or more cells are detected and discarded). Just

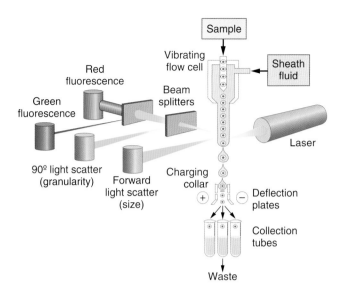

Figure 1–14. Fluorescence-activated cell sorting (FACS). Antibodies tagged with red or green fluorescent molecules and specific each for one of two different cell surface proteins (e.g., CD4 and CD8) are used to label a population of cells (e.g., a population containing the CD4 and or the CD8 protein on their surface). The labeled cells pass into a vibrating flow cell, from which they emerge within individual fluid droplets. The droplets are excited by a laser beam. Forward-scattered laser light, side-scattered laser light, and red and green fluorescent light from the droplet are measured. Based on these measurements, individual droplets will be given a positive (+) or negative (−) charge, and then diverted to collection tubes via charged deflection plates. *(Modified from Roitt IM, Brostoff J, Male D. Immunology, 5th ed. St. Louis: Mosby Year-Book, 1998.)*

before the cells enter the nozzle, each cell is illuminated by a laser beam that causes any cell-associated dye to fluoresce. Forward- and side-scattered light is also measured. Based on these measurements, individual droplets are given either no charge (i.e., empty droplets or droplets with clumps of cells) or a positive or negative charge, and are then deflected (or not, if uncharged) by a strong electric field, which sends them to a particular sample collector.

The FACS device can be used simply to measure the characteristics of a population of cells (**cytometry**), or to sort and isolate subpopulations of cells (**cell sorting**). Earlier devices had a single laser source, and four light detectors, one each for forward scatter (a measure of cell size), side scatter (cellular granularity), and red or green fluorescence. Coupled with the use of red and green fluorescently tagged monoclonal antibodies directed against particular surface proteins, devices such as these played a large role in working out the role of various populations of precursor cells in the process of lymphocyte differentiation.

Second- and third-generation instruments now use as many as 3 lasers, and can detect and sort cells based on as many as 12 different fluorescent colors. FACS analysis is quite useful in studies of, for example, cytokine production by individual T-cell populations, expression

of activation markers, and apoptosis induction in cell population subsets. FACS not only uses monoclonal antibodies for staining of surface markers, but it can also be used to sort and clone hybridoma cells present at low frequency in a postfusion population. This can permit the rescue of rare hybridoma clones expressing useful monoclonal antibodies that might otherwise be lost to overgrowth by nonproducing hybrids. When coupled with reporter gene constructs, such as those expressing proteins tagged with GFP, rare cells expressing the reporter can be identified and captured for further growth and analysis. Additional applications of flow cytometry are discussed in Chapter 2.

Subcellular Fractionation

Subcellular fractionation is a set of techniques that involve cell lysis and centrifugation. These techniques were intimately involved in the discovery, over the course of the last half of the preceding century, of all the various compartments, membrane structures, and organelles that are now known to make up the internal structure of a cell (see Box 1–1). They are also part of the working repertoire of any contemporary cell biologist.

Cell Lysis

Cells can be lysed in any of a variety of ways; the optimum method depends on the cell or tissue type, and the intent of the investigator. One common way to gently break open tissue culture cells, for example, is a device called a *Dounce homogenizer*. A **Dounce homogenizer** consists of a glass pestle with a precision-milled ball at the end; the dimensions of the ball are such that it slides tightly into a special tube in which the cell suspension is contained. Several up and down strokes of the pestle suffice to break open the majority of the cells while leaving nuclei and most organelles intact.

Centrifugation

After cell lysis, centrifugation can then be used to separate the various components of the cellular homogenate, based on their particular size, mass, and/or density. In one common approach, used for the rough fractionation of a homogenate, the lysate is centrifuged in a stepwise fashion at progressively greater speeds and longer times, collecting pelleted material after each step. A low-speed spin will pellet unbroken cells and nuclei; centrifugation of the supernatant from the low-speed spin at a higher, intermediate speed and for a longer duration will bring down organelles such as the mitochondria; centrifugation of the supernatant from the intermediate speed pellet at yet greater speeds and even longer durations will pellet microsomes (ER) and other small vesicles (Fig. 1–15). This type of procedure is termed **differential**

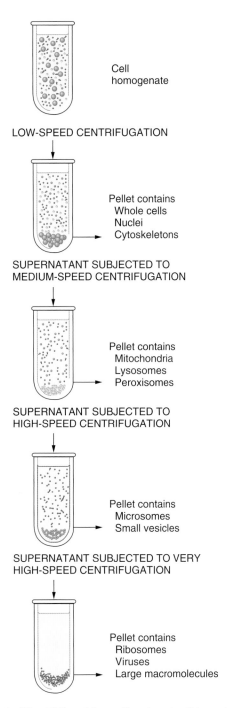

Figure 1–15. **Differential centrifugation.** A cell lysate is placed in a centrifuge tube, which, in turn, is mounted in the rotor of a preparative ultracentrifuge. Centrifugation at relatively low speed for a short time (800 g/10 minutes) will suffice to pellet unbroken cells and nuclei. The supernatant of the low-speed spin is transferred to a new tube, and centrifuged at a greater speed and longer time (12,000 g/20 minutes) will pellet organelles (mitochondria, lysosomes, peroxisomes); centrifugation of that supernatant at high speed (50,000 g/2 hours) will pellet microsomes (small fragments of endoplasmic reticulum and Golgi membranes); centrifugation at very high speeds (300,000 g/3 hours) will pellet free ribosomes or viruses or other large macromolecular complexes. *(Modified from Alberts B, et al.* Molecular Biology of the Cell, *4th ed. New York, NY: Garland Science, 2002.)*

centrifugation. Differential centrifugation can be usefully applied to separate subcellular components that differ greatly in size or mass. But the pelleted materials thus obtained are usually contaminated with many different components of the cell; in the case of Dounce homogenates, for example, the low-speed nuclear fraction contains not only unbroken cells but also large sheets of plasma membrane wrapped around the nuclei; mitochondrial pellets contain lysosomes and peroxisomes.

Further purification or more detailed analyses can be obtained by two other techniques of centrifugation: **rate-zonal centrifugation** (also known as **velocity sedimentation**) and **equilibrium density gradient centrifugation** (sometimes called **isopycnic density gradient centrifugation**). In both of these techniques, an aliquot of cellular material (whole-cell lysate or a resuspended pellet from differential centrifugation) is added as a thin layer on top of a gradient of some dense solute such as sucrose.

In the case of **rate-zonal centrifugation** (Fig. 1–16, A), the sample is layered on a relatively shallow sucrose gradient (e.g., 5–20% sucrose), and then spun at an appropriate speed (based on the size and mass of the material in the sample); in this case, cellular material is not pelleted; instead, the centrifugal field is used to separate materials based on their size, shape, and density; the shallow sucrose gradient serves simply to stabilize the sedimenting material against convective mixing. After the sample components have been resolved based on their sedimentation velocity (but typically before any of the material has actually formed a pellet on the bottom of the centrifuge tube), the centrifuge is stopped, the bottom of the tube is pierced, and sequential fractions of the resolved material are collected for assay. In this way, for example, ribosomes and polyribosomes were first isolated and characterized. The velocity at which a particle moves during centrifugation can be characterized by a number called its "**sedimentation coefficient,**" often expressed in **Svedbergs (S)**. The value of S is a function of the mass, buoyant density, and shape of an object. Large and small mammalian ribosomal subunits, for example, have sedimentation coefficients of 60S and 40S, respectively, whereas the whole ribosome has a sedimentation coefficient of 80S.

In the previously described applications of centrifugation, objects are separated based largely on their relative mass and size. Alternatively, cellular materials can be resolved based on their **buoyant density.** Various proteins, for example, can differ widely in molecular mass, but all proteins have approximately the same buoyant density (approximately 1.3 g/cm³); carbohydrates have densities of approximately 1.6 g/cm³; RNA has a density of about 2.0 g/cm³; membrane phospholipids have densities on the order of 1.05 g/cm³; and cellular membranes, composed of both lipid and protein,

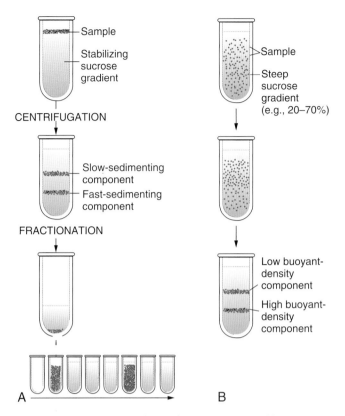

Figure 1–16. Rate-zonal centrifugation versus equilibrium density gradient centrifugation. **A:** In rate-zonal centrifugation, the sample is layered on top of a shallow sucrose gradient. During centrifugation, the various components in the sample then move toward the bottom of the tube based on their sedimentation coefficients. After resolution of the components, the bottom of the plastic tube is pierced and fractions are collected. **B:** Equilibrium density centrifugation resolves components in the sample based on their molecular density. The sample is either layered onto or incorporated into a steep sucrose gradient; during centrifugation, individual components move in the centrifugal field until they reach a density in the gradient that is identical to the buoyant density of the sample component. At this point, each component stops moving and forms a band in the gradient. *(Modified from Alberts B, et al. Molecular Biology of the Cell, 4th ed. New York, NY: Garland Science, 2002.)*

have densities of approximately 1.2 g/cm³. These differences in intrinsic molecular densities permit the resolution of a variety of cellular substituents by the technique of **equilibrium density gradient centrifugation** (see Fig. 1–16, B). Here again, the sample would be layered on top of a gradient of dense solute. For resolving cellular membranes and organelles, the solute would be sucrose, and a 20% to 70% sucrose gradient typically would be used, generating densities ranging from 1.1 to 1.35 g/cm³. For resolving proteins and nucleic acids, higher density gradients made with cesium chloride would be used. During centrifugation over the course of several hours, cellular components migrate in the tube until they reach a point in the density gradient equal to their

own buoyant density, at which point they cease moving and form a disk or "band" at their equilibrium position in the gradient. Rough ER membranes, smooth ER membranes, lysosomes, mitochondria, and peroxisomes all have unique buoyant densities, for example, and are readily separated from each other by this method.

THE TECHNIQUES OF PROTEOMICS AND GENOMICS ARE DISCUSSED IN LATER CHAPTERS

Identification of the functions of the many novel proteins revealed by the Genome Project is, of course, not the only project of contemporary cell biology. Other important research goals to which cell biologists are making contributions include a deeper understanding of the molecular basis of cancer and embryologic development, stem cell properties and function, and perhaps most formidably of all, the "neural correlates of consciousness," to use the phrase of Francis Crick. In these open-ended kinds of investigations, a number of other techniques in addition to the ones described in this chapter are required. Two of the most important such techniques are **mass spectrometry** and **microarrays,** or "gene chips" as the latter is sometimes called. Both of these techniques are discussed in Chapter 5.

An important component of the solution to these problems is a subdiscipline called **Systems Biology,** whereby one seeks to understand all the interrelations between individual signaling pathways and genetic regulatory mechanisms, and how they function as an integrated whole to produce the overall behavior of the cell. A systems biology attitude is integral to the approach we have taken to all the topics discussed in this text.

SUMMARY

Cell biologists have many powerful and sophisticated tools to deploy in their investigations of the function of uncharacterized cellular proteins. Microscopy techniques, in the forms of fluorescence microscopy, EM, and AFM, are among the most useful of these tools, as are the allied techniques of immunology. Tissue culture techniques provide a source of defined, uniform cell types for protein expression and analysis, and flow cytometry technology permits rapid and extremely sensitive analysis of cell populations. Epitope tagging of the proteins encoded by cloned complementary DNA molecules permits their efficient affinity purification, especially in conjunction with the standard techniques of subcellular fractionation and liquid chromatography. Two-dimensional gel electrophoresis and Western blotting are powerful analytic methods for resolving and characterizing complex mixtures of proteins.

Suggested Readings

General

Alberts B, Johnson A, Lewis J, Raff M, Roberts K, Walter P. In: Manipulating proteins, DNA, and RNA. *Molecular Biology of the Cell*, 4th ed. New York, NY: Garland Science, 2002.

Fluorescence Microscopy

Giepmans BNG, Adams SR, Ellisman MH, Tsien RY. The fluorescent toolbox for assessing protein location and function. *Science* 2006;312:217–224.

Tsien RY. The green fluorescent protein. *Annu Rev Biochem* 1998;67:509–544.

Antibodies

Harlow E, Lane D. *Using Antibodies. A Laboratory Manual.* Cold Spring Harbor, NY: Cold Spring Harbor Laboratory Press, 1999.

Electron Microscopy

Hayat MA. *Principles and Techniques of Electron Microscopy*, 4th ed. Cambridge: Cambridge University Press, 2000.

Atomic-Force Microscopy

Fotiadis D, Scheuring S, Muller SA, Engel A, Muller DJ. Imaging and manipulation of biological structures with the AFM. *Micron* 2002;33:385–397.

Hörber JKH, Miles MJ. Scanning probe evolution in biology. *Science* 2003;302:1002–1005.

Cell Culture

Davis JM, ed. *Basic Cell Culture: A Practical Approach.* Oxford, England: IRL Press, 1994.

Flow Cytometry

Herzenberg LA, Parks D, Sahaf B, Perez O, Roederer M, Herzenberg LA. The history and future of the fluorescence activated cell sorter and flow cytometry. *Clin Chem* 2002;48:1819–1827.

Subcellular Fractionation

Howell KE, Devaney E, Gruenberg J. Subcellular fractionation of tissue culture cells. *Trends Biochem Sci* 1989;14:44–48.

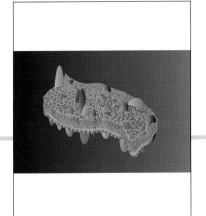

Chapter 2

Cell Membranes

Biological membranes are complex mixtures of lipids and proteins. They form barriers between different cellular compartments and define inside from outside. Without this separation, cells would not be able to function, and as such, proper membrane structure and function is essential for life as we know it.

Phospholipids, important building blocks of the membrane, spontaneously arrange themselves such that their polar head groups face the water and the hydrocarbon fatty acid tails face the hydrophobic core (Fig. 2–1). The forces that determine this arrangement are collectively called the *hydrophobic effect*. It can be argued that the hydrophobic effect is perhaps the most important single factor to organize molecules of living matter into complex structural entities such as cellular plasma membranes or organelles. It is equally important in the formation of detergent micelles and a number of other phenomena that occur in aqueous solutions.

If you have witnessed the beading of oil and its separation from the water in the kitchen, you have witnessed the hydrophobic effect. Water is a polar molecule in which the hydrogen is partially electropositive and oxygen partially electronegative. Because of their polar nature, water molecules tend to form clusters by hydrogen bonding to each other.

Hydrocarbons or lipids dropped into an aqueous solution disrupt the hydrogen bonding of the water molecules. The coalescing of the hydrocarbons is driven by reestablishment of the hydrogen-bonding pattern of the water molecules. Phospholipid molecules are amphipathic, meaning they have hydrophilic (water-loving) polar head groups and hydrophobic (water-fearing) fatty acyl chains. When placed in water, these amphipathic phospholipids try to bury their hydrophobic fatty acyl chains away from water by forming either spherical micelles in which the fatty acyl chains face the center of the sphere and the polar head groups are at its surface or bilayers, which are sheets of two phospholipid monolayer or leaflets (see Fig. 2–1). The bilayers eventually must form a spherical vesicle so that there is no hydrophobic edge facing the water. Depending on the shape of the phospholipid molecules, different structures are preferred, including structures in which the polar head groups are close together with the hydrophobic chains sticking outward. Such structures can occur in the hydrophobic cores of bilayers.

Alternatively to the formation of micelles and bilayer structures in water, phospholipids can form a monolayer at the air–water interface, with their hydrocarbon tails facing the air and polar head groups in the water. Dutch pharmacists Gorter and Grendel observed this phenomenon in 1925. They extracted the lipids from human red cell membranes and placed them on a water surface. When the phospholipids were compressed with a movable barrier, the surface area covered by the phospholipids was twice the surface area of the erythrocyte membranes from which they were extracted. Based on this experiment, they concluded that the lipids in the red cell membrane must be organized in a double layer, a bilayer. In 1965, Alec Bangham and colleagues at the

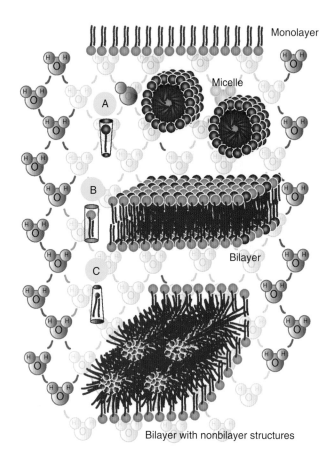

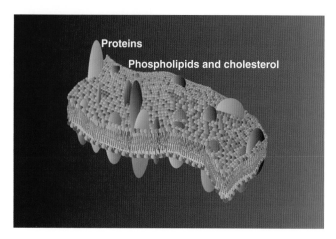

Figure 2–2. The principle of the fluid mosaic model of biological membranes as proposed by Singer and Nicolson. In this model, globular integral membrane proteins are freely mobile within a sea of phospholipids and cholesterol.

Figure 2–1. The hydrophobic effect drives structural rearrangement of lipids, including the formation of bilayers. Phospholipids placed in water would potentially disrupt the hydrogen bonding of water clusters. This causes the phospholipids to bury their nonpolar tails by forming micelles, bilayers, nonbilayer structures, or monolayers. The driving force of the hydrophobic effect is the tendency of water molecules to maximize their hydrogen bonding between the oxygen and hydrogen atoms. Which of the lipid structures is preferred depends on the lipids, as well as the environment. The shape of the molecules (size of the head group and characteristics of the side chains) can determine lipid structure. Molecules that have an overall inverted conical shape, such as detergent molecules (**A**), form structures with a positive curvature, such as micelles. Cylindrical-shaped lipid molecules (**B**) preferentially form bilayer structures. When lipid molecules have an overall conical shape (**C**), structures with a negative curvature are preferred, such as the hexagonal (HII) phase. Any combination of these structures can be found in lipid mixtures.

Agricultural Research Council Institute of Animal Physiology in Cambridge, United Kingdom, showed that phospholipids can self-assemble into small bags. These microscopic phospholipid vesicles or liposomes share some of the properties of cell membranes. In 1974, Efraim Racker and Walther Stoeckenius of Cornell University incorporated bacteriorhodopsin and an ATPase into a lipid membrane and were able to generate ATP from light, showing that simple (artificial) membranes that mimic biological membranes can be generated in the laboratory. In 1972, Singer and Nicolson proposed the fluid mosaic model of membranes (Fig. 2–2). The

basic principles of this model suggest that membrane proteins are embedded within the bilayer, with hydrophobic portions of the proteins buried within the hydrophobic core of the lipid bilayer and hydrophilic portions of the protein exposed to the aqueous environment. These embedded membrane proteins were believed to be mobile, likened to "icebergs within a lipid sea." Notably, the simple presentation of membrane proteins and the "ball-and-stick" representation of the extremely diverse and dynamic nature of the phospholipids is at best a cartoon-like presentation of the biological membrane. Figure 2–16 shows a more detailed model of one membrane protein embedded in a lipid bilayer with water on both sides.

Most of the points of this bilayer model remain intact today. However, the structure is much more complex. Proteins found only on the periphery of the membrane also play important functions. Membranes are continuously remodeled; lipid–lipid, protein–protein, and lipid–protein interactions in the membrane, as well as with constituents on either side of the membrane, determine membrane function. Although movements in the plane of the bilayer and across the bilayer are highly dynamic, these movements are not random. Not all membrane proteins are freely mobile, and similarly, lipids aggregate in specific areas, leading to a heterogeneous lipid and protein distribution in the plane of the bilayer. Both lipids and proteins are also highly asymmetrically distributed across the bilayer. The maintenance of both protein and lipid composition and organization play a crucial role in lipid–protein interaction and functionality of the bilayer.

All biological membranes, including plasma and organelle membranes of mammalian cells, have a similar basic structure and function. They contain a lipid bilayer and serve as a selective permeability barrier separating cellular compartments and inside from outside. Consequently, the composition of the extracellular fluid can differ from that of the cytoplasm, and the polar molecules within the cytoplasm can differ from those within the lumen of the endoplasmic reticulum (ER), the Golgi complex, or mitochondria. This chapter focuses primarily on the plasma membrane.

MEMBRANE LIPIDS

The Lipid Composition of Human and Animal Biological Membranes Includes Phospholipids, Cholesterol, and Glycolipids

The cellular plasma and organelle membranes within the body contain 40% to 80% lipid. Among these lipids, the phospholipids are the most prevalent. Four of the major phospholipids found in human and animal membranes are the glycerophospholipids phosphatidylcholine (PC), phosphatidylserine (PS), and phosphatidylethanolamine (PE), and the sphingosine-based phospholipid sphingomyelin. Many minor species exist in the membrane, such as phosphatidylinositol (PI) or phosphatidic acid (PA). These latter species may be minor in their abundance, but they have important physiological functions. The polar head group for the phospholipids can be choline, serine, ethanolamine, or inositol, linked by a phosphate ester bond to carbon 3 of the glycerol or sphingosine backbone (Fig. 2–3). The hydrophobic portion of the glycerophospholipids contain two hydrocarbon fatty acyl chains linked to carbons 1 and 2 of the glycerol backbone, or in sphingomyelin, a fatty acid bound to the sphingosine back-

bone. The charge of the polar head group differs based on its nature and the pH of the surrounding medium. At neutral pH, PS contains a net negative charge. PC, in contrast, has both a negative (on the phosphate) and positive (on the choline) charge, and therefore behaves as a Zwitterion at neutral pH. These different phospholipid species are not only distinguished by their head groups or backbone (glycerol or sphingosine). They also differ by the way the fatty acyl chains are bound to the glycerol backbone, which can be via an ether, vinyl ether, or ester bond. In addition, the fatty acyl chains can be 14 to 26 carbon atoms in length with 0 to 6 double bonds.

In human and other animal cellular plasma membranes, the phospholipids contain fatty acids that typically contain an even number of carbon atoms (16, 18, or 20). Often, one of the fatty acids (in particular on the carbon 2 position in glycerophospholipids) is unsaturated (contains at least one double bond). The saturated fatty acid is straight and flexible, whereas the unsaturated fatty acid (which most typically contains one cis double bond) has a kink at the site of the double bond.

These fatty acyl tails interact with each other by Van der Waals forces. This interaction depends significantly on the level of unsaturation of the acyl chains. As an example, when phospholipids have two palmitic acids (16 carbon atoms, 0 double bonds) as side chains, they will interact much more strongly with their neighbors as compared with species that have 1 palmitic acid and 1 oleic acid (18 carbon atoms, 1 double bond).

As can be expected, all these molecules, or "molecular species," have different chemical and physical chemical characteristics. Their relative abundance in a

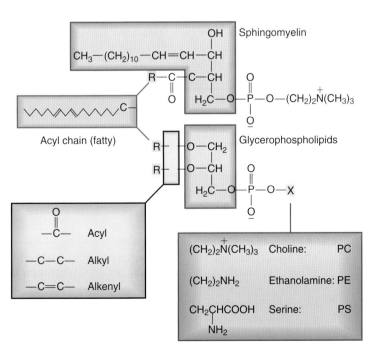

Figure 2–3. Structure of phospholipids. All phospholipids have a polar, hydrophilic head group and nonpolar, hydrophobic hydrocarbon tails. Glycerophospholipids are characterized by their glycerol backbone. Long carbon chains connected to the first and second carbon of glycerol provide the hydrophobic part of the molecule. The phosphate and additional head group structure provide the hydrophilic portion of the molecule. In sphingomyelin, the backbone is sphingosine. A long-chain fatty acid provides the second hydrophobic tail. Note that both phosphatidylcholine and sphingomyelin have a choline-containing polar head group.

membrane will determine membrane characteristics. Although all possible combinations could lead to the presence of thousands of different molecules randomly distributed in a given biological bilayer, this is not the case. The molecular species composition is well defined and specific for each cell type and membrane. As an example, approximately 25% of the phospholipids in the human red cell membrane are PC. Of PCs, approximately 10% is diacylated (an ester bond on both the sn-1 and sn-2 glycerol position) with two palmitic acids. This molecule is shown as an example in Figure 2–4, containing two palmitic acid moieties (16 carbon atoms in length, 0 double bonds). The relative level of this particular molecule is maintained over the life of the red cell and changes will result in abnormalities in the function of the membrane. In the human red cell, more than 90% of the glycerophospholipid molecular species of both PC and PS is of the diacyl variety. In contrast, PE, which makes up about 25% of the glycerophospholipids, contains approximately 40% of molecular species with a vinyl ether on the sn-1 position. An example of one of these plasmalogen phospholipids (1-palmitenyl, 2-linoleoyl PE) is shown in Figure 2–4.

Cholesterol is also a major component of biological membranes (see Fig. 2–4). The structure of cholesterol indicates that it is amphipathic, with a polar hydroxyl group and a hydrophobic planar steroid ring. Cholesterol intercalates between the phospholipids, with its hydroxyl group near the polar head groups and its steroid ring parallel to the fatty acid chains of the phos-

pholipids and perpendicular to the membrane surfaces, and modulates membrane phospholipid structure. The diverse characteristics of membrane phospholipids lead to significantly different behavior of these molecules in pure lipid bilayers. As an example, dipalmitoyl phosphatidylcholine (DPPC) exhibits a solid or gel-like behavior at room temperature because of the strong van der Waals interactions of the palmitic acid side chains with its neighbors in a bilayer. This is not the case for palmitoyl-oleoyl phosphatidylcholine (POPC), in which this interaction is much less between the oleic acid–palmitic acid pair with its neighbors, and a bilayer of this lipid has a more fluid behavior at room temperature. When DPPC and POPC are mixed together, their different interactions will lead to segregation of the two species in the bilayer, like ice shelves in the Arctic Ocean. When equimolar amounts of cholesterol are added, it will alter the van der Waals interactions between the acyl groups, and the segregation is abolished. Hence, one could argue that cholesterol has a major function as a "species mediator" in the lipid bilayer.

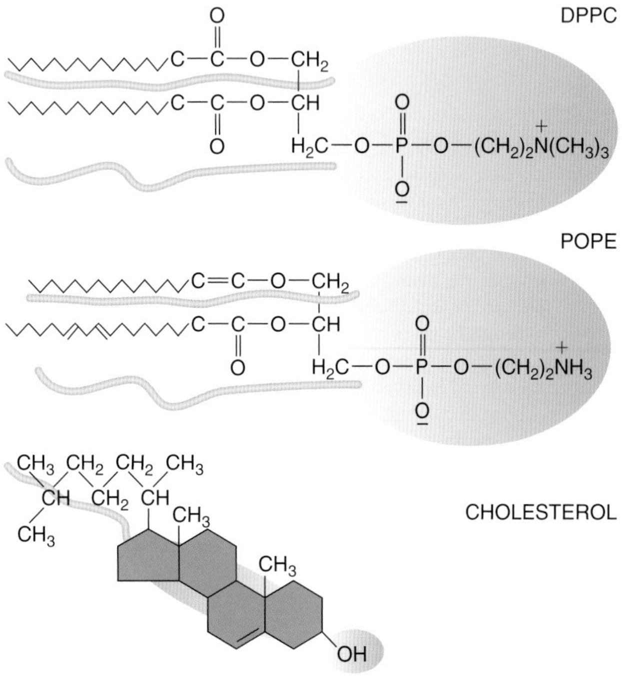

In addition to the major components of the lipid bilayer that determine its overall structure, many minor lipid components are present and essential for the normal function to the cell membrane. As an example, plasma membranes contain many different molecular species of glycolipids, lipids with attached sugar residues. The sugar residues of plasma membrane glycolipids almost always face the outside of the cell; that is, they have an asymmetric distribution, being found only in the outer leaflet of the bilayer. In human and animal cells, glycolipids are produced primarily from ceramide and are referred to as glycosphingolipids. These lipids can be neutral glycolipids with 1 to 15 uncharged sugar residues, or gangliosides with 1 or more negatively charged sialic acid sugars. The glycolipids are important in cell–cell interactions, contribute to the negative charge of the cell surface, and play a role in immune reactions.

Together it will be clear that the lipid bilayer is a complex mixture of many lipid molecular species. These molecules, and thereby the membranes that they build, are generated during development of the cell, by *de novo* synthesis of the different components, as well as by remodeling of the phospholipids after they are assembled in the membrane.

Membrane Lipids Undergo Continuous Turnover

The lipids in membranes are continuously turned over. The breakdown and reassembly may be important for structural rearrangements or may be induced as a repair process. Examples include endocytosis or exocytosis, processes that involve fusion, and processes of membranes that affect the structure of the cellular plasma

Figure 2–4. Structure of the glycerophospholipids, dipalmitoyl-phosphatidylcholine (DPPC), 1-palmitoyl, 2-linoleoyl plasmalogen phosphatidylethanolamine (POPE), and cholesterol.

membranes. Hence, systems are in place to facilitate not only this membrane flow but also remodeling and restoration of the cell membrane. This process is just one example of the continuous flow of lipids between different membrane fractions such as the ER, the Golgi complex, and the plasma membrane. Despite such dynamic exchange processes, these membranes have distinct compositions, which are important for their individual function. Therefore, to counteract this potentially randomizing mixing process of lipids, systems are in place to remodel and restore lipid composition and organization. The precise nature of these systems that would elucidate how they "sense" and act so specifically is poorly understood.

Another example of lipid turnover is the oxidative alteration and repair that has to take place in all membranes, as illustrated in Figure 2–6. Our ability to use oxygen for life comes at a high price. Oxygen will attract electrons whenever it can, and unpaired electrons in reactive oxygen species (ROS) are highly reactive. Polyunsaturated fatty acyl groups are vulnerable for attack by ROS, and despite a complex system of antioxidants, including vitamin E in the lipid bilayer, ROS will react with the double bonds, inserting a polar

oxygen moiety. This will alter the normal packing in the bilayer and compromise its normal barrier function. Phospholipases will recognize this defect in membrane packing and cleave the ester bond of the fatty acid (see Fig. 2–7). The resulting "lysophospholipid" with a free alcohol needs to be restored to a phospholipid by a reacylation process. Fatty acid is activated by binding coenzyme A (CoA) to the carboxyl group, forming a thioester at the expense of ATP. The enzyme responsible for this step, acyl CoA synthase (ACSL), thereby generates a molecule with energy stored in the thioester bond. This acyl CoA molecule is then used by a next enzyme, acyl CoA-acyltransferase (LAT), which generates a new ester bond between the fatty acid and the lysophospholipid to generate a new phospholipid and regenerate CoA.

Given that ROS-induced damage is random in nature, the composition of the bilayer needs to be restored to its original composition, and a diverse pool of fatty acids is available to choose from (depending on your diet), this reaction must be highly selective. A complex family of enzymes in many different isoforms is involved in these reactions. Many of these family members are currently poorly defined. A much better understanding

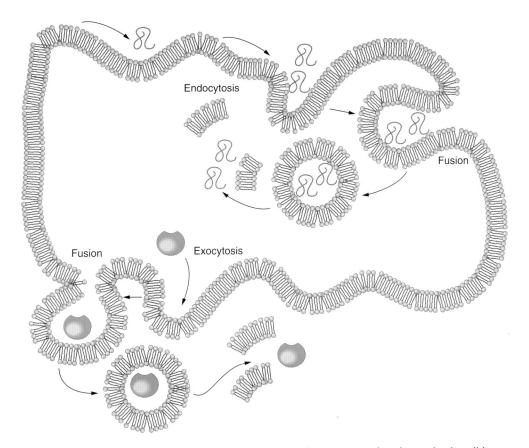

Figure 2–5. Endocytosis and exocytosis. Particles and other entities can be taken up by the cell by an active process called *endocytosis*. The plasma membrane rearranges its lipids and encloses the particle to be taken up. As a last step, the membrane fuses and closes. Lipids in the membrane have to be remodeled to restore the lipid bilayer to its original composition. Examples are resorption processes in the gut, or phagocytosis. Exocytosis is a similar process in the reverse direction. Examples are secretion of enzymes and hormones and release of neurotransmitters.

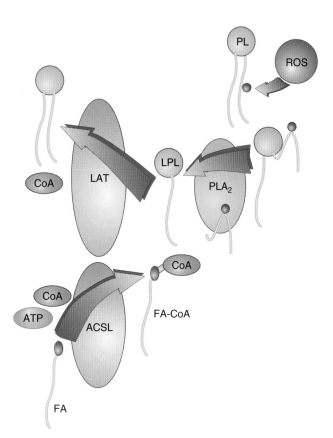

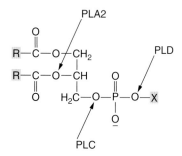

Figure 2-6. Repair of an oxidatively damaged phospholipid. Reactive oxygen species (ROS) oxidize unsaturated fatty acid (FA) in phospholipids (PL). Phospholipase A_2 recognizes and hydrolyzes PL to lysophospholipid (LPL). FAs are activated to acyl coenzyme A (FA-CoA) by acyl CoA synthetase (ACSL) using ATP. FA-CoA and LPL are used by LPL acyl CoA-acyltransferase (LAT) to form phospholipids, releasing CoA for the next cycle.

Figure 2-7. Phospholipases hydrolyze phospholipids. The ester bond hydrolyzed by phospholipases determines the nomenclature of these enzymes. Phospholipase A_2 (PLA$_2$), phospholipase D (PLD), and phospholipase C (PLC) are shown.

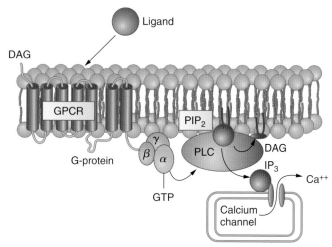

Figure 2-8. G-protein–mediated signal transduction. A ligand binds to a G-protein–coupled receptor (GPCR) in the membrane. This, in turn, activates a phospholipase C, which hydrolyzes phosphatidylinositol biphosphate (PIP$_2$) to form diacylglycerol (DAG) and inositol triphosphate (IP$_3$). IP$_3$ acts to increase cytosolic calcium as part of a signal transduction cascade.

ester bond by adding water. Phospholipases are a large family of proteins with different locations, sizes, and characteristics. They can be grouped based on the ester bonds that they are able to cleave. Figure 2–7 shows a number of examples. Phospholipase D cleaves the ester bond between the phosphate moiety and the head group. Hence, if it hydrolyses PC, the choline group will be released and PA will result from this action. Phospholipase C cleaves at the third carbon in the glycerol backbone to generate diacylglycerol (DAG). When a similar enzyme cleaves sphingomyelin, ceramide is formed. Hydrolysis at the second carbon by phospholipase A_2 will result in a lysophospholipid (see Fig. 2–6).

The products of these reactions can have important physiological consequences. The release of arachidonic acid by phospholipase A_2 action is an important step in the synthesis of leukotrienes, thromboxanes, and similar molecules that perform important cellular signal functions in processes as inflammation. Hydrolysis of PC to PA by phospholipase D, followed by phospholipase A_2 hydrolysis, will generate lysophosphatidic acid (LPA), a powerful lipid mediator involved in many processes, including wound healing. The action of phospholipase C will generate DAG, which in turn has a function in signal transduction. Similar effects are found when products from sphingomyelin, such as ceramide (by sphingomyelinase C), are generated. Another important product is sphingosine 1-phosphate, which has structural similarities to LPA and is similarly involved in signal transduction pathways. When the target of phospholipase C is PIP$_2$, the phosphorylated form of phosphatidylinositol, the released phosphorylated inositol (IP$_3$) acts as an important mediator in cellular signal transduction. One simple step in lipid-mediated signal transduction is illustrated in Figure 2–8. G-protein–

of their actual function is essential, not only to understand the normal repair processes, but also to understand the pathology when this system is not functioning properly.

In addition to structural rearrangements and repair processes, membrane lipids are also metabolized as normal steps in physiological processes. As an example, this section examines the action of phospholipases. The ester bonds in phospholipids are vulnerable for hydrolysis by enzymes that will catalyze the breaking of the

coupled receptor (GPCR) acts on binding of ligands to its extracellular portion. GPCRs are known to play a crucial role in the development and progression of major diseases such as cardiovascular, respiratory, gastrointestinal, neurological, psychiatric, metabolic, and endocrinological disorders. GPCRs can be divided into groups distinguished by the G protein to which they bind. They will trigger different pathways, including the activation of phospholipase C, which in turn generates bioactive molecules including DAG and IP_3.

Together phospholipid turnover plays an important role in the maintenance of membrane phospholipid composition and many important physiological pathways.

Membrane Lipids Are Constantly in Motion

In addition to the complex composition of lipid bilayers, it is essential to realize that membranes are highly dynamic systems, often defined by terms such as *fluidity, mobility,* and *packing.* Membrane phospholipids are capable of several types of motion within the biological membrane (Fig. 2–9).

The phospholipids can rotate rapidly around a central long axis. The fatty acid chains of phospholipids are flexible and will "wiggle" in all directions, with greatest flexion toward the center of the hydrophobic bilayer core. Phospholipids can move laterally across a biological membrane at a rate of approximately 1×10^{-8} cm^2/sec at 37°C, which means that they exchange places with their nearest neighbor about 10^7 times per second. They can move (flip and flop) across the bilayer. In the absence of proteins, this flip-flop is a slow process as a polar entity, the head group, has to transfer through an apolar layer (the fatty acyl chains). However, in the presence of proteins, this transfer can be facilitated in both directions. The overall movement of membrane components is often described as *fluidity,* a rather vague term that describes how a particular molecule or sets of molecules can move around, wobble, and the like. The mobility of any membrane component is governed by molecular interactions including the van der Waals interactions between the neighboring molecules, be it lipids or proteins. Because the composition is complex, the interactions that govern the "packing" of the molecules can differ significantly from neighbor to neighbor. Therefore, overall description of movement can be deceiving because it will differ significantly from molecule to molecule. One example that illustrates how packing can change mobility is the finding that cholesterol, which alters the packing of the phospholipids (see earlier), also tends to slow their lateral mobility. Despite the finding that phospholipids move rapidly in the plane of the bilayer, this is not a random process. Both lipid and protein organization will lead to domains in the membrane where certain lipids and proteins are enriched. As an example, these microdomains, or "rafts," are involved in specific physiological processes such as signal transduction (see Fig. 2–8). It makes intuitive sense to aggregate different proteins and lipids closely together when they have to act in concert. Alternatively, this means that if this organization is disrupted, these processes will not proceed properly. The rafts in plasma membranes are, in general, enriched in molecules such as sphingomyelin, saturated glycerol phospholipids, and cholesterol. Lowering of the cholesterol content of the membrane tends to "dissolve" these rafts, leading to an altered function of the membrane. It will be obvious that all lipids are in a tightly organized equilibrium in the bilayer, but also with the environment. Although the hydrophobic effect forces these molecules in the direction of bilayer structures, this still is an equilibrium and not an absolute "all or nothing" distribution. Depending on the nature of the lipid, they will be able to exchange between bilayers. When the acyl chains are shorter than 16 carbon atoms, the hydrophobicity decreases such that a significant fraction will not stay in the bilayer. Hence, mammalian plasma membranes contain few molecular species with shorter carbon chains because the stability of the membranes would be compromised. Free fatty acids, lysophospholipids, and cholesterol can much more easily leave the membrane and exchange with other bilayers. In particular, the transfer of long-chain phospholipids between membranes can be facilitated by lipid exchange proteins, the cargo vessels for lipid transport. They bind lipids such that the hydrophobic portions are kept away from the water phase. Some of these proteins can be quite specific with respect to the lipids they bind, whereas others are less specific. Serum albumin that reaches concentrations of about 2% will bind a large variety of lipid molecules and as such can function as a transporter.

Lipoproteins are structures (as the name implies) that are made up of lipids and specific proteins, with a function to transport these molecules from the liver to other spots in the organism where they are needed. Changes in any of these pathways can affect the lipid composi-

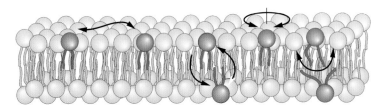

Figure 2–9. **The motion of phospholipids within the lipid bilayer.** Phospholipids are able to move in the plane of the bilayer at very fast rates, they are capable of transbilayer movement (flip-flop), they can rotate rapidly around a central axis, and the fatty acyl tails undergo constant flexion.

tion of cells, including the plasma membranes. As an example, blood cells exchange lipids with lipoproteins. Therefore, changes in lipoproteins will affect the membranes of these cells. Individuals with significant liver problems are often characterized by changes in their blood cell membranes as the result of changes in lipoproteins. Cells may result with too high concentrations of cholesterol, which in turn will affect their function. One could argue that plasma membranes have significant amounts of cholesterol to start with and a bit extra would not be so bad. Small variations are indeed tolerated, but they are bound to well-defined limits; outside these limits, the altered lipid protein interactions that occur can have dire consequences for the cell.

Clinical Case 2-1

Timothy Doyle was a 41-year-old, white, unemployed carpenter who was brought by the police to Boston City Hospital the day after Thanksgiving. He had been found asleep at midnight on a bench in the Boston Common. Though initially difficult to arouse, he was able to give the emergency department (ED) intern the simple history: he "had a few drinks after the turkey." His vital signs were normal. On physical examination, he was a very thin, almost emaciated man who looked much older than his stated age. He was extremely pale and obviously jaundiced. His heart and lungs were clear. His abdomen was protuberant, and there was a radiation of prominent veins surrounding his umbilicus, which the intern's resident called a "caput medusa." There was also an abdominal fluid wave indicative of ascites. Though difficult to assess because of the fluid, the spleen appeared to be enlarged about twice its size by ballottement. The liver was easier to feel, with a distinct edge and nodular surface, which descended four finger-breadths below the ribcage with deep inspiration. On neurological examination, Mr. Doyle was oriented in time and place and recognized the ED staff. However, when asked to extend his arms forward with his eyes closed, he had both a coarse tremor and a repetitive slow rhythmic flexing, or "liver flap," of his wrists. Limited laboratory data obtained in the ED included a hemoglobin level of 6.8 g, with a reticulocyte count of 19%, indicating a significant hemolytic anemia. The white cell count was 11,000, and the platelet count was substantially reduced to 70,000. The blood smear showed that the majority of the red cells were "spur" cells carrying multiple irregular spiky extensions of their normally smooth membranes.

The bilirubin was 8.7, with 3.5 indirect and 5.2 direct reacting, and his dark urine was positive for bile. The alkaline phosphatase was 50, and the aspartate transferase was 270, whereas the alanine transferase was 115. The prothrombin time gave an International Normalized Ratio (INR) of 3.7. Because he was apparently stable, the intern gave Mr. Doyle an intravenous injection of vitamin K for his elevated prothrombin time, and also a vitamin B cocktail in an effort to avert the development of delirium tremens. The intern also requested that the blood bank cross and match six units of blood for Mr. Doyle. Unfortunately, at 4 A.M., the monitoring ED nurse was called stat to find Mr. Doyle vomiting up massive amounts of bright red blood, which had already saturated his sheets. His blood pressure was 60/40 mm Hg, and his pulse thready at 45. Despite an immediate injection of epinephrine and a call to the blood bank for the matched blood, Mr. Doyle died from exsanguination before the blood arrived.

Cell Biology, Diagnosis, and Treatment of Cirrhosis of The Liver

Mr. Doyle, who was well known to the ED team, had advanced Laennec cirrhosis secondary to intractable alcoholism. The marked elevation of his prothrombin time, together with his depressed platelets, made him particularly susceptible to gastrointestinal bleeding, which is typical of patients with increased portal pressure from end-stage liver disease. The striking caput medusa sign suggested that this increased pressure had already induced the fragile esophageal varices, which eventually bled so catastrophically. Of all the typical laboratory tests indicating his extensive liver cell damage, the most alarming was the finding of the spur cell hemolytic anemia. This unusual manifestation of distorted cholesterol balance in both his plasma and red cell membrane lipids almost always heralds imminent death in patients with liver failure.

In addition to the heterogeneous distribution in the plane of the bilayer, phospholipids also transfer from the outer to the inner monolayer and back. Again, this flip-flop is not random; it is orchestrated by proteins in the bilayer. Figure 2–10 shows the normal distribution of phospholipids in the human red cell membrane. The choline-containing phospholipids (PC) and sphingomyelin (SM) are mainly found in the outer monolayer, whereas the amino phospholipids are predominantly (PE) or exclusively (PS) found in the inner monolayer. This highly asymmetrical plasma membrane distribution is typical for most mammalian cells. PS (and PE) is actively transferred from the outer to inner monolayer by an aminophospholipid translocase, or flippase, that consumes one Mg-ATP for each PS molecule transported. Two isoforms of this flippase, family members of the P-ATPases, recently have been identified in the red cell. Their structure in the membrane needs to be established, and it is unknown whether these two forms have different functions. They are able to transport

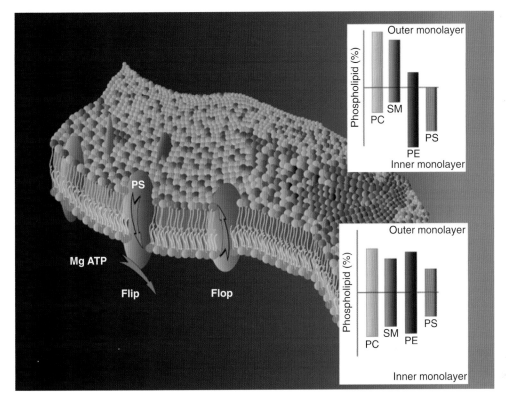

Figure 2-10. The normal distribution of phospholipids in the human red cell membrane. The choline-containing phospholipids (phosphatidylcholine [PC] and sphingomyelin [SM]) are mainly found in the outer monolayer, whereas the amino phospholipids are predominantly (phosphatidyletha-nolamine [PE]) or exclusively (phosphatidylserine [PS]) found in the inner monolayer. Deactivation of the flippase and activation of a scrambling process will lead to the exposure of PS on the surface of the cell.

aminophospholipids, but may do so at different rates for either PE or PS or for different molecular species of these classes of lipids.

In addition to active movement from the outer to inner monolayer, several proteins have been identified that can transport lipids from the inner to outer monolayer by either a directional, active (ATP-consuming) process or a bidirectional scrambling process. ATP-binding cassette (ABC)-containing transporters have a special role in this regard. Important members of the ABC transporter superfamily are the multidrug resistance (MDR) ATPase and the ABCA1, which is involved in lipoprotein metabolism.

The MDR ATPase was initially discovered when oncologists realized that tumors were often resistant to a broad range of distinct anticancer drugs. The reason for this resistance was that tumor cells often overexpress the MDR ATPase, which uses energy of ATP hydrolysis to pump hydrophobic molecules out of the cell. Because a broad range of anticancer drugs are hydrophobic, the effective concentration was difficult to establish. Normally, the MDR ATPase is expressed in liver, kidneys, and intestines, where it is thought to pump toxic substances into the bile, urine, and intestinal lumen, respectively. Therefore, liver cancer (hepatoma) is, in part,

difficult to treat because the affected cells are resistant to a wide range of chemotherapeutic drugs.

Both the inward movement of PS and the scrambling of the bilayer such that PS is exposed are highly controlled in the cell. Even when the scrambling is activated, as long as the flippase is active, PS exposure will be minimal. However, when both the flippase is deactivated and scrambling of the lipids across the bilayer is proceeding, PS will be exposed on the surface of the cell, with important physiologic consequences. During platelet activation, exposure of PS is essential because this will form the docking site for hemostatic factors such as the prothrombinase complex. The proteins in this complex assemble on the PS surface and prothrombin is cleaved to thrombin, an essential step in blood coagulation. In contrast, random, unwanted exposure of PS on plasma membranes leads to a prothrombotic state and imbalance of the normal hemostatic processes. Exposure of PS is also an important trigger for recognition and removal of cells. Early in apoptosis or programmed cell death, PS is exposed on the surface of the cell. Macrophages recognize this abnormal surface and will engulf the cell as it undergoes its apoptotic program, ensuring that the cell is processed before it expels into the environment its potentially noxious breakdown products. Therefore, PS exposure has become a powerful indicator of the onset of apoptosis, and the whole process of programmed cell death is obviously of high importance in tissue remodeling. The processes that turn off the flippase and turn on scrambling are only partly understood but may involve signal transduction pathways that can be activated inside the cell or be triggered from outside the cell. Abnormal regulation of this system may lead to unwanted PS exposure and premature removal of the cell. In contrast, removal of cells is a normal, if well-regulated, event in tissue development. This highly regulated programmed death is essential, and dysregulation in this program may lead to cancerous cells. In these processes, movement of calcium across membrane bilayers appears to be a common theme in the regulatory processes. As an example, in the human red cell, the calcium-ATPase, an active calcium transporter, pumps out calcium against a strong concentration gradient (see later for a description of ion transport). Inside the red cell, the calcium concentration is kept at nanomolar levels, whereas in plasma, it is approximately 2 mM. When the cell becomes "leaky" for calcium, either by natural causes or induced by calcium ionophore to the membrane, calcium levels will increase. In particular, this will be the case when the calcium-ATPase is unable to pump it out efficiently anymore. This will, in turn, induce scrambling and may down-regulate the flippase and PS will be exposed. In cases where the flippase does not work optimally anymore, either because it is damaged or the cell has run out of ATP, PS will be rapidly exposed and the cell is recognized and removed. Hence, calcium fluxes may play an important role in the onset of apoptotic processes and the removal of cells that are not able to deal with this properly.

In addition to an asymmetrical distribution of phospholipids based on their head groups, species will move and distribute asymmetrically based on their fatty acyl chains. The inner monolayer is often enriched in fatty acyl chains with higher levels of unsaturation. In addition, cholesterol, a major component in the bilayer, exhibits different properties with respect to transbilayer movement as compared with phospholipids. Because the polar head group of cholesterol is a small hydroxyl group, cholesterol (unlike phospholipids) can readily flip-flop from inner to outer monolayer and back, regardless of the presence of proteins. Therefore, cholesterol distributed on both sides of the bilayer can move across the bilayer in response to shape changes within the plasma membrane.

Together, the dynamic movement of lipids in the plane and across the bilayer, combined with enrichment in certain domains in the plane of the bilayer or across the bilayer, plays an important role in protein–lipid interaction and therefore physiologic function. Many details on the significance of the heterogeneous or asymmetric distribution of proteins and phospholipids across the membrane need to be explored. Nevertheless, the clues of this dynamic organization point at a simple significance: It allows the two sides of the membrane to be functionally distinct and gives opportunities for proteins and enzymes in distinct pathways to act more efficiently.

Membrane Protein–Lipid Interactions Are Important Mediators of Function

From the preceding discussion, it should be clear that lipid composition, organization, and lipid–protein interactions are essential for the structure, function, biogenesis, and trafficking of membranes. Lipid–protein interactions are at the heart of many biological processes such as lipolysis, blood coagulation, signal transduction, and the maintenance of proper cytosolic composition. The Human Genome Project did not directly address lipids. The protein mechanisms that maintain them are encoded in DNA, not the lipids or their macromolecular organization. Nevertheless, the wealth of information that has become available through the major advances in understanding the genome has only further increased interest in lipid–protein interactions.

The discovery through genome analysis of the abundance of membrane proteins, the identification of the many specific functions that lipids have, and the detection of lipids at specific sites in atomic structures of membrane proteins are just three issues that will likely lead to a better understanding at the molecular level of how biomembranes function. Bridging the specifics of the world of membrane lipids and the special features and difficult handling of proteins that are present in membranes has been, and will continue to be, extremely challenging. Nevertheless, a better understanding of how these interactions drive the functionality of membranes will be essential to understanding the function of cells.

Integral and Peripheral Membrane Proteins Differ in Structure and Function

The two basic types of membrane proteins, integral and peripheral, are distinguished by their operational definition, based on the ability to extract them from the lipid bilayer. Integral membrane proteins are embedded in the lipid bilayer and can be removed only by disrupting the bilayer. To do so, detergents are used. These amphipathic molecules disrupt the bilayer by forming mixed phospholipid-detergent micelles. In these micelles, the integral membrane proteins are coated by the hydrophobic domains. Peripheral membrane proteins can be removed from membrane without dissolving the bilayer. Most frequently, these peripheral proteins are removed by shifting the ionic strength or pH of the aqueous solu-

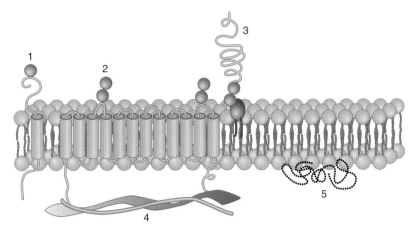

Figure 2–11. **Integral and peripheral membrane proteins.** Integral and peripheral membrane proteins can interact with the lipid bilayer in many different ways. The following situations are presented: *(1)* a single-pass glycosylated integral membrane protein (note that a single α-helical segment of the protein crosses the bilayer); *(2)* a multipass glycosylated integral membrane protein (this structure is found in transporters and membrane channels); *(3)* an integral membrane protein for which the protein itself does not enter the bilayer; instead, it is covalently linked by sugars to phosphatidylinositol; *(4)* a peripheral membrane protein associated by ionic interactions to an integral (or another peripheral) membrane protein; and *(5)* a peripheral membrane protein associated with the polar head groups of phospholipids by an ionic interaction.

tion, thereby dissociating the ionic interactions of the peripheral protein with either phospholipid polar head groups or other membrane proteins.

Various subtypes of integral and peripheral membrane proteins exist (Fig. 2–11). Some integral membrane proteins are transmembrane and make a single pass through the membrane. This type of protein usually has a hydrophilic section containing charged and polar amino acids in the aqueous environment outside the cell, a hydrophobic stretch of 20 to 25 nonpolar amino acids forming an α-helix within the hydrophobic core of the bilayer and a hydrophilic portion within the aqueous interior of the cell.

An example of a single-pass transmembrane protein is glycophorin A, the major sialoglycoprotein in the erythrocyte plasma membrane. Other proteins make multiple α-helical passes through the membrane. Transporters and ion channel are examples of multipass transmembrane proteins. Unlike single-pass transmembrane proteins, multipass transmembrane proteins can contain polar, and even charged, amino acids within the bilayer core. These polar amino acids, when facing one side of the α-helix, contribute to the formation of aqueous pores. An example of an important multipass transmembrane protein is the erythrocyte anion-exchange protein called *band 3*. Band 3 is responsible for the one-for-one exchange of HCO_3^- for Cl^- across the erythrocyte membrane that allows the release of CO_2 in the lungs. The protein makes several α-helical passes through the membrane, which contains a small hydrophilic COOH terminus and a longer hydrophilic

NH_2-terminal domain extending into the cytoplasm. The NH_2-terminal domain has binding sites for glycolytic enzymes, hemoglobin, and regions that link band 3 to the cytoskeleton. The carbohydrate moieties associated with band 3 are on the outside surface of the red blood cell membrane.

Notably, all transmembrane proteins are integral membrane proteins, but not every integral membrane protein is a transmembrane protein. Some proteins are linked to the bilayer by covalently bound fatty acids or phospholipids. Examples are the so-called glycosylphosphatidylinositol (GPI)-anchored proteins. These proteins are linked to sugar moieties that, in turn, are linked to PI, which is (as all phospholipids are) embedded in the hydrophobic core with its acyl chains. An example of this class of proteins is acetylcholinesterase, an enzyme essential for the breakdown of acetylcholine.

This structure, by which the protein itself is outside the bilayer but firmly connected to the bilayer by its phospholipids, gives these proteins a relatively high mobility compared with other proteins. With the large head that sticks out in the water phase, the mobility may be less than that of lipids, but it is much faster than proteins with amino acid chains embedded in the bilayer. Therefore, it is logical to expect such structures for proteins that must be able to move rapidly, such as certain receptors or enzymes. Many proteins also use fatty acids to increase their interaction with the hydrophobic portion of the bilayer. Fatty acids (mainly myristic or palmitic acid, 14 and 16 carbon atoms long, respectively) are covalently bound to amino acid moieties. One example is aquaporin, the water transporter. The peripheral proteins can attach to the membrane surface by ionic interactions with an integral membrane protein (or another peripheral membrane protein) or by interaction with the polar head groups of the phospholipids.

MEMBRANE PROTEIN ORGANIZATION

Similar to lipids, membrane proteins are capable of lateral movement within the plane of the membrane. A quantitative measurement of protein lateral mobility can be obtained by fluorescence recovery after photo bleaching. In these studies, a surface protein is fluorescently labeled. A focused laser beam bleaches a small selected region of the plasma membrane, decreasing fluorescence in that spot. The fluorescence of the bleached area returns with time because unbleached, labeled surface molecules diffuse into it. The percentage recovery in the bleached spot is proportional to the fraction of that specific integral membrane protein that is mobile. The rate of recovery of fluorescence allows a diffusion coefficient to be calculated. From this approach, we know that integral proteins within artificial lipid vesicles diffuse at a rate of 10^{-9} to 10^{-10} cm^2/sec, whereas in a biological membrane, they diffuse at a rate of 10^{-10} to 10^{-12} cm^2/sec. The slower movement in the biological membrane is caused by protein–protein and lipid–protein interactions. One example is band 3, the anion transporter of the red cell membrane. A major fraction of these molecules cannot move laterally within this membrane, but these molecules are rapidly mobile when purified and placed within an artificial lipid vesicle. The reason is that the band 3 molecules are restricted by their interaction with the cytoskeleton on the cytoplasmic membrane surface. Therefore, although band 3 is capable of movement, its lateral mobility is restricted by its interaction with the membrane skeleton. Another example was indicated earlier, for GPI-linked proteins. In contrast with band 3, they are linked to the bilayer only by a lipid interaction and will be able to move much more rapidly. Nevertheless, their protein moiety will "bump" into proteins that are not moving, decreasing their mobility compared with lipids.

Membrane proteins are not randomly distributed in the plane of the bilayer. They aggregate in areas depending on their function. Similar to lipids, integral membrane proteins can also rotate along their axis within the membrane, but in contrast with lipids, they do not flip-flop from one leaflet to the other. Proteins also have an asymmetric distribution. Spectrin is always associated with the inner cytoplasmic leaflet of the erythrocyte membrane. Glycophorin A always has its NH$_2$-terminal domain outside the red cell and its COOH-terminal domain within the cytoplasm. Because proteins cannot flip-flop from one leaflet of the bilayer to the other, their asymmetry is more absolute. Similar to the glycolipids, the carbohydrate portion of glycoproteins has an asymmetric distribution across biological membranes. The sugar residues are almost always found on the outside of the membrane. Oligosaccharides are attached to glycoproteins by N-linkage to asparagine residues or linkage to serine or threonine. The carbohydrate moieties of glycolipids and glycoproteins, as well as glycosaminoglycans, which are oligosaccharides bound together by small protein cores, make up a fuzzy coat observed on electron microscopy of the outer surface of the plasma membrane. This fuzzy coat is often referred to as a **glycocalyx**, meaning "sugar chalice."

Optical Technologies Such as Microscopy and Flow Cytometry Have Revolutionized the Study of Membranes

A myriad of techniques has been developed over the years to study membranes. Proteins, lipid, or sugar components can be labeled to identify certain cells, as well as to study the function of different membrane building blocks. The optical technologies currently in use to visualize these extremely small entities started about 300 years ago with the development of the science of microscopy by a pioneer in cell biology.

Antony van Leeuwenhoek was a tradesman of Delft, the Netherlands. He had no higher education or university degrees and knew no languages other than his native Dutch. However, in a letter of June 12, 1716, he describes:

My work, which I've done for a long time, was not pursued in order to gain the praise I now enjoy, but chiefly from a craving after knowledge, which I notice resides in me more than in most other men. And therewithal, whenever I found out anything remarkable, I have thought it my duty to put down my discovery on paper, so that all ingenious people might be informed thereof.

He did not "invent" the microscope, as is often thought, but his simple yet very powerful "magnifying glasses" allowed him to significantly increase magnification. Importantly, he described and communicated his observations in detail. He discovered and described bacteria, free-living and parasitic microscopic protists, blood cells, sperm cells, microscopic nematodes and rotifers, and much more. His studies, which were widely circulated, exposed an entire world of microscopic life to the awareness of scientists.

Today, light microscopy is joined by electron microscopy, fluorescent microscopy, and many other technologies that pry information out of cells. Because every research project in cell biology, particularly studies of

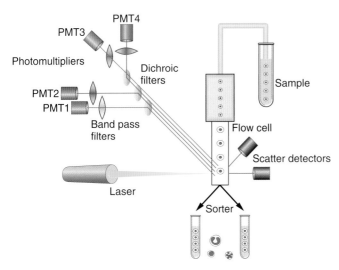

Figure 2–12. Flow cytometry. Cells enter the flow cell in a single file. When they enter the path of the laser, laser light is scattered and fluorescence is detected. The scatter detectors and photomultiplier tubes (PMTs) relate their signals to a computer for analysis of the "events" that are detected by the laser. Criteria can be defined that determine whether these events should be separated (sorted) from the population.

biomembranes, is likely to use flow cytometry, a short description and examples are provided here.

Since the 1980s, flow cytometers have developed from complex machines that could easily fill a room to desktop workhorses for cell biology. Similar to microscopy, this technology enables the analysis of single cells. Although this technique does not generate a regular picture of a cell, it does define the optical properties of a cell. In contrast with microscopy, many millions of cells are analyzed in a short time. This renders valuable information on cell populations or rare events in that population.

Flow cytometry is well suited for clinical applications because samples such as peripheral blood cells are already in suspension and flow analysis can be performed on small samples sizes, which is an important consideration for clinical samples.

Figure 2–12 shows the basic principle of flow cytometric analysis. Many variations on this theme are currently available, but in all, the plumbing of a flow cytometer is designed to let cells move in a single line through the beam of a laser. The laser light is scattered by the cell and the scatter characteristics of this "event" are recorded by detectors that measure the amount of light detected at different angles. Because scatter will change with volume and internal structure of the cell, this "event" can be used to describe the cell.

Before loading into the cytometer, the cells can be incubated with fluorescently labeled probes (such as antibodies or dyes) that recognize cellular molecules of interest. When the laser hits these fluorophores, they

will be exited. When the excited fluorochromes in the cells return to their natural state and emit energy, a set of optics collects the emitted light and sends it to filters that separate the emission spectra. Different bands of light go to different detectors, called *photomultiplier tubes* (PMTs). The data from the PMTs are handled by electronics and a computer processes it so that a record exists of how much light was emitted by each event (cell). By analyzing the data, it can be concluded whether that particular event did have the fluochrome attached and how much was attached. Because many different fluorochromes can be detected at the same time, a complex picture arises that describes if and how many probe molecules did bind to the cell, a direct reflection of, for example, the proteins or lipids on the surface that were targeted.

The development of fast desktop computers has significantly improved the capabilities of these instruments. The intense data flow and analysis in real time, as well as evaluation of the data sets after collection, requires significant computing power, which currently can be performed easily with commercially available desktop or laptop computers. The streamlined cytometers and data analysis capabilities are not the only improvements. A larger number of commercially available antibodies and probes for cellular molecules, as well as more fluorescent dyes and advances in laser optics, are helping to make flow cytometry more accessible to researchers and clinicians.

With the increase in available fluorescent labels and antibodies for cell-surface parameters, as well as intracellular compounds, the flow cytometers are increasingly used for multiparameter analyses. In addition, the machines incorporate more lasers and detectors so many more fluorochromes can be excited and detected at once. Typically, one or two lasers intersect a sample to excite two to four fluorescent reporters. More sophisticated machines now have 3 to 4 lasers that commonly detect 8 colors and can be configured to detect 11 colors.

Another major application of flow cytometry uses the cell-sorting capabilities of these instruments. Based on the information gathered from a particular event, the sorter can send the cell on a separate path, allowing collection of a subset of cells from a population. High-speed sorters enable more cells to be analyzed so that the best cells that would be useful for the study at hand are selected away from the rest. They can subsequently be used for biochemical analysis, RNA expression, or to develop high-throughput screens for drugs. The simplest example uses sorters to separate live cells from dead cells and/or different types of cells that may arise in culture or are co-isolated from primary tissue. With a high-speed sorter, primary or transfected cells, sorted between live and dead cells, can be sorted at a rate of about 10,000 cells/sec.

Linking markers to cell-surface proteins are routine measures to identify cell populations in cell biology research. Identification of compatibility markers is essential in clinical evaluations such as before bone marrow transplant. This technology can also be used to measure lipid structure and movement (Fig. 2–13). In the experiment in Figure 2–13, the effect of sulfhydryl reagents on the activity of the flippase (see Fig. 2–10) is evaluated.

A population of red cells is biotinylated, creating a linker for streptavidin on its surface. A second population is treated with N-ethylmaleimide (NEM), a sulfhydryl reagent. The two cell populations are mixed in a 1:1 ratio. From here on, the biotinylated population acts as the nontreated control and can be distinguished by adding streptavidin, fluorescently labeled with allophycocyanin (APC), to the mixture. In flow cytometry, these cells will show up in the fluorescent channel for the streptavidin (APC) probe. NBD-PS is added to the mixture of NEM-treated and nontreated cells. This compound is a fluorescent PS derivative, which can be identified in the nitrobenzoxadiazole (NBD) channel because its fluophore has a different wavelength compared with APC. Hence, we have set up an experiment with dual colors. Following a similar logic, approaches can be taken with more than two colors. NBD-PS will rapidly incorporate from the water phase into the outer monolayer, driven by the hydrophobic effect of both cell types. When the flippase is active, NBD-PS will be transported rapidly to the inner monolayer. After a short incubation time, the cells are incubated with bovine serum albumin (BSA). NBD-PS still present in the outer monolayer will be extracted BSA, a lipid binding protein. In cells where the flippase was active, fluorescence will stay high, as NBD-PS has moved to the inner monolayer and is not available for BSA extraction. In cells where the flippase is inactivated, NBD-PS will stay on the outside, will be

extracted with BSA, and the fluorescence of these cells will decrease. Therefore, the more fluorescent the cells are, the more activity of the flippase exists.

Let us look at the results of this experiment and the use of flow cytometry. We count a few million cells and use the scatter channels (see Fig. 2–12) to only "look" at the intact red cells. This way we exclude false results from label that has moved from outside to inside in a hemolyzed (leaky) cell, cell debris, or other entities positive for either APC (the streptavidin label) or NBD (the PS label). All these events will have scatter characteristics that make them different from intact red cells. In a properly conducted experiment, this background should be low and virtually all events should be intact red cells.

Figure 2–13A shows a dot plot. Each dot represents an event as measured by the laser. Because we have chosen to look at the red cell scatter channel, each dot represents a red cell. On the y-axis, streptavidin (APC) fluorescence is plotted. Hence, cells in the top two (left and right) quadrants are biotinylated red cells with a high fluorescence in the "APC channel." Similarly, the cells in the bottom two quadrants are negative for APC, because they do not bind streptavidin since no biotin is present on the surface.

Before BSA treatment, both the blue (APC-positive) and orange (APC-negative) events are high in NBD fluorescence (right two quadrants) as the result of NBD-PS in the membrane. Treatment with BSA (+) shows that NBD-PS is extracted from the APC-negative population, and they move to a lower fluorescence (green events). In contrast, most of the APC-positive events will stay positive in the NBD channel (overlay blue and red events), and only a few cells (red events) move to a lower fluorescence in the NBD channel. This simple

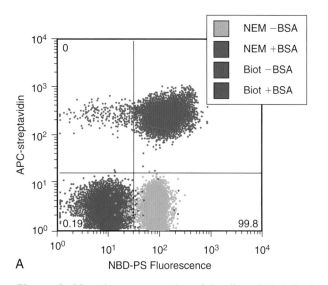
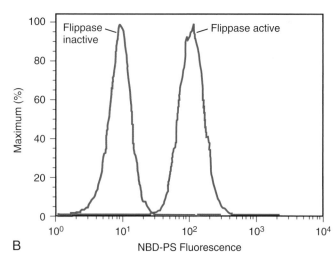

Figure 2–13. Flow cytometry data of the effect of N-ethylmaleimide (NEM) on the ability of the cell to transport NBD phosphatidylserine (PS). **A:** Dot plot of cells treated without (Biot) or with NEM. Fluorescence of NBD-PS was recorded before (–BSA) and after (+BSA) treatment. **B:** Cells with an active flippase are identified as high-fluorescence cells in the NBD channel after bovine serum albumin treatment.

experiment shows that treatment of the cells with NEM will inhibit the flippase. NBD-PS stays in the outer monolayer and is extracted with BSA. Figure 2–13B shows a histogram of the same data. The cells with an active flippase (red curve) are clearly distinct from the population with inhibited flippase activity (green curve).

In this example of an *in vitro* experiment, the action of the flippase is evaluated and shows that modification of a sulfhydryl group will inactivate its activity. Can we measure the activity of the flippase in red cell pathology, or the loss of phospholipid asymmetry and the exposure of PS? Indeed, this can be accomplished. As indicated, exposure of PS has become a routine measurement in apoptosis research. In the following section, the use of flow cytometry is used to define the exposure of PS in subpopulations of sickle red cells.

Important Changes in Membrane Phospholipids Occur in Sickle Cell Disease

A disease that affects humans was well known to the African tribes for the past 5000 to 10,000 years. In 1910, the Chicago-based physician James B. Herrick described, "Peculiar elongated and sickle-shaped red corpuscles in a case of severe anemia."

It was another 40 years before it became apparent what the underlying molecular mechanism of the presence of these "sickled cells" was. In the late 1940s, Linus Pauling perfected his technique to separate proteins and was able to show that hemoglobin collected from a sickle cell patient had a different protein mobility in his separation compared with normal hemoglobin. Pauling's article "Sickle Cell Anemia: A Molecular Disease," published in the journal *Science* in 1949, set the stage for molecular medicine that we are familiar with today. Pauling made the (first) direct link between an abnormal protein and a disease. A few years later, after the discovery of the DNA structure by Watson and Crick, it became apparent that a simple point mutation from a T to an A in one codon of the β-globin locus on chromosome 16 led to a switch from glutamic acid to valine on the sixth position of β globin. This simple mutation results in abnormal behavior of hemoglobin. The protein polymerizes under low oxygen, resulting in the "sickle shapes" that Herrick described. Importantly, the presence of the hemoglobin mutation has a significant effect on the cell, particularly the cell membrane, and its interaction with all other components in blood, including the endothelial cell layer of the vessel wall. The changes in the membrane of red cells in patients with sickle cell disease will determine the premature demise of these cells, play a role in the prothrombotic state, and lead to an increased interaction of these cells with each other and other blood cells; these combined effects may play a role in vasoocclusive crisis, that is, the blockage of a vessel with the ischemia/reperfusion that results from it. One of the altered characteristics of the sickle cell membrane is the exposure of PS in a subpopulation of the cells (Fig. 2–14). Changes in sickle cell membrane proteins are discussed in Chapter 3.

Clinical Case 2-2

Wendell Washington is a 17-year-old black boy who was admitted to the Sacramento City Hospital on June 2 for "excruciating total body pain." He had just graduated as the valedictorian of his high school class and was on a weekend trip with his class to Lake Tahoe to celebrate graduation and the Memorial Day holiday. The weather was unseasonably warm. In the middle of the picnic, he felt feverish and extraordinarily thirsty, despite drinking large amounts of water. By late afternoon, all his joints had started to hurt, and he was asking the school nurse/chaperone for pain medication. A painful cough had also developed. By 10 P.M., he was exquisitely uncomfortable and anxious, and he asked to be taken to the hospital. Recognizing that he was seriously ill, the nurse immediately called for ambulance transport with oxygen and arranged for Wendell to be taken to the Sacramento hospital rather than to Reno, which was appreciably closer. On admission to the ED, Wendell presented as an obviously dehydrated thin black teenager in acute distress. He was crying out in pain and complaining that his legs, ribs, and abdomen hurt more than he could bear. He could not be examined until he had received 5 mg morphine sulfate intravenously.

His blood pressure was 110/60 mmHg, his pulse 144 and regular, his respirations 24, and his temperature 41.2°C. Pertinent findings on physical examination were dry, pale mucous membranes, an extensive area of dullness and rales covering most of his right upper lung field, and markedly tender legs bilaterally. He also was guarding his abdomen, but there were no masses or focal tenderness. Neither the liver nor the spleen could be palpated, and there was no splenic dullness by percussion. A stat chest film confirmed extensive right upper and midlobar pneumonia, and a sputum sample obtained with difficulty showed gram-positive diplococci. Immediate blood analysis showed a white blood cell count of 21,000 with 92% polys, a hemoglobin level of 8.6 g, and a platelet count of 210,000. Blood smears showed a shift to the left and Dohle bodies in the polys, with 4+ poikilocytosis and anisocytosis in the red cells. There was also an 8% reticulocytosis and a 10% population of irreversibly sickled cells in the air-dried smears. A scanty urine specimen showed a specific gravity of 1.010 and

Continued

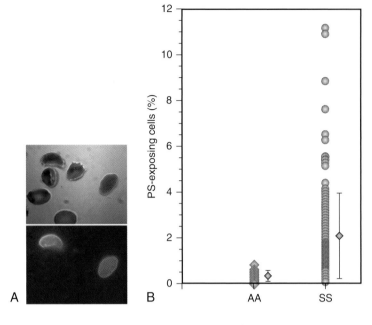

Figure 2–14. **A:** An example of sickle cells incubated with fluorescent annexin V and observed by fluorescent microscopy. The top micrograph is in normal light; the bottom is in fluorescence. Some cells bind fluorescently labeled annexin V. This protein, originally purified from human placenta, binds to phosphatidylserine (PS) on the surface of cells in the presence of calcium. In the example, this protein was labeled with a fluophore that can be excited with a laser. Therefore, flow cytometry can be used to identify these fluorescent red cells in the population. When they bind annexin, they expose PS on their surface. **B: The percentage of red cells that expose PS, as measured by flow cytometry.** In these cells, the flippase is inactive and phospholipids are scrambled, likely as the result of a (transient) increase in cytosolic calcium. This subpopulation of cells is readily recognized by macrophages, similar to apoptotic cells. That they are observed in peripheral blood is an indication that these cells are generated so rapidly that the normal removal system cannot keep up to remove them. In platelets, PS exposure is essential for the normal action of the hemostatic system. Obviously, exposure of PS on the surface of red cell could lead to an imbalance in the normal pathways, and indeed, PS-exposing red cells are related to the prothrombotic state in these patients, as well as their risk for stroke.

extensive microscopic hematuria. Wendell was given high-flow oxygen by mask, whereas major efforts to hydrate him with intravenous fluids were begun. He was immediately started on intravenous antibiotics for his presumed pneumococcal pneumonia, a sample was sent to the blood bank for typing and matching of six units of blood, and he was promptly transfused. A history subsequently obtained from Wendell's mother disclosed that he was the second of her six children with sickle cell disease. Previously, he had had only mild problems, with no more than one yearly admission for a moderately painful crisis as he grew up. Although he had limited his athletic activities in school, he had done well academically and was highly regarded by his classmates.

Cell Biology, Diagnosis, and Treatment of Sickle Cell Disease

Sickle cell anemia is a serious chronic hemolytic anemia. The disease is due to the homozygous inheritance of mutant β-globin genes that contain a substitution of valine for glutamic acid at the sixth amino acid position. Ever since the path-breaking biochemistry of Pauling, it has been recognized as the quintessential molecular disease. Wendell had the bad fortune to inherit the aberrant sickle globin gene from each of his heterozygous parents. As such, he was vulnerable to the hypoxia-induced conformational change in his hemoglobin that converts a normally globular and highly soluble protein into bundles of rigid, elongated, insoluble fibers. From this basic defect flow virtually all of the secondary problems of

his red cells and his physiology. These include the entrapment of red cells in the microcirculation in regions of low oxygen tension, the exquisite painful crises when this entrapment induces anoxia in bones, the gradual infarction of the spleen and the immunologic defects consequent to that, and the infarction of the renal papillae, which removes the ability to concentrate urine. This, in turn, makes patients hypersensitive to dehydration. In some severe cases, perhaps from secondary membrane stickiness, there is a tendency toward cerebral infarction and cumulative brain damage. Ironically, both low oxygen tension and dehydration increase the likelihood of *in vivo* sickling. The elevated altitude of Lake Tahoe, together with the dehydration caused by the weather and Wendell's fever, induced his sickling crisis. The pneumonia that rapidly infected his lungs, perhaps because of his limited immune response, added to his hypoxia and threatened to establish a lethal hypoxia-sickling-hypoxia cycle. Fortunately, the good judgment of the nurse in promptly sending Wendell to a responsive ED at a lower altitude saved his life.

Together, flow cytometry is a powerful technique to probe presence of proteins or lipids on the surface (or inside) of large populations of cells. Although not described here, fluorescent probes exist that report on the presence of certain ions in the cell. These probes can be used in fluorescent microscopy to evaluate ion fluxes, and also in flow cytometry to define populations of cells that exhibit altered transport characteristics (transport characteristics are described in detail in the following sections).

The Cell Membrane Is a Selective Permeability Barrier That Maintains Distinct Internal and External Cellular Environments

As we have discussed aspects of membrane structure, let us examine some of the functions of the plasma membrane. As it separates inside from outside, transport of molecules or ions across this barrier is important. This section reviews different modes of transport—diffusion, osmosis, and active transport. Other modes of transport are also possible, such as endocytosis and exocytosis (see Fig. 2–5). Small, uncharged molecules can pass through a lipid bilayer by simple diffusion (Fig. 2–15, A [part I]). For example, small gaseous molecules, such as O_2, CO_2, and N_2, and small uncharged polar molecules, such as ethanol, glycerol, and urea, can simply diffuse down their concentration gradient across the

lipid bilayer. The underlying mechanism is simple. Particles of a substance dissolved in liquid or gas solvent are in continuous random (Brownian) movement. They tend to spread from areas of high concentration to areas of low concentration until the concentration is uniform throughout the solution. If these areas of high and low particle concentration are separated by a permeable membrane, particles on the high- or low-concentration side will move to the opposite side of the membrane in a given time. However, because there are many more particles on the high-concentration side, the net flux will be from high to low concentration. This will continue until the concentrations on both sides of the membrane are the same. Although the particles will continue to move across the bilayer, no further *net* diffusion will occur. When one would label such a compound, for example, with a radioactive label, diffusion can be observed without a net increase of the actual compound on either side. Such experiments can be revealing with respect to the compartments in a cell that are in equilibrium for certain compounds. Given the situation in Figure 2–15A (part I), in which small molecules are present in two solutions at concentrations outside (C_o) and inside (C_i) the cell ($C_o > C_i$), separated by a mem-

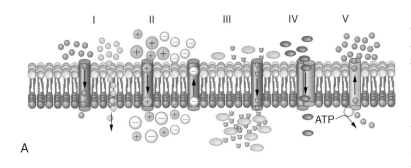

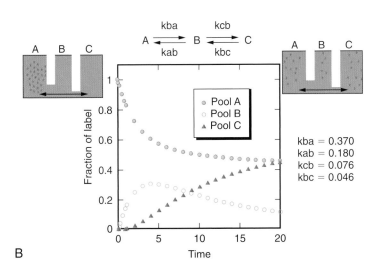

Figure 2–15. **A:** Examples of transport across a biological membrane. *(I)* Diffusion of small hydrophilic or hydrophobic particles driven by a concentration gradient. *(II)* Diffusion of hydrophilic or charged particles driven by a voltage gradient. *(III)* Osmosis, diffusion of solute driven by a concentration gradient of a nonpermeable compound. *(IV)* Facilitated diffusion. *(V)* Active transport against a concentration gradient. **B: Diffusion among the three compartments A, B, and C.** At time zero, label is added to follow the diffusion between the different compartments. In time the fraction of label will decrease in compartment *A*, as compared with start. Similarly, the concentration in the other two compartments will increase. Equilibrium will be established and the size of the compartments can de deducted from the concentration of label in each.

brane permeable to the molecule, but not to solvent, Fick's first law of diffusion states that the net rate of diffusion is $J = -DA\, dc/dx$.

This equation tells us that the rate of diffusion will be proportional to the surface area of the membrane *(A)*, the concentration gradient across the membrane *dc/dx*, and a factor called the **diffusion coefficient** *(D)*, which is a measure of the rate at which the molecule can permeate the membrane. The minus sign in this equation is because diffusion is positive in the direction of higher to lower concentration. *D* is proportional to the speed that the diffusing molecule can move in the surrounding medium. As an example, for spherical molecules, this is described by Einstein's equation: $D = KT/6prn$, where K is equal to the Boltzmann constant, T is the absolute temperature in kelvins (K), r is equal to the molecular radius, and n is the viscosity of the medium.

In other words, diffusion is inversely proportional to the radius of the diffusing molecule and to the viscosity of the surrounding medium. Because the molecular mass of large molecules is proportional to r^3, D will be inversely proportional to (molecular mass)$^{1/3}$. A molecule with a mass that is 64 times bigger will move 4 times slower. Importantly, this describes the movement of a particle between *two* compartments (inside and outside). When considering the diffusion of a molecule *through* a biological membrane, one must also consider the bilayer as a compartment of a "size" that depends on the particle that moves between the outside and inside aqueous compartments. This becomes obvious if we think about the lipid solubility of the diffusing molecule. The permeability of the plasma membrane to a particular molecule increases with its partition coefficient (the movement *into* and *out* of the lipid bilayer). For molecules with the same partition coefficient (same affinity to go into the lipid layer), there is decreasing permeability, with increasing molecular mass. Molecules that are small and soluble in the lipid core, such as O_2, CO_2, N_2, ethanol, glycerol, and urea, will diffuse down their concentration gradient through the bilayer. Diethylurea, which has a *50-fold* greater partition coefficient than urea, diffuses through the membrane at approximately *50* times the rate. For a substance diffusing through a biological membrane, its concentration at the outer face of the bilayer will be bC_o, in which b represents the partition coefficient. Its concentration at the inner face will equal bC_i. The concentration gradient within the membrane will be $dc/dx = b(C_o - C_i)/dx$. The more lipid soluble the substance, the larger the partition coefficient, b; therefore, the larger the effective concentration gradient within the membrane is. If we substitute the concentration gradient across a biological membrane into Fick's equation, the result is $J = Db/dx*A(C_o - C_i)$, with Db/dx referred to as the permeability coefficient, or P. The permeability coefficient for a diffusing molecule is proportional to the partition coefficient, b,

and to the diffusion coefficient within the membrane, D, but is inversely proportional to the distance across the membrane, dx. Therefore, Fick's equation applied to biological membranes simplifies to $J = -PA(C_o - C_i)$.

Another way to explain this is that the membrane is a third compartment in the diffusion process, and the "size" is dependent on the thickness of the bilayer and the affinity of the compound to be in that compartment. Compounds equilibrate between the "three" compartments, as indicated in Figure 2–15B. Figure 2–15B describes the movement of a compound in time when added to compartment A. The constants "k" describe the rate of movement among compartments A, B, and C. Although particles keep moving, the concentrations will equilibrate. Such experiments can be used to access both diffusion rates and partitioning between different compartments.

Fick's equation applies to small, uncharged molecules. The diffusion of charged molecules across biological membranes is determined not only by the concentration gradient but also by the membrane potential; that is, the movement of small, charged molecules across the membrane is related to the electrochemical gradient, not simply to the chemical gradient.

Water Movement across Membranes Is Based on Osmosis

The movement of water across a lipid bilayer has long been a difficult issue to understand. The whole reason of a membrane is to separate two aqueous compartments, and its formation is driven by the hydrophobic effect (see Fig. 2–1). A major development in the understanding how water can move across a lipid bilayer came with the discovery of aquaporin (Fig. 2–16), for

Figure 2–16. The movement of water through a membrane, facilitated by the water channel aquaporin (www.ks.uiuc.edu/Research/aquaporins).

which Peter Agre received the Nobel Prize in Chemistry in 2003, which he shared with Roderick MacKinnon for his studies of ion channels. Aquaporins are membrane water channels that play critical roles in controlling the water contents of cells. They are widely distributed in all kingdoms of life, including bacteria, plants, and mammals. In humans, more than 10 different aquaporins have been identified. Several diseases, such as nephrogenic diabetes insipidus and congenital cataracts, are connected to the impaired function of these channels. Aquaporins form tetramers in the cell membrane and facilitate the transport of water and, in some cases, other small solutes across the membrane. Interestingly and quite paradoxical, these water pores are completely impermeable to charged species, such as protons. This seems paradoxical because protons are usually transferred readily through the water phase. This remarkable property is critical for the conservation of membrane's electrochemical potential.

Osmosis is the flow of water across a membrane from a compartment where solute concentration is lower to one in which the solute concentration is higher. In Figure 2–15A (part IV), a semipermeable membrane (permeable to water) separates a solution with a low concentration (top) from a high concentration (bottom). Water flow will occur from top to bottom. The amount of pressure that would have to be applied on the bottom compartment to keep water from entering is called the **osmotic pressure**. The osmotic pressure of a solution depends on the number of particles in solution; therefore, it is referred to as a **colligative property**. Because it depends on the actual number of particles, the degree of ionization must be taken into account. In calculating osmotic pressure, one molecule of NaCl yields two particles, whereas Na_2SO_4 yields three particles. By Van't Hoff's law, osmotic pressure can calculated as follows:

$$Osmotic\ pressure = iRTc$$

where

- i = number of ions formed by dissociation of a solute molecule
- R = ideal gas constant
- T = absolute temperature (measured in K)
- c = molal or molar concentration

At physiologic concentrations of solutes, such as NaCl, the values obtained for osmotic pressure differ from theoretical values based on Van't Hoff's law. Therefore, a correction factor called the **osmotic coefficient** (Φ) is inserted into the Van't Hoff formula rendering: Osmotic pressure (π) = $\Phi * iRTc$. The factor Φ approaches a value of 1 as the solution becomes increasingly dilute. The term in the equation $\Phi * ic$ is referred to as the **osmolar concentration** with units of osmoles per liter (Osm/L).

To calculate the osmotic pressure (at 37°C) of saline (154 mM NaCl) (Φ = 0.93) solution, we find that π = $(0.93) \times (2) \times (8.2 \times 10^{-2}) \times (310\ K) \times (0.154) = 7.28$ atm. To calculate the osmolarity of this solution: $\Phi * ic$ = $(0.93) \times (2) \times (0.154) = 0.286$ Osm/L or 286 mOsm. When solutions are compared, they can be defined as hypoosmotic, hyperosmotic, or isoosmotic. A solution that has a lower osmotic pressure than 154-mM NaCl at 37°C is defined as *hypo*osmotic compared with the NaCl solution. In contrast, the 154-mM NaCl solution is *hyper*osmotic compared with that particular system. If the osmotic pressures of two solutions are equal, they are called **isoosmotic solutions.**

Cellular membranes are highly permeable to water as it moves rapidly through the aquaporin water channels. The driving force of osmosis is to maintain isoosmotic conditions across the membrane. Hence, if the NaCl concentration would drop on one side of the membrane, water would move rapidly to the other side to regain that equilibrium. Let us look at an example. Erythrocytes are useful when discussing the osmotic properties of cells because they behave as almost perfect osmometers. At 154 mM NaCl, human erythrocytes have their normal shape and volume (Fig. 2–17). Therefore, this concentration of NaCl (286 mOsm) is termed **isotonic.** Importantly, this can differ in different species. In the mouse, isotonicity for murine red cells is at 330 mOsm. The erythrocyte shrinks in more concentrated solutions (**hypertonic**) as water will leave the cell. The erythrocyte swells in more dilute (**hypotonic**) solutions. Under these conditions, water enters the cell and it changes shape and starts to swell. Obviously, this swelling can only go that far. The cells change their shape from their discoid form to spherical, and when the erythrocyte reaches 1.4 times its original volume, the membrane cannot handle the increasing pressure. The cell bursts (hemolyses), releasing hemoglobin and other cytosolic components. Not all cells in the cell population behave identically. Some cells reach their critical volume earlier (at higher osmolarities) and other cells later (at lower osmolarities). The relation between the osmolarity and hemolysis, also called the *osmotic fragility curve*, is shown in Figure 2–17.

The intracellular substances within the erythrocyte that produce its cytosolic osmotic pressure include the O_2-carrying protein hemoglobin, K^+, organic phosphate, Cl^-, and many more. In total, the tonicity of the red cell cytosol is 286 mOsm; hence, the medium must have an osmolar concentration of 286 mOsm and an osmotic pressure of 7.28 atm at 37°C to maintain the normal volume of the cell. The major osmotic particle in the extracellular fluid or plasma that surrounds the red cell is the Na^+ ion. Sodium is maintained at high concentrations in the extracellular fluid by Na/K-ATPase (see later). Therefore, if these pumps, and thereby the ion balance, are altered, the cell may swell or shrink, altering its normal characteristics.

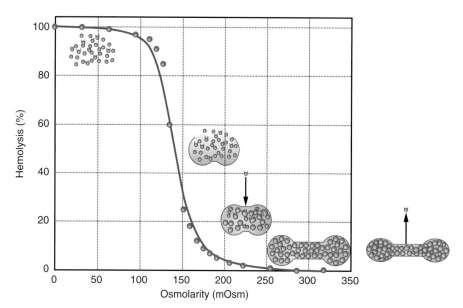

Figure 2-17. The red cell as an osmometer. When red cells are placed in a buffer of isotonicity (290 mOsm), they have their normal volume. Increase in tonicity will lead to the loss of water and the cell shrinks. Decrease in tonicity will lead to the influx of water and the cells swell, increasing the pressure in the cells. At 150 mOsm, approximately 50% of the cells hemolyze, rapidly progressing to 100% hemolysis at lower tonicity of the surrounding medium.

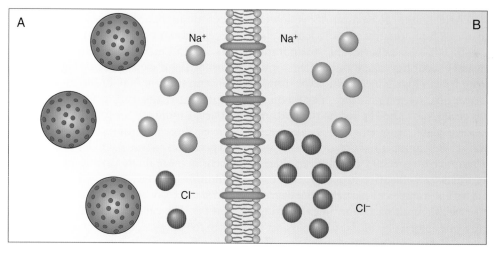

Figure 2-18. The Donnan effect. A semipermeable membrane will allow the diffusion of Na^+ and Cl^- ions, but not the negatively charged protein, in chamber A. Initially, there is an equal number of cations and anions in chambers A and B, which means that the Cl^- is at much greater concentration in B than in A to balance the anionic charge of the impermeable protein. With time, Cl^- will flow down its concentration gradient from B to A. To maintain electroneutrality, Na^+ would also move from B to A. However, this would cause the osmotic concentration to be greater in chamber A than in chamber B; therefore, water moves from B to A.

DONNAN EFFECT AND ITS RELATION TO WATER FLOW

Cells contain many negatively charged ions in the cytoplasm, such as protein and RNA, that cannot diffuse through the membrane. These impermeable anions cause a redistribution of permeable anions in a manner that Donnan and Gibbs predicted (Fig. 2–18).

The two compartments, A and B, are separated by a semipermeable membrane. Both compartments have NaCl. Compartment A contains a protein with a net negative charge that cannot diffuse through the membrane. The total concentration of cations and anions are equal in compartments A and B. The Cl^- initially will be at a higher concentration in compartment B than in A. Because the membrane is permeable to Cl^-, it will move down its chemical gradient toward compartment A. Na^+ will follow Cl^- into compartment A to maintain electroneutrality on side A. After these events, the concentration of particles will be higher in compartment A compared with compartment B. For single cations and anions with identical valence, their distribution across the semipermeable membrane at equilibrium can be expressed by the Donnan equation: $[Na]_A[Cl]_A = [Na]_B[Cl]_B$. Because the total number of particles in compartment A will be greater at equilibrium than in compartment B, water will flow toward compartment A. For this same reason, the Donnan effect tends to cause water to flow into cells but is balanced by the active outward pumping of ions by Na/K-ATPase.

FACILITATED TRANSPORT

Both diffusion and osmosis are passive transport properties across biomembranes, governed by mechanisms as described earlier. However, many molecules may exhibit very low permeability coefficients *(P)*; therefore, they diffuse slowly, slower than needed to maintain functionality of the cells. Hence, this process needs to be facilitated.

Molecules can cross biological membranes far more rapidly than would be predicted by Fick's equation because of a process called **facilitated diffusion.** Transmembrane proteins play an essential role in this process. As an example, let us consider the uptake of glucose into the erythrocyte. The cell needs glucose for its metabolic processes. The concentration of glucose in blood plasma (approximately 5 mM) is much higher than the concentration of glucose within the erythrocyte. However, because of its large size, glucose cannot simply diffuse through the lipid bilayer (Fig. 2–19). It is moved rapidly across the erythrocyte membrane, down its concentration gradient, by associating with a glucose transporter or permease that changes conformation and allows passage of glucose.

As indicated in Figure 2–19B, the rate of uptake of glucose is saturable. At high concentrations of glucose, all of the erythrocyte glucose transporters are occupied so that a maximal velocity is reached (V_{max}). The K_m represents the concentration of glucose at which the rate of uptake $V = \frac{1}{2}V_{max}$. It is a measure of the affinity of glucose to bind to its transporter. For example, whereas the K_m for D-glucose is 1.5 mM, the K_m for its stereoisomer L-glucose is greater than 3000 mM. This illustrates that facilitated diffusion, unlike simple diffusion, is highly (stereo) specific. The glucose transporter will bind D-glucose, but it binds L-glucose poorly, if at all. Other six-carbon sugar molecules that are similar in structure to D-glucose can be transported by the glucose transporter, but they have a greater K_m value. For example, the K_m values for D-mannose and D-galactose are 20 and 30 mM, respectively. These sugars can competitively inhibit the uptake of D-glucose into erythrocytes. The curve in Figure 2–19B fits the Michaelis–Menton equation for enzymatic activity: $V = V_{max}/(1 + K_m/C)$, where *C* is the concentration of substrate. Although the transporter is not an enzyme, it carries out a function similar to that of an enzyme. Instead of chemically converting a substrate, as an enzyme does, transporters move a substrate across a biological membrane in a saturable manner. The glucose transporter is a multipass transmembrane protein (see Fig. 2–11) with 12 α-helical transmembrane segments. Multipass transmembrane proteins have a higher proportion of polar amino acids within the bilayer core than do the single-pass ones. The binding of D-glucose to an extracellular domain of the glucose transporter is thought to cause a conformational change in the protein, which allows the polar amino acids within the bilayer core to hydrogen-bond with the hydroxyl groups of glucose,

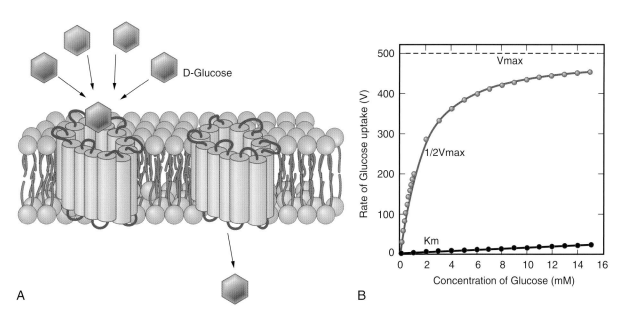

Figure 2–19. Glucose transport into red blood cells: facilitated diffusion. **A:** A glucose transport protein (permease) for glucose. **B:** Glucose moves down its concentration gradient into red blood cells at a rate much faster than would be predicted by simple diffusion through the lipid bilayer *(black line)*. The plot for rate of glucose uptake versus external glucose concentration is hyperbolic *(red line)*. The rate increases with external glucose concentration until it reaches a maximal velocity (V_{max}). The K_m is the concentration of external glucose at which half maximal velocity is reached.

thereby facilitating its movement down its concentration gradient. Facilitated diffusion is sometimes called **passive transport** because it requires no external source of energy. D-Glucose simply moves down its concentration gradient.

ACTIVE TRANSPORT

Active transport and selective permeability of the plasma membrane for ions creates large differences in the ionic composition of the cytosol and extracellular fluid surrounding mammalian cells. As an example, the Na^+ concentration is maintained at approximately 10- to 20-fold greater concentration outside than inside cells, whereas K^+ is at a 20- to 40-fold greater concentration inside the cells. Whereas the calcium concentration in plasma is about 2 mM, inside the red cell, it is maintained at nanomolar concentrations by a pump that uses energy (ATP) to expel calcium from the cells. Similarly, the enzyme Na/K-ATPase uses the energy of ATP hydrolysis to pump three Na^+ ions out of the cell against their electrochemical gradient and two K^+ ions into the cytosol against their electrochemical gradient (Fig. 2–20). The Na/K-ATPase is an important example of **primary active transport,** for which energy derived from ATP hydrolysis directly moves molecules across membranes against their electrochemical gradient.

The extracellular domains of the α subunit of mammalian Na/K-ATPases contain binding sites for two K^+ ions and the inhibitor ouabain, a compound related to the cardiac glycoside digitalis. The intracellular domains contain binding sites for three Na^+ ions, ATP, and a phosphorylation site. The α subunit is autophosphorylated by ATP in a single aspartate residue, conformationally converting the ATPase from a form that transports K^+ to a form that transports Na^+. At the outer surface of the membrane, binding of K^+ promotes hydrolysis of the phosphate group from the α subunit. Cleavage of the phosphate converts the carrier back to a form that preferentially transports K^+.

That three Na^+ ions are pumped out and two K^+ ions enter the cell with hydrolysis of one ATP molecule is important in several ways. First, Na/K-ATPase is defined to be electrogenic because each cycle leads to one net positive charge to the outside surface of the membrane, thereby contributing in a small way to the development of the membrane potential. Second, three osmotic particles are pumped out, whereas only two osmotic particles are pumped in for each cycle of ATP hydrolysis. This counteracts the swelling of the cells induced by the large impermeable anions within cells. Similar ATPases pump Ca^{2+} out of cells or back into the sarcoplasmic reticulum of muscle, or pump protons into the lumen of the stomach. These are called the *P class of ATPases* because they contain nearly identical sequences surrounding the phosphorylated aspartate. Similarly, the flippase moves lipid components against its concentration gradient hydrolyzing one ATP per PS molecule transported (see Fig. 2–10).

SECONDARY ACTIVE TRANSPORT

The movement of Na^+ down its electrochemical gradient can be coupled to the movement of another molecule against its gradient. This is referred to as **secondary active transport,** because the Na^+ electrochemical gradient is maintained by Na/K-ATPase. The transported molecule and cotransported molecule can move in the same direction, which is called **symport,** or they can move in opposite directions across the membrane, which is referred to as **antiport** (Fig. 2–21).

An example to illustrate such cotransport is the uptake of amino acids and glucose from the diet. Glucose is transported against its concentration gradient from the intestinal lumen into the intestinal epithelial cell across the apical membrane by cotransport with sodium (Fig. 2–22).

The glucose can then be transported down its concentration gradient at the basolateral membrane into the blood by facilitated diffusion. The process at the apical membrane is conducted by a glucose-Na^+ symport

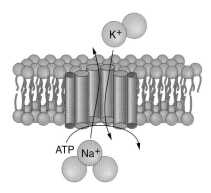

Figure 2–20. The Na/K-ATPase is an electrogenic pump. It moves three Na^+ ions out of the cell and two K^+ ions into the cytoplasm at the expense of ATP hydrolysis to ADP and inorganic phosphate.

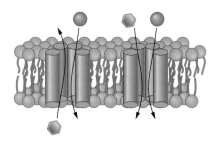

Figure 2–21. **Coupled transport.** The transport of one molecule across a biological membrane can be coupled by the transport protein to the movement of another molecule. If the movement of both molecules is in the same direction, the cotransport is referred to as *symport*. If the molecules are being moved in opposite directions, the cotransport is referred to as *antiport*.

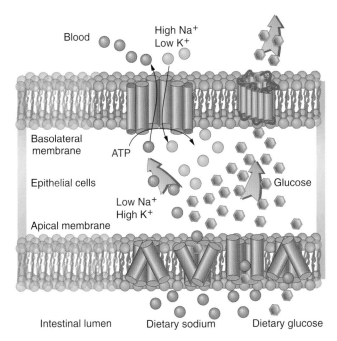

Figure 2–22. **Transport of glucose into and out of intestinal epithelial cells.** Glucose enters from the intestinal lumen through the apical membrane of the epithelial cells by a glucose Na⁺ symport transporter. The binding of one Na⁺ molecule and one glucose molecule to separate sites on the extracellular domain of a glucose Na⁺ symport protein causes the transporter to change conformation. The change of conformation creates a channel through which Na⁺ and glucose can be transported into the cytosol. The symport protein then returns to its original conformation. The Na⁺ ions are pumped back out of the cell by Na/K-ATPase, located on the basolateral membrane. Glucose then exits the cell by facilitated diffusion, by permeases located in the basolateral membrane.

protein. The binding of Na⁺ and glucose to the symport protein causes a conformational change that allows both Na⁺ and glucose to enter the cell together. The energy that moves glucose against its chemical gradient is derived from Na⁺ moving down its electrochemical gradient. This process is indirectly linked to the ATP hydrolysis, which fuels the pumping of Na⁺ out of the epithelial cell. Similarly, amino acids are transported against their concentration gradient. The Na/K-ATPase that pumps Na⁺ out of the cell is located at the basolateral surface. Thus, glucose, amino acids, and Na⁺ enter at the apical surface and exit the epithelial cell into the blood at the basolateral surface.

Another example is the antiport of sodium and calcium ions, which plays an important role in cardiac muscle contraction. The uptake of Ca²⁺ into cardiac muscle triggers contraction. The accumulated Ca²⁺ is then moved out of the cardiac muscle cell by an antiport protein, which is powered by the simultaneous movement of Na⁺ down its electrochemical gradient into the heart muscle cells. Again, the Na⁺ electrochemical gradient is maintained by Na/K-ATPase; therefore, this is another example of secondary active transport. Drugs such as digoxin (Crystodigin, Lanoxin) and ouabain

increase the force of heart muscle contraction. These drugs inhibit Na/K-ATPase, causing an increase in intracellular Na⁺. This dissipates the Na⁺ electrochemical gradient and inhibits the Na⁺-Ca²⁺ antiport. This results in increased intracellular Ca²⁺ and, consequently, stronger heart contraction.

ION CHANNELS AND MEMBRANE POTENTIALS

The hydrophobic environment of the phospholipid membrane bilayer is virtually impermeable to hydrated ions in aqueous solution. Ions are immiscible with the nonpolar hydrocarbon fatty acid chains of membrane phospholipids and, therefore, are unable to cross the lipid bilayer by simple diffusion. Permease proteins allow transporting of ions and other hydrophilic substances across membranes, but another major class of transport proteins works differently and highly efficiently. These transporters form aqueous ion channels that traverse the lipid bilayer. Ion channels are distinguished physiologically from carrier proteins by the high velocity of ion flux they allow across the membrane, without expending energy. Whereas permease-mediated transport operates at a maximum of 10⁵ ions/sec, ions flow at a rate 100 to 1000 times faster through channels. This mediates not only rapid electrical signaling of nerve impulses and muscle contraction but also many biological responses common to nonexcitable cells.

An important principle is that ion channels determine the rate but not the direction of ion flow across cell membranes. Movement of a nonionic solute through a channel is strictly passive, determined only by its concentration gradient across the membrane. For an ion, the direction of flow depends on both the chemical concentration gradient *and* the electrical potential across the membrane.

The concept of the ion channel is 150 years old. However, a better understanding with respect to their basic properties was not gained until the 1960s. A major breakthrough came with the discovery that ion channels could be artificially introduced into cell membranes by treating them with small hydrophobic proteins, termed **ionophores.** Made primarily by different fungi, some ionophores have been used as antibiotics, but many more have become important tools for the cell biologist.

The loss of phospholipid asymmetry can be induced by activating phospholipid scrambling by increasing cytosolic calcium. In the laboratory, this can be achieved by treating cells with the calcium ionophore A23187. The addition of this compound to membranes makes them highly permeable to calcium. The calcium pump, which normally keeps intracellular calcium low by actively pumping calcium out at the expense of ATP, is completely overwhelmed as the ionophore equilibrates calcium across the bilayer at near to diffusion rates.

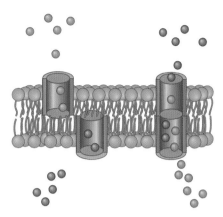

Figure 2–23. Structure of the helix and transmembrane channel formed by gramicidin A. Two gramicidin A peptides dimerize head to head to span the lipid bilayer.

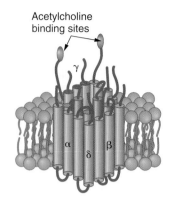

Acetylcholine binding sites

Figure 2–24. A model of the acetylcholine receptor. The pentameric configuration of the receptor is typical of many other members of the ligand-gated ion channel family. Two of the subunits are identical and three are different; each of the two α subunits contains an extracellular binding site for acetylcholine.

Gramicidin A, as another example, is a linear polypeptide with 15 amino acids that, like all ionophores, is readily miscible with the phospholipid bilayer. This ionophore, when inserted into the membrane, forms an unusual α-helix. Alternating D-and L-amino acid residues in the molecule orient the polar (hydrophilic) carbonyl oxygen atoms and amide nitrogens of the peptide bonds toward the hollow center of the helix, where they form the wall of the channel pore. Hydrophobic amino acid side chains radiate outwardly from the helix to anchor the ionophore in surrounding lipid (Fig. 2–23). Two helical molecules of gramicidin A are thought to align end to end, to completely span the lipid bilayer. The different flow rates of various ion species across membranes punctuated with these simple channels indicate that gramicidin pores prefer small cations to anions, and that partial dehydration is necessary for ions to traverse the channel in single file, as they do through many animal membrane channels.

Whereas channel-forming proteins expressed by mammalian cells also form helices in membranes, they differ fundamentally from ionophores such as gramicidin A. Because membrane-spanning proteins in eukaryotes consist solely of L-amino acids, their α-helix lacks a central pore. The carbonyl oxygens and amide nitrogens of the polypeptide backbone hydrogen-bond to each other, and thus are prevented from interacting with water. As a result, these α-helices are naturally hydrophobic and highly suited for association with the lipid bilayer. By organizing several membrane-spanning domains such that hydrophilic amino acid residues point toward the aqueous channel and hydrophobic residues face the lipid bilayer, channels can be formed as illustrated for the acetylcholine receptor in Figure 2–24.

Many different ion channels have been reported to date. Channels are often glycoproteins that contain several α-helical membrane-spanning regions, flanked by hydrophilic portions protruding into the extracellular space and cytoplasm. Given the high flux of ions through these channels, regulation of their permeability status appears to be important to maintain proper ion balance across the membrane. Indeed, ion channels, in general, exist in two or more conformations, including a brief but stable open state and a stable closed state. Different types of stimuli can control the opening and closing, or **gating,** of ion channels. Three major varieties of rapidly gated ion channels and a fourth slowly gated type, the gap junction, can be distinguished (Fig. 2–25).

Voltage-gated channels are responsible for the propagation of electrical impulses over long distances in nerve and muscle, and they open specifically in response to a change in the electric field that exists across the plasma membrane of cells at rest. **Ligand-gated** ion channels are insensitive to voltage change, but they are opened by the noncovalent, reversible binding of a chemical ligand. These substances include neurotransmitters or drugs that bind to the extracellular portion of the receptor. Alternatively, an intracellular second messenger or enzyme interacting with the cytoplasmic face of the channel can also influence its conformational state. Ligand-gated ion channels enable rapid communication between different neurons and between neurons and muscle or glandular cells across synapses. A few cell types have **mechanically gated** ion channels for which the opening is controlled by cellular deformation. A fourth class of ion channel, the **gap junction,** enables ions to flow between adjacent cells without traversing the extracellular space. Gap junctions are not rapidly gated, but rather open and close in response to changes in the intracellular concentration of Ca^{2+} and protons. An example of the ligand-gated class is the channel gated by acetylcholine on skeletal muscle cells. Many different neurotransmitters other than acetylcholine operate elsewhere, primarily by binding to transmitter-gated ion channels. As a rule, each of these ligands has its own specific ion channel receptor that, when opened,

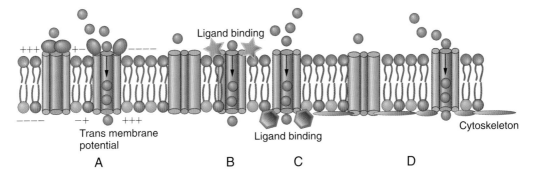

Figure 2–25. Classes of ion channels stimulated by different gating mechanisms. Ion channels are distinguished according to the signal that opens them. **A:** Voltage-gated channels require a deviation of the transmembrane potential. Ligand-gated receptors respond to the binding of a specific ligand, either (**B**) an external neurotransmitter molecule or (**C**) an internal mediator such as a nucleotide or ion. **D:** Mechanically gated channels can sense movement of the cell membrane linked by cytoskeletal filaments to the channel protein. Each effector causes an allosteric change that opens the channel, thereby causing an ion flux across the membrane.

is selectively permeable to a certain ion. These channels are restricted to special junctions, termed **synapses,** between nerve terminals and apposed stretches of the muscle cell membrane. Acetylcholine released from a stimulated nerve terminal diffuses across a cleft in the synapse and functions at the muscle cell as a neurotransmitter molecule that binds to external sites on the acetylcholine receptor channel protein. The binding of two acetylcholine molecules to the α subunits evokes a conformational change that briefly opens the channel for about 1 millisecond before the channel recloses, still remaining bound to acetylcholine (Fig. 2–26).

Once the channel is closed, acetylcholine dissociates from the channel, which returns to an unbound conformation. Lining the circumference of the channel pore is one of four α-helical–spanning regions in each subunit that contains more hydrophilic amino acid residues than do the others. Unlike the multimeric construction of these ligand-gated channels, voltage-gated ion channels for Na^+ and Ca^{2+} are constructed from a single large polypeptide chain. The interconnected domains are thought to arrange as a tetramer surrounding the voltage-gated ion channel. Differing somewhat from Na^+ and Ca^{2+} channels is the formation of the voltage-gated K^+ channel, which consists of four identical subunits. Despite these differences, structures of K^+, Ca^{2+}, and Na^+ channels in the membrane are strikingly similar. In these voltage-gated receptors, one of the membrane-spanning helices is thought to function as the actual voltage sensor of depolarization that causes a conformational change to open the channel pore.

Ion channels in mammalian cell membranes are selective for the type of ions that flow through them. The acetylcholine receptor at the neuromuscular junction is permeable to small cations (K^+, Na^+, Ca^{2+}), not anions. Others channels are permeable to a single cation, either K^+, Na^+, or Ca^{2+}. Similarly, channels permeable only to

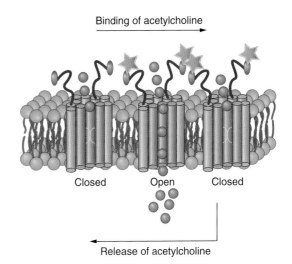

Figure 2–26. Three conformational states of the acetylcholine-gated ion channel. The binding of two acetylcholine molecules alters the protein conformation to open the channel pore. However, the effect is only transient; the pore soon closes with acetylcholine still bound to the receptor sites. Once the ligand dissociates from the receptor, the channel can return to a closed but receptive conformation.

Cl^- have been identified. The voltage-gated K^+ channel, for example, is too small to pass Ca^{2+}. Interestingly, even though Na^+ is the smaller ion, this channel shows a 100-fold greater selectivity for K^+ than for Na^+. Hydration of cations in solution appears to play an important role. Because Na^+ is smaller than K^+ in ionic diameter, its charge density and electric field are stronger. Hence, Na^+ interacts more strongly with surrounding water molecules compared with K^+. This reduces its mobility and the larger shell of electrostatically bound water molecules results in a larger "virtual" diameter such that hydrated Na^+ cannot permeate the smaller-diameter K^+ channels. Although ion channels are most often recognized in the role of electrical signaling in nerve and muscle cells, they exist, to some degree, in all cells to mediate other functions. The selectivity of a cell membrane for permeant ions depends on the relative proportions of various types of ion channels. A single neuron can have as many as four different ion-selective channels. The most common one among neural and non-neural cell types is termed the **K⁺ leak channel** because its opening does not require a specific gating stimulus. These channels enable all cells in the body to maintain a voltage difference across their plasma membrane, the **membrane potential.**

The Membrane Potential Is Caused by a Difference in Electric Charge on the Two Sides of the Plasma Membrane

Cells maintain slightly more negative than positive ions in the cytosol and more positive than negative ions in the extracellular fluid (Fig. 2–27A). Membranes with their hydrophobic lipid bilayer are poor conductors of ionic current, and the transmembrane voltage across the plasma membrane resembles an electrical capacitor. As a result, an accumulation of negative charges along the cytosolic side of the membrane attracts positively charged ions on the extracellular side of the bilayer. The voltage gradient that arises across the membrane (5.0 nm) is nearly 200,000 V/cm. Therefore, membrane potential and transmembrane ionic gradients provide a driving electrical force for many biological processes.

A membrane potential V_m is defined as $V_m = V_i - V_o$, where V_i is the voltage inside the cell and V_o is the voltage outside the cell. Because V_o is arbitrarily set to zero, V_m in the undisturbed, or **resting,** cell becomes negative as the result of the slightly negative net ionic charge of the cytoplasm. Membrane potential is primarily based on four ion species: K^+, Na^+, Cl^-, and organic anions (A^-), such as amino acids and other metabolites. Of these, Na^+ and Cl^- are concentrated in the extracel-

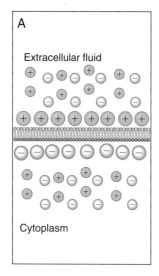
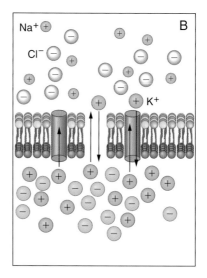

Figure 2-27. **A: The separation of charge across the cell membrane forms a membrane potential.** A net excess negative charge inside the membrane and a matching net excess positive charge outside the membrane form a transmembrane potential difference that is maintained across an impermeable lipid bilayer. Charge on either side of the membrane is concentrated in a thin layer (<1 nm thick) and formed by an extremely small percentage of the total ions in the cell. **B: Opposing forces regulate K⁺ flux across the plasma membrane.** Membrane potential is primarily based on four ion species: K⁺, Na⁺, Cl⁻, and organic anions (−), such as amino acids and other metabolites. The resting membrane potential of a cell permeable only to K⁺ depends on the passive diffusion of K⁺ out of the cell down its concentration. If left unchecked, K⁺ efflux would eventually create an excess negative charge in the cell (an overbalance of organic anions) and a buildup of [K⁺]ₒ, were it not for an electrical driving force moving K⁺ in the opposite direction. An equilibrium results when these two opposing forces counterbalance each other.

lular fluid, whereas K+ and A− are preponderant inside the cell. The active extrusion of Na+ from the cell by Na/K-ATPase maintains the osmotic balance of the cytosol by preventing the influx of water that would otherwise occur. In exchange for pumping three Na+ ions out of the cell, two K+ ions enter the cytosol to counterbalance organic anions that do not permeate the plasma membrane.

Neurons or muscle cells have many Na+ channels in their membranes. They remain closed, however, when the cell is at rest. Therefore, Na+ extruded by Na/K-ATPase cannot readily reenter the cell down its steep concentration gradient. Only nongated K+ leak channels remain open. As a result, K+ will tend to passively leak out of the cell through the K+ leak channels down its steep concentration gradient until the force of outward diffusion is counterbalanced by a second and opposing inward electrical force created by the attraction that organic anions in the cytosol have for K+ (see Fig. 2–27B).

Together, the concentration gradient and the voltage gradient for a particular ion across a membrane determine the net electrochemical gradient that drives the flow of that particular ion species through a membrane channel. When these forces balance, the electrochemical gradient is zero, and no net flow of this ion occurs across the membrane. Given the concentration of an ion inside and outside the cell, the voltage necessary to achieve this equilibrium, termed the **equilibrium potential,** can be calculated from the Nernst equation. For K+ this would be:

$$E_k = \frac{RT}{ZF} \; \ln \frac{[K^+]_o}{[K^+]_i}$$

where E_k is the value of the equilibrium potential of K+, R is the gas constant ($2 \, cal \, mol^{-1} \, K^{-1}$), T is the absolute temperature in K, F is the Faraday constant ($2.3 \times 10^4 \, cal \, V^{-1} \, mol^{-1}$), and $[K^+]_o$ and $[K^+]_i$ are the concentrations of K+ outside and inside of the cell, respectively, such as K+. The charge $(Z) = +1$, for a monovalent cation such as K+, and at 37°C, RT/ZF is 27 mV. By using typical values for $[K^+]_o$ and $[K^+]_i$ in mammalian tissue, the equation renders $E_k = -96$ mV.

The equilibrium potential for any of the other ion species across the membrane can be similarly calculated. Ultimately, the resting membrane potential (V_R) for a cell is determined by permeability of the membrane to specific ions and their concentrations inside and outside the cell. The Goldman equation is a summation of contributions of fluxes of different ions, determined by their concentrations and permeabilities (P).

$$V_R = \frac{RT}{F} \; \ln \frac{P_K [K^+]_o + P_{Na} [Na^+]_o + P_{Cl} [Cl^-]_i}{P_K [K^+]_I + P_{Na} [Na^+]_i + P_{Cl} [Cl^-]_o}$$

Based on the individual contributions, this equation can be simplified drastically. Because Ca^{2+} does not contrib-

ute significantly, it is not included. Chloride equilibrates across the membrane through nongated Cl− channels, but most remains in the extracellular fluid to counterbalance nonpermeable intracellular anions. That is, if the permeability to one ion far supersedes that of others ($P_K \gg P_{Cl}$, P_{Na}), the Goldman equation is reduced to the Nernst equation for that ion. When the cell is at rest, the ratio of open K+ leak channels to Na+ channels is high, making the cell much more permeable to K+ than to Na+. The contribution of Na+ influx to V_R is minimal, and a balance of −80 to −90 mV for V_R is attained by mammalian cells, a value *near* E_K and far from E_{Na}.

Together, for a nerve or muscle cell with multiple ion-selective channels in its membrane, V_R is affected by the permeability of the membrane for each diffusible ion species (K+, Na+, Cl−) and the concentration of each inside and outside the cell. Consider, for instance, the glial cell, having only nongated K+ leak channels in its membrane. V_R is essentially equal to E_k. What would happen when we open a few Na+ channels in the glial cell membrane, as illustrated in Figure 2–28. Because Na+ is more concentrated outside than inside the cell, it will flow passively into the cell through the opened Na+ channels. The electrical force of attraction of a slightly electronegative membrane potential generated by the Na/K-ATPase and E_K (−96 mV) will also attract Na+ into the cell. The magnitude of the latter can be calculated from the Nernst equation: $E_{Na} = +67$ mV. This tells us that E_{Na} is 163 mV away from the V_R established by E_K alone (−96 mV). As a result, both the electrical and chemical forces work in the same inward direction to form a strong electrochemical gradient driving Na+ into the cell. The Na+ influx should then depolarize the cell; that is, it should reduce the charge separation across the membrane by making the interior less electronegative relative to outside of the cell. Indeed, if K+ efflux and the smaller Na+ influx were allowed to continue unchecked across the plasma membrane of a neuron, transmembrane gradients for both ions would eventually dissipate, because $[K^+]_i$ would plummet and $[Na^+]_i$ would gradually increase, thereby reducing the V_R. Opposing this effect is Na/K-ATPase, which continues to pump three Na+ ions out of the cell for every two K+ ions pumped in, and thereby maintains the V_R constant. Because there is a net transfer of positive charge out of the cell, the pump is said to be **electrogenic,** creating the slight excess of negative charge inside the cell membrane. Thus, passive channel-mediated ion fluxes occurring by simple diffusion are balanced by active fluxes that require energy. A steady state between the two processes is reached where the *net* ion flux across the membrane is zero to define the resting membrane potential (Fig. 2–28).

Any disturbance to a cell that increases the membrane permeability for an ion will drive the membrane potential away from V_R in the direction of the equilibrium potential for that ion. These transient deviations

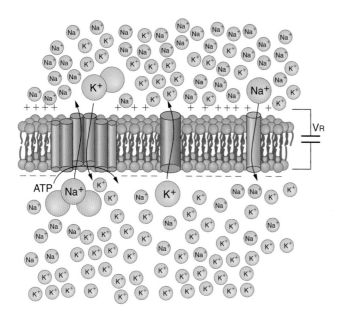

Figure 2-28. Passive and active fluxes maintain the resting membrane potential. The cell at rest maintains a steady state whereby Na⁺ influx and K⁺ efflux defined by passive diffusion is balanced by active transport of these ions in the opposite direction by Na/K-ATPase.

from V_R and the opening of gated ion channels that cause them are the basis of electrical signals that convey information not only along and between nerve cells but also from nerve to muscle.

Neurons convey signals along their length by generating passively spreading local potentials and self-propagated action potentials. A local potential arises when the cell membrane is stimulated to become more permeable to certain ions than it is at rest. If Na⁺ influx results, the voltage drop across the membrane is reduced and the signal is said to be **depolarizing.** Alternatively, enhanced Cl⁻ influx will increase the voltage drop in some neurons and **hyperpolarize** the cell relative to its resting state. When current enters the neuron through open membrane channels, it diffuses in all directions for a distance that depends on the intrinsic properties of the neuron. The voltage signal decreases exponentially as it travels away from the site of entry as ions will leave the cell by nongated K⁺ leak channels. This rapid decrease in the spreading wave of depolarization is sufficient to convey electrical signals toward the cell body along relatively short input fibers called **dendrites.** However, such signals would dissipate well before leaving the cell along its **axon,** which is typically a much longer structure (Fig. 2–29).

Action Potentials Are Propagated at the Axon Hillock

"Nerve impulses" are transmitted along most neurons as local potentials that travel toward the cell body through dendritic processes and propagate away from

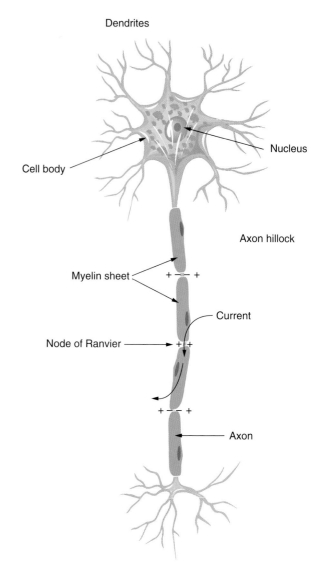

Figure 2-29. Prototypical neuron. Incoming local potentials from dendrites converge at the cell body and reach the axon at its origin. The action potential, typically triggered at a region known as the *axonal hillock,* is propagated down the length of the axon, where it will initiate a chain of events leading to neurosecretion from the terminal branches.

the cell body along the axon as **action potentials.** The action potential is typically triggered at a membrane segment where the axon emanates from the cell body, a region called the **axon hillock.**

In this region, electrical signals encounter voltage-gated Na⁺ channels at sufficient density to trigger the action potential. At appropriate current load, a few local voltage-gated Na⁺ channels in the axon will open, allowing a small amount of Na⁺ to enter the axon down its steep electrochemical gradient. Increased intracellular Na⁺ further depolarizes the cell and recruits more voltage-sensitive channels to open, resulting in more Na⁺ entry. This progressive, self-amplifying mechanism for depolarization reaches a limit when the membrane potential (V_R) has shifted from about −90 mV for mammalian myocytes and neurons to about −50 mV, when the equilibrium potential for Na$^+$ (E_{Na}) is nearly reached. Current will diffuse longitudinally along the axon. In this way, what started as a local depolarization reaches the threshold to initiate a new action potential at a neighboring membrane segment.

Two mechanisms next engage to return the same membrane patch to its original resting potential. First, opened voltage-gated Na$^+$ channels rapidly close in the depolarized membrane because they are conformationally unstable in the open state. Once closed again, the

channels remain in an inactivatable state until after the membrane is repolarized. Similarly, as described for the acetylcholine receptor (see Fig. 2–26), voltage-gated Na$^+$ channels can exist in three different conformations: closed, but activatable; open; and closed, but inactive. In this case, not a ligand but voltage drives the transition of recruited channels through each of these states (Fig. 2–30).

The second mechanism contributing to the decay of the action potential is the opening of voltage-sensitive K$^+$ channels. These channels respond more slowly to depolarization than do voltage-gated Na$^+$ channels and

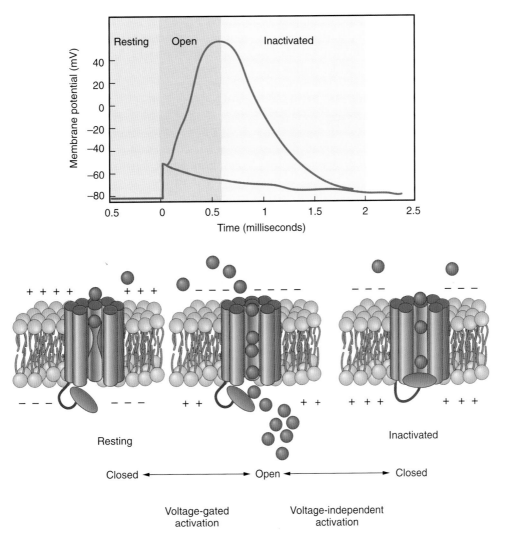

Figure 2–30. Three conformation states of the voltage-gated Na$^+$ channel. In response to a brief pulse of current, depolarization of a mammalian neuron to a threshold of about −50 mV *(red curve)* triggers the opening of the voltage-gated Na$^+$ channel by opening an activation gate formed by the channel protein lining. Ions flow and the membrane potential increases *(blue curve)*. The open state is metastable and is inactivated rapidly by a separate inactivation-gating mechanism provided by the cytosolic portion of the channel protein. After reaching a maximum of +50 mV, the membrane potential declines to its resting state because Na$^+$ channels inactivate, and the efflux of K$^+$ ions through nongated leak channels continues unabated. Another action potential is not possible until the Na$^+$ channels have returned to the closed but activatable state. In the absence of voltage-gated Na$^+$ channels, the modest depolarization evoked by the current stimulus would have immediately begun to decay *(green curve)*.

are not opened until the action potential is near its peak. With a membrane potential of −50 mV, a strong electrochemical gradient for K+ efflux builds inside the cell and is released when the voltage-sensitive K+ channels transiently open. As a result, the sudden loss of positive charge repolarizes the cell more quickly. The voltage-gated Na+ channels are closed now and the voltage-sensitive K+ channels aid strongly the nongated K+ leak. The return of the membrane potential toward the K+ equilibrium potential closes voltage-gated K+ channels and allows the inactivated Na+ channels to regain their activatable state. Thus, by accelerating the decay of the action potential and membrane repolarization, voltage-gated K+ channels reduce the length of a refractory period (<1 millisecond) before another action potential can be triggered at the same site. Thus, the action potential self-propagates as a transient depolarizing wave form racing toward the axonal terminal at speeds in vertebrates ranging from 1 to 100 m/sec, depending on the type of axon.

The all-or-none nature of the action potential overcomes the problem of keeping a proper signal over relatively long distances. However, just as with your cable television signal, DSL modem, or the electrical power grid, additional properties of the axon determine the efficiency at which these nerve impulses travel. As the action potential propagates from one patch of axolemma to the next, local current diffuses within the axon to depolarize neighboring segments to the threshold value. The effective distance over which this occurs is determined by the same principles that govern any flow of electricity. In this case, important parameters are the internal resistance (r_i) of the axonal cytosol and the membrane resistance (r_m) to current (K+) escaping through open channels. Both a decrease in r_i and an increase in r_m will allow larger currents to flow longer distances inside the axon, thereby charging the capacitance of neighboring membranes at a faster rate. Consequently, a mere doubling of the axonal diameter (reducing r_i) can significantly speed the conduction rate of nerve impulses. Although r_m would actually decrease as a result of doubling the axonal circumference, and thereby the number of open ion channels per segment, this is compensated by a fourfold increase in cross-sectional area of the axon, which reduces r_i by a factor of 4. The net effect of enlarging the axonal diameter is to increase the propagation velocity of the action potential. Although such rapid signaling along giant axons evolved by the squid works well, it would not be sufficient in humans, because it would require a spinal cord the size of a tree trunk. In vertebrates, considerable savings of energy and space is achieved by insulating many axons with a **myelin sheath.**

The formation and maintenance of myelin is the task of two glial cell types: oligodendrocytes in the central nervous system and their counterparts in peripheral nerves, the Schwann cells. These cells generate enormous quantities of flattened plasma membrane and wrap it around axons in a concentric fashion to form a biochemically specialized sheath up to 200 layers in thickness. Every myelinating Schwann cell invests a single axon to form a segment of sheath, termed an **internode,** that occupies approximately 1 mm of axon length. By contrast, a single oligodendrocyte extends branches from its cell body that flatten and expand to cover as many as 40 different axons.

The extremely high lipid-to-protein ratio and the stacking of the individual myelin membranes results in a high transmembrane resistance (r_m), greatly improving the electrical characteristics of the axon. Between one internode of myelin and the next are regions of axon, varying in length from 0.5 to 20 mm, termed **nodes of Ranvier** (see Fig. 2–29). Almost all membrane channels, including voltage-gated Na+ channels, are confined to nodes of Ranvier along a myelinated axon. Current entering the axon through Na+ channels at nodes of Ranvier spreads with greater efficiency than would otherwise occur in a nonmyelinated axon, because the increased r_m minimizes current leakage.

Moreover, in a nonmyelinated axon, the buildup of opposite charges on either side of the membrane occurs along its entire length. For a myelinated axon, this is confined to the nodes of Ranvier. Having the lower capacitance, myelinated axons require the influx of fewer positive charges to reduce the transmembrane potential to the threshold for an action potential. Hence, in addition to speeding the nerve impulse, myelination drastically reduces the amount of energy needed for axonal conduction. In the unmyelinated axon, Na/K-ATPase is needed to restore the membrane potential along the entirety of the fiber. In the myelinated axon, this occurs only at the depolarized nodes of Ranvier.

Because myelin speeds the conduction of nerve impulses, diseases or toxins that injure the myelin sheath or myelin-producing cells can cause significant neurologic problems. Axons stripped of their myelin sheath conduct impulses slowly or not at all. Conduction deficits resulting from damage to the brain and spinal cord are particularly severe because oligodendrocytes, unlike Schwann cells, fail to regenerate sufficiently in most circumstances. Moreover, damage to a single oligodendrocyte is of greater physiological consequence because it produces internodes of myelin that surround segments of several different axons. Multiple sclerosis is the most prevalent, and is the prototype, of demyelinating diseases that affect the central nervous system. The symptoms of this disease that are usually manifested clinically are the result of dysfunctional sensory and motor neuron systems that are most dependent on rapid neurotransmission.

SUMMARY

Membranes form the boundaries of cells and cell organelles. They separate inside from outside and allow compartmentalization of DNA, RNA, proteins, and other molecules or ions. Without membranes, life as we know it would not be possible. The prokaryotic plasma membrane often surrounded by a protective cell wall encloses a single cytoplasmic compartment. The eukaryotic cell is surrounded by a plasma membrane that defines its outer boundaries. Inside this cell we find cell organelles, including the nucleus, mitochondria, lysosomes, peroxisomes, Golgi apparatus, and ER, all of which by themselves are defined by membranes that allow them to perform their specific functions. In a multicellular organism, cells differentiate to develop particular features. Mammalian cells are different depending on the tissue that they form (e.g., heart, kidney, liver). Moreover, within a certain tissue we find cells that are radically different depending on their function.

All cells in a human have descended from the same fertilized egg and are generated with the same genetic information embedded in DNA. The variation in gene expression that determines the characteristics of each cell in each tissue also determines the characteristics of each membrane. Although plasma membranes or membranes of cell organelles differ with their different functions, their basic structure is similar. They all form a barrier between water-containing compartments and are specific with respect to the compounds that they let permeate. Therefore, whether a cell is (or will become) a neuron or kidney cell, the basic structure of its membranes is similar, whereas the actual composition of the lipids and proteins that form the biological membranes will differ drastically depending on the specific function of the membrane. This variation in composition gives the different cell types their typical structure and shape and allows them to communicate with other cells and transport specific ions, proteins, and other compounds across the membrane.

This chapter describes a number of characteristics of cell membranes that form the basis for the function of membranes in all mammalian cells with their dizzying variety of characteristics and function.

Suggested Readings

de Kruijff B. Biomembranes. *Biochim Biophys Acta* 2004; 1666(1–2):1–290.

Embury S, Hebbel R, Mohandas N, Steinberg M, eds. *Sickle Cell Disease, Basic Principles and Clinical Practice*. New York: Raven Press, 1994.

Hoffman R, Benz JE, Shattil S, Furie B, Cohen H, Silberstein L, McGlave P. *Hematology, Basic Principles and Practice*. New York: Elsevier, 2005.

Shapiro H. *Practical Flow Cytometry*. New York: Wiley-Liss, 2005.

Singer SJ, Nicolson GL. The fluid mosaic model of the structure of cell membranes. *Science* 1972;175:720–731.

Tanford C. *The Hydrophobic Effect*. New York: John Wiley & Sons, 1980.

Yeagle P, ed. *The Structure of Biological Membranes*. London: CRC Press, 1992.

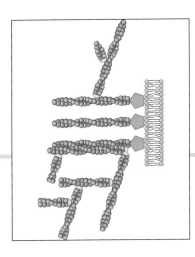

Chapter 3

Cytoskeleton

An intriguing feature of eukaryotic cells is the ability of extracts that contain cytosol, devoid of organelles, to roughly maintain the shape of the cell and even to move or contract, depending on how the extracts are prepared. This maintenance of structure by the cytosol arises from a complex network of protein filaments that traverse the cell cytoplasm, called the **cytoskeleton**. The cytoskeleton is not simply a passive feature of the cell that provides structural integrity; it is a dynamic structure that is responsible for whole-cell movement, changes in cell shape, and contraction of muscle cells—it provides the machinery to move organelles from one place to another in the cytoplasm. In addition, recent studies have provided evidence that the cytoskeleton is the master organizer of the cell's cytoplasm, furnishing binding sites for the specific localization of ribonucleic acids (RNA) and proteins that were once thought to diffuse freely through the cytoplasm.

Amazingly, the many activities of the cytoskeleton depend on just three principal types of protein assemblies: **actin filaments, microtubules,** and **intermediate filaments** (IFs). Each type of filament or microtubule is formed by a specific association of protein monomers. The dynamic aspects of the cytoskeletal structures arise from accessory proteins that control the length of the assemblies, their position within the cell, and the specific-binding sites along the filaments and microtubules for association with protein complexes, organelles, and the cell membrane. Thus, although the protein filaments and microtubules define the cytoskeleton, the partici-

pation of accessory or regulatory proteins conveys its diverse activities. This chapter discusses the structures built from the interaction of proteins with the individual cytoskeletal assemblies, beginning with an examination of actin filaments. The initial focus is on their well-defined role in muscle cell contraction; then their participation in the membrane skeletal complex and structures formed in nonmuscle cells is described. Through a discussion of cell motility, we consider how the different components of the cytoskeleton work together as an integrated network that leads to an essential cellular function. Finally, the IFs and microtubule components of the cytoskeleton are discussed.

MICROFILAMENTS

Actin-Based Cytoskeletal Structures Were First Described in Muscle Tissue

Actin, first isolated from skeletal muscle, was originally thought to be a protein found exclusively in muscle tissue. However, actin is a component of all cells, representing 5% to 30% of the total protein in nonmuscle cells. Although present in all eukaryotic cells, actin isolated from nonmuscle cells is different from that found in skeletal muscle. Six different isoforms of actin have been described in human and animal cells: α-skeletal, in skeletal muscle; α-cardiac, in heart muscle; α-vascular, in smooth muscle of the vasculature; γ-enteric, in smooth muscle of the viscera; and β-cytoplasmic and γ-cytoplasmic, preponderantly in nonmuscle cells. Actin is an extremely conserved protein, with greater than 80% identity of amino acid sequence between the different isoforms. The major difference in amino acid sequence occurs at the NH_2-terminal end of the actin isoform and appears to have little effect on the rate of actin monomer polymerization into filaments, but it is essential for the association of specific actin-binding and regulatory proteins (see detailed discussion later in this chapter).

Many of the other protein components that are common to actin-based cytoskeletal structures in all cells were also first isolated from muscle tissue. In muscle, these proteins demonstrate a rigorous organization, forming the specialized contractile machinery of the muscle cell. Therefore, we examine the role of actin filaments and associated proteins in muscle cells first to lay the groundwork for our understanding of actin-based structures in nonmuscle cells.

Skeletal Muscle Is Formed from Bundles of Muscle Fibers

The organization of the skeletal muscle from the gross level to the molecular level is depicted in Figure 3–1. Skeletal muscle is composed of long, cylindrical, multinucleated cells that can be several centimeters in length. The individual muscle fibers are surrounded by a delicate, loose connective tissue, called the **endomysium,** that carries the capillary network of blood supply for the muscle. Bundles of the individual muscle fibers are grouped together, forming the muscle fasciculi, which are bounded by a layer of connective tissue, the **perimysium.** The fasciculi are grouped to form the definitive muscle tissue that is covered by a thick, tough connective tissue layer, the **epimysium.** The three connective tissue layers of muscle tissue contain fibers of collagen and elastin and differ from each other primarily by their thickness. Skeletal muscle causes specific movements of the body by their attachment to tendons that are usually attached to the skeleton or bone.

The Functional Unit of Skeletal Muscle Is the Sarcomere

Each skeletal muscle cell, or myofiber, contains many bundles of regularly arranged filaments, called *myofibrils.* It is the highly structured arrangement of filaments within the **myofibrils** that give skeletal muscle its characteristic striped or striated appearance. Skeletal muscle is the best biological example of the relation of structure, as viewed through the microscope, with function. Longitudinal sections of skeletal muscle, viewed under the light and electron microscopes, demonstrate an

ordered banding pattern (Fig. 3–2). These are called the **A band, I band,** *and* the **Z disk** *or* **Z line.**

The A band is the dark-staining region of the myofilaments, and it contains the thick filaments, composed of the protein myosin II, as well as overlapping thin filaments. The light-staining I band contains the thin filaments, of which the main protein component is actin. The Z disk appears as a dark line that bisects the I band. In electron micrographs of skeletal muscle, the dark-staining A band is observed to have distinct regions, termed the *H band* and *M line.* The H band is a zone of lighter staining within the central region of the A band, which is bisected by a dark-staining M line. This region of the A band is where the assembly of the myosin thick filaments occurs.

The segment of the myofibril between two Z disks, containing a complete A band and two halves of adjoining I band regions, is called the **sarcomere.** The sarcomere is the functional contractile unit of the myofibril. The myosin thick filaments mark the A band, which is equidistant from the two Z disks of the sarcomere. The thin filaments of the sarcomere are joined to the Z disk and extend through the light-staining I band region and partially into the A band, where they interdigitate with the myosin thick filaments. The Z disk functions to anchor the thin filaments of the sarcomere. Cross sections through different portions of the sarcomere provide additional information about the organization of the thick and thin filaments (see Fig. 3–1). A cross section through the I band shows only thin filaments, arranged in a hexagonal pattern. A section through the H zone of the A band demonstrates only thick filaments, whereas a section through the M band zone of the A band shows a network of coiled filaments, representing the assembly of the bipolar myosin thick filaments. The segment of the A band at which the thin filaments interdigitate with the myosin thick filaments shows that each thick filament is surrounded by six thin filaments. This arrangement of thick and thin filaments is an essential structural feature of the sarcomere and is required for the sliding of filaments during contraction.

Thin Filaments Are Built from the Proteins Actin, Tropomyosin, Troponin, and Tropomodulin

All eukaryotic cells appear to contain filaments 7 to 8 nm in diameter, called **microfilaments,** that are polymers of the protein actin. These filaments, referred to as **filamentous** or **F-actin,** are built from polymerization of a globular actin monomer, called **G-actin,** which has a relative molecular mass (M_r) of 43,000 (43 kDa). Each F-actin microfilament appears as two helically intertwined chains of G-actin monomers for which a complete turn of the helix occurs over a distance of 37 nm, or 14 G-actin monomers (Fig. 3–3).

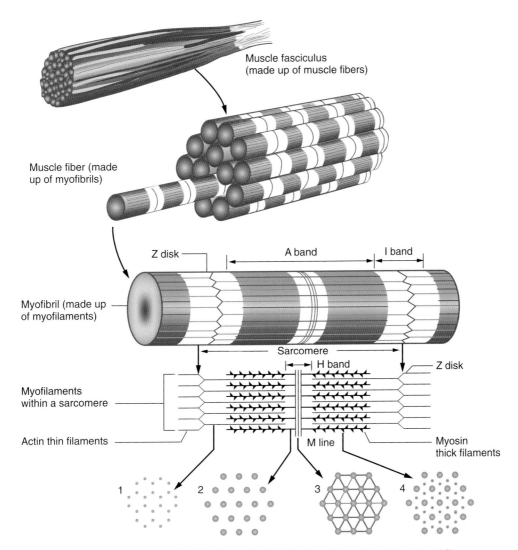

Figure 3–1. Organization of a skeletal muscle. A skeletal muscle consists of bundles of fibers called *fasciculi.* Each fasciculus consists of a bundle of long, multinucleated muscle fibers that are the cells of the muscle tissue. Within the muscle cells are the myofibrils, which are composed of highly organized arrangements of myosin II (thick) and actin (thin) filaments. The extreme structural organization of the myofilaments is the basis for the striated appearance of skeletal muscle. The myofilaments are organized into the functional units of skeletal muscle, the sarcomere, which extends from one Z disk to the next. Actin thin filaments extend from the Z disk *(light-staining I band)* toward the center of the sarcomere, where they interdigitate with the myosin thick filaments *(dark-staining A band).* Cross sections through the sarcomere near the Z disk *(1)* show the ~8-nm actin thin filaments, whereas sections in the regions of the A band *(4)* demonstrate that each –15-nm thick filament is surrounded by a hexagonal array of six actin thin filaments. Sections through the sarcomere near the center in the segment of the A band referred to as the *H band* show the organization of the myosin thick filaments *(2),* whereas cross sections through the center of the H band demonstrate a network of filaments that participate in the assembly of the thick filaments to form the M line *(3). (Modified from Bloom W, Fawcett DW. A Textbook of Histology, 10th ed. Philadelphia: WB Saunders, 1975.)*

Each G-actin monomer must have an adenosine triphosphate (ATP) molecule bound to polymerize onto an actin filament. ATP hydrolysis is slow for actin monomers but greatly accelerates once the monomer is incorporated into the actin filament. If ATP and Mg^{2+} (or physiologic salt concentrations) were added to G-actin at a high enough concentration, it would spontaneously polymerize to F-actin. The polymerization would have several stages (Fig. 3–4).

First would be a lag phase when three G-actin monomers form an actin trimer, which can then serve as a seed or nucleation site for the polymerization of G-actin monomers onto the actin filament during the polymerization phase. Finally, a steady-state phase is reached, during which the rate of addition of G-actin monomers onto the filament equals the rate at which these monomers leave the filament. Actin microfilaments have a polarity, with a fast-growing plus (+) end and a slow-

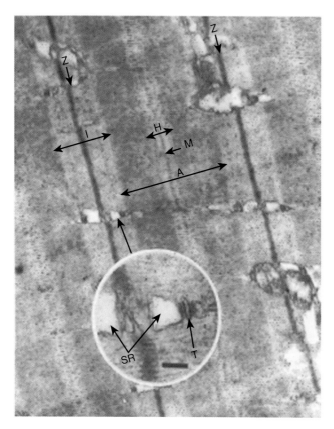

Figure 3–2. Electron microscopy of skeletal muscle. A longitudinal section through a skeletal muscle cell demonstrates the regular pattern of cross-striations derived from the myofibrils. As shown in this low-magnification electron micrograph, the skeletal-muscle cell has many myofibrils aligned in parallel. In this repeating structure, one can easily discern the Z disk. Scale bar = 0.3 μm for A, I, and H bands and M line of the sarcomere. (inset) Terminal cisternae of the sarcoplasmic reticulum *(SR)* and associated transverse tubule *(T)*. *(Courtesy Dr. Phillip Fields.)*

growing minus (–) end. At each end of the actin filament, there is a critical concentration of G-actin, at which the rate of addition to that end matches the rate of monomer removal from the same end. For the plus end, this concentration of G-actin is approximately 1 μM; it is 8 μM at the minus end. Therefore, at concentrations of G-actin between 1 and 8 μM in the presence of ATP and Mg^{2+}, a treadmill is formed by which actin is being added to the plus end and subtracted from the minus end. If no energy was supplied to this system, this would be a perpetual motion machine, which is thermodynamically impossible. However, shortly after the addition of each G-actin monomer to the actin filament, ATP is cleaved to adenosine diphosphate (ADP), with release of inorganic phosphate (Pi). This raises several interesting questions. First, the concentration of actin within the cytoplasm of muscle and nonmuscle cells is greater than 100 μm, suggesting that almost all of the actin within the cells of your body would be filamentous actin (because this is far above the critical concentration for the plus or minus end). However, most cells have a mechanism for maintaining a pool of G-actin monomers. Why is this the case? Second, does treadmilling occur within the living cell? The answer is yes, as we will discuss later in this chapter.

Although F-actin is the preponderant protein of the skeletal muscle thin filament, these filaments also contain other proteins, tropomyosin, troponin, and tropomodulin (Fig. 3–5).

Tropomyosin is a long, rod-shaped molecule (~41 nm in length), so-called because of its similarities with myosin, specifically the rodlike tail domain of the myosin molecule. Tropomyosin is formed from a dimer of two identical subunits. The individual subunit polypeptides are α-helical, and the two α-helical chains wind around

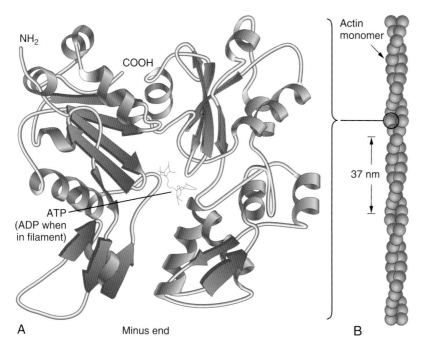

Figure 3–3. Structure of globular (G) and filamentous (F) actin. **A:** G-actin is a 43-kDa monomer with four structural domains. ATP and ADP bind to G-actin within the groove separating domains 1 and 3. **B:** F-actin is a helical filament composed of polymerized G-actin monomers *(spheres)*. The filament undergoes a complete turn of the helix every 14 G-actin monomers, or 37 nm.

each other in a coiled coil to form the rigid, rod-shaped molecule. Tropomyosin binds along the length of the actin filament, lining the grooves of the helical F-actin molecule, thereby stabilizing and stiffening the filament.

Another major accessory protein of the skeletal muscle thin filament is troponin. Troponin is a complex of three polypeptides: troponins T (TnT), I (TnI), and C (TnC). These polypeptides are named for their apparent functions within the troponin complex: TnT for its

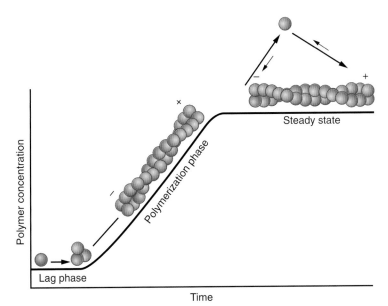

Figure 3-4. Polymerization of actin. The polymerization of actin occurs in three stages: (1) a lag phase in which an actin trimer nucleation site is formed; (2) a polymerization phase, during which G-actin monomers are added preferentially at the plus end of the actin filament; and (3) a steady state, at which actin monomers are being added at the plus end at the same rate they are being removed at the minus end.

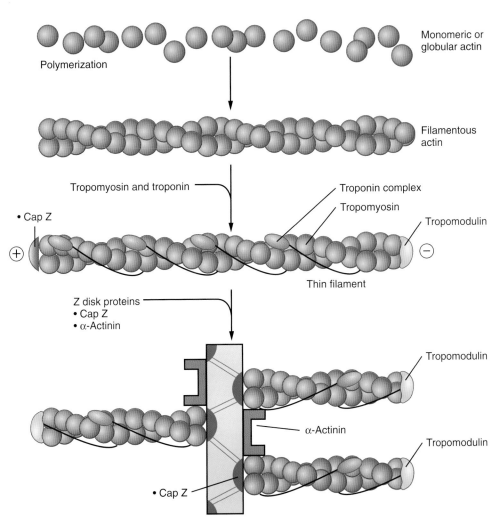

Figure 3-5. Formation of actin thin filaments and their arrangement in the sarcomere. Globular actin monomers polymerize through head-to-tail association to form the helical filamentous (F-actin) form of actin. Thin filaments are built from the specific association of the F-actin filaments with the rodlike tropomyosin molecule, which lines the grooves of the actin filament, and the troponin polypeptide complex. In the sarcomere, the thin filaments are anchored at the Z disk through their interactions with binding proteins, principally cap Z and α-actinin. The exact structure of the Z disk is unknown; the protein interactions shown in this diagram are based on the in vitro capabilities of the isolated cap Z and α-actinin proteins. As illustrated, the specific protein interactions of the Z-disk proteins immobilize the thin filaments at their plus (+) ends, thereby maintaining the polarity of the actin thin filaments in the sarcomere. The minus (−) ends are capped by tropomodulin.

tropomyosin binding, TnI for its inhibitory role in calcium regulation of contraction (see later discussion), and TnC for its calcium-binding activity. The troponin complex is elongated, with the I and C subunits forming a globular head region and the T polypeptide forming a long tail domain. The tail domain, formed from the T subunit, binds with tropomyosin, which is thought to position the complex on the actin thin filament. Because there is only one troponin complex for every seven actin monomers in an actin filament, the positioning of the complex by the specific interactions of the T subunit with the tropomyosin molecule is critical for its ability to regulate contraction.

Tropomodulin (a 43-kDa globular protein) binds tropomyosin and caps the minus end of the actin filament, thereby regulating the length of the actin thin filaments within the sarcomere.

Thick Filaments Are Composed of the Protein Myosin

Myosin was also first described in muscle cells but is now known to be a ubiquitous component of nonmuscle cells. The major form of myosin found in most cells, including skeletal muscle, is referred to as **myosin II** because it contains two globular heads or motor domains. Myosin II has an M_r of approximately 460 kDa, with two identical heavy chains of M_r of 200 kDa, which form a coiled-coil helical tail and two globular heads (Fig. 3–6).

For a coiled-coil helix to form, the myosin heavy chains must have a heptad amino acid repeat sequence—a, b, c, d, e, f, g, a, b, c, d, e, f, g—with hydrophobic amino acids in positions a and d. Because an α-helix makes a complete turn every 3.5 amino acids, such a repeat would create a hydrophobic stripe that slowly rotates around the helix. To bury this hydrophobic

stripe away from the aqueous environment, two such α-helices would wind around each other into a coiled coil. The myosin molecule also contains two pairs of light chains with M_r of 20 and 18 kDa. These light chains are found associated with the myosin heads. If purified myosin is proteolytically cleaved with the enzyme papain, the globular heads (called SF1 *fragments*) can be separated from the myosin tails (see Fig. 3–6). The myosin tails brought to physiologic ionic strength and pH will spontaneously form thick filaments, similar to those found in skeletal muscle. The SF1 heads contain all of the myosin ATPase activity required for muscle contraction. If the purified heads are added to preformed F-actin and viewed by electron microscopy, the SF1 fragments look like arrowheads that all face in one direction. The pointed end of the arrowheads face the minus, or slow-growing, end of the filament, and the barbed end faces the plus, or fast-growing, end (Fig. 3–7).

The polarity of actin filament interaction with myosin SF1 fragments (see Fig. 3–7) is important for muscle contraction. Thick filament formation arises from association of the tail or rod segment of the myosin molecule, as demonstrated by the aggregation of isolated tail domains produced by proteolytic cleavage of myosin II. The association of the myosin II heavy chain dimers is due to hydrophobic interactions of the rodlike tail segments, and the formation of filaments depends on interactions between the coiled myosin II tail domains. In muscle, the rodlike fibrous tails of 300 to 400 myosin II dimers pack together to form the bipolar 15-nm-diameter-thick filaments. This association of the myosin II tail segments results in the formation of a filament that has a bare central zone composed of an antiparallel array of myosin II tails (Fig. 3–8).

The globular myosin II head segments protrude from the filament at its terminal regions in a helical array, with a periodicity of 14 nm. Thick filaments then display a high degree of structure. They are symmetric about the bare central zone with the polarity of the filament determined by the arrangements of the globular head segments, which are reversed on either side of this central zone.

Accessory Proteins Are Responsible for Maintenance of Myofibril Architecture

In vertebrate skeletal muscle, the structural orientation of the thick and thin filaments is crucial for contraction. Thus, the maintenance of this structure is important for muscle function. Several proteins (but probably not all that are necessary) that interact with the thick and thin filaments and play a role in the maintenance of myofibril structure have now been identified.

The thin filaments terminate and are anchored to the Z-disk structures of the sarcomere. This immobilizes the

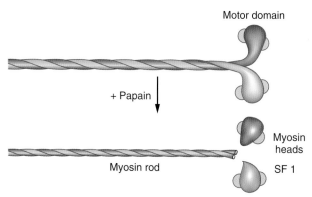

Figure 3–6. **Structure of myosin II and its cleavage by papain.** Myosin II is a 150-nm-long fibrous protein, with two globular heads. Treatment of myosin II with the proteolytic enzyme papain releases the two myosin heads, or SF1 fragments, from the myosin rod.

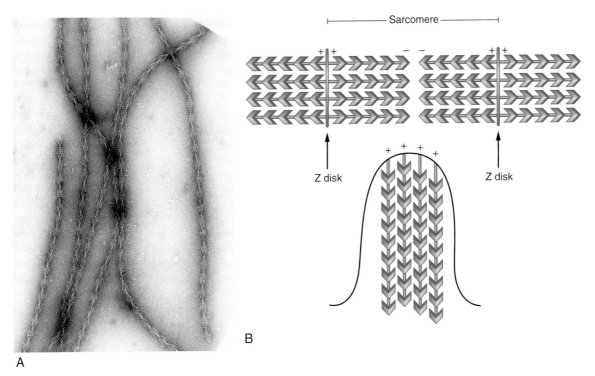

Figure 3–7. **Actin filaments have a polarity.** The polarity of actin filaments can be visualized by labeling with myosin SF1 fragments. **A:** This is an electron micrograph of *in vitro*–formed actin filaments that have bound myosin SF1 fragments. The myosin fragments bind to the actin filaments, demonstrating their polarity. The myosin heads look like arrowheads that all point to the minus (–) ends of the actin filament, the barbed ends facing the plus (+) ends of the filaments. *(Courtesy Dr. Roger Craig, University of Massachusetts.)* **B:** In a sarcomere, the barbed or plus (+) ends are attached to the Z disk. When actin filaments are bound to the cytoplasmic surface of the plasma membrane, it is the plus end that is associated with the membrane. The example shown here is the attachment of actin filaments to the tip of the microvillus.

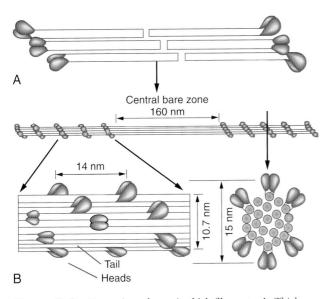

Figure 3–8. **Formation of myosin thick filaments. A:** Thick filament formation is initiated by the end-to-end association of the rodlike tail domains of myosin II molecules. **B:** This results in the formation of the bipolar thick filament, with globular heads at either end separated by a 160-nm central bare zone consisting of myosin II tail domains. At the filament ends, the myosin globular head domains protrude from a 10.7-nm diameter central core at intervals of 14 nm. The successive myosin heads rotate around the fiber, which forms a filament containing six rows of myosin head domains to contact the adjacent thin filaments of the sarcomere.

thin filaments with their plus ends at the Z disk and their minus ends extending to the center region of the sarcomere. Therefore, a **sarcomere unit** (defined as the distance between two adjacent Z-disk structures) contains actin filaments that extend from each Z disk and exhibit polarity that is opposite on either side of the central region of the sarcomere. Cap Z, a two-subunit protein (M_r 32,000 and 36,000) that binds selectively to the plus ends of actin filaments, is one of the proteins that helps with the anchoring of thin filaments to the Z disk. Because it binds to the fast-growing or plus end of the actin filament, cap Z is thought to prevent growth and depolymerization of F-actin, causing the filaments of the myofibril to be very stable structures. Its localization at the Z disk suggests that cap Z may assist in the immobilization of the thin filaments (see Fig. 3–5), perhaps by interactions with other proteins of the Z disk. The major component of the Z disk is the protein α-actinin, a fibrous protein composed of two identical subunits (M_r 190,000). The NH$_2$-terminal domain of α-actinin bears a strong resemblance to NH$_2$-terminal domains of other cytoskeletal proteins (principally members of the spectrin supergene family) that function to bind and cross-link actin filaments. It is the NH$_2$-terminal domain of α-actinin that provides the ability of this protein to bind tightly to the sides of actin filaments, allowing the bundling together of adjacent thin filaments at the Z disk. Although the exact structure of the Z disk is unknown, evidence now indicates that it contains two sets of overlapping actin filaments of opposite polarity that originate in the two sarcomeres adjacent to the Z disk, and the thin filaments are anchored to the disk structure by interactions with proteins such as cap Z and α-actinin. As mentioned earlier, tropomodulin, a 43-kDa protein, binds to, and caps, the minus slow-growing ends of actin filaments, regulating their length.

In skeletal muscle, there are mechanisms that maintain the relative position of the myofilaments and regulate the length of the polymerized filaments. Two proteins, titin and nebulin, appear to be important for these functions (Fig. 3–9).

Titin, a large fibrous protein, appears to connect the thick filaments to the Z disk. Titin is the largest protein described to date (~3.5 million daltons), and it contains a long series of immunoglobulin-like domains. It functions to keep the myosin thick filaments centered within the sarcomere structure. It may act as an elastic band to keep filaments in an appropriate orientation, also inhibiting the structural deterioration of the sarcomere during muscle contraction. Another large fibrous protein, nebulin, forms a long, inextensible filament that extends from the Z disk to the minus end of the thin filaments. Nebulin contains a 35-amino acid repeating actin binding motif. Because of their exacting length and their repeating association with the actin filaments, nebulin fibers may regulate the number of actin monomers that polymerize into thin filaments and aid in the formation of the regular geometric pattern of thin filaments during muscle formation.

Muscle Contraction Involves the Sliding of the Thick and Thin Filaments Relative to Each Other in the Sarcomere

Measurements of sarcomere and A and I band lengths from electron micrographs of contracted and resting muscle firmly established the mechanism of muscle contraction: The sliding of actin thin and myosin thick filaments passed each other within the sarcomere unit. These measurements demonstrated that the lengths of the individual filaments do not change as a muscle contracts; yet, the distance between two adjacent Z disks becomes shortened in contracted muscle relative to relaxed muscle. When the length of a sarcomere decreases in contracted muscle, the I band region shortens, whereas the length of the A band remains unchanged (Fig. 3–10).

Because the lengths of the thick and thin filaments do not change, the change in length of the I band could occur only if the thin filaments were to slide past the thick filaments. Therefore, the reversed polarity of the

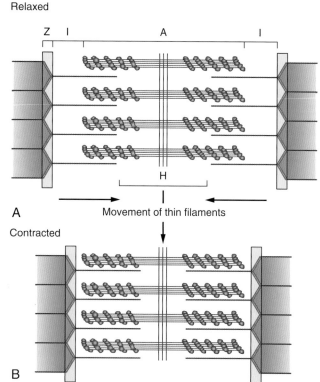

Figure 3–10. Sliding filament model of muscle contraction. Muscle contraction occurs by the sliding of the myofilaments relative to each other in the sarcomere. **A:** In relaxed muscle, the thin filaments do not completely overlap the myosin thick filaments, and a prominent I band exists. **B:** With contraction, movement of the thin filaments toward the center of the sarcomere occurs, and because the thin filaments are anchored to the Z disks, their movement causes shortening of the sarcomere. The sliding of thin filaments is facilitated by contacts with the globular head domains of the bipolar myosin thick filaments.

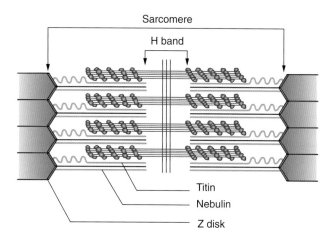

Figure 3–9. Titin and nebulin: accessory proteins of the skeletal muscle sarcomere. The location of the proteins titin and nebulin within the sarcomere is shown. Titin, a large protein that has elastic properties and links the myosin thick filaments to the Z disks, helps maintain their location in the sarcomere. Nebulin, a large filamentous protein anchored at the Z disk, is in close apposition to the actin thin filaments. Their close association with the thin filaments suggests that the nebulin fibers serve to organize the actin filaments of the sarcomere.

thick and thin filaments relative to the center line of the sarcomere (defined by the M line) would cause a shortening of the sarcomere during contraction by the sliding of thin actin filaments, which are attached to the Z disk, past the thick myosin filaments toward the center of the sarcomere. This model of muscle contraction, called the **sliding filament model,** was first proposed in 1954 and led to the dissection of molecular mechanisms of contraction.

Adenosine Triphosphate Hydrolysis Is Necessary for Cross-Bridge Interactions with Thin Filaments

Skeletal muscle contraction requires the interactions of myosin II head groups with the thin filaments. These interactions are governed by binding and hydrolysis of the high-energy molecule ATP by the ATPase activity resident in the globular myosin II head domain. The ATP-driven interactions between myosin II and actin are illustrated in Figure 3–11.

When a myosin II head binds a molecule of ATP, it causes a weakening of the myosin–actin interaction. A dissociation of the myosin II head group binding to the thin filament occurs (step 1). The cleavage of ATP to ADP and Pi creates an "activated" myosin II head that has undergone a change in structure, facilitated by the flexible hinge regions of the molecule, such that the myosin II head is perpendicular with an adjacent actin thin filament (step 2). The conversion between these two stages is reversible, because the ADP and Pi remain

bound to the myosin II head, and the energy released from ATP hydrolysis is stored in the strained bonds resulting from the rotation of the myosin II head group. The activated myosin II molecule then comes into contact with a neighboring actin subunit, and this binding triggers the release of Pi, which, in turn, strengthens the myosin–actin interaction (step 3). This strong binding causes a conformational change in the myosin II head, generating a "powerstroke," which pulls the actin filament relative to the fixed myosin II filament, resulting in contraction (step 4). The product of this step is the so-called rigor complex, in which the actin–myosin linkage is inflexible, and the thick and thin filaments cannot move past each other. If no ATP is available to the muscle (e.g., after death), the muscle will remain rigid, owing to the tight myosin–actin interactions. This condition is referred to as **rigor mortis.** Under normal circumstances, a molecule of ATP will displace the bound ADP, causing release of the actin filament from the myosin head group, effectively relaxing the muscle and returning to step 1 of the cycle. The hydrolysis of the newly bound ATP then prepares the muscle for further rounds of myosin–actin interactions.

Each cycle of myosin–actin interaction would result in movement of an actin thin filament by about 10 nm. A coordination of multiple myosin II head group interactions to provide a concerted movement of filaments and a mechanism by which these interactions are regulated must exist to achieve the rapid rates of contraction for intact muscle fibers. Each thick filament is formed from the aggregation of multiple myosin II rod domains,

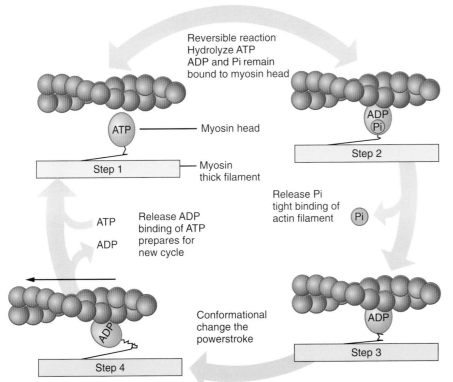

Reversible reaction
Hydrolyze ATP
ADP and Pi remain
bound to myosin head

ATP — Myosin head

Step 1

Myosin
thick filament

ATP — Release ADP
binding of ATP
ADP — prepares for
new cycle

ADP
Pi

Step 2

Release Pi
tight binding of
actin filament — Pi

Conformational
change the
powerstroke

ADP

Step 4

ADP

Step 3

Figure 3–11. Illustration of the ATP-driven myosin–actin interactions during contraction. The binding of ATP to a myosin head group causes release from the actin filament (step 1). The hydrolysis of ATP to ADP+Pi readies the myosin head to contact an actin filament (step 2). The initial contact of the myosin with an actin filament causes the release of Pi and a tight binding of the actin filament (step 3). This tight binding induces a change in conformation of the myosin head, such that it pulls against the actin filament, the powerstroke (step 4). This change in conformation is accompanied with the release of ADP. The binding of an additional ATP causes a release of the actin filament and a return of the myosin head to a position ready for another cycle.

which results in a bipolar filament, with each side of the filament containing approximately 300 to 400 head groups protruding in a spiral fashion. This arrangement provides multiple contacts of a thick filament (called **cross-bridge interactions**) with a thin filament. Along the length of the thin filament, there will be myosin II cross-bridges at various points in the myosin–actin cycle (see Fig. 3–11), such that the collective actin cross-bridged contacts ensure the smooth and rapid movement of the thin filament relative to the thick filament. Each myosin head group cycles about five times per second, sliding the myosin thick filaments and actin thin filaments past each other at a rate of ~15 μm/sec. Sarcomeres can shorten by about 10% in length in about 20 milliseconds.

To effectively coordinate the sliding of filaments from entire groups of myofibrils, leading to muscle contraction capable of producing mechanical work, a transient increase in calcium regulates the interactions of myosin II head group cross-bridges at the cellular level. The calcium-based regulation of muscle contraction occurs by overcoming a block of myosin–actin interactions by the troponin-tropomyosin complexes on the thin filament. The specific interactions involved in this regulation are discussed in the following section.

Calcium Regulation of Skeletal Muscle Contraction Is Mediated by Troponin and Tropomyosin

When myosin is mixed with filaments made from purified actin, the myosin ATPase activity is stimulated to its maximal activity, independent of calcium addition to the reaction. If thin filaments, which contain actin, tropomyosin, and troponin, are added to purified myosin, the stimulation of the myosin ATPase activity is wholly dependent on the presence of calcium. The basis of calcium-dependent hydrolysis of ATP in this reaction is a reversal of the inhibition of the actin–myosin interaction caused by the position of tropomyosin and troponin on the thin filaments (Fig. 3–12).

Each rodlike tropomyosin molecule contacts seven actin monomers and lines the grooves of the F-actin helix. Bound to a specific site of each tropomyosin molecule is the troponin complex that comprises three polypeptides: TnT, TnI, and TnC. The elongated TnT molecule (M_r 37,000) binds the COOH-terminal region of tropomyosin and links both TnI and TnC to the tropomyosin. TnI (M_r 22,000) binds TnT, as well as actin, and in concert with tropomyosin, causes a change in the conformation of F-actin such that it interacts only weakly with myosin head groups. This weak interaction cannot activate the myosin ATPase activity. Together with TnI, the TnC ($M_r = 20,000$) subunit forms a globular domain of the troponin complex. TnC, the calcium-binding subunit, has a structure and function similar to that of the intracellular calcium receptor protein calmodulin. The binding of calcium ions at all four of the calcium-binding domains of TnC releases the TnI tropomyosin inhibition of actin activation of the myosin ATPase, thereby allowing contraction of the myofibril. The binding of calcium by TnC results in a shift or movement of the tropomyosin toward the center of the actin helix, which exposes a region of the actin monomer,

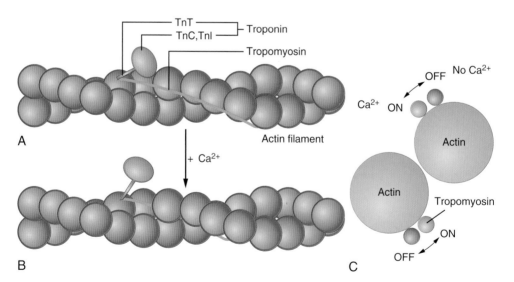

Figure 3–12. Diagram of Ca²⁺-mediated movements of troponin and tropomyosin filaments during muscle contraction. **A:** In the relaxed muscle, the tropomyosin filament is bound to the outer domains of seven actin monomers along the actin filament. The troponin complex is bound to the tropomyosin by the rod-shaped troponin T (TnT) polypeptide. **B:** In the presence of Ca²⁺, troponin C (TnC) binds the calcium, causing the globular domain of troponin (TnC and troponin I [TnI]) to move away from the tropomyosin filament. **C:** This movement permits the tropomyosin to shift to a position that is farther inside the groove of the helical actin filament, allowing the myosin heads to make contact with the released sites of the actin monomers.

allowing the binding of myosin head groups in such a way that activation of the myosin ATPase activity occurs. The hydrolysis of ATP permits cycling of cross-bridge interactions and the sliding of filaments. The myosin-activating sites of F-actin are blocked by the troponin-tropomyosin complex in the resting, but not in the active, state of the myofibril. Thus, the contraction of skeletal muscle is regulated by the concentration of intracellular calcium ions.

Intracellular Calcium in Skeletal Muscle Is Regulated by a Specialized Membrane Compartment, the Sarcoplasmic Reticulum

In the resting or relaxed state, the concentration of calcium ions in skeletal muscle cells is low. Thus, to have contraction and relaxation cycles of muscle, a mechanism must exist by which the internal calcium ion concentration is regulated. Moreover, the concerted contraction of a muscle to produce work depends on the simultaneous contraction of all of its constituent myofibers and their myofibrils. Therefore, the rapid changes in calcium ion concentration that are needed along the entire length of the myofibril for contraction must be maintained by mechanisms other than simple diffusion, which would be too slow for simultaneous contraction of myofibrils in skeletal muscle cells. To deliver calcium in a uniform fashion throughout the muscle cell, there is a special membrane-bound tubule system, derived from the endoplasmic reticulum (ER) in these cells.

Electron microscopy of skeletal muscle shows a network of smooth membranes, called the *sarcoplasmic reticulum* (SR), surrounding the myofibrils. The SR forms a network of membrane-limited tubules and cisternae that surround the outer regions of the A band of each myofibril (Fig. 3–13). In addition, the SR forms a more regular structure, called the **terminal sac** or **terminal cisternae**, which is a membrane-limited channel that surrounds the A-I junction of each individual myofibril. The terminal cisternae are in close proximity to a specialized channel formed from delicate invaginations of the sarcolemma (plasma membrane of the muscle cell), called the **transverse tubules** (t-tubules). The t-tubules, in association with terminal cisternae from adjacent myofibrils, form a triad structure (see Fig. 3–13). These structures are important for the coupling of external stimuli (e.g., signals from motor neurons) to muscle contraction.

The SR forms a membranous compartment that occupies 1% to 5% of the total muscle volume and serves as a reservoir of calcium ions sequestered away from the myoplasm and myofibrils. For its role in maintenance of calcium ion concentration, the SR membrane contains numerous proteins for the transport of calcium,

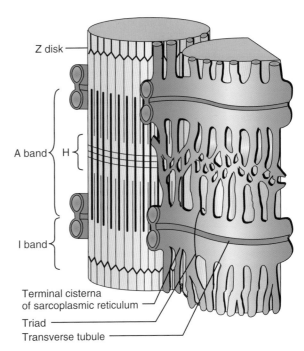

Figure 3–13. Diagram of part of a skeletal muscle fiber, illustrating the organization of the sarcoplasmic reticulum (SR) and transverse tubule (t-tubule) networks. The SR is a specialized smooth endoplasmic reticulum (ER) that in muscle serves as a store for Ca^{2+} ions. The SR forms a membranous tubule network that surrounds the myofibrils. At the A-I band junctions of the sarcomere, the SR forms a more regular channel, referred to as the terminal cisternae. Two terminal cisternae are separated by a second tubule system, the t-tubules, which are special invaginations of the sarcolemma. These three membrane-bound tubules form a structure known as the triad: a t-tubule flanked on either side by a terminal cisternae of the SR, at the region of the A-I junction of the sarcomere. (*Modified from Cormack DH. Ham's Histology, 9th ed. Philadelphia: JB Lippincott, 1987.*)

including a Ca^{2+}-ATPase protein that pumps calcium from the cytosol into the lumen of the SR. For each 1 mol of ATP hydrolyzed by the ATPase activity of the calcium pump, 2 mol of calcium are sequestered into the lumen of the SR. This active transport mechanism is responsible for the maintenance of the low calcium ion concentration in resting muscle. The stored calcium is released from the SR into the sarcoplasm as the action potential spreads along the sarcolemma. The action potential, stimulating Ca^{2+} release, travels through the t-tubule system. A voltage-sensitive protein sensor located in the t-tubule membrane, termed the **dihydropyridine-sensitive receptor** (DHSR), feels the action potential and translates its presence to the SR through direct interaction with an SR calcium channel, the **ryanodine receptor**. These proteins, the DHSR and ryanodine receptor, are analogous to proteins found in other cells whose function is to release calcium from internal stores, the so-called IP$_3$ (inositol 1,4,5-triphosphate) receptor pathway (described in Chapter 8). The large complex formed by these proteins in muscle when viewed in the electron microscope is often referred to as

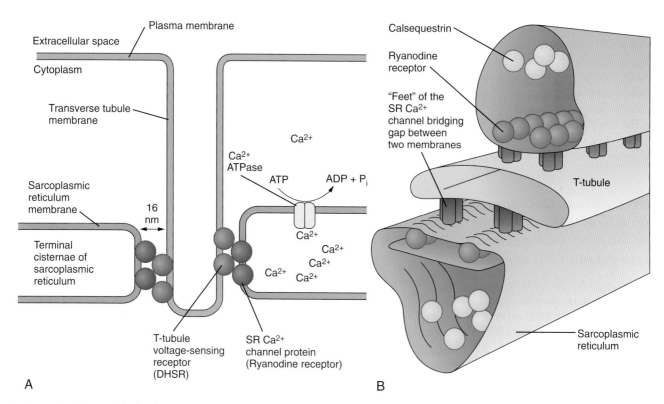

Figure 3–14. Model of Ca²⁺ ion regulation by the sarcoplasmic reticulum (SR) in muscle. **A:** An illustration of the association of the transverse tubules (t-tubules) and the terminal cisternae of the SR. The SR Ca²⁺ channel is shown to make direct contact with the voltage-sensing Ca²⁺ channel of the t-tubule. When depolarized, the t-tubule voltage-sensing protein (DHSR) undergoes a change in conformation and, because of its close association with the SR Ca²⁺ channel, causes the SR channel to open and release calcium to the cytoplasm. This Ca²⁺ release occurs with essentially no delay because of the direct interaction of the DHSR and the Ca²⁺ channel of the SR, the ryanodine receptor. The Ca²⁺ ions in the cytoplasm are returned to the lumen of the SR by the Ca²⁺-ATPase pump in the SR membrane. **B:** View of the t-tubule and SR terminal cisternae associations. The t-tubules and SR terminal cisternae are in close proximity; "feet" of the SR channel protein are shown bridging the gap between the t-tubule and SR membranes. Inside the lumen of the SR is the protein calsequestrin that weakly binds the internalized Ca²⁺ ions, reducing the effective internal concentration of free Ca²⁺ ions. *(A: Modified from Agnew WS. Nature 1988;344:299–303. B: Modified from Eisenberg BR, Eisenberg RS. Gen Physiol 1982;79:1–17.)*

"feet" on the SR (Fig. 3–14). The net effect is the release of a pulse of calcium into the sarcoplasm by the transit of an action potential. The released calcium stimulates contraction through binding the troponin complex on the thin filaments. After contraction, the calcium is actively transported into the lumen of the SR by Ca²⁺-ATPase, returning the muscle to the relaxed state (see Fig. 3–14).

Within the lumen of the SR are proteins that function to bind and store the internalized calcium ions (see Fig. 3–14). The best-characterized example is calsequestrin. Although the binding affinity of Ca²⁺ by calsequestrin is low, each molecule of the protein binds 40 to 43 Ca²⁺ ions. Thus, calsequestrin, together with other proteins that have similar properties, effectively reduced the SR luminal concentration of calcium from 20 to 30 mM (if all the Ca²⁺ ions were free in solution) to about 0.5 mM. The result of binding Ca²⁺ ions in the SR lumen is to greatly reduce the concentration gradient against which the membrane Ca²⁺ pump must act.

Three Types of Muscle Tissue Exist

In the preceding section, we focused on the contractile apparatus found in skeletal muscle. Two other major types of muscle are present in the vertebrates. Cardiac muscle forms the walls of the heart and is also found in walls of the major vessels adjacent to the heart. Smooth muscle is found in the hollow viscera of the body (e.g., the intestines) and in most blood vessels. All three types of muscle use actin-myosin structures for contraction by a sliding filament mechanism. However, some fundamental differences exist in the structural organization of the contractile apparatus and the regulation of contraction in the different muscle cells.

Myocardial Tissue: Striated Muscle Built from Individual Cells

Cardiac tissue consists of long fibers that, like skeletal muscle, exhibit cross-striations under the light microscope. The striated appearance of cardiac muscle derives from the highly organized arrangement of actin and myosin filaments of the contractile apparatus. Although cardiac muscle is similar in appearance to striated skeletal muscle, two main histologic criteria distinguish these two muscle types.

The first criterion is the positioning of the nuclei within the cells. In skeletal muscle, nuclei are located at the periphery of the cell, just under the sarcolemma, whereas in cardiac muscle, the nuclei are found at the central regions of the cell. Thus, cardiac cells have a bare or cleared zone surrounding the nucleus, the perinuclear space, which arises from the myofilaments arranging themselves such that they detour around the nuclear compartment.

The second major criterion that distinguishes cardiac from skeletal muscle is the appearance of dark staining disk structures in cardiac muscle, the **intercalated disk.** These are specialized junctional complexes that separate one cardiac muscle cell from another. Thus, cardiac muscle fibers are built from an arrangement of single cells, unlike skeletal muscle fibers that are built from the fusion of individual cells into a multinucleated fiber. Although at the light microscopic level the intercalated disk appears as straight lines, demarcating one cell from another, the view of these structures in the electron microscope reveals that they take an irregular steplike path, such that part of this cell–cell junction is horizontal and part is longitudinal. Therefore, the individual cardiac muscle cells interdigitate with each other, forming the myocardial muscle fibers (Fig. 3–15). This arrangement of cell–cell contact allows myocardial muscle to contain straight fibers and fibers that branch to effectively construct a hollow organ capable of pumping blood.

Different regions of the intercalated disk contain specific junctional complexes (see Fig. 3–15). In the transverse (or vertical) sections of the intercalated disk, two junctional complexes exist. The first is the **desmosomes,** which are sometimes referred to as the **macula adherens.** These junctions function as the "spot welds" that hold the adjacent cardiac cells together. In this region of the plasma membrane is a second type of junctional complex, termed the **fascia adherens,** which functions to connect the thin filaments of adjacent cells and to hold them in register with the myosin thick filaments (see later). In the longitudinal regions of the intercalated disk are junctional contacts called **gap junctions** (sometimes referred to as nexus). These contacts allow the cardiac cells to exchange small cytoplasmic solutes. The gap junction contacts permit electrical coupling of the cardiac muscle cells, such that synchronization of contraction exists among these cells.

Some of the muscle cells within the heart, the Purkinje fibers, are specialized to carry electrical impulses. These cells are grouped into bundles that form two branches, one to each ventricle. Histologically, these cells are larger and more irregular in shape than are the surrounding cardiac muscle cells. The Purkinje cells contain large glycogen deposits and have smaller bundles of myofibrils at their periphery. These special conducting fibers are responsible for the final distribution of electrical stimulus to the myocardium.

The Contractile Apparatus of Cardiac Muscle Is Similar to That of Skeletal Muscle

Cardiac muscle owes its striated appearance to the arrangement of thick and thin filaments that make up the contractile apparatus. Electron micrographs of

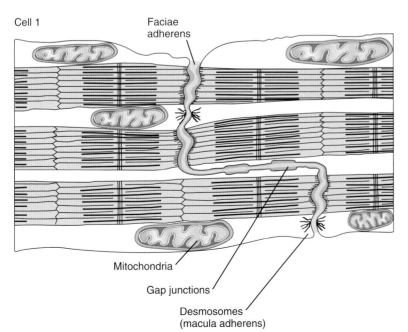

Cell 1
Faciae adherens
Mitochondria
Gap junctions
Desmosomes (macula adherens)

Figure 3–15. Diagram of an intercalated disk between two cardiac muscle cells. The intercalated disk is a steplike structure that allows the interdigitation of cardiac muscle cells. In the transverse sections of this structure are the desmosomes, which hold the cells together, and the junctional complexes, the fasciae adherens, which function as Z-disk structures to anchor actin thin filaments from adjacent cells. In the longitudinal sections of the intercalated disk are the gap junctions. These junctional complexes allow communication between the cells such that adjacent cardiac cells are coupled electrically.

cardiac muscle reveal a banding pattern of myofibrils similar to that observed for skeletal muscle. Like skeletal muscle, these bands are referred to as the A band, I band, and Z disk. The dark-staining A band is the region of the myofilament that contains the thick filaments composed of myosin and overlapping thin filaments. The I band contains the thin actin filaments and is bisected by the Z disk, to which the actin filaments are anchored. One notable difference in the structure of myofilaments in cardiac cells, compared with skeletal muscle cells, is the termination of some actin thin filaments at the region of the intercalated disk (Fig. 3–16).

The fascia adherens complex in the transverse segment of the intercalated disk functions to anchor actin thin filaments at the cell periphery. Although the molecular details of how this junctional complex binds and arranges actin thin filaments are unknown, these complexes function as a Z disk, in that they maintain the exacting arrangement of six actin filaments surrounding each myosin thick filament.

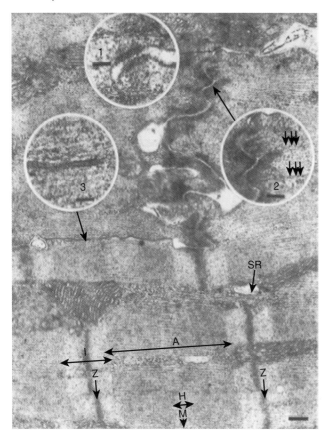

Figure 3–16. Electron micrograph of cardiac muscle. This electron micrograph of cardiac muscle cells shows the regular arrays of the myofibrils into the sarcomeres. In this arrangement, one can easily see the Z disk; the A, I, and H bands; and the M line of the sarcomere structure. The inserts show a higher magnification of the junctional compartments of the intercalated disk: *(1)* the macula adherens or desmosomes, *(2)* the fasciae adherens, and *(3)* gap junctions. Notice the thin filaments that terminate in the fascia adherens complex. Scale bar = 0.2 μm; (insets) 0.05 μm. *(Courtesy Dr. Phillip Fields.)*

Cardiac muscle thick filaments are made from a cardiac isoform of myosin II, which has a subunit structure similar to that found in skeletal muscle. Cardiac myosin II has two heavy chains of approximately 200,000 M_r that assemble by association of a rodlike tail domain and fold into a globular head domain at their NH_2 terminus. There are four light chains, two pairs of M_r 18,000 to 20,000, with one polypeptide from each set bound with each head segment of the molecule. Associated with the globular head domain of cardiac myosin II is an actin-activated ATPase activity that functions in cross-bridge formation and contraction. However, the isozymes of myosin expressed in cardiac muscle have a lower ATPase activity than those in skeletal muscle. Familial hypertrophic cardiomyopathy is caused by defects in the cardiac β myosin (at or near the head or motor domain) or in myosin light chains, troponin, or tropomyosin. This disease affects approximately 2 of every 1000 people, with the clinical outcome being enlargement of the heart and cardiac arrhythmias.

The thin filaments of cardiac muscle are built from actin, tropomyosin, and troponin. Although these proteins form the same complex as that found in skeletal muscle, they are different from the polypeptides found in their skeletal muscle counterpart; that is, they are cardiac-specific isoforms. Cardiac muscle thin filaments exhibit the same stoichiometry and structure as those discussed for skeletal muscle. Thus, in cardiac muscle, there is an arrangement of six thin filaments surrounding each thick filament, and contraction or cross-bridge formation in cardiac muscle is regulated by Ca^{2+} by the thin filament-based troponin-tropomyosin complex. Defects in cardiac α-actin cause familial dilated cardiomyopathy. Patients with dilated cardiomyopathy demonstrate a defective left and/or right systolic pump function leading to cardiac enlargement and hypertrophy, which leads to early heart failure.

The Smooth-Muscle Cell Does Not Contain Sarcomeres

Smooth muscles are made of individual cells that can vary notably in size, from 20 μm in length in the walls of the small blood vessels to 200 to 300 μm in length in the intestine. The smooth-muscle cell is characterized by its fusiform shape. The cells are thickest at their midregion and taper at each end. Smooth muscles are built from sheets of cells that are linked together by various junctional contacts that serve as sites of cell–cell communication (e.g., gap junctions) and mechanical linkages. Cells of the smooth muscle are active in the synthesis and deposition of connective tissue matrix, which serves to embed the cells and acts in limiting the distension of the hollow viscera.

Smooth-muscle cells do not contain a highly ordered array of thick and thin filaments; thus, they do not appear striated. In electron micrographs of smooth muscle, numerous dense staining regions, known as **dense bodies,** are found throughout the cytoplasm of the cell. The major protein component of the dense body is the actin-binding protein, α-actinin, which indicates that they serve as the functional equivalent of a skeletal muscle Z disk. Indeed, actin thin filaments are found anchored to the dense bodies. Two proteins, desmin and vimentin, belonging to the IF class of proteins (discussed later), are expressed at high levels in smooth-muscle cells. The filaments formed from these proteins are prominent in these cells and appear to serve as links between the dense bodies and the cytoskeletal network of the cell. These links aid in contraction by maintenance of the dense body positioning (Fig. 3–17), allowing movement of the cell by an inward pulling of the plasma membrane.

The Contractile Apparatus of Smooth Muscle Contains Actin and Myosin

Actin and myosin can be isolated from smooth-muscle cells, and *in vitro,* these proteins demonstrate a sliding filament mechanism for contraction. However, the regu-

lation of contraction in smooth muscle follows a path very different from that observed for striated muscle. The thin filaments of smooth and striated muscle have similar structures, except that the calcium regulatory protein troponin is not present in smooth muscle. The cellular content of actin and tropomyosin is greater in smooth muscle than in striated muscle (by about twofold). This, in combination with a reduced quantity of myosin in smooth muscle compared with striated muscle, produces a greater ratio of thin-to-thick filaments in smooth muscle (–12 thin per 1 thick) than that observed in the striated muscles (–6 thin per 1 thick).

Smooth muscle contains numerous thin filaments that approximately align along the long axis of the cell. These thin filaments are embedded into the cytoplasmic densities (dense bodies) and exhibit the same polarity relative to their attachment points. The thin filaments have their plus (+) ends at the dense body and their minus (–) ends extending into the cellular cytoplasm. Thus, although the filaments in smooth muscle are not as highly organized as those found in striated muscle, the polarity of actin thin filaments in smooth muscle is such that contraction by myosin cross-bridge cycling would cause a pulling of dense bodies toward one another. This is critical for smooth-muscle contraction, in that it would cause an inward pulling of the plasma

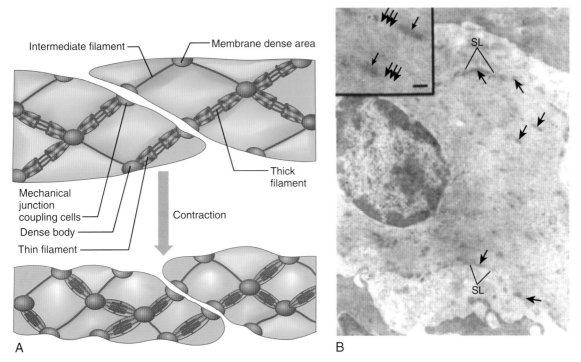

A   B

Figure 3–17. Organization of cytoskeletal and myofilament elements in smooth muscle. A: Smooth-muscle cells contain small contractile elements that are not organized as in striated muscle. Numerous actin thin filaments are anchored into dense bodies within the smooth-muscle cytoplasm, which is the functional equivalent of the striated muscle Z disk. Intermediate filaments of desmin and vimentin form linkages between the dense bodies and the cytoskeleton of the cell. These links are important for contraction, which pulls the plasma membrane inward and changes the shape of the cell. B: In this electron micrograph of a smooth-muscle cell, the dense bodies are seen throughout the cell and near the sarcolemma *(SL).* At higher magnification (insert), myofilaments *(small arrows)* are observed as they emanate from the dense bodies *(large arrows).* Scale bar = 0.29 pm. *(B: Courtesy Dr. Phillip Fields.)*

membrane, creating force generation by essentially reshaping the cell. A change in shape of several coupled cells would generate the force of smooth-muscle contraction.

Myosin isolated from smooth muscle has properties different from those of striated muscle. Similar to skeletal muscle, smooth-muscle myosin consists of two heavy chains and four light chains. Two polypeptides of myosin light chains are associated with each globular head domain of the smooth-muscle myosin. However, smooth-muscle myosin will form filaments under only certain conditions. When myosin isolated from smooth-muscle cells is dephosphorylated, it remains fully soluble. Analysis of soluble myosin by sedimentation assays and electron microscopy shows that the dephosphorylated myosin folds up into a compact unit, with the tail domain reaching toward the globular head domain. In this configuration, the isolated myosin resists the formation of thick filaments, and its actin-activated ATPase activity is essentially blocked. On phosphorylation of the 18-kDa light chain of smooth-muscle myosin by the enzyme myosin light chain kinase (MLCK), the tail segment is released from the head segment (Fig. 3–18). The resulting released myosin tails can form bipolar

thick filaments. Moreover, the freeing of the tail domain to form bipolar filaments allows the activation of the head domain ATPase (permitting cross-bridge formation).

Smooth-Muscle Contraction Occurs via Myosin-Based Calcium Ion Regulatory Mechanisms

Smooth-muscle cells lack troponin, the Ca^{2+} regulatory protein found in the thin filaments of striated muscle, yet micromolar increases in intracellular Ca^{2+} concentrations are required for smooth-muscle contraction to occur. The Ca^{2+} regulation of smooth-muscle contraction occurs by changes in the phosphorylation state of the myosin molecule. The regulation of smooth-muscle contraction is said to be myosin based. When stimulated, the Ca^{2+} concentration in the smooth-muscle cytoplasm increases, and the released Ca^{2+} first encounters the Ca^{2+}-binding protein, calmodulin. Calmodulin is present in all cells, and it is referred to as a **modulator protein**. The calmodulin molecule lacks enzymatic activity but exerts its effects by binding Ca^{2+}, and the Ca^{2+}-

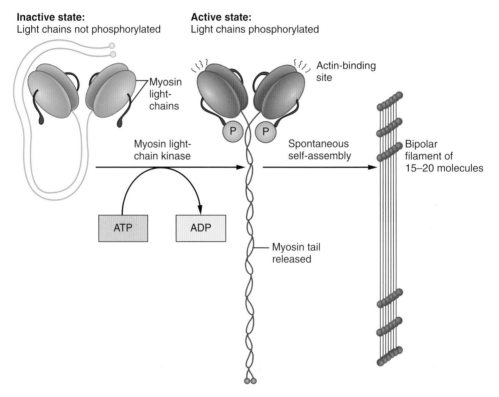

Figure 3–18. Model for assembly of smooth-muscle myosin thick filaments. Dephosphorylated myosin isolated from smooth-muscle cells is in an inactive state and does not readily form thick filaments because of the conformation of the tail domain that binds with the globular head domain. The phosphorylation of the 18-kDa light chain of myosin has two effects: It causes a change in the conformation of the myosin head, exposing its actin-binding site, and it releases the myosin tail from its inactive conformation, allowing the myosin molecules to assemble into bipolar thick filaments. *(Modified from Alberts B, Bray D, Lewis J, et al. Molecular Biology of the Cell, 2nd ed. New York: Garland Publishing, 1989.)*

calmodulin complex is then able to bind with other proteins and modulate their activity. One such calmodulin-regulated protein is smooth-muscle myosin light chain kinase (SmMLCK). Without Ca^{2+}-calmodulin, the SmMLCK is in an inactive state. After the binding of Ca^{2+}-calmodulin, SmMLCK is active and phosphorylates the 18-kDa regulatory light chain of smooth-muscle myosin II (Fig. 3–19).

This phosphorylation permits myosin II to aggregate into thick filaments and allows cross-bridge formation of the thick filaments with the thin filaments of the smooth muscle. Thus, the phosphorylation of myosin light chains is an obligatory event for cross-bridge formation and cycling in smooth muscle. During relaxation, Ca^{2+} ion concentration decreases, and a net dephosphorylation occurs. The reduction in intracellular Ca^{2+} concentration causes an inactivation of the SmMLCK (by a reversal of Ca^{2+}-calmodulin binding). The regulatory activity of the requisite phosphatase enzyme(s) that dephosphorylates the myosin light chain is not well defined.

Smooth-Muscle Contraction Is Influenced at Multiple Levels

Because it can be stimulated by a variety of sources—neuronal and hormonal inputs—smooth-muscle contraction can be regulated by several mechanisms. These include regulation by cyclic adenosine monophosphate (cAMP), diacylglycerol, and the protein caldesmon (see Fig. 3–19). Each of these pathways effects a negative regulation; they serve to maintain a relaxed state of the smooth muscle. For example, activation of β-adrenergic receptors on smooth-muscle cells causes an increase in intracellular cAMP levels, which, in turn, activates cAMP-dependent protein kinase. One of the targets for cAMP-dependent protein kinase in smooth muscle is SmMLCK, and phosphorylation of the MLCK results in a lower affinity of the kinase for the Ca^{2+}-calmodulin complex. As a result, the SmMLCK does not phosphorylate myosin, and the myosin (and smooth muscle) remains in its relaxed state. Other hormones relax smooth muscle by activation of protein kinase C, which is mediated by Ca^{2+} and 1,2-diacylglycerol. The activation of protein kinase C allows it to phosphorylate SmMLCK, causing it to remain in an inactive state.

In addition to hormonal regulation of contraction, smooth-muscle cells contain Ca^{2+}-binding proteins that interact with the actin thin filaments, thereby affecting contraction. Caldesmon is an elongated calmodulin-binding protein. In the absence of Ca^{2+}, caldesmon will bind to the actin filaments of smooth muscle, restricting the ability of actin and myosin to interact. In the presence of increased Ca^{2+} concentrations, the Ca^{2+}-calmodulin complex binds with caldesmon, causing a release of the protein from the thin filaments. Thus,

the Ca^{2+}-calmodulin complex modulates contraction in smooth muscle by affecting myosin head group phosphorylation, in addition to releasing the caldesmon block on actin thin filaments. This dual control by Ca^{2+}-calmodulin allows the cell to regulate the duration and frequency of contractions.

Actin-Myosin Contractile Structures Are Found in Nonmuscle Cells

In nonmuscle cells, the actin/myosin ratio is about $100:1$. Thick filaments and microfilaments form within the cytoplasm, but they are in equilibrium with pools of nonpolymerized myosin and G-actin. Although the nonmuscle thick filaments are shorter than those of skeletal muscle, and the myosin and actin filaments do not form the highly structured array found in skeletal muscle, they are still responsible for contraction in nonmuscle cells. Figure 3–20 gives two examples.

During telophase, the last stage in mitosis, a contractile ring forms on the cytoplasmic membrane surface at the cleavage furrow. This contractile ring contracts (like a belt pulled tightly around the waist), forming a cleft between two cells that are separating. Just before telophase, actin filaments begin to form at the site that will become the cleavage furrow. In addition, free myosin begins to polymerize at the same site and, together with actin and the actin-binding protein α-actinin, forms the contractile ring. Because the actin filaments that are attached to the plasma membrane have mixed polarity, the short myosin filaments can use the energy of ATP hydrolysis to cause a contraction that pulls the dividing cell into a dumbbell shape. Before cell division, the actin and myosin filaments rapidly depolymerize.

A second example of nonmuscle contraction is pulling on the plasma membrane, created by stress fibers formed within fibroblasts. Fibroblasts are cells that synthesize and are in contact with extracellular matrix proteins throughout much of the connective tissue that surrounds the organs of your body. The plasma membrane of the fibroblast makes contact with extracellular matrix proteins, both within a tissue culture dish and within the body's connective tissue, at sites called **focal contacts** or **adhesion plaques.** At these contact sites, an integral transmembrane protein of the integrin family binds an extracellular matrix protein, such as fibronectin, at the outer cell surface; this interaction pulls outward on the plasma membrane. The integrins are heterodimers that contain various isoforms of α and β subunits, and the distinct combination determines the specificity of binding various extracellular matrix proteins. The fibroblast is not pulled apart because the same integrin binds actin bundles called **stress fibers** at the cytoplasmic membrane surface. The stress fibers contain interdigitated actin bundles of mixed polarity that are linked

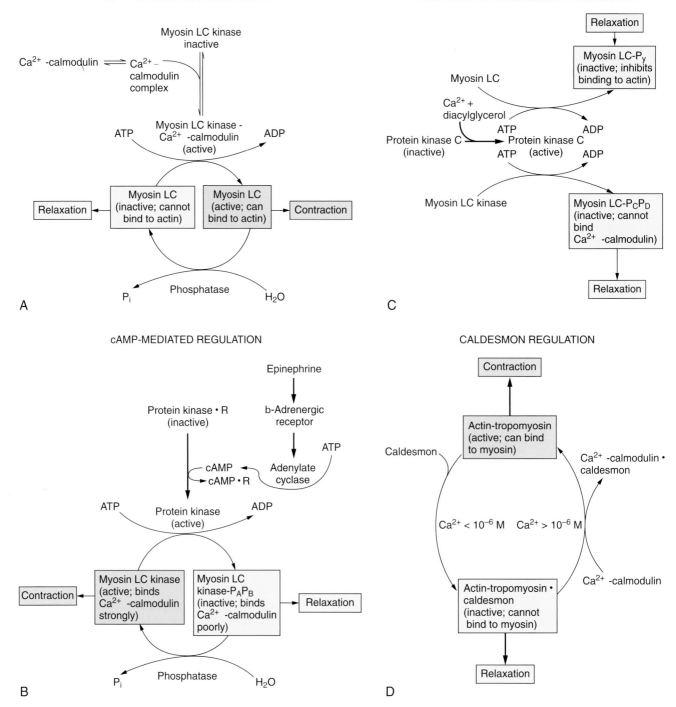

Figure 3–19. Mechanisms that regulate smooth-muscle contraction and relaxation. **A:** Regulation by Ca^{2+}-calmodulin: As intracellular Ca^{2+} increases, excess Ca^{2+} is bound by calmodulin, and the Ca^{2+}-calmodulin complex binds to and activates myosin light chain (LC) kinase. The activated kinase phosphorylates the regulatory LC of myosin at site X, which leads to contraction. As the intracellular Ca^{2+} concentration declines to less than 0.1 μM, there is a dissociation of the Ca^{2+}-calmodulin complex from myosin LC kinase, rendering it inactive. Under these conditions, the myosin LC phosphatase, which is not dependent on Ca^{2+} for activity, dephosphorylates myosin, causing relaxation. **B:** Regulation by cyclic adenosine monophosphate (cAMP): Stimulation of β-adrenergic receptors by catecholamines, such as epinephrine, causes the stimulation of adenylate cyclase and an increase in intracellular cAMP concentrations. This stimulates the cAMP-dependent protein kinase that phosphorylates myosin LC kinase at sites A and B near the calmodulin-binding domain of the molecule. This causes the myosin LC kinase to have a lower affinity for calmodulin, rendering it inactive, such that it does not phosphorylate the regulatory light chain of myosin, causing relaxation. Dephosphorylation of the myosin LC kinase restores its ability to bind Ca^{2+}-calmodulin for contraction. **C:** Diacylglycerol-mediated regulation: Diacylglycerol and Ca^{2+} stimulate the activity of protein kinase C, which phosphorylates myosin LC kinase at sites different from those of the cAMP-dependent kinase (sites C and D). In addition, protein kinase C phosphorylates the regulatory LC of myosin at a position different from the myosin LC kinase. Both of these events render the proteins inactive and cause relaxation. **D:** Caldesmon regulation: At low concentrations of Ca^{2+} (> 1 μM), caldesmon binds to tropomyosin and actin, inhibiting the binding of myosin, thereby keeping the muscle in a relaxed state. When the intracellular Ca^{2+} concentration increases, the Ca^{2+} is bound by calmodulin and the Ca^{2+}-calmodulin complex binds with caldesmon, releasing it from the actin filament and allowing contraction. *(Modified from Adelstein RS, Eisenberg, E. Annu Rev Biochem 1980;49:92–125; and Rasmussen H, Takuwa Y, Park S. FASEB J 1983;1:177–185.)*

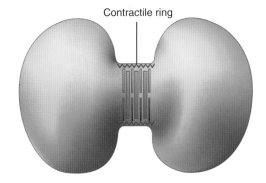

Contractile ring

Figure 3-20. **Nonmuscle actin and myosin have contractile functions.** Two examples of a contractile function for nonmuscle actin and myosin are demonstrated. An assembly of actin and myosin creates a contractile ring (**top**) that draws in the center of a cell, leading to cell division. A simplified presentation of stress fibers (**bottom**) that interact with the plasma membrane at focal contacts, and because of the contractile activity of actomyosin, cause flattening of substrate-attached fibroblasts.

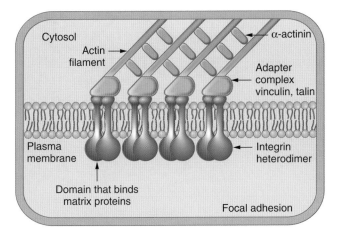

together in a parallel array. The stress fibers bind the integrin through adapter proteins, called **talin** and **vinculin,** as well as a plus end–capping protein. Again, it is the plus end of the actin bundle that binds end-on to the plasma membrane at the focal contact, as well as to plus end–capping proteins on that side of the fibroblast that is not attached to the substrate. The stress fibers also contain short myosin filaments, and they exert a contractile force on the actin bundles, which results in an inward pulling on the plasma membrane that counteracts the outward pull of the extracellular matrix and, in culture, leads to flattening of the fibroblast. The stress fibers rapidly assemble in response to fibroblast attachment to a substrate, and they rapidly depolymerize when the cells are detached. The depolymerization of actin bundles causes the cells to round up. A third example is the actin and myosin filaments that associate with the adhesion belt, characteristically located below the tight junction of an epithelial cell. Adjacent cells in an epithelial cell layer are held 15 to 20 nm apart by a Ca^{2+}-dependent transmembrane protein called *uvomorulin* or *E-cadherin.* Uvomorulin also binds, through the actin-binding proteins, α-actinin and vinculin, to the sides of actin filament bundles that form an adhesion belt around the cytoplasmic membrane surface. Myosin filaments and this circumferential F-actin contract, thereby mediating an important process in human development: the folding of epithelial cells into tubes. In the

neural plate, this contraction causes an apical narrowing, which leads to the plate rolling up to form the neural tube during human development.

Members of the Myosin Supergene Family Are Responsible for Movement of Vesicles and Other Cargo Along Actin Tracks in the Cytoplasm

We now know that a large family of myosins exists. The human genome contains 40 myosin genes. What all of these myosins have in common is a conserved motor domain, but they vary in their tail domains, which allows for the diversity of their functions. It will be instructive to look at four myosins (I, II, V, and VI) (Fig. 3–21).

After the discovery of muscle myosin (myosin II), which is involved in contraction, the next myosin described was myosin I. This myosin I was so named because it has only one motor head group and also has a short tail. Myosin I has the ability to walk along an actin filament, with the energy source being ATP hydrolysis, toward the plus end. In so doing, it can carry vesicular cargo through the cytoplasm, helping to establish intracellular organization. Many more myosins were then discovered and named for the order of their

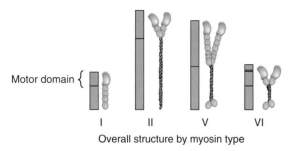

Motor domain {

I   II   V   VI

Overall structure by myosin type

Figure 3–21.   **Four members of the myosin family.** We demonstrate the domain structure of myosin I, II, V, and VI. All have a common N-terminal motor domain shown in blue and a variable C-terminal domain shown in red that allow for their different functions. All of the myosins move toward the plus ends of actin filaments except for myosin VI, which moves toward the minus end. This is due to the small break (amino acid sequence change) in the structure of the motor domain (shown in white).

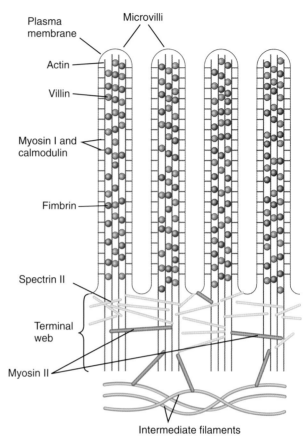

Plasma membrane
Microvilli
Actin
Villin
Myosin I and calmodulin
Fimbrin
Spectrin II
Terminal web
Myosin II
Intermediate filaments

Figure 3–22.   **Bundled actin filaments have a structural function within microvilli.** Actin bundles are attached at their plus ends to the tip of microvilli. The actin filaments are bundled by the proteins villin and fimbrin, and the bundles are attached to the side walls of the microvilli plasma membrane by association with myosin I and calmodulin. Within the terminal web, nonerythroid spectrin (spectrin II) and myosin II link adjacent actin bundles to each other and to intermediate filaments.

discovery (myosin III to XVIII to date). Myosin V has two heads and is involved in vesicular and organelle transport. Myosin VI is the only member of the myosin family that moves toward the minus end of actin filaments, because of a break in the conserved sequence of its motor domain. The diversity of directional movement and tail domains allows for different cargo to be moved throughout the cell cytoplasm on actin tracks.

Bundles of F-Actin Form a Structural Support for the Microvilli of Epithelial Cells

Epithelial tissue, which lines the surfaces of the body, the internal organs, body cavities, tubes, and ducts, contains absorptive cells with numerous fingerlike projections, called *microvilli,* on their apical surface. These **microvilli** increase the surface area of the apical plasma membrane of the epithelial cell, thereby permitting a greater absorption of important nutrients. The microvilli, which are approximately 80 nm wide and 1 μm long, need a stable cytoskeletal scaffolding to maintain their shape and upright position. A stable and highly structured core of 20 to 30 bundled actin filaments, which run parallel to the microvilli and attach to the cytoplasmic surface of the plasma membrane, serves as this scaffolding (Fig. 3–22).

The actin filaments are bundled by two proteins, named *fimbrin* and *villin.* Actin-bundling proteins are characterized by having two binding sites for F-actin. As they bind to the sides of actin filaments in a helical staircase, they group the filaments into parallel bundles. Villin has an interesting second function: At Ca^{2+} concentrations greater than 10^{-6} M, villin becomes an actin-severing protein. (This class of proteins is discussed later in this chapter.) The actin bundles are attached at their plus end to the tip of the microvilli plasma membrane by undefined proteins. The lateral attachments of the

actin bundles to the side wall of the microvilli's plasma membrane are through a complex that contains calmodulin and myosin I (minimyosin). The core bundles of microvillar actin filaments end just below the surface of the apical plasma membrane in a region of the epithelial cell, called the *terminal web* because it contains a meshwork of actin filaments, actin-binding proteins, and IFs. The actin cross-linking protein, spectrin II (or nonerythroid spectrin), and short myosin filaments run perpendicular to and attach adjacent actin core bundles. These attachments of the core bundles to spectrin II and myosin are thought to hold the microvilli upright. Spectrin II also cross-links the actin core bundles to IFs.

The Gel-Sol State of the Cortical Cytoplasm Is Controlled by the Dynamic Status of Actin

The cytoplasm of human and animal cells has regions that have the characteristics of a pseudoplastic gel and other regions that liquefy into the sol state. Gel-sol transformations of the cytoplasm are essential for altering the shape of cells and controlling their movement. The gel-sol conversion within the cytoplasm is regulated by the dynamic state of actin and its interaction with actin-binding proteins.

For example, in the cortical cytoplasm just below the plasma membrane, there is a thick, three-dimensional matrix of actin filaments that excludes organelles from this region of the cell cytoplasm. Long actin filaments tend to self-associate, causing a highly viscous solution. In the cortical cytoplasm, however, these actin filaments are cross-linked into a three-dimensional meshwork by long, fibrous, actin cross-linking proteins. The two most prevalent actin cross-linking proteins are spectrin II (nonerythroid spectrin) and filamin, both of which are long fibrous proteins with two well-separated actin-binding sites at their ends. On occasion, it is essential that a region of the cortical cytoplasm becomes liquefied. For instance, when a macrophage contacts a bacterial cell, the cortical actin network must locally disassemble so that the cell surface can restructure to engulf the microorganism. This is conducted by a local increase in the cytoplasmic Ca^{2+} concentration, to 10^{-6} M, which stimulates a Ca^{2+}-sensitive, actin-severing protein, called **gelsolin,** to cut the actin filaments into short protofilaments. In the process, the gelsolin molecule binds to the plus end of the severed actin filaments and caps that end. Gelsolin is removed from the plus end of actin filaments by association with phosphatidylinositol-4,5-bisphosphate (PIP_2). (This is discussed further later in the Regulation of Actin Dynamics section.)

Another protein that causes severing of actin filaments, by a different mechanism, is cofilin. Cofilin is a member of a family of proteins that are called *actin depolymerizing factors.* Cofilin binds to G-actin and the sides of actin filaments tightening the helix and increasing the torque. As a result, these more strained actin filaments are more easily severed, by mechanical stress, producing new plus ends for nucleation of actin growth.

Although the concentration of actin within nonmuscle cells is 50 to 200 μM, far greater than the critical concentration for the plus and minus ends of F-actin, only 50% of the actin is in polymerized form in most cells. The actin within nonmuscle cells is in a dynamic state, undergoing polymerization and depolymerization as required. The reason for the pool of G-actin within nonmuscle cells is a group of small actin-binding proteins of the thymosin family. Thymosin β_4 is a 5-kDa protein that binds to G-actin. G-actin bound to thymosin cannot hydrolyze or exchange its ATP, nor can it bind to the plus or minus end of F-actin. When rapid polymerization of actin is required, for example, at the leading edge of a motile cell, then activated profilin competes with thymosin for binding to G-actin. The profilin–actin complex then binds to the plus end of actin filaments, followed by the dissociation of profilin. Profilin binds to the second and fourth domains of G-actin on the opposite side from the ATP-binding cleft. The binding of profilin causes a conformational change in G-actin opening further the cleft region, leading to more rapid ATP-ADP exchange. Localized release of G-actin with ATP bound, in turn, promotes rapid polymerization. Profilin is activated by phosphorylation and binding to inositol phospholipids. Figure 3–23 provides a summary of thymosin and profilin function.

Cell Motility Requires Coordinated Changes in Actin Dynamics

The movement of cells, or **cell motility,** is essential for human development. Many cells are moving during embryogenesis. Growth cones, at the leading edge of motile axons, move toward their synaptic targets; macrophage and neutrophils move toward sites of infection; and fibroblasts migrate through connective tissue. Cell motility is also important in cancer biology. Cancer cells from primary tumors can crawl from the primary tumor sites to invade neighboring tissues and enter blood vessels. Therefore, understanding how cells move is important.

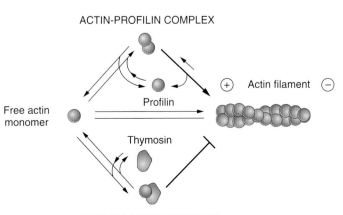

Figure 3–23. Role of thymosin and profilin in the dynamics of polymerization and depolymerization of actin. Summary of the interactions of thymosin and profilin with actin.

ACTIN-PROFILIN COMPLEX

Free actin monomer

Profilin

Thymosin

(+) Actin filament (−)

ACTIN-THYMOSIN COMPLEX

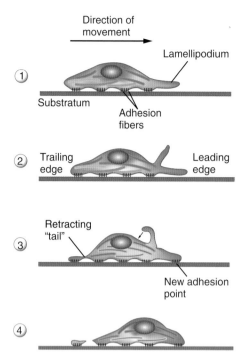

Direction of movement

Lamellipodium

① Substratum — Adhesion fibers

② Trailing edge — Leading edge

③ Retracting "tail" — New adhesion point

④

Figure 3–24. The steps in cell motility. The cell to the left (**1**) is moving toward the right. The cell first extends a flat, sheetlike projection of its plasma membrane (lamellipodia; [**2**]). The projection then attaches to the substrate or travels back toward the cell soma (**3**). The protrusions then adhere to the substrate and form sites of attachment for actin filaments to the focal contact (**4**). The tension on the actin filaments cause the cell to be pulled forward, leaving behind remnants of the tail region.

STATIONARY:
Polymerization and retrograde flow balanced

Myosin

F-actin

G-actin

Lamellipodium

Myosin-driven retrograde flow of actin

Substrate

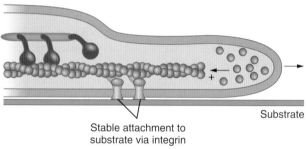

MOVING FORWARD:
Cell attachment resists retrograde flow, resulting in extension

Stable attachment to substrate via integrin

Substrate

Figure 3–25. Protrusion in a motile cell is based on the balance of actin polymerization at the leading edge and retrograde flow of actin filaments toward the rear. (*Modified from Becker et al.* The World of the Cell, *6th ed. Pearson, Benjamin Cummings, 2006.*)

The steps in cell movement are protrusion of the plasma membrane on the leading edge, attachment to the substrate, tension on the actin filaments, and retraction of the tail of the cell (Fig. 3–24).

The protrusions of the plasma membrane include lamellipodia, which are sheetlike projections, and filopodia or microspikes, which are pointed, narrow projections at the leading edge of the cell. In the lamellipodia and filopodia, two processes are occurring. There are the rapid polymerization of actin filament at the plus end, which is associated with the membrane and is pushing it forward, and the retrograde flow of the actin filaments toward the rear of the cell based on a myosin II–based contractile process. When polymerization at the plus end is more rapid than retrograde flow, then the cell moves forward. When the two are balanced, the cell is stationary (Fig. 3–25).

The lamellipodia contain branched actin filaments with their plus ends associated with the plasma membrane. The association to the cytoplasmic surface of the plasma membrane is through a protein family called *formins*. The formins have two actin-binding domains and can therefore remain attached to the actin filament while moving toward the plus end as new actin monomers are added. The branching is due to the Arp 2/3 nucleating complex. This complex contains two actin-related proteins (Arp 2 and 3), with 45% sequence iden-

tity when compared with actin, five other smaller proteins, and a nucleation-promoting factor. The Arp 2/3 nucleating complex can bind to the sides of existing actin filaments at its minus end and nucleate plus end growth of a new actin filament. Branches formed by the Arp 2/3 complex are at a 70-degree angle relative to the original filament. Therefore, the lamellipodia is being forced forward by a branched actin meshwork with its plus ends pushing out the plasma membrane (Fig. 3–26).

Filopodia, in contrast, are formed by bundled parallel actin filaments with their plus end polymerizing in association with the filopodia tip. The actin filaments are bundled within the filopodia by the actin bundling protein called fascin. The result is a stiff microspike that serves as a feeler as the cell moves forward.

The lamellipodia must now make an attachment to the surface through interaction of the integrins with extracellular matrix proteins on the outside of the cell and linker or adapter proteins on the cytoplasmic side. These linker proteins (talin, vinculin, α-actinin) attach actin filaments to the focal contact site as previously described.

An alternative form of cell motility is amoeboid movement. This form of motility is used by amoeba, slime molds, and leukocytes. In amoeboid movement, the cell extends pseudopodia (derived from the Greek meaning "false feet"). The basis for movement is conversion of a thick, gelatinous, actin-crosslinked ectoplasm at the periphery of the cytoplasm into a fluid

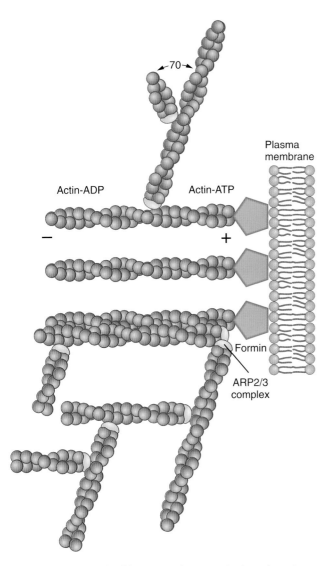

Figure 3–26. Role of formins and Arp 2/3 in the polymerization, branching, and membrane association of actin filaments within the lamellipodia.

endoplasm in the direction of the protrusion. Then the endoplasm congeals into ectoplasm, thereby turning the tip of the pseudopodia into a pseudoplastic gel. A series of this gel-sol-gel-sol interconversion forces the cell forward. Meanwhile, at the rear of the cell, the gel-sol interconversion allows the rear of the cell to retract. Severing of actin filaments by gelsolin, due to increased local calcium concentrations, causes the gel-sol conversion. The reverse sol-gel conversion is caused by actin polymerization and cross-linking.

Cells in the body and in embryos demonstrate directional movement. This directional movement is based on diffusible molecules that are recognized by the cell surface. These diffusible molecules are called chemoattractants or chemorepellants based on whether the cell moves toward or away from this particular directional cue. The process of directed movement based on chemical gradients is called *chemotaxis*. An example of chemotaxis is the movement of a neutrophil toward a

bacterial infection. Receptors exist on the surface of neutrophils that allow them to detect low levels of N-formylated peptides. Because prokaryotes, and not eukaryotes, produce proteins with an N-terminal methionine, these peptides can be derived from only a bacterial source. The receptors on the neutrophil surface allow this white blood cell to move in the direction of the bacterial infection.

Inhibitors of Actin-Based Function

Cytochalasins, a group of chemicals excreted by various molds, block cell movement. The cytochalasins bind to the plus end of microfilaments; block further polymerization; and inhibit cell motility, phagocytosis, microfilament-based trafficking of organelles and vesicles, and the production of lamellipodia and microspikes. Lantrunculin, extracted from a sea sponge, binds to and stabilizes actin monomers, the result being a net depolymerization of actin filaments. Lantrunculin has similar effects on actin-based function as the cytochalasins have. Phalloidin, an alkaloid isolated from the toadstool *Amanita phalloides*, stabilizes microfilaments and does not allow depolymerization. These chemicals also block cell movement, indicating that both actin filament assembly and disassembly are required for cell motility.

ACTIN-BINDING PROTEINS

Figure 3–27 and Table 3–1 summarize the functions of the various proteins that interact with actin. We have discussed intracellular actin-based assemblies (e.g., within the cytoplasm of the cell); now we consider two mechanisms by which actin filaments bind to the cytoplasmic surface of the cell membrane. First, we discuss the binding through the **ERM family** and then associations found in the **spectrin membrane skeleton**.

The ERM Family Mediates End-on Association of Actin with the Cytoplasmic Surface of the Plasma Membrane

The ERM family (ezrin, radixin, and moesin), of protein 4.1–related proteins, attaches actin filaments to the plasma membrane in many cell types. The C-terminal domain of activated ERM proteins binds the side of preformed actin filaments, and the N-terminal domain binds the cytoplasmic domain of transmembrane proteins such as CD44 (the receptor for hyaluronan) (Fig. 3–28).

Defects in the ERM protein, called merlin, leads to the human genetic disease called *neurofibromatosis*, where multiple benign tumors develop in auditory nerves and other parts of the nervous system. The ERM interactions are regulated by phosphorylation or binding to PIP_2.

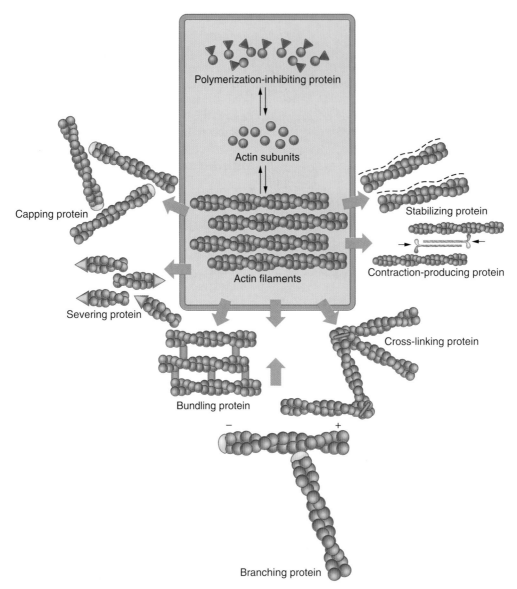

Figure 3–27. Various roles of actin-binding proteins. Summary of the ways in which various actin-binding proteins regulate the cellular organization of actin. (Modified from Widnell CC, Pfenninger KH. *Essential Cell Biology*. Baltimore: Williams & Wilkins, 1990.)

TABLE 3–1. Actin-Binding Proteins	
Protein	**Functions**
Tropomyosin	Stabilizes filaments
Fimbrin, α-actinin, villin	Bundles filaments
Formin, Arp 2/3	Nucleates polymerization and forms branches
Filamin	Cross-links filaments
Spectrin I/II	Cross-links filaments in membrane skeleton
Gelsolin	Fragments filaments
Myosin II	Slides filaments in muscle
Myosin I	Moves vesicles on filaments
Cap Z	Caps plus ends of filaments
Profilin, thymosin	Binds actin monomers

SPECTRIN MEMBRANE SKELETON

The spectrin membrane skeleton, first described in erythrocytes, but now known to be a ubiquitous component of nonerythroid cells, is essential for maintaining cellular shape and membrane stability and for controlling the lateral mobility and position of transmembrane proteins within biological membranes.

The Structure and Function of the Erythrocyte Spectrin Membrane Skeleton Are Understood in Exquisite Detail

The spectrin membrane skeleton was first described, and is best understood, in the mammalian erythrocyte. The spectrin membrane skeleton of the human erythro-

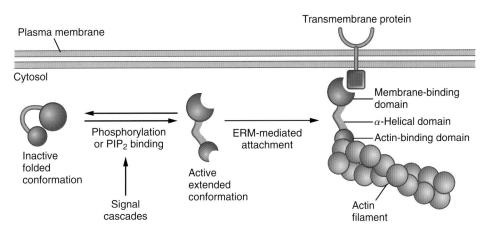

Figure 3–28. ERM proteins serve to link the plus end of actin filaments to the plasma membrane. When in their folded inactive form, the ERM proteins cannot associate with the plus end of actin filaments or the cytoplasmic domain to transmembrane proteins such as CD44. However, they become activated by phosphorylation or association with PIP₂, and the unfolding of the ERM proteins allows binding to transmembrane proteins and actin filaments.

cyte maintains the biconcave shape of the erythrocyte, gives it its properties of elasticity and flexibility, stabilizes the plasma membrane, and controls the lateral mobility of integral membrane proteins. These are important properties for an 8-μm diameter biconcave disk that must continuously deform as it passes through capillaries as small as 2 μm in diameter.

The major proteins of the skeleton are **spectrin I**, **actin**, and **protein 4.1** (a nomenclature based on migration on sodium dodecyl sulfate-polyacrylamide gel electrophoresis [SDS-PAGE]). Spectrin I is composed of two large subunits of approximately 280 (α) and 246 kDa (β). The simplest form of spectrin is an antiparallel αβ heterodimer; however, on the cytoplasmic surface of the erythrocyte membrane, it is an (αβ)2 tetramer, formed by head-to-head interaction of two heterodimers. Each end of the spectrin tetramer contains an actin-binding site, and spectrin I cross-links the actin filaments into a two-dimensional meshwork that covers the cytoplasmic surface of the plasma membrane. The actin filaments are short, approximately 14 actin monomers long (~33 nm); therefore, they are called **actin protofilaments.** The actin protofilaments are stabilized by **tropomyosin,** the minus end capped by **tropomodulin,** and each protofilament binds six spectrin tetramers, forming a hexagonal array. The spectrin-F-actin complex is stabilized by protein 4.1 and adducin, which also binds to the ends of the spectrin tetramer. The spectrin skeleton is attached to the bilayer by two types of linkages. A peripheral protein, called **ankyrin,** binds to the spectrin β subunit toward the junctional end of the heterodimers and links spectrin to the cytoplasmic NH₂-terminal domain of band 3. Protein 4.1, in addition to stabilizing the spectrin–actin interaction, binds to a member of the glycophorin family, thereby serving as a link to the bilayer. The structure of the membrane skeleton is shown in Figure 3–29.

Phosphorylation of protein 4.1 and adducin by A kinase and C kinase, respectively, down-regulates the formation of the spectrin-4.1-actin and spectrin-adducin-actin ternary complexes. Recently, Goodman and colleagues have demonstrated that α-spectrin has an E2/E3 ubiquitin conjugating and ligating activity that can ubiquitinate itself near the tail regions of the tetramer, as well as various other target proteins. Ubiquitination of the C-terminal tail region of α-spectrin also down-regulates the affinity of the spectrin-4.1-actin and spectrin-adducin-actin ternary complexes. As discussed later, this becomes important in red blood cells from patients with sickle cell disease, in whom the level of ubiquitination of spectrin is substantially diminished.

Because the spectrin membrane skeleton is responsible for the normal biconcave shape of an erythrocyte, genetic defects in these proteins cause abnormal red cell shapes and stability. Hereditary spherocytosis (HS) is a common hemolytic anemia in white populations, in whom the erythrocytes are spherical and fragile (Fig. 3–30).

All patients with the common dominant form of HS show a small-to-moderate reduction in spectrin content, sometimes because of genetic defects in ankyrin, band 3, or both. A small number of HS subjects have a defective spectrin molecule that cannot bind protein 4.1 and, therefore, cannot form a stable spectrin 4.1-actin complex.

Clinical Case 3-1

Douglas Richmond is a 16-year-old British student at a prestigious Eastern boarding school. He wanted very much to try out for the football team, but the school nurse insisted that he see the consultant physician first, because "his eyes were yellow." Douglas told

Continued

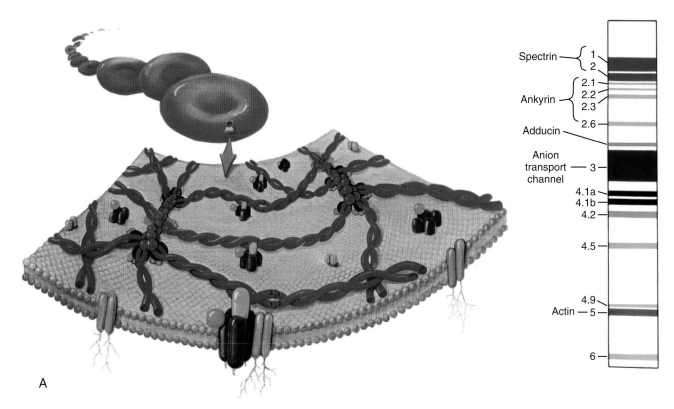

A

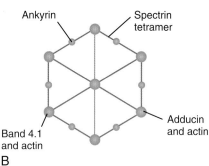

Ankyrin   Spectrin
                 tetramer

Band 4.1   Adducin
and actin   and actin

B

Figure 3-29. Protein interactions in the erythrocyte's spectrin membrane skeleton. **A:** Schematic drawing of the protein interactions within the membrane skeleton with matching color-coded SDS PAGE of RBC proteins to the right. **B:** Schematic diagram of spread membrane skeleton examined by negative-staining electron microscopy. We see a hexagonal lattice of junctional complexes containing actin protofilaments and protein 4.1, cross-linked by spectrin tetramers (Sp4), three-armed spectrin molecules (Sp6), and double spectrin filaments (2Sp4). Ankyrin is attached to spectrin filaments 80 nm from their distal end. *(A: Modified from Goodman SR, Zagon IS. Brain Res Bul 1984;13:813–832; B: Modified from Liu SC, Derick LH, Palek J. J Cell Biol 1987;104:527–536.)*

the doctor that he felt completely fine and had no symptoms at all—he just wanted to play American-style football. He did admit that his eyes had been a little yellow for years, but he never thought about it because his brother and sister also had a little yellow color in their eyes.

On physical examination, the doctor found him to be a well-developed, intelligent young man who couldn't understand "what all the fuss was about"; he played soccer actively at home. Douglas's vital signs were normal, his sclerae were distinctly yellow, as were the creases in his palms. His heart and lungs were normal. On abdominal examination, he had slight upper right quadrant tenderness, which he said he had had for several years. In the left upper quadrant, the doctor detected a distinct mass that moved down two fingerbreadths from the rib margin when Douglas took a deep breath. The doctor was confident that this was Douglas's spleen, which he estimated to

be twice normal size. The rest of the examination was unremarkable except for some dermatitis on the inside aspect of the boy's ankles, which Douglas ascribed to minor trauma from his soccer-playing in England.

The doctor asked Douglas if he knew he had an enlarged spleen. Douglas told him he didn't, but that that was interesting because his older brother had to have his spleen out in England because "it burst after he wracked up his MG three years ago." On further questioning, Douglas revealed that his brother and sister had always been anemic. He said that his parents were fine, but he thought his father had had his gall bladder out 6 years ago, just before his 40th birthday.

Douglas's initial laboratory tests showed a minimally reduced hemoglobin level of 11.8, a reticulocyte count of 16%, and normal white cell and platelet counts. Interestingly, the mean cell hemoglobin concentration (MCHC), automatically calculated by

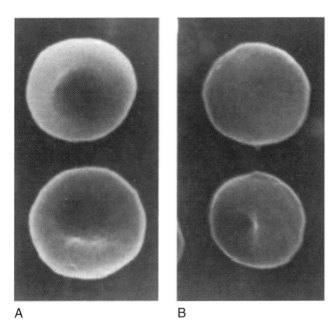

A   B

Figure 3–30. Scanning electron microscopy of normal and hereditary spherocytosis (HS) erythrocytes. **A:** Biconcave erythrocytes from a normal subject. **B:** A spherocyte and stomatospherocyte from an HS subject. (From Goodman SR, Shiffer K. *Am J Physiol* 1983;244:C134–141, by permission.)

the laboratory cell-counting machine, was elevated, at 39. The blood smear showed many small, dark, apparently dense red cells without the usual central pallor.

The serum bilirubin was elevated, at 3.2, and this was mostly the so-called indirect fraction. There was no bilirubin in the urine. The immunology laboratory reported that hepatitis and Coombs' tests were negative.

Because of these results, the doctor asked for an additional osmotic incubation test on Douglas's blood. That result showed an osmotic fragility curve that was markedly shifted to the right, with more than 50% of Douglas's red cells hemolyzing in 0.5 g% saline.

When Douglas returned for his report, the doctor told him that he had HS—the most common hereditary hemolytic anemia of western Europeans. He explained that he was very well compensated, and just barely anemic. Reassuringly, he anticipated that Douglas would continue to do well as long as he took a little bit of supplemental folic acid each day. He said that he probably had bilirubin gallstones like his father because of his increased blood turnover, but this required no special attention now. However, he strongly advised Douglas not to play any contact sport, because he might injure his enlarged spleen, which would require urgent surgery.

Douglas thanked him and elected to try out for the chess team.

Cell Biology, Diagnosis, and Treatment of Spherocytosis

HS is an autosomally dominant hemolytic disorder. The primary physiologic defect is a loss of the redundant lipid membrane, which allows the normal red cell to be so remarkably deformable. This deformability is essential for the red cell to squeeze through the labyrinthine interstices of the microcirculation. Specific genetic defects in any one of several proteins of the red cell membrane skeleton, which normally supports and stabilizes the lipid membrane, are responsible for the membrane loss. As the cells circulate, they are progressively depleted of membrane and become close to spherical. This makes them much less deformable, which, in turn, makes them vulnerable to entrapment in the microcirculation, especially in the spleen.

In vitro, the membrane loss can be nicely demonstrated by the increased osmotic fragility, because the cells act as perfect osmometers and their reduced membrane cannot accommodate the swelling that occurs in hypotonic saline.

The high MCHC of the red cells in this disorder is unique. The dark small cells certainly do look dense by microscopy, and this is confirmed by density gradient centrifugation. This interesting phenomenon is not fully understood, but it is likely to be related to some aberrant membrane transport process.

In hereditary elliptocytosis, in which the red cells are elliptical and fragile, the most prevalent defect is a spectrin dimer that cannot form tetramers.

The primary genetic defect in sickle cell anemia is in the hemoglobin molecule, but a subset of sickle cells lock into an irreversibly sickled cell, which is a major factor leading to the sickle cell crisis. The molecular bases of the irreversibly sickled cell are: (1) a posttranslational modification in β actin, in which a disulfide bridge is formed, leading to actin filaments that depolymerize slowly; and (2) diminished ubiquitination of spectrin, which leads to spectrin-4.1-actin and spectrin-adducin-actin ternary complexes that disassemble slowly. The result is a "locked" membrane skeleton leading to a cell that cannot change shape (Fig. 3–31).

Spectrin Is a Ubiquitous Component of Nonerythroid Cells

Until 1981, spectrin and the membrane skeleton were thought to be components found only in erythrocytes. That year, Goodman and coworkers demonstrated that spectrin-related molecules were ubiquitous components of nonerythroid cells. This led to the important question of the function of these nonerythroid spectrin molecules.

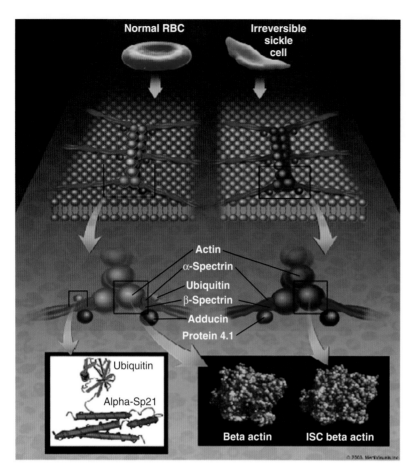

Figure 3–31. Molecular basis of the irreversible sickle cell (ISC). The inability of the ISC and its membrane skeleton to remodel is due to the inability of actin protofilaments to disassemble (indicated by the darkening of the green monomers) and the lack of α-spectrin ubiquitination leading to tightened spectrin-4.1-actin and spectrin-adducin-actin ternary complexes (indicated by the darkening of the red spectrin tails). *(From Goodman SR. Cell Mol Biol 2004;50:53–58, by permission.)*

Not only spectrin, but also ankyrin, protein 4.1, and band 3 analogues were found lining the membranes of nonerythroid cells. Two mammalian α spectrin genes and five β spectrin genes have been identified. Furthermore, two isoforms of nonerythroid spectrin have been extensively characterized. One isoform contains α-spectrin, linked to an alternately spliced form of erythroid β-spectrin (β-spectrin 1Σ2). This isoform is found in brain, skeletal muscle, and cardiac muscle. The second isoform contains a nonerythroid α- and β-spectrin that shares approximately 60% sequence identity with erythroid spectrin. This isoform, called **spectrin II**, is the product of a distinct set of genes and is the most universal form of spectrin. The spectrins I and II, which are found in nonerythroid cells, line the cytoplasmic surface of the plasma membrane and organelle membranes, and probably control the membrane contour and stability. However, nonerythroid spectrins are also multifunctional cross-linkers within the cytoplasm of nonerythroid cells and tissues. Spectrin II also cross-links actin rootlets within the terminal web region of the epithelial cell cytoplasm. Within neurons, spectrins I and II link actin filaments to microtubules, neurofilaments, organelles, and synaptic vesicles within the cytoplasm. The synaptic vesicle–spectrin interaction within the presynaptic terminal is central to synaptic transmission. Spectrins and their binding partners within nonerythroid cells play functions as diverse as tethering small spherical synaptic

vesicles to the active zone of the presynaptic plasma membrane; regulating the release of calcium from internal stores; controlling membrane traffic between the ER, Golgi, and plasma membrane; and bringing DNA repair enzymes in contact with damaged DNA.

Spectrins I and II, α-Actinin, and Dystrophin Form the Spectrin Supergene Family

The complete sequences for spectrins I and II have been determined. Both α and β subunits contain triple-helical repeat units of approximately 106 amino acids, separated by flexible nonhelical regions throughout most of their sequence. These repeats share approximately 20% to 40% sequence identity. Interestingly, the NH_2- and COOH-terminal ends of α- and β-spectrin I and II do not contain the typical repeat structure. Furthermore, a 140-amino acid stretch at the NH_2 terminus of the β subunit has been demonstrated to represent the actin-binding domain of spectrin (Fig. 3–32).

Whereas spectrins I and II share only 60% sequence identity throughout the α and β sequence, they are more than 90% identical in the actin-binding domain.

α-Actinin (an actin-bundling protein) and dystrophin (the protein missing in subjects with Duchenne's muscular dystrophy [DMD]) have sequences that are highly related to spectrin I and II.

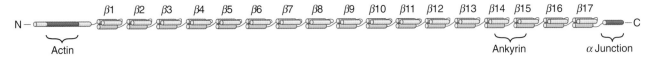

β–Spectrin II

A

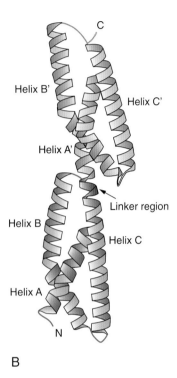

B

Figure 3–32. Structure of β-spectrin II. A: β-Spectrin II is presented as an example of the structure of the members of the spectrin supergene family. β-Spectrin II has a nonhelical actin-binding domain at its NH₂ terminus. There are 17 triple-helical spectrin repeats separated by flexible hinge regions. The COOH terminus contains a nonhelical region involved in association of the α-spectrin subunit. **B:** Detailed structure of one triple-helical repeat is presented. *(A: From Ma Y, Zimmer WE, Riederer BM, et al. Mol Brain Res 1993;18: 87–99, by permission.)*

Clinical *Case 3-2*

Robbie Franklin is an 11-year-old boy who was brought to the hospital in a wheelchair complaining of dyspnea. Over the past 3 days he has developed a painful productive cough and a fever to 39.7°C.

On physical examination, he is a thin but normally developed child, who looks younger than his stated age. He has steel braces from his ankles to his thighs on both legs and severe kyphoscoliosis of his thoracic spine. His blood pressure is normal, but his pulse is elevated to 112 mmHg. His lung examination shows a region of dullness to percussion, with sticky rales by auscultation and a harsh, grating rub on inspiration in the right lower lobe.

It is difficult to take a history from him because of dysarthric speech and possible mental slowness. His

mother says that Robbie appeared fine until he was about 3 years old. He then started to fall frequently when playing with his sisters, and his leg muscles became so weak that he had to "climb up himself" to get upright. Four years ago, he began to develop painful contractures in his legs, and 2 years ago, he required braces to limit those contractures. One year ago, to try to control his advancing weakness, he was started on oral prednisone (40 mg/day) therapy. He never had pulmonary problems previously.

Significant laboratory examinations showed a white cell count of 19,000 with 85% polys and a shift to the left. Many of the white cells contained Dohle bodies. His red cell and platelet counts were normal. His serum creatine kinase level was 3400 [sic]—consistent with serious, persistent, muscle necrosis.

Continued

Cell Biology, Diagnosis, and Treatment of Duchenne's Muscular Dystrophy

Robbie's primary disease is DMD. This is an X-linked recessive disorder that presents in young boys as progressive weakness of the proximal limb muscles. It is due to a genetic abnormality in the sarcolemmal protein dystrophin. Dystrophin is a large, elongated, spectrin-like protein, which normally links laminin in the extracellular matrix to F-actin within the cytoplasmic membrane skeleton. It is essential for the maintenance of muscle membrane structure and function. The abnormal dystrophin of Duchenne's dystrophy does not properly complete this important transmembrane linkage, so membrane stability fails, and eventual muscle fiber necrosis with marked seepage of muscle creatine kinase ensues. It is also possible that similar instability of membranes in neurologic structures is related to the mental difficulties that many of these patients have.

Robbie currently also has acute bacterial lobar pneumonia. This frequently occurs in Duchenne's patients as scoliosis (caused by muscle imbalance along the spinal column) and weakened intercostal muscles impinge on respiratory function. It is also likely that the steroid therapy given in an effort to delay the muscle weakness has contributed to his susceptibility to bacterial infection.

Robbie's prognosis is poor. It is likely that he will have progressive weakness, worsening contractures, and recurrent pneumonias for the next several years. He will probably succumb to one of the pneumonias by the age of 20.

α-Actinin, a 190-kDa dimer, is composed of two identical antiparallel subunits. Dystrophin is an 800-kDa homodimer, with two antiparallel 400-kDa subunits. Both α-actinin and dystrophin contain the spectrin triple-helical repeat units, with about 10% to 20% identity with the spectrin repeats. Both proteins contain a nonhelical region at their NH_2 terminus, with 60% to 80% identity with the actin-binding domain of β-spectrin. This finding was important in determining the function of dystrophin and the cause of DMD. Because of its sequence identity with the actin-binding domain of spectrin, dystrophin was proposed, and since demonstrated, to function in anchoring actin filaments to the plasma membrane in skeletal muscle. The common structure found for spectrins I and II, α-actinin, and dystrophin has led to the concept that they are descendants of a common ancestral gene, and thus to their being called the **spectrin supergene family**.

REGULATION OF ACTIN DYNAMICS

We mentioned earlier that contacts between spectrin and actin are regulated by the phosphorylation of protein 4.1 and adducin, as well as the ubiquitination of spectrin within the red blood cell membrane skeleton. The regulation of actin dynamics with cells that contain a cytoskeleton is far more complex.

As discussed in Chapter 8, phosphatidyl inositol and its derivatives are involved in cell signaling cascades. Several activated kinases convert phosphatidyl inositol into PIP_2. PIP_2 then is a key regulator of actin microfilament polymerization, depolymerization, attachment of actin binding proteins, and cleavage by gelsolin. PIP_2 associates with profilin at the cytoplasmic surface of the plasma membrane, preventing its association with actin. Hydrolysis of PIP_2 by a signal-activated phospholipase C releases profilin from the membrane, allowing it to bind G-actin. This promotes ADP-ATP exchange with subsequent polymerization of G-actin to F-actin (Fig. 3–33).

Gelsolin and cofilin, which sever actin filaments, are both inhibited by PIP_2. Once PIP_2 is released from these proteins, or hydrolyzed, filaments are severed, increasing the number of plus ends available, thereby stimulating actin polymerization.

WASP (the Wiskott–Aldridge syndrome protein; WASP is an inherited immune system disorder characterized by low levels of platelets and white blood cells) and Scar (or WAVE) bind to Arp 2/3, activating it and promoting actin polymerization. WASP and Scar also bind to PIP_2 and Rho family proteins.

The Rho protein family includes Cdc 42, Rac, and Rho, which are monomeric G proteins. These GTPases cycle between an active guanosine triphosphate (GTP)-bound form and an inactive guanosine diphosphate (GDP)-bound form. Activation of Cdc 42 causes activation of Scar, leading to its binding the ARP 2/3 complex and rapid actin polymerization and bundling. This results in the formation of filopodia or microspikes. Activation of Rac causes activation of WASP and PI(4)P 5-kinase, leading to the formation of lamellipodia and membrane ruffles. The kinase generates a form of PIP_2 that uncaps plus ends of F-actin, leading

PROFILIN RELEASE BY PIP_2 CLEAVAGE

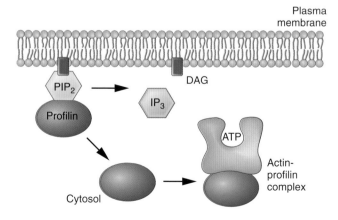

Figure 3–33. Regulation of the profilin-actin interaction by PIP_2.

to polymerization. Activation of Rho promotes bundling of actin filaments, with myosin II thick filaments, forming stress fibers interacting with focal contacts. Rho activates ROCK (Rho kinase), which phosphorylates myosin II light chains. This phosphorylation stimulates the association of the myosin heads with actin filaments, resulting in contraction.

Therefore, the Rho family of GTPases regulates the state of actin and is critical to actin-based cell motility. The effect of Rac, Rho, and Cdc 42 on a fibroblast is shown in Figure 3–34.

INTERMEDIATE FILAMENTS

Intermediate filaments (IFs) are 10 nm in diameter and, therefore, intermediate in thickness between microfilaments and myosin thick filaments, or microfilaments and microtubules. Although much work is required to determine the functions of this ropelike filament, their role appears to be primarily structural; that is, the major function of IFs is to provide resistance to mechanical stress placed on the cell. IFs within muscle cells link together the Z disks of adjacent myofibrils. Neurofilaments within the axon serve as a structural support to resist breakage of these long, slender processes, and they increase in number as the caliber of an axon increases during development. IFs of epithelial cells interconnect spot desmosomes, thereby stabilizing epithelial sheets.

A Heterogeneous Group of Proteins Form Intermediate Filaments in Various Cells

The protein monomers that constitute IFs differ from the components of microfilaments and microtubules (see

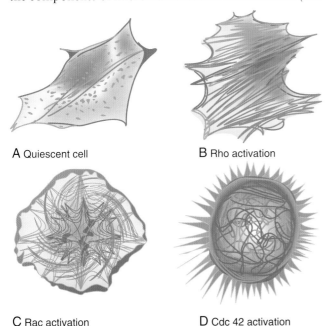

A Quiescent cell

B Rho activation

C Rac activation

D Cdc 42 activation

Figure 3–34. Effect of Rho family activation on actin organization in fibroblasts.

discussion later in this chapter) in several important ways. The IFs in various human and animal cells are composed of a heterogeneous group of proteins, but microfilaments are always composed of actin, and microtubules are always composed of tubulin. The IF subunits are fibrous proteins, although both G-actin and tubulin are globular. Almost all of the IF subunits are incorporated into stable IFs within various cells, whereas the same cells contain a substantial pool of unpolymerized G-actin and tubulin. No energy in the form of ATP or GTP hydrolysis is required for IF polymerization. IFs have no polarity, whereas microfilaments and microtubules have plus and minus ends. IFs are composed of a heterogeneous class of subunits (Table 3–2).

The keratin filaments found in epithelial cells always contain an equal number of subunits of acidic (type I) and neutral basic (type II) cytokeratins. In humans, there is a genetic disease, epidermolysis bullosa simplex, that arises from mutations in the keratin genes expressed in the basal cell layer of the epidermis. This disrupts the normal network of keratin filaments in these cells, and people afflicted with these keratin gene mutations are keenly sensitive to mechanical injury; even a gentle squeeze can cause disruption of this cell layer and blistering of the skin.

Vimentin, desmin, glial fibrillary acidic protein (GFAP), and neurofilament light chain (NF_L) are capable of forming homopolymeric IFs, but when together in a single cell (e.g., muscle cells, glial cells), they may copolymerize. However, in an epithelial cell where cytokeratins can be coexpressed with vimentin, they do not copolymerize, but instead, form separate IFs. Within axons and dendrites, NF_L, neurofilament medium chains (NF_M), and neurofilament heavy chains (NF_H) copolymerize to form the neurofilaments. The cell-type specificity of IF proteins has been useful to pathologists, who use fluorescent IF-type–specific monoclonal antibodies to identify the tissue of origin of metastatic cancer cells. The nuclear lamina is composed

TABLE 3–2.	Intermediate Filaments of Human Cells	
Intermediate Filament	**Subunits (M_r)**	**Cell Type**
Keratin filaments	Type I acidic keratins Type II neutral/basic keratins (40–65 kDa)	Epithelial cells
Neurofilaments	NF_L (70 kDa) NF_M (140 kDa) NF_H (210 kDa)	Neurons
Vimentin-containing filaments	Vimentin (55 kDa) Vimentin + glial fibrillary acidic protein (50 kDa)	Fibroblasts Glial cells
	Vimentin + desmin (51 kDa)	Muscle cells
Nuclear lamina	Lamins A, B, and C (65–75 kDa)	All nucleated cells

NF_L, neurofilament light chain; NF_M, neurofilament medium chain; NF_H, neurofilament heavy chain.

of the IF-related proteins lamin A, lamin B, and lamin C. These proteins and the square lattice that they form on the inner nuclear envelope are discussed in Chapter 5 (see the section dealing with the nucleus).

How Can Such a Heterogeneous Group of Proteins All Form Intermediate Filaments?

It is truly remarkable that proteins of the IF class, which range in M_r from 40 to 210 kDa, are all capable of forming IFs. The molecular basis for this common morphology is shown in Figure 3–35.

All IF proteins contain a subunit-specific NH₂ terminus of variable size, a homologous central α-helical region of approximately 310 amino acids (with 3 nonhelical gaps), and a subunit-specific COOH terminus of variable size. Only the homologous 310-amino acid α-

helical region is a portion of the 10-nm IF core. The variable regions extend from the core and are responsible for cross-linking IFs to other cytoskeletal structures. In the formation of the IFs, the first step is that the 310-amino acid α-helical region of two monomers wind around each other into a parallel coiled coil. The IF proteins contain the heptad repeat within the 310-amino acid α-helical region required for coiled-coil formation. Next, two dimers link side by side in an antiparallel conformation to form a tetramer. Because the tetramers have an antiparallel conformation, the IFs have no polarity. The IF tetramers attach laterally to each other in a staggered array until there are 8 tetramers (32 monomers) making up the wall of the IF. The eight tetramers are wound to form the ropelike structure of the IF. Microfilaments have actin-binding proteins to allow their association with other cytoskeletal structures, and microtubule-associated proteins (MAPs) play a similar function for microtubules. There are also specific cross-linking proteins for IFs, such as filaggrin,

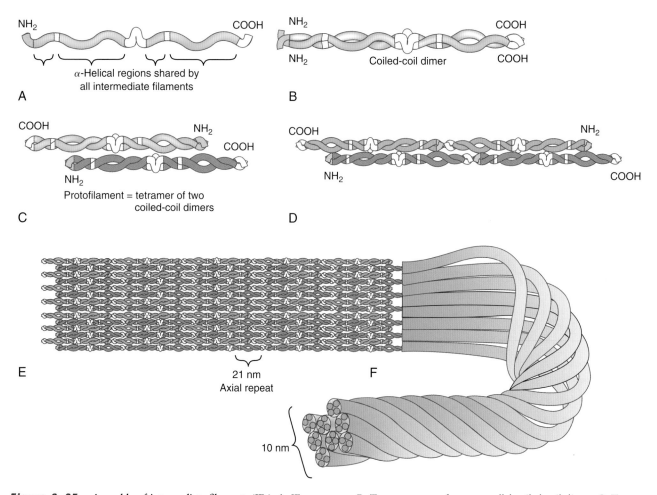

Figure 3–35. Assembly of intermediate filaments (IFs). **A:** IF monomers. **B:** Two monomers form a parallel coiled-coil dimer. **C:** Two dimers form an antiparallel tetramer by side-to-side interaction. **D:** The two dimers forming a tetramer are staggered, which allows the formation of higher order structure. **E:** The tetramers continue to associate in a helical array up to eight tetramers (protofilaments) wide. **F:** The intermediate filaments became longer and wound into a ropelike structure. *(Modified from Alberts B, Bray D, Lewis J, et al. Molecular Biology of the Cell, 3rd ed. New York: Garland Publishing, 1994.)*

which bundles keratin filaments, and plectin, which bundles vimentin-containing IFs, and neurofilaments that bundle to each other, as well as to microtubules and microfilaments. The variable COOH-terminal regions of NF heavy and medium subunits bundle neurofilaments together, giving a structurally stable core to axons and dendrites.

The importance of these IF linkages is demonstrated in people who have a mutation in plectin. The result of such a mutation is a human disease that combines the phenotypes of epidermolysis bullosa simplex, muscular dystrophy, and neurodegeneration.

In anaphase cells, the IFs normally form a tight weave around the nucleus and then spread in wavelike fashion toward the plasma membrane. If the microtubules of an anaphase cell are depolymerized with colchicine (Col-Benemid) or demecolchicine (Colcemid), the IFs collapse around the nucleus; obviously, the IFs are highly integrated with microtubules. When antibodies against spectrin II were microinjected into fibroblasts, the IF network again collapsed around the nucleus, even though there was no obvious effect on microtubules or microfilament stress fibers. This suggests that spectrin II may also play an important role in linking IFs to other cytoskeletal structures. Indeed, immunoelectron microscopy has demonstrated such a role for spectrin II within the terminal web of epithelial cells and within the axons and dendrites of mammalian neurons. All of the studies described earlier suggest an association of IFs with the nuclear envelope. In addition, IFs attach to the plasma membrane by interactions with ankyrin and with spectrin II.

the tubulin molecules are added to the growing microtubule to form the 13 protofilaments. The individual protofilaments are organized such that α and β subunits alternate along the length of the protofilament, which provides a microtubule with an inherent polarity. The tubulin dimer α subunits, for each protofilament, all face the minus slow-growing end. The β subunits all face the plus end (Fig. 3–36).

Tubulin is one of the most highly conserved proteins known. The significance of this is currently unclear, but it is presumed that this results from the many essential functional subdomains within the tubulin molecule. Not only are there regions that are necessary for subunit interactions during microtubule assembly, but tubulin also contains regions for GTP binding, for interacting with MAPs, and sites for binding to several different drugs. Pharmacologic agents that are bound by tubulin, such as colchicine, vinblastine sulfate (Velban), nocodazole, and paclitaxel (Taxol), disrupt the normal dynamic behavior of microtubules. Because proper microtubule functioning is essential for spindle formation and cell division, microtubule inhibitors are commonly used for cancer chemotherapy.

Microtubules Undergo Rapid Assembly and Disassembly

Cytoplasmic microtubules are labile structures that have the capacity to undergo rapid assembly and disassembly. This characteristic is important for many microtubule functions. For example, cytoplasmic microtubules must be broken down rapidly as the cell enters mitosis. Like-

MICROTUBULES

Microtubules Are Polymers Composed of Tubulin

Microtubules are the third type of cytoskeletal structure, and they have been implicated in a variety of cellular phenomena, including ciliary and flagellar motility, mitotic and meiotic chromosomal movements, intracellular vesicle transport, secretion, and several other cellular processes. Their principal component is the protein tubulin, a heterodimer composed of nonidentical α and β subunits, with each subunit having an M_r of nearly 50 kDa. In addition, several other proteins are associated with microtubules, and it is these accessory proteins that are responsible for many of the characteristics of microtubule-based motility. The MAPs are described later in this section. Through the electron microscope, microtubules are seen as hollow cylinders with an outer diameter of 24 nm. When viewed in cross section, the wall of each microtubule is seen to be composed of 13 tubulin dimers, which represent 13 protofilaments composed of tubulin subunits. As microtubules assemble,

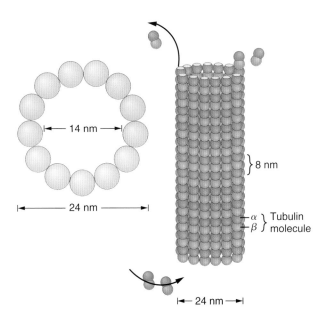

Figure 3–36. Morphology of cytoplasmic microtubules. Schematic representations of microtubules in cross and longitudinal sections.

wise, the disassembled microtubules must be re-formed to assemble the mitotic spindle. The ability of the spindle microtubules to be broken down by the cell during mitosis also appears to be essential for chromosomal separation. If dividing cells are cultured in the presence of Taxol, a drug that blocks microtubule disassembly, chromosomal segregation is blocked. The ability of the microtubule cytoskeleton to reorganize rapidly may be important for many other cellular events, such as cell migration and the establishment of cellular polarity.

Similar to actin filaments, growing microtubules have an inherent structural polarity. This polarity occurs because of the orientation of the tubulin subunits in the microtubule polymer. When growing microtubules are analyzed *in vitro,* subunits add to one end of the elongating polymer faster (the plus end) than to the other end (the minus end) (Fig. 3–37).

Inside cells, however, the minus end of the microtubule is capped, owing to its association with the centrosome complex; therefore, only the events that occur at the plus end will be considered. Current ideas concerning microtubule dynamics focus on the binding of GTP by tubulin subunits during microtubule assembly and the subsequent hydrolysis of the bound GTP to GDP (see Fig. 3–37).

For the tubulin dimer to add to an elongating microtubule polymer, the tubulin α and β subunits must each bind GTP. The GTP-tubulin can then add to the growing end of a microtubule, and some time after adding to the microtubule, the bound GTP associated with the β subunit is hydrolyzed to GDP. In effect, this results in the presence of either a small GTP- or GDP-tubulin cap on the end of a microtubule, with the remainder of the microtubule polymer being composed of GDP-tubulin. As long as a microtubule continues to grow rapidly, tubulin subunits will be added to the tubule faster than the nucleotide can be hydrolyzed, and the GTP cap will remain intact. This is important because tubulin subunits add to GTP-capped microtubules much more efficiently than they bind to GDP tubules. If the rate of microtubule assembly slows, GTP hydrolysis can catch up, the GTP cap will be lost, and the entire length of the microtubule polymer will comprise GDP-tubulin. The GDP-capped microtubules are unstable and tend to lose GDP-tubulin subunits from the end of the microtubules, which results in microtubule shortening. Moreover, this rate of loss is rapid and is referred to as a catastrophe. If the hydrolysis of GTP-tubulin becomes slower than GTP-tubulin addition, then the same microtubule can resume growth; this is referred to as a "rescue." This formation and catastrophic breakdown of microtubules is called **dynamic instability.** Dynamic instability provides a partial explanation for how a cell is able to reorganize its microtubule cytoskeleton so rapidly.

By Capping the Minus Ends of Microtubules, the Centrosome Acts as a Microtubule-Organizing Center

Unlike other cytoskeletal filaments, which appear to be nucleated and oriented haphazardly throughout the cytoplasm, cytoplasmic microtubules are all nucleated by the centrosome complex. If cultured cells are fixed and processed for antitubulin immunofluorescence microscopy, a starlike array of microtubules is observed that originates near the nucleus and radiates throughout the cytoplasm (Fig. 3–38).

At the focal point of the astral microtubule array is the centrosome. Ultrastructurally, the centrosome is composed of a centriole pair and an osmiophilic cloud of amorphous material, called *pericentriolar material,* which surrounds the centrioles (Fig. 3–39). The individual centrioles of the centriole pair are oriented at right angles to each other, and each centriole comprises nine triplets of short microtubules (0.4–0.5 μm in length).

Experimental analysis shows that the microtubule nucleating capacity of the centrosome complex is contained within the pericentriolar material and not in the centriole pair. The centrosome component that nucleates microtubules is a **γ-tubulin ring complex** found within the pericentriolar material.

The γ-tubulin ring complex is composed of γ-tubulin plus several accessory proteins. γ-Tubulin is a special-

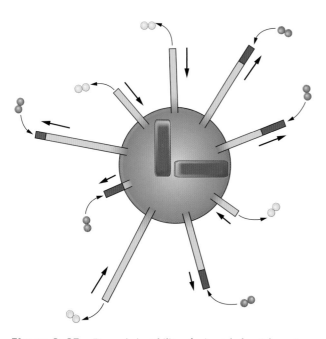

Figure 3–37. Dynamic instability of microtubules. Schematic demonstrating dynamic instability. The GDP-tubulin regions of microtubules are shown as yellow regions, and GTP-tubulin microtubule caps are shown as purple regions. Microtubules that are assembling from the centrosome contain GTP caps, whereas those that are catastrophically disassembling contain only GDP-tubulin.

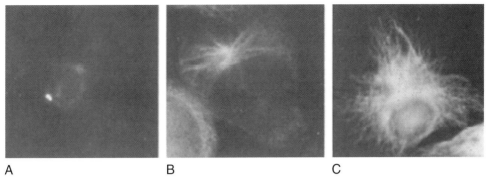

A   B   C

Figure 3-38. **The centrosome nucleates cellular microtubules.** Antitubulin immunofluorescent staining of cultured mammalian cells demonstrating that the centrosome is the microtubule-organizing center in mammalian cells. **A:** Mammalian cells were experimentally treated so that all of the microtubules were disassembled. Only the microtubule-containing centrosomes could be identified. **B:** When the experimental treatment was reversed, a starlike array of microtubules began to form off the centrosome. **C:** With time, the microtubules elongated until they eventually filled the cytoplasm. *(Courtesy R. Balczon)*

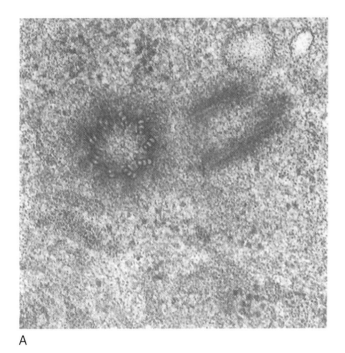

A

Figure 3-39. Morphology of the centrosome complex. **A:** Electron micrograph of a centrosome complex. The centrioles of a centriole pair are oriented at right angles to each other. The centrioles are composed of nine short-triplet microtubules and are surrounded by weakly staining pericentriolar material. **B:** γ-Tubulin ring complex nucleates microtubule polymerization from the centrosome. (**A:** *Courtesy R. Balczon*)

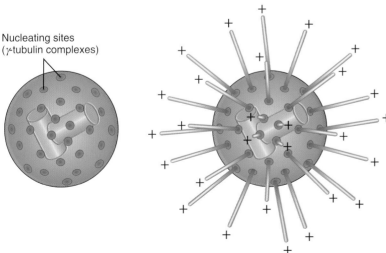

Nucleating sites
(γ-tubulin complexes)

B

ized member of the tubulin superfamily of proteins that is localized specifically to the pericentriolar material. It has been demonstrated that γ-tubulin forms a nucleation complex within the centrosome that allows microtubule formation. Moreover, by binding to the minus ends of microtubules, the γ-tubulin within the centrosome caps these microtubule ends. This means that all assembly and disassembly phenomena must occur at the plus ends of microtubules. In addition, the capping of the minus ends of the microtubules provides an explanation of how microtubule formation could occur inside a cell. Like actin microfilaments, free microtubules *in vitro* have an assembly, or plus end, and a disassembly, or minus end. Microtubules will assemble *in vitro* only if the concentration of tubulin is so high that tubulin subunits are adding to the assembly end more quickly than tubulin monomers are dissociating from the disassembly end. Apparently, the concentration of tubulin inside a cell is below this critical concentration necessary for spontaneous assembly. Therefore, the rate of tubulin loss from the minus end of a microtubule that is free in the cytoplasm would exceed the rate of addition of tubulin subunits to the plus end. As a result, microtubules cannot form freely in the cytoplasm and must be nucleated by the centrosome. By capping the minus end of the microtubule, the presence of a centrosome permits a cell to maintain its cytoplasmic tubulin concentration at levels that are too low to support spontaneous microtubule assembly. Because tubulin levels are so low, the only microtubules that can form in a cell under physiologic conditions are those capped at their minus ends by their associations with a centrosome.

The Behavior of Cytoplasmic Microtubules Can Be Regulated

The picture that develops when considering dynamic instability is one of a cytoplasm that is constantly changing because of the rapid turnover of microtubules. At any one instance, many microtubules would be rapidly growing, whereas others would be quickly and catastrophically disassembling. Although true to some degree, in reality, the average life span of cytoplasmic microtubules is about 10 minutes. Microtubules can exist for periods that are longer than might be expected because the cell has several mechanisms for stabilizing cytoplasmic microtubules.

One adaptation that cells can use for stabilizing microtubules is to cap the plus end. Because cytoplasmic microtubules are capped on their minus ends by the centrosome, if they were capped on their plus ends, they would not have a free end available for disassembly. This mechanism is, in fact, used by cells during mitotic spindle assembly. As a cell enters into mitosis, numerous microtubules begin to form off the centrosomes. Most of these microtubules disassemble rapidly because of

dynamic instability. However, some of the growing microtubules make contact with the kinetochore regions of the mitotic chromosomes and are capped on their plus ends by the proteins of the kinetochore. This microtubule capping selectively stabilizes these microtubules and is an important event in spindle morphogenesis. Other molecular and biochemical modifications can result in the increased stability of cytoplasmic microtubules. These changes in microtubule behavior can be caused either by posttranslational modifications of tubulin or by the interaction of microtubules with any one of several MAPs. The principal posttranslational modification of tubulin in cellular microtubules is the removal of the COOH-terminal tyrosine of the α-tubulin subunit by a detyrosinating enzyme that is present in cells. This detyrosination occurs after tubulin has been incorporated into a microtubule and results in the stabilization or maturation of cytoplasmic microtubules. However, detyrosination may have no direct effect on microtubule kinetics; microtubules that are formed *in vitro*, using either tyrosinated or detyrosinated tubulin, show no differences in their inherent stability. Therefore, it is conceivable that detyrosination may act as a signal that induces the binding of a second protein to microtubules, which results in the increased stability that is observed in detyrosinated cytoplasmic microtubules. When a detyrosinated microtubule is disassembled, a cellular cytoplasmic enzyme is responsible for adding a tyrosine back onto the COOH terminus of the α-tubulin polypeptide.

The interaction of tubulin with MAPs also results in considerable modifications in the behavior of microtubules. A protein is characterized as being a MAP if it binds to and copurifies with microtubules during their isolation from cellular homogenates. The individual types of MAPs appear to vary among cell types. However, considerable information on the functions of MAPs has been obtained by studying those that have been isolated from neural tissue. Two major classes of MAPs have been identified in neurons: the **high M_r MAPs,** a small family of proteins of 200 to 300 kDa; and the **tau proteins,** a group of polypeptides of 40 to 60 kDa. Experimental analysis has demonstrated that these proteins bind to tubulin monomers and assist with microtubule nucleation. In addition, they appear to be involved in the tight bundling of microtubules that is characteristic of the microtubular configurations seen in nerve axons and at other selected cellular sites (Fig. 3–40).

Other types of MAPs can also regulate microtubule length and polymerization rate. A pool of free tubulin dimers can be found within cells. One reason is the protein **stathmin,** which binds to tubulin dimers and prevents their addition to microtubules. The stathmin–tubulin interaction is dissociated on phosphorylation of stathmin. **Katanin** is a microtubule-severing protein and detaches them from the microtubule-organizing center.

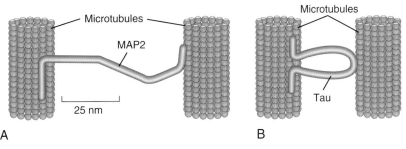

Figure 3–40. High-molecular-weight microtubule-associated proteins (MAPs) and tau bundle microtubules. Because MAP2 has a longer cross-linking domain than tau, it creates tighter bundles of microtubules.

The severing activity of katanin is ATP dependent and results in rapid depolymerization of released microtubules.

Microtubules Are Involved in Intracellular Vesicle and Organelle Transport

One of the important functions of microtubules is the intracellular transport of organelles and vesicles. For the microtubule cytoskeleton to fulfill this role, microtubules must have a means of generating a force that allows such motile behavior to occur. Experimental analysis has allowed the identification and purification of the ATPases that appear to be involved in force generation.

Two families of microtubule-dependent ATPases have been identified that appear to be important in cytoplasmic transport. One of these families is kinesin and kinesin-related proteins (KRPs). The original kinesin, isolated from squid axons, is a large multi-subunit protein that is involved in translocating vesicles along microtubules from the minus, or centrosome, end toward the distal plus ends. This allows the transport of vesicles from deep within the cytoplasm, where they are produced by budding from the Golgi apparatus, to the cell cortex where secretion can occur. In neurons, it allows for the transport of cargo from the soma to the presynaptic terminal of axons.

We now know that there are many families of kinesins and KRPs where the only common feature is a conserved motor domain. The original kinesin had two heavy chains, where the motor domain was close to the N terminus, and two light chains. Its structure is remarkably similar to myosin II. Other members of the kinesin and KRP family that have the motor domain at or near the N-terminus move vesicles toward the plus ends of microtubules (an example is KIF 1B). Members of the kinesin and KRF families that have the motor domain close to the C terminus move cargo toward the minus ends of microtubules (an example is KIF C2). KIF2 is a member of an unusual family of kinesins that have their motor domain toward the center of the heavy chain and

cannot translocate along microtubules. Instead, this family, named catastrophins, binds to the end of microtubules and stimulates dynamic instability. KIF 1B is a member of a family of monomeric kinesins that translocates cargo toward the plus end of microtubules (Fig. 3–41).

The other family of enzymes that are involved in cellular motile events is the cytoplasmic form of the ciliary enzyme dynein. Cytoplasmic dynein is a high M_r multi-subunit protein complex that translocates structures along microtubules from their plus to their minus ends (Fig. 3–42). Microtubules appear to play a relatively passive role in most types of intracellular movement, with the active roles being performed by the microtubule-dependent ATPases. A good analogy for visualizing these events would be to consider a railroad: the microtubules would serve as the tracks, and the locomotory forces responsible for transporting the vesicular cargo would be generated by ATPases, such as kinesin and dynein.

How does the cargo associate with kinesin or dynein? Membrane-associated motor receptors (MAMRs) interact with the tails of kinesin family members. One example of a MAMR is the amyloid precursor protein (APP). The abnormal processing of APP has been linked to Alzheimer's disease. Cytoplasmic dynein interacts with membrane vesicles through the dynactin complex (Fig. 3–43).

Cilia and Flagella Are Specialized Organelles Composed of Microtubules

Cilia and flagella are specialized cellular appendages that extend from the surfaces of several different cell types. Cilia are prominent in the respiratory tract and on the apical surface of the epithelial cells that line the oviduct. In the respiratory tract, cilia are involved in clearing mucus from the respiratory and nasal passages, whereas those that line the oviduct are involved in transporting ova toward the uterus. The major type of flagellated cell in humans is the spermatozoon. For the mature sperm cell, the beating flagellum provides the force that

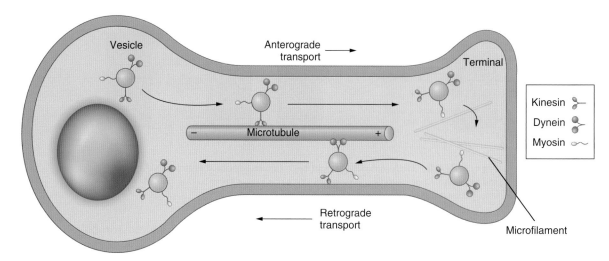

Figure 3–41. Vesicle transport along microtubule tracks. Schematic demonstrating how vesicles might be transported along cytoplasmic microtubules by micro-tubule-dependent ATPases. Microtubules are polar structures with defined plus and minus ends. Vesicles are thought to be transported in the anterograde direction (from the minus to the plus end of a microtubule) by certain members of the kinesin family. Vesicles and organelles are thought to be translocated in the retrograde direction (from the plus to the minus end of the microtubule) by cytoplasmic dynein and other kinesin-related proteins.

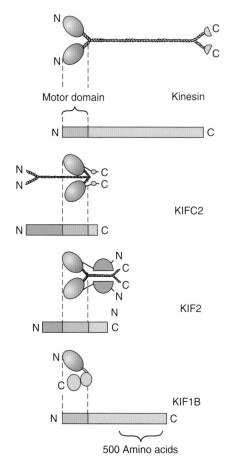

Figure 3–42. Kinesin and kinesin-related proteins.

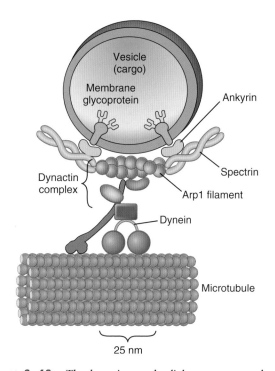

Figure 3–43. The dynactin complex links cargo to cytoplasmic dynein. The dynein tail region is linked by accessory proteins to an Arp 1 filament that is associated with actin protofilaments within a spectrin membrane skeleton linked to the cargo via ankyrin.

allows the sperm to swim. Cilia and flagella are similar ultrastructurally. At the core of one of these organelles is the axoneme, a complex structure composed of microtubules and various other proteins that allows ciliary and flagellar bending to occur. When

viewed in cross section, the axonemal microtubules are arranged in a distinctive nine-plus-two array (Fig. 3–44).

The term **nine-plus-two** refers to the orientation of the microtubules that comprise the axoneme. In axo-

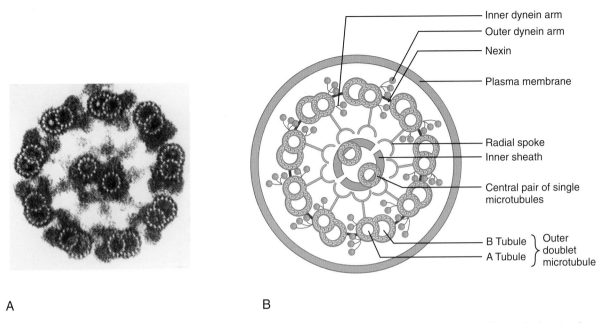

Figure 3–44. Axonemal nine-plus-two array of microtubules and other axoneme proteins. **A:** An electron micrograph showing the nine-plus-two organization of microtubules in an axoneme. **B:** Schematic showing the organization of the proteins in an axoneme. (**A:** *Courtesy Dr. W.L. Dentler;* **B:** *modified from Alberts B, Bray D, Lewis J, et al. Molecular Biology of the Cell, 2nd ed. New York: Garland Publishing, 1989.*)

nemes, two complete central microtubules (the central pair) are surrounded by a circumferential ring of nine doublet microtubules. The outer doublets are arranged so that each doublet pair is composed of one complete microtubule (the A tubule), which consists of 13 protofilaments, and an incomplete microtubule (the B tubule), which is composed of only 11 protofilaments. The B tubule shares a portion of the A tubule wall. The nine-plus-two array of microtubules traverses the axoneme, extending from the specialized centrioles, called **basal bodies,** which are located at the base of the cilium or flagellum, all of the way out to near the tip of the cilium or flagellum.

In addition to the microtubules, several other important proteins can be found in axonemes. These accessory proteins are absolutely essential for normal ciliary function. Extending from the A tubule of each doublet toward the B tubule of the neighboring doublet are two proteinaceous arms (see Fig. 3–43). These arms are actually the enzyme dynein, the ATPase that is responsible for ciliary and flagellar motility. Also extending between the A and B tubules of the neighboring doublet is a protein called *nexin*. This attaches neighboring doublets to each other. Finally, a radial spoke extends off the A tubule of each doublet and makes contact with an electron-dense sheath surrounding the central pair of microtubules, thereby connecting the doublet microtubules to the central pair. The dynein arms, nexin links, and spoke proteins exhibit a periodicity along the entire length of the axoneme. In addition to these prominent proteins, numerous other minor proteins are present in the axoneme.

Axonemal Microtubules Are Stable

Most cellular microtubules are labile structures that can be assembled and disassembled rapidly. Axonemal microtubules, however, are stable structures that resist breakdown. One of the modifications of axonemal microtubules that may contribute to this increased stability is the enzymatic acetylation of a lysine residue on α-tubulin subunits. Cytoplasmic microtubules, which are turning over rapidly, are nonacetylated. Like detyrosinated tubulin, acetylated tubulin shows *in vitro* kinetic behavior similar to unmodified tubulin, suggesting that the acetylation of axonemal α-tubulin subunits may serve as a signal for other proteins that bind to the microtubule wall to stabilize the axonemal tubules.

Microtubule Sliding Results in Axonemal Motility

Axonemes can be isolated from several sources quite easily. This is accomplished by shearing cilia or flagella from cells, selectively removing the residual plasma membrane that surrounds the axonemes, and extracting the axonemes in a buffer that contains a mild detergent. When isolated sperm flagellar axonemes are incubated in a buffer that contains ATP, the axonemes will continue to beat in a relatively normal fashion; therefore, all of the information that is required for ciliary and flagellar motility is contained within the structure of the axoneme alone. Biochemical and genetic studies have been used to dissect the mechanism of axonemal motility.

When isolated axonemes are treated with a proteolytic enzyme, the nexin cross-links and radial spokes are selectively digested, whereas the microtubules and dynein arms remain intact. If these protease-treated axonemes are incubated with ATP, the axoneme elongates up to nine times its original length. Microscopic analysis has shown that this is because the nine outer doublet microtubule pairs actively slide past one another. That these treated axonemes lack cross-links and spokes suggests that the dynein arms are the ATPase that drives axonemal motility. Moreover, this result suggests that nexin and the spoke protein are able to convert the activity of dynein into the bending that results in ciliary and flagellar motility. In fact, this is true. When isolated axonemes are low-salt extracted, the dynein arms are released, whereas the microtubules, nexin links, and spoke proteins remain intact. Such extracted axonemes will not beat when ATP is added. However, if ATP is added to the salt extract that contains the dynein arms, the ATP is actively hydrolyzed. Therefore, the following mechanism can be visualized for ciliary and flagellar activity (Fig. 3–45).

In the presence of ATP, dynein undergoes a conformational change. The net effect of this change is that the dynein arm releases from the B tubule of the adjacent microtubule doublet pair and then reattaches to that same doublet pair farther down the length of the doublet. This cycle is repeated when another ATP binds to the dynein arm. However, this "walking" of the dynein arms down the length of the microtubule wall is resisted by the nexin cross-links and radial spokes, and these two structures convert the sliding of the adjacent microtubule pairs into a bending motion.

Such a complex pathway must be tightly regulated because at any one instance, dynein arms on one region of an axoneme must be active, whereas in other areas, the dynein arms must be relaxed for ciliary and flagellar beating to occur.

Several human diseases are the result of mutations in one of the genes that encodes axonemal proteins. As would be expected, male individuals with such conditions are sterile because their sperm are immotile. In addition, patients afflicted with one of these conditions show chronic respiratory tract problems because the respiratory cilia are unable to clear mucus from the bronchiole and nasal passages.

Microtubules and Motor Proteins Are Responsible for the Function of the Mitotic Spindle

Mitosis and the mitotic spindle are discussed in relation to the cell cycle in Chapter 9. Here, we discuss the role of microtubules, MAPs, and motor proteins in the proper functioning of the mitotic spindle. As mitosis begins, the centrosomes and chromosomes have dupli-

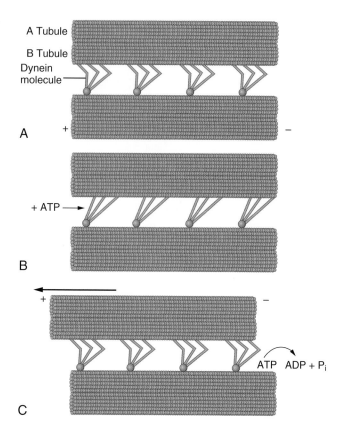

Figure 3–45. Dynein cross-bridge cycles lead to the bending of cilia and flagella. **A:** Schematic showing how the binding of ATP causes a conformational change in the structure of the ATPase dynein. **B:** The ATP binding causes dynein to release from the wall of the adjacent tubule of a microtubule pair and then reattach to that same microtubule pair farther down the length of that microtubule doublet. **C:** Cleavage of ATP to ADP results in a force that causes the two microtubule pairs to slide past one another. *In vivo*, the sliding of the microtubule doublets relative to one another is converted to bending by the nexin cross-links and spoke proteins.

cated, and during this prophase, the nucleus breaks down, the chromosomes condense, and then the chromosomes become separated. The interphase microtubules disassemble based on an increase in catastrophic microtubule depolymerization and decreased rescue events. As a result, new microtubules are formed with their plus ends associating with the newly condensed chromosomes. At prometaphase, newly formed microtubules growing from the two centrosomes become attached to the kinetochore of the condensed chromatids, which move to the center of the microtubule array. The microtubules growing from the two centrosomes have opposite polarity, or are antiparallel in polarity. At metaphase, the chromosomes are aligned at the center of this antiparallel set of microtubules. A second class of microtubules is the interpolar microtubules that overlap at the center of the spindle, and the third class is the astral microtubules that polymerize from the centrosomes or spindle poles growing away from the kinetochore and interpolar microtubules. At anaphase A, the sister chromosomes separate, moving toward the spindle poles, and at anaphase B, the spindle poles separate farther with a formation of a central spindle. Finally, at telophase, cytokinesis occurs and the nuclear envelope re-forms.

As shown in Figure 3–46, the role of microtubules, MAPs, and motor proteins in the process described earlier is beginning to be understood. As discussed earlier in the chapter, microtubules grow from the γ-tubulin ring complex and undergo catastrophic shrinkage when the plus end contains a GDP cap while being

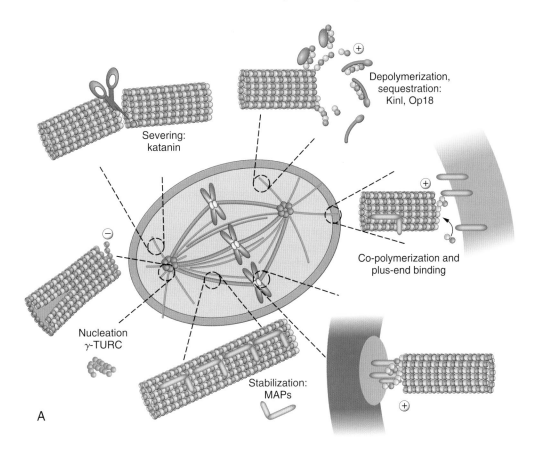

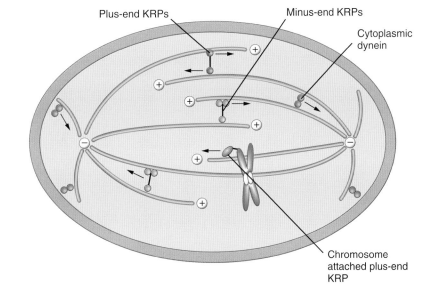

Figure 3–46. The role of microtubules, microtubule-associated proteins (MAPs), and motor proteins in mitosis. **A:** The role of various classes of MAPs in the dynamics of the microtubules of the mitotic spindle. **B:** The role of various classes of kinesin-related proteins and cytoplasmic dynein in mitosis. (*Modified from Gadde, S. and Heald, R. Current Biology 2004;14:R797–R805.*)

rescued from shrinkage when a GTP cap exists. Catastrophins (like KinI), members of the KRP family, stimulate catastrophic shrinkage, whereas MAPs such as MAP-2 and tau protein stabilize microtubules. End-binding MAPs such as EB1 and CLIP 170 are thought to attach microtubules to kinetochores and cell membranes by linking to proteins called APC and CLASP. Katanin (see earlier discussion) severs microtubules, creating new GDP capped ends and releasing the microtubules from the centrosome. Stathmin/Op18 bind to tubulin dimers, stimulating GTP hydrolysis and maintaining a pool of nonpolymerized tubulin. As shown in Figure 3–46B, motor proteins play several important roles in the mitotic process. Tetrameric plus end kinesin family members (BimC and Eg5 family) bind microtubules of opposite polarity, causing spindle pole separation. Minus end KRPs cross-link microtubules and focus astral microtubule minus ends at the spindle poles. Chromosome-attached plus end KRPs are involved in chromosomal attachment to the microtubules and movement of chromosomes toward the metaphase plate. The minus end cytoplasmic dynein serves as a "feeder" moving microtubules toward the spindle poles where they are fed into the "chipper" KinI family members that chew the plus ends of the microtubules in Pacman fashion. Cytoplasmic dynein is also thought to move kinetochore microtubules in a plus end direction toward the chipper. Although many players remain to be determined, we are now starting to see how microtubule dynamics, MAPs, and motor proteins work together to cause the detailed movements seen in mitosis.

SUMMARY

The cytoskeleton is responsible for contraction, cell motility, movement of organelles and vesicles through the cytoplasm, cytokinesis, establishment of the intracellular organization of the cytoplasm, establishment of cell polarity, and many other functions that are essential for cellular homeostasis and survival. It accomplishes these tasks through three basic structures: 7- to 8-nm diameter microfilaments composed of actin, 10-nm IFs with cell-specific composition, and 24-nm outer diameter microtubules composed of tubulin dimers. The cytoskeleton is a dynamic structure where the three major filaments and tubules are under the influence of proteins that regulate their length, state of polymerization, and level of cross-linking. Members of the myosin family move vesicles along actin microfilaments with specific directionality, whereas members of the kinesin/KRP and dynein families move cargo along microtubule tracks and play essential roles in the formation and function of the mitotic spindle. Both forms of translocation require ATP hydrolysis. Contraction caused by the interaction of myosin heads with actin filaments also derives its energy from ATP hydrolysis.

The membrane skeleton was first described in erythrocytes, where it maintains the cellular biconcave shape and gives the cell its properties of elasticity and flexibility. Spectrins are now known to be present in all eukaryotic cells, and these membrane skeletons play roles as diverse as controlling membrane traffic, DNA repair, calcium release from internal stores, and synaptic transmission.

Suggested Readings

Intermediate Filaments

Herrmann H, Aebi U. Intermediate filaments and their associates: multi-talented structural elements specifying cytoarchitecture and cytodynamics. *Curr Opin Cell Biol* 2000;12:79.

Microfilaments

Etienne-Manneville S, Hall A. Rho GTPases in cell biology. *Nature* 2002;420:629–635.
Pollard TD, Borisy GG. Cellular motility driven by assembly and disassembly of actin filaments. *Cell* 2003;112:453–455.
Spudich JA. The myosin swinging cross-bridge model. *Nat Rev Mol Cell Biol* 2001;2:387–392.
Yin HL, Janmey PA. Phosphoinositide regulation of the actin cytoskeleton. *Annu Rev Physiol* 2003;65:761–789.

Microtubules

Bornens M. Centrosome composition and microtubule anchoring mechanisms. *Curr Opin Cell Biol* 2002;14:25–34.
Endow SA. Kinesin motors as molecular machines. *BioEssays* 2003;25:1212–1219.
Gadde S, Heald R. Mechanisms and molecules of the mitotic spindle. *Curr Biol* 2004;14:R797.
Gundersen GG, Gomes ER, Wen Y. Cortical control of microtubule stability and polarization. *Cum Opin Cell Bid* 2004;16:1–7.

Spectrin Membrane Skeleton

Hsu YJ, Goodman SR. Spectrin and ubiquitination: a review. *Cell Mol Biol* 2005;(suppl 51):OL801–OL807.

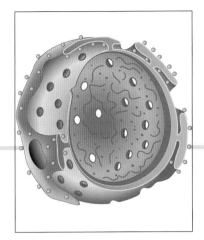

Chapter 4

Organelle Structure and Function

Prokaryotic organisms such as bacteria, with dimensions in the 1- to 2-μm range, are surrounded by an external membrane (two for gram-negative bacteria) and usually have no internal membranes. In contrast, eukaryotic mammalian cells have dimensions in the 10- to 60-μm range and contain complex membrane-bound subcellular organelles essential for cell functions. Organelle membranes have one surface facing the cytosol and the other facing the interior, or lumen, of the organelle. This compartmentalization defines a volume that contains different types and amounts of molecules than those present in the cytosol. Some organelles have two membranes, creating an additional inter-membrane space between them.

Figure 4–1 presents an overview of the subcellular organelles of a typical mammalian cell, including the plasma membrane that surrounds the cell. Most cells in the human body are part of a solid organ and contain plasma membranes specialized to contact the plasma membranes of adjacent cells. This specialization results in two distinct types of plasma membrane that, although they form a continuous layer around the cell, have different protein and lipid compositions that do not mix freely. Typically, one extracellular surface of the plasma membrane faces a cavity, such as the lumen of the intestine, and is called the *apical membrane*. Projections called microvilli extend out from the apical face into the cavity space. The plasma membrane surface that faces adjacent cells on the side, and material on the opposite side of the apical face, is called the *basolateral plasma membrane*. Specialized junctions between the cells prevent proteins and lipids in apical and basolateral plasma membranes from diffusing past the junctions and mixing. Because the shapes of cells in a solid organ are not symmetric, they are referred to as polarized cells. In contrast, cell types such as red blood cells and lymphocytes that are not part of a solid organ have only one type of plasma membrane and are not polarized.

The nuclear membrane contains two distinct lipid bilayers, an inner and outer membrane (see Fig. 4–1). There are specialized structures called *nuclear pores* that form a channel connecting the inside of the nucleus (the nucleoplasm) with the cytoplasm. The outer nuclear membrane is an extension of the endoplasmic reticulum (ER), the membrane with the largest surface area in the cell. There are two types of ER: the rough ER (RER) and the smooth ER (SER). These two ER forms have different functions (see later).

The Golgi complex is a set of slightly cupped pancake-shaped membranes in a stack typically containing four to seven members. The stack has a forming face termed the *cis*-Golgi network (CGN) that receives newly bio-synthesized membrane from the RER. There are intermediate layers that contain the *cis*- and medial Golgi cisternae, and a maturing face called the *trans*-Golgi cisternae. Adjacent to the *trans*-Golgi cisternae is the terminal part of the Golgi complex called the *trans*-Golgi network (TGN) from which membrane departs for either the plasma membrane or an endosome, eventually reaching the lysosome. The lysosome is an

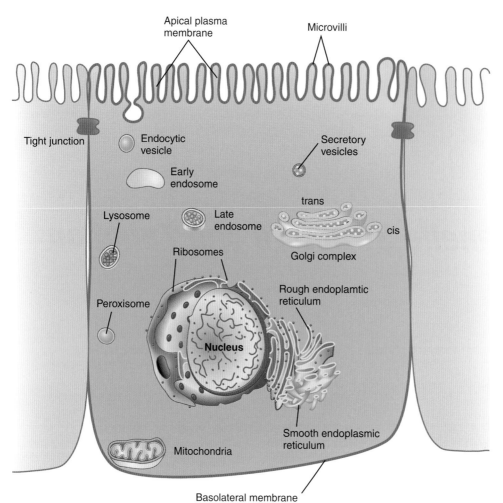

Apical plasma membrane

Microvilli

Tight junction

Endocytic vesicle

Early endosome

Lysosome

Late endosome

Ribosomes

Peroxisome

Nucleus

Mitochondria

Secretory vesicles

trans

cis

Golgi complex

Rough endoplamtic reticulum

Smooth endoplasmic reticulum

Basolateral membrane

Figure 4–1. Overview of **plasma membrane and subcellular organelles of mammalian cell.** The subcellular organelles of a typical mammalian cell include the nucleus (surrounded by a double membrane), the rough endoplasmic reticulum (RER), the smooth endoplasmic reticulum (SER), the Golgi complex, secretory vesicles, various endosomes, lysosomes, peroxisomes, and mitochondria (contains an inner and an outer membrane). This example shows a simple polarized cell with apical and basolateral plasma membranes and the tight junctions that separate them. Other junctions between cells that are not shown here are discussed in Chapter 6.

organelle that contains hydrolytic enzymes specialized for degrading proteins, polysaccharides, lipids, DNA, and RNA.

The plasma membrane generates endocytic vesicles that fuse together to form the early endosome. The early endosome becomes the late endosome, and much of the membrane internalized in endocytic vesicles is economically recycled back to the plasma membrane from endosomes, avoiding the need to synthesize new membrane. The late endosome receives membrane vesicles made by the TGN and also makes vesicles that return membrane to the TGN.

All of the organelles described in the preceding paragraphs are connected by pathways of membrane flow that involve membrane vesicles or membrane tubules that form at one membrane and then fuse with the next membrane in the pathway. Most proteins and lipids are first incorporated into a membrane in the ER, move to the Golgi complex, and then travel on to the plasma membrane, an endosome, or a lysosome. Membrane traffic from the ER out toward the plasma membrane is called *anterograde transport*. Membrane traffic in the reverse direction, called *retrograde transport*, also occurs to recover membrane components removed by anterograde transport, maintaining homeostasis. Because the

different organelles have different functions, they each have different protein and lipid compositions necessary to support their functions. How membranes maintain their unique compositions despite robust anterograde and retrograde membrane traffic is not well understood. Interestingly, two organelles in mammalian cells, the mitochondria and peroxisomes, are not stations in these pathways of membrane flow. Mitochondria contain two membranes, an inner and an outer membrane. Peroxisomes are bounded by a single membrane. Because mitochondria and peroxisomes are not connected to pathways of membrane flow, specialized mechanisms exist to deliver proteins and lipids to them.

A typical mammalian cell makes about 20,000 different proteins and one-half to two-thirds of these stay in the cytosol where they are synthesized. Proteins that do not stay in the cytosol pass through a membrane en route to secretion, become a membrane protein, or are transported across a membrane to another aqueous subcellular compartment (such as the nucleus or the matrix of the mitochondria). Intense studies have been attempted to understand how proteins are sorted and delivered to their final destinations, and one general concept that has emerged is that the information for where a protein will end up is encoded in the gene for the protein, expressed as amino acid sequences in the protein that are "address signals." Reading the address signals and transporting the proteins to their final destinations use different complex cellular machines, depending on where the protein ultimately goes in the cell. An overview of protein trafficking in the cell is shown in Figure 4–2.

Subcellular organelles are intimately related to many pathologic conditions that cannot be appreciated without an understanding of organelle cell biology. In the rest of this chapter, the structure and function of subcellular organelles is presented beginning with the nucleus and moving to the ER and outward, following the innate pathways of membrane flow that connect the organelles. A discussion of mitochondria and peroxisomes concludes this chapter.

THE NUCLEUS

The largest and most visible subcellular organelle in most mammalian cells is the nucleus, bounded by the double nuclear membrane called the *nuclear envelope*. Within the nuclear envelope is the nucleoplasm that contains the cell's DNA (except for mitochondrial DNA). The nucleus also contains the nucleolus, one or more dark substructures readily seen by light microscopy that are not encircled by a membrane. The nucleoli are sites where ribosomal RNA (rRNA) is synthesized and begins to assemble into ribosomes. Evidence also exists that the nucleoplasm contains a fibrous network called the *nuclear matrix* that may provide a scaffold for chromatin binding. The synthesis of DNA (replication) and RNA (transcription) occur in the nucleus (see Chapter 5 for further discussion). A diagram of the nuclear envelope is shown in Figure 4–3A.

The inner nuclear membrane faces the nucleoplasm and contacts an underlying nuclear lamina, a network made from fibrous proteins called *lamins*. The nuclear lamina helps shape the nuclear envelope and is a key element in the breakdown of the nuclear envelope at mitosis. During mitosis, the nuclear membrane

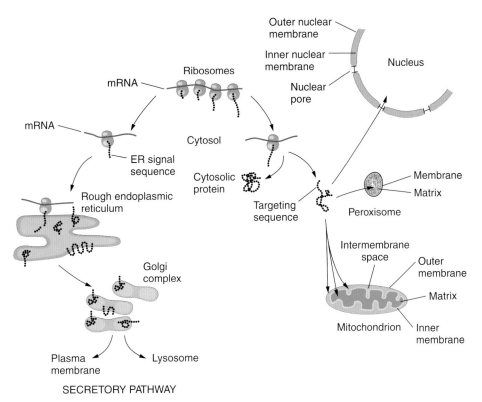

Figure 4–2. Organelles and protein sorting in a mammalian cell. Proteins that emerge from cytosolic ribosomes are of two types: those that contain a peptide signal sequence that specifies the binding of the ribosome to the rough endoplasmic reticulum (RER), and those that do not contain this signal. Proteins made by ribosomes that bind the RER are then translocated partially (for integral membrane proteins) or completely (for secretory proteins) through the membrane of the RER. Other signals encoded in the amino acid sequences of the proteins specify where the proteins ultimately end up, with locations ranging from the RER to other internal membranes to the plasma membrane to the cell exterior. Proteins that are not made on membrane-bound ribosomes either stay free in the cytosol or contain protein signals that are posttranslationally decoded to send the protein to the cytoplasmic face of an existing membrane, to mitochondria, to peroxisomes, or to the interior of the nucleus. *(Adapted from Lodish H, Berk A, Matsudaira P, et al. Molecular Cell Biology, 5th ed. New York: W. H. Freeman and Company, 2004.)*

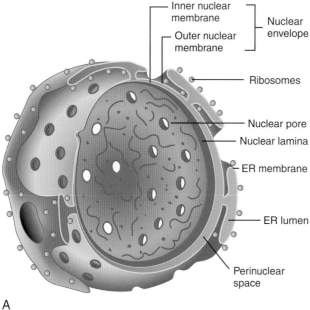

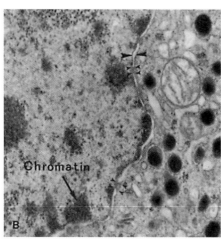

Figure 4–3. Relationship of the nuclear envelope with cellular structures. A: Diagram shows the double membrane that surrounds the nuclear compartment. The inner nuclear membrane is lined by the fibrous protein meshwork of the nuclear lamina. The outer nuclear membrane is contiguous with the membrane of the endoplasmic reticulum (ER). As illustrated in the diagram, the outer nuclear membrane often has ribosomes associated with it that are actively synthesizing proteins that first enter the region between the inner and outer nuclear membranes, the perinuclear space, which is contiguous with the lumen of the ER. The double membrane of the nuclear envelope contains pores that have regulated channels for passage of material between the cytoplasm and the nucleoplasm. B: Electron micrograph of a nucleus from a luteal cell. *Large arrowheads* denote the inner and outer nuclear membranes of the nuclear envelope, which contains the nuclear pores *(small arrows). (Electron micrograph provided by Dr. Phillip Fields.)*

disassembles, releasing chromosomes for distribution into daughter cells at division, followed by reassembly to re-form the nucleus.

The perinuclear space is between the inner and outer nuclear membranes; in most cells, this space is about 20 to 40 nm across (see the large arrowheads in Fig. 4–3B). The outer nuclear membrane is in continuity with the ER so the perinuclear space is an extension of the ER

lumen. The outer nuclear membrane in many cells contains bound ribosomes, as does the surface of the RER, which participate in translocating proteins into or through the membrane. The nuclear envelope contains holes where the inner and outer membranes fuse together to create pores (see the small arrowheads in Fig. 4–3B). rRNA, transfer RNA (tRNA), and messenger RNA (mRNA) made in the nucleus must be exported to the cytoplasm, and proteins made in the cytoplasm must be imported into the nucleus. This bidirectional transport occurs through the nuclear pores (see Chapter 5 for a detailed description of bidirectional transport and nuclear function).

ENDOPLASMIC RETICULUM

The SER and RER together form the ER, the largest membrane system in mammalian cells. No discontinuity exists in the luminal space enclosed by the SER and RER, but the two membranes nevertheless are morphologically and functionally distinct. The SER most often appears by electron microscopy as tubules and cross-sectional vesicles without any unusual surface features (Fig. 4–4). The RER is studded with ribosomes engaged in the synthesis of secretory and integral membrane proteins, and it is the bound ribosomes that give the RER its rough appearance in electron micrographs (see Fig. 4–4).

Smooth Endoplasmic Reticulum

The SER has a variety of functions that are often more prominent in certain cell types whose roles require an enhanced SER ability. Four common functions are the mobilization of glucose from glycogen, calcium storage, drug detoxification, and the synthesis of lipids. Glucose is stored as the polymer glycogen in close proximity to the SER, especially in liver, kidney, and intestinal cells that specialize in glucose homeostasis. To be released for use, individual glucose units are excised from glycogen and converted to glucose-1-phosphate, which is then converted to glucose-6-phosphate. However, to be exported from the cell for use by other cells, the glucose must traverse the plasma membrane; to do this, the phosphate must be removed. The enzyme that removes phosphate, glucose-6-phosphatase, is an SER-bound protein, prominent in liver, kidney, and intestine, which are organs that are glucose reservoirs. Type 1 glycogen storage disease (Von Gierke disease), one of about a dozen diseases that affect glycogen metabolism, is due to a genetic deficiency of glucose-6-phosphatase. Patients with this disease can store glycogen but cannot break it down, and with time glycogen accumulates, enlarging the liver. The disease causes chronic low blood sugar, abnormal growth, and is frequently fatal.

The SER is a storage site for calcium within cells. Calcium is pumped into the SER by active transport and

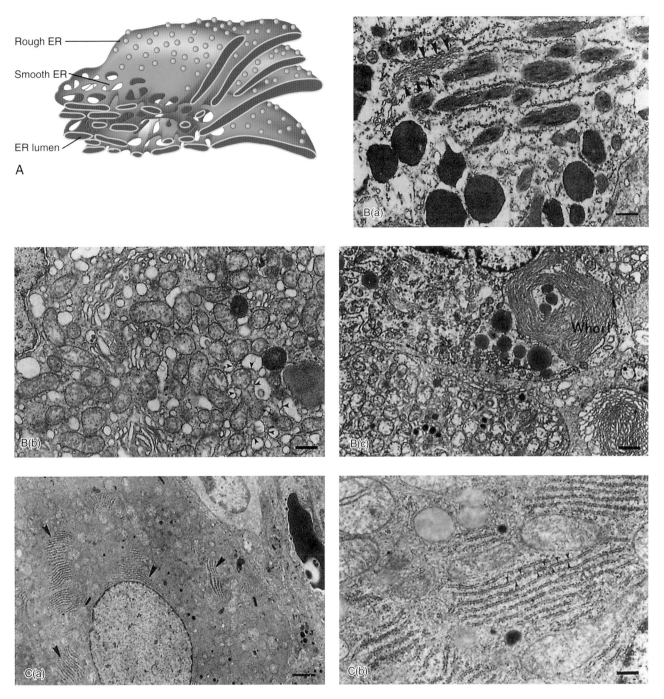

Figure 4–4. Structure of the rough and smooth endoplasmic reticulum (ER). **A:** The rough endoplasmic reticulum (RER) consists of oriented stacks of flattened cisternae studded with ribosomes on the cytoplasmic surface. The luminal space is 20 to 30 nm. The smooth ER (SER) often appears as 30- to 60-nm-diameter membranous tubes that have connections to the RER so the two ER membranes share the same luminal space. The SER has no associated ribosomes. **B:** Electron micrographs of SER from a mammalian luteal cell. The SER can be found in various forms. At least three types are presented in the following micrographs: *(a)* Lamellar stacks of SER are observed (between the *arrowheads*) (small luteal cell from pregnant cow corpus luteum). *(b)* The cytoplasm of the cell may be filled with SER that appears as empty vesicles (*small arrowheads* inside the vesicles). This is characteristic of active steroid-secreting cells (large luteal cell from pregnant cow corpus luteum). *(c)* SER may be observed in spirally arranged cisternae (whorl) (small luteal cell from a pregnant cow corpus luteum). Scale bars = 0.39 μm (a, b); 1.31 μm (c). **C:** Electron micrographs of RER from a mammalian large luteal cell (21-day pregnant rat). *(a)* Stacks of RER *(arrowheads)* are observed throughout the cytoplasm of the cell. *(b)* At a higher magnification, ribosomes *(small arrowheads)* can be observed lining cytosolic membrane surfaces of the RER. Scale bars = 1.6 μm (a); 0.27 μm (b). *(Electron micrographs in **B** and **C** provided by Dr. Phillip Fields.)*

released in response to hormonal signals. This is particularly important in muscle cells where the SER is so prominent it has a special name, the sarcoplasmic reticulum. Calcium is released in response to signaling pathways initiated on neurotransmitter binding to the cell-surface receptors.

Cytochrome P450s are a large family of enzymes resident in the membrane of the SER that use oxygen and nicotinamide adenine dinucleotide phosphate (NADPH) to hydroxylate a wide variety of substrates, including steroids and drugs. Hydroxylation often increases the solubility of hydrophobic drugs, facilitating clearance from the body, and selected cytochrome P450 enzymes are up-regulated in response to different drugs. This up-regulation can be large enough to cause dramatic expansion of the SER membrane. For example, chronic barbiturate use leads to expansion of the SER caused by induction of detoxifying cytochrome P450 enzymes. The increased inactivation of the drug requires larger barbiturate doses to achieve an effect, which is part of the addictive spiral in chronic users. Carcinogens, such as polycyclic aryl hydrocarbons, are also hydroxylated by SER-associated cytochrome P450 enzymes, which frequently enhances their carcinogenic activity.

Phospholipids, ceramide, and sterols are primarily synthesized in mammalian cells by enzymes in the ER, usually associated with the cytoplasmic leaflet of the SER. Exceptions to this include mitochondria that make selected phospholipids and peroxisomes that can biosynthesize cholesterol and some other lipids. The initial step in phospholipid synthesis is the condensation of two molecules of fatty acyl coenzyme A (CoA) with glycerol phosphate to make phosphatidic acid (Fig. 4–5). Each molecule of fatty acyl CoA is added separately, enabling the cell to control the type of fatty acid esterified to the 2 and 3 positions of the glycerol, with position 2 often containing an unsaturated fatty acid. Free fatty acids in the cytosol are usually bound to a fatty acid–binding protein and are converted to the fatty acyl-CoA derivatives that are substrates for the acyl-transferase enzymes in the cytosolic side of the membrane. A phosphatase removes the phosphate from phosphatidic acid to make diacyl glycerol, and in the polar head group, either cytidine-diphosphoethanolamine (CDP-ethanolamine) or cytidine-diphosphocholine (CDP-choline) is added (see Fig. 4–5).

Phospholipids are assembled in the cytoplasmic leaflet of the ER and must then be translocated to the other half of the bilayer to distribute a particular phospho-

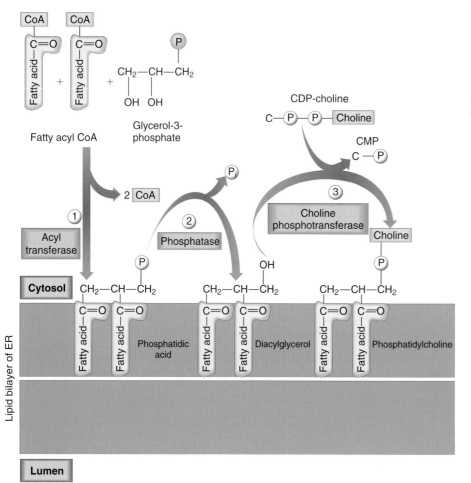

Figure 4–5. Synthesis of phospholipids in the smooth endoplasmic reticulum (SER). Diagram presents the pathway for synthesis of phosphatidylcholine from fatty acyl-coenzyme A (CoA), glycerol-3-phosphate, and cytidine-diphosphocholine (CDP-choline).

lipid between the two monolayers. The spontaneous flipping of phospholipids from one monolayer to the other is extremely slow and proteins termed *flippases* have evolved that catalyze the flipping of specific lipids. Phospholipids are often asymmetrically distributed in the two halves of a bilayer; for example, phosphatidylcholine and sphingomyelin are predominantly in the luminal face (or topologically equivalent extracellular face) of the membrane, whereas phosphatidylethanolamine and phosphatidylserine are mainly on the cytosolic face. Because the distribution of a phospholipid on the two sides of the membrane depends on the type of flippase present, it is believed that the asymmetric distribution of phospholipids in the two halves of a membrane is achieved by control of flipping, although it is unclear how the process is regulated to achieve the diversity of lipid asymmetry that is observed with different membranes.

Ceramide, the precursor of phosphosphingolipids and glycosphingolipids, is synthesized in the ER from serine and palmitoyl CoA. Phosphosphingolipids are also made in the ER. Glycosphingolipids, such as gangliosides, are made when ceramide reaches the Golgi complex and is glycosylated on the luminal face of the Golgi complex by glycosyl transferases. Glycosphingolipids are found only on the extracellular (luminal) side of membranes, suggesting that there are no flippases for this type of lipid.

The committed step in cholesterol synthesis, the production of mevalonate, is catalyzed by 3-hydroxy-3-methylglutaryl-CoA reductase (HMG-CoA reductase), an integral membrane proteins of the SER. Other enzymes involved in the process of making cholesterol, as well as metabolically modifying cholesterol, are also ER residents. Although initially made on the cytosolic side of the SER, cholesterol is found on both sides of the membrane, and evidence exists that cholesterol flippases catalyze the flipping.

Not only are many lipids unevenly arranged in two halves of the bilayer, but most membranes maintain a unique lipid composition. For example, the ER of mammalian cells is typically 50% or more phosphatidylcholine, with less than 10% each for sphingomyelin and cholesterol, whereas the plasma membrane contains less than 25% phosphatidylcholine and more than 20% each for sphingomyelin and cholesterol. Thus, once a lipid is incorporated into the ER, it must not only be transported to other membranes, but the characteristic compositions of the destination membranes must be maintained. There are three basic thought processes on how this occurs. One is vesicle-mediated transport whereby vesicles bud from the donor membrane and fuse to a target membrane, thus moving lipids from one membrane to the other. Known pathways of vesicular transport whereby vesicles originating from the ER move sequentially through the Golgi apparatus and on to endosomes or the plasma membrane have been

reported (see later). Ceramides that require glycosylation in the Golgi complex to make glycosphingolipids are believed to use this pathway. Cholesterol, however, can go from the ER to other membranes bypassing the Golgi complex, and evidence exists for a type of vesicle that buds from the ER carrying cholesterol and selected phospholipids that transports lipids to other membranes without passing through the Golgi complex. A second idea in lipid transport is that lipid transfer proteins directly extract a lipid from a membrane and shield the lipid in a hydrophobic pocket while the protein diffuses through the cytosol and deposits the lipid in an acceptor membrane. The third idea in lipid transport is that the ER makes transient contact with another membrane, passing lipids from the ER to the other membrane.

Rough Endoplasmic Reticulum

The RER is where secretory and membrane proteins (other than mitochondrial and peroxisomal proteins in mammalian cells) first pass through or enter a membrane. The RER contains machinery that, in cooperation with ribosomes, selects secretory and membrane protein from all the other proteins being synthesized in the cell and targets them to the RER. This process includes the binding of ribosomes to the RER, which gives the RER its defining rough appearance in electron micrographs. Many other RER functions are related to the processing of secretory and membrane proteins. These include glycosylation, folding, quality control, and the degradation of proteins that do not pass quality-control standards.

The Translocation of Secretory Proteins across the Rough Endoplasmic Reticulum Membrane Requires Signals

In the 1970s, cell biologists noticed that mature secretory proteins (proteins that had left the cell by secretion) had a slightly smaller molecular weight than when the proteins were translated in a cell free protein synthesis system containing ribosomes, tRNA, mRNA, and all the other factors necessary for protein synthesis in a test tube. This difference in molecular weight was traced to the presence of extra amino acids at the N-terminal end of secretory proteins that were cleaved off to form the mature protein during the secretory process. Evidence from electron microscopy and biochemical experiments showed that secretory proteins entered the lumen of the RER en route to the cell exterior, and it was proposed that these extra amino acids were the signal to select secretory proteins from all other proteins and somehow divert them into the RER. This was called the *signal hypothesis,* and the extra amino acids at the N terminus are called the *signal peptide* or *signal sequence*. When signal sequences from different proteins were compared, several common features were found (Fig. 4–6).

The overall length of N-terminal signal peptides is usually 16 to 30 amino acids, and these peptides contain 1 or more positive amino acids on the N-terminal side of a continuous core of 6 to 12 hydrophobic residues. There is also a trailer after the hydrophobic sequence, and it later became clear that certain amino acids in this region direct the signal peptidase to remove the signal peptide from secretory proteins by proteolytic cleavage. Except that the core residues are hydrophobic, there is no particular amino acid sequence that distinguishes one signal peptide from another; rather, the recognition of signal sequences by the secretory machinery is based on hydrophobicity, not specific amino acids.

The signal hypothesis was tested in cell free assays that reconstituted the transport of secretory proteins across the RER membrane. The reaction mixtures contained the mRNA for a secreted protein and all other components necessary to translate the mRNA, including ribosomes prepared from cytosol (i.e., the ribosomes were from the soluble pool, not those bound to the RER). In addition, it was possible to add microsomes, small intact vesicles that are derived from the RER when cells are homogenized. The experimental design is illustrated in Figure 4–7.

When the protein was made in the absence of microsomes, it contained the signal peptide and was not glycosylated. If microsomes were added after synthesis, the signal peptide was not cleaved, the protein remained outside of the microsomes, and no carbohydrate was added. The presence of the protein outside of the microsomes was demonstrated by adding proteases that digested the protein. However, when the microsomes were present at the same time that protein was being synthesized, the signal sequence was cleaved, the protein was glycosylated, and the completed protein was

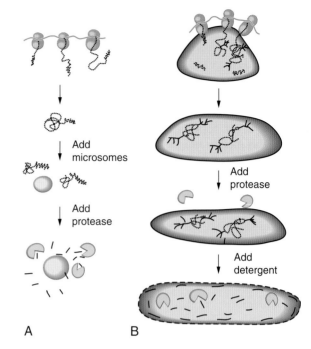

Figure 4–7. Testing the signal hypothesis in cell free protein synthesis assays. A: Messenger RNA (mRNA) for a secretory protein is translated in a test tube until the protein is completed and released from the ribosome. If microsomes (vesicles derived from the RER) are added after translation, the protein does not enter the microsomes, the signal peptide remains intact, and no carbohydrate is added to the protein. If a protease is added, the protein is digested because it is not protected inside the microsomes.
B: mRNA for a secretory protein is translated in the presence of microsomes. Ribosomes making the protein bind the microsomes and translate the protein to the interior of the microsomes. Analysis of the proteins indicates that the signal peptide is cleaved and the protein is N-glycosylated. If a protease is added, the protein is not digested because it is sheltered within the microsomes. However, if a detergent is added to disrupt the membrane, the protease and the protein can come into contact and the protein is degraded. *(Adapted from Lodish H, Berk A, Matsudaira P, et al. Molecular Cell Biology, 5th ed. New York: W. H. Freeman and Company, 2004.)*

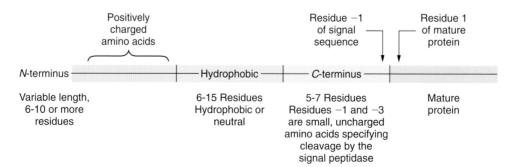

Figure 4–6. Features of signal peptides. Signal peptides typically have three functional regions. The N terminus has a variable length and contains positively charged amino acids at the C-terminal side. The middle region has 6 to 15 hydrophobic or neutral amino acids. The C-terminal region has 5 to 7 residues, and positions −1 and −3 (where the first amino acid residue of the mature protein is +1) are small, uncharged amino acids important for proteolytic removal of the signal peptide by signal peptidase.

released inside the microsomes, deduced from the observation that it was protected from added proteases that could not enter the microsomes. When the microsomal membrane was dissolved with detergents, the protein inside was released and digested with proteases.

The results from these experiments, as well as many others, were incorporated into a model with the following features: (1) Secretory proteins are translocated across the ER membrane cotranslationally, while new amino acids are being polymerized into the protein (there are exceptions, especially in yeast). (2) The signal peptide is cleaved off during translocation, before synthesis is complete. (3) Certain types of carbohydrate are cotranslationally added to the growing polypeptide on the luminal side of the RER. (4) Ribosomes from the soluble pool participate in synthesis and translocation of secreted proteins, indicating that there are not two different types of ribosomes, one for the synthesis of cytoplasmic proteins and a second for secreted proteins.

Because secretory proteins begin to cross the RER membrane before their synthesis is complete, there must be a mechanism to recruit ribosomal complexes engaged in making secretory proteins to the RER. Continued biochemical studies of the cell free synthesis of secretory proteins identified a soluble factor that when added to the cell free system arrested the elongation of secretory proteins, but not cytoplasmic proteins. Because the only known feature that distinguishes ribosomes making secretory proteins from those making cytoplasmic proteins is the signal peptide on the secretory protein, the factor was called *signal recognition particle* (SRP). SRP arrests the elongation of secretory proteins just as the signal peptide emerges from the ribosome, which gives the ribosomal complex time to bind to the RER. Elongation arrest is relieved when the ribosome complex associates with the RER and translocation proceeds. SRP was initially thought to be one or more proteins, but subsequent work revealed a surprise: purified SRP contained not only 6 polypeptides, but also a 300-nucleotide small RNA molecule. Structural studies of SRP are incorporated into the model shown in Figure 4–8.

The polypeptides are designated P9, P14, P19, P54, P68, and P72, where the number indicates the protein molecular weight divided by 1000. P54 is a guanosine triphosphate (GTP)–binding protein and also contains the site that interacts with the signal sequence via a groove lined with hydrophobic residues called *methionine bristles*. Other SRP subunits contribute to translational arrest by interacting with the ribosome, or are required for protein translocation. Because SRP contains both RNA and protein like a ribosome, some consider it a detachable ribosomal subunit.

The discovery of SRP motivated the search for an SRP receptor on the RER. The receptor was found and is a heterodimer containing two integral membrane proteins that bind GTP. Thus, two GTPases, the P54 subunit

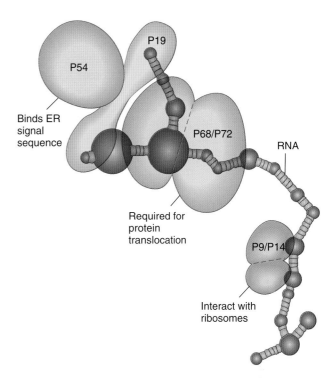

Figure 4–8. Structure of signal recognition particle (SRP). SRP contains six proteins (P1, P14, P19, P54, P68, and P72), plus a single copy of a 300-nucleotide RNA. Notice that the RNA is folded into an extensive secondary structure. *(Adapted from Lodish H, Berk A, Matsudaira P, et al. Molecular Cell Biology, 5th ed. New York: W. H. Freeman and Company, 2004.)*

of SRP and the SRP receptor, are involved in recruiting to the RER ribosomes that are making secretory proteins. The GTPase cycle is believed to increase the fidelity of the recruitment process, ensuring that only ribosomes making nascent secretory proteins bind the SRP receptor and proceed to the next step, eventually establishing a connection with the RER. By analogy with the GTPase function in codon/anticodon selection of protein synthesis, the GTP-bound forms of P54 and the SRP receptor are believed to monitor the fit of sites that bind one another, and if the affinity of binding is high and the complex lasts long enough, GTP is hydrolyzed to guanosine diphosphate (GDP) and the process proceeds. Complexes that do not have a strong affinity are released before GTP hydrolysis occurs, rejecting them before the next step is taken.

Because the signal sequence contains a hydrophobic core, it was originally proposed that this core inserts into the RER lipid bilayer and the rest of the polypeptide followed to translocate the protein across the membrane without the need for a protein pore. However, it is now clear that there is a protein pore in the membrane and its opening and closing is highly regulated. The pore is called the *translocon*, and seminal studies in identifying the protein constituents of the translocon combined yeast genetics, where mutated pore components were obtained, with biochemical experiments, where secretory proteins trapped part way through the pore were chemically cross-linked to adjacent pore proteins. The three main proteins of the translocon complex are Sec61α, an integral membrane protein that spans the membrane 10 times, and two other subunits, Sec61β and Sec61γ. These proteins have been purified and the translocon reconstituted in defined lipid vesicles that are active in translocating secretory proteins across the vesicle membrane. From these studies, it is believed that the energy source for thrusting a protein through the membrane comes mainly from the energy used to elongate proteins during protein synthesis.

High-resolution structural studies of the pore associated with ribosomes suggest that the aqueous channel in the pore center has a diameter of about 2 nm, although there is evidence that the size of the pore may change depending on what is passing through it. This channel would breach the ER membrane and allow cytosolic and luminal molecules to mix if the opening and closing of the channel were not highly regulated. There are two competing theories on how the pore is gated. One suggests that the translocon constituents do not assemble until a ribosome is recruited to direct assembly and form a tight seal around the pore, blocking free access through

the pore. Alternatively, evidence exists that proteins on the luminal side of the RER form a gate that closes the pore until the ribosome forms a seal on the cytoplasmic side, followed by opening the gate on the luminal side. Gating of the pore is more complicated than it may appear at first because, as discussed later in this chapter, the pore is also gated laterally in the plane of the membrane to allow membrane proteins in the pore to move sideways into the bilayer. A model summarizing the translocation of secretory proteins through a membrane is depicted in Figure 4–9.

On emerging from a ribosome in the cytosol, the signal peptide of the secretory protein interacts with SRP. Elongation of the nascent polypeptide is arrested until the ribosomal complex binds to the ER via the SRP receptor. The nascent chain is transferred to the channel of the translocon, and the signal peptide inserts into the translocon as a loop with the N terminus on the cytoplasmic side. During this process, GTP is hydrolyzed to GDP, releasing SRP into the cytosol to participate in another round of signal sequence recognition. The signal peptide is removed by the signal peptidase before protein synthesis is complete so that continued elongation of the polypeptide chain pushes the N terminus of the cleaved

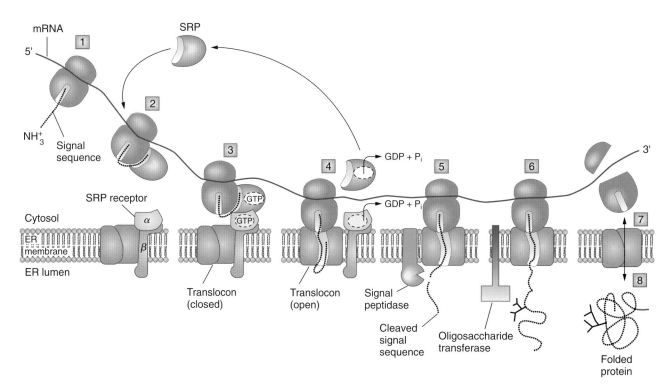

Figure 4–9. Model for translocation of secretory proteins across the endoplasmic reticulum (ER) membrane. When the signal peptide emerges from ribosomes, the signal recognition particle (SRP) binds to the hydrophobic amino acid region of the signal peptide. This transiently delays further elongation of the secretory protein until the SRP/ribosome complex binds to the SRP receptor on the rough endoplasmic reticulum (RER) via interactions between the SRP and the receptor. Once contact with the translocon is sufficient, GTP is hydrolyzed and the SRP dissociates from the complex, which is now available to target another ribosome to the RER. Elongation of the protein resumes and the peptide inserts into the translocon as a loop with the N terminus on the cytoplasmic side. As elongation proceeds, the loop extends to the luminal side and the signal peptidase cleaves the signal sequence. As elongation continues, the oligosaccharide transferase covalently adds the preassembled N-linked oligosaccharide to Asn residues in the appropriate context. *(Adapted from Lodish H, Berk A, Matsudaira P, et al. Molecular Cell Biology, 5th ed. New York: W. H. Freeman and Company, 2004.)*

protein farther into the lumen of the RER. Proteins are also glycosylated during elongation by addition of a preformed polysaccharide to certain asparagine residues, a process that is covered in more detail later in this chapter. The signal hypothesis and insights into how proteins are targeted to and across membranes has been so influential in cell biology and medicine that, in 1999, a major contributor in this area, Dr. Günter Blobel, was awarded the Nobel Prize in Medicine and Physiology for his work.

The Insertion of Proteins into Membranes Requires Stop-Transfer Anchor Sequences

As explained in Chapter 2, single-pass integral membrane proteins are anchored to the bilayer by a membrane-spanning α-helix that contains predominantly hydrophobic or neutral amino acids. Single-pass membrane proteins may be oriented with their N terminus on the luminal side and their C terminus on the cytoplasmic side of the membrane, or the reverse. How do the membrane-spanning domains get into the membrane? An extension of the signal hypothesis provides the answer. If a typical secretory protein containing an N-terminal signal peptide also contained an internal hydrophobic membrane-spanning anchor, then during SRP-mediated translocation of the peptide through the translocon, the membrane-spanning domain would eventually emerge from the ribosome and enter the translocon. If transfer through the translocon stopped at this point, and the translocon opened laterally, the membrane-spanning α-helix could move into the bilayer in the correct orientation to anchor the protein. Because translocation stops when the hydrophobic membrane anchor spans the translocon, membrane-spanning sequences of this type are called stop-transfer anchor sequences. Cell-free studies on the synthesis of model membrane proteins in the presence of microsomes suggested this simple extension of the signal hypothesis was correct in producing single-pass membrane proteins that had their N termini in the lumen of the ER and their C termini in the cytosol. These proteins have a typical cleavable N-terminal signal peptide, just like secretory proteins. Thus, the two features that dictate the luminal N-terminal and cytosolic C-terminal orientation are an N-terminal cleavable signal peptide and an internal stop-transfer anchor sequence. Membrane proteins with these features are called type I integral membrane proteins (Fig. 4–10A).

But what about proteins whose orientation is the opposite of type I proteins? This mystery was solved when it was realized that membrane-spanning domains could themselves act as signal peptides recognized by SRP, thus targeting a membrane protein to the RER by the same machinery that secretory proteins used. The only difference is that membrane-spanning domains do not contain the amino acid signal that directs the signal peptidase to cleave signal sequences, so membrane-spanning domains are not cleaved. Integral membrane proteins of this type have no N-terminal signal peptide, but they do have an internal membrane-spanning domain that serves the purpose of the signal peptide, including one or more positively charged residues on the N-terminal side of the membrane-spanning domain, which is important (see later). As the membrane-spanning domain emerges from the ribosome, SRP binds and targets the complex to the RER via the SRP receptor. The positively charged N-terminal side of the membrane-spanning domain stays in the cytosol and the protein loops into the translocon, just as with proteins that contain a typical signal peptide. The result is that the N terminus of the protein that has already emerged from the ribosome stays on the cytoplasmic side. As the protein is elongated, the C terminus eventually emerges from the ribosome and passes completely through the translocon. The translocon opens laterally and the protein moves out into the bilayer with its N terminus in the cytoplasm and C terminus in the lumen of the ER. Integral membrane proteins that enter the membrane by this mechanism, termed *type II proteins*, have no N-terminal signal peptide, but they do have an internal membrane-spanning domain that has two functions, a signal peptide function and a membrane anchor function. Thus, type II proteins are said to have a signal-anchor sequence (see Fig. 4–10B).

In addition to type I and II single-pass membrane proteins, there is a type III. Type III proteins have a signal-anchor sequence but no N-terminal signal peptide, just like type II proteins; yet, the orientation is a luminal N terminus and a cytosolic C terminus, which is just the opposite of a type II membrane protein. The one structural difference researchers noticed was that the cluster of positive charges usually found adjacent to the N-terminal side of the signal-anchor sequence was not there; instead, there was a cluster of positive charges adjacent to the C-terminal side of the signal anchor. By a mechanism that is not well understood, the side of the membrane anchor that is positively charged has a strong tendency to stay in the cytoplasm, perhaps because the membrane potential is negative inside, attracting the positive charges. Type III membrane proteins initially insert into the translocon as a loop such that the C-terminal side of the anchor stays in the cytoplasm and the N-terminal side goes through, threading the already completed N-terminal part of the peptide through the membrane. As elongation proceeds, the C-terminal part of the protein is synthesized and stays in the cytosol (see Fig. 4–10C).

Many proteins span the membrane multiple times, from 2 to more than 10 times, termed *multipass membrane proteins*, or type IV integral membrane proteins. The insertion and orientation of these proteins can be understood as combinations of type I, II, and III

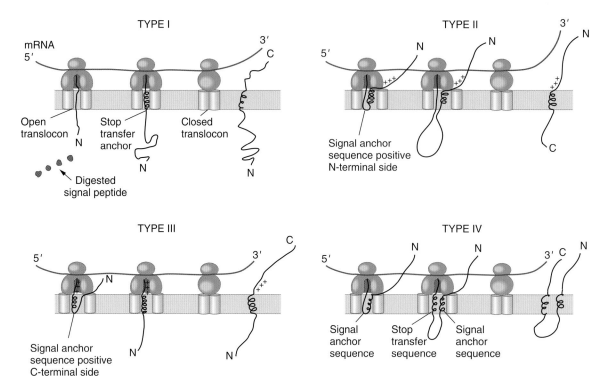

Figure 4–10. Type I, II, III, and IV membrane proteins. Type I proteins contain a typical N-terminal signal peptide that directs ribosomes to bind the rough endoplasmic reticulum (RER). The proteins insert as a loop (see Fig. 4–9), and the signal is removed and digested. Elongation of the protein continues to push the N terminus of the protein into the lumen of the RER until a second hydrophobic region, termed a *stop-transfer anchor*, emerges into the translocon. This sequence stops further transfer through the translocon, and the remainder of the protein is elongated on the cytosolic side of the membrane. The stop-transfer sequence causes the translocon to open laterally (the mechanism for this is not understood), releasing the protein into the bilayer to make an integral membrane protein anchored by the hydrophobic membrane-spanning domain. Type II and III proteins do not contain an N-terminal signal sequence; rather, the internal hydrophobic domain that will eventually span the membrane serves as both the signal sequence (that binds to signal recognition particle [SRP]) and the membrane anchor. Type II proteins contain a cluster of positively charged amino acids on the N-terminal side of the anchor and insert into the membrane with a loop as if the signal anchor were a normal signal peptide, resulting in the N terminus of the protein on the cytosolic side. Type III proteins have the cluster of positively charged amino acids on the C-terminal side of the signal anchor, which changes the orientation of the protein in the membrane by an uncertain mechanism. One possibility is that the protein still inserts as a loop with the N terminus on the cytosolic side, but that passage of the anchor through the membrane is blocked by the positive charges, and the N-terminal domain then slips through the translocon. Type IV proteins have multiple membrane-spanning domains that alternate as signal-anchor and stop-transfer sequences. Two membrane-spanning domains are shown, but the insertion mechanism can accommodate many more to produce proteins with 10 or more membrane-spanning domains.

mechanisms. For example, suppose a protein has two membrane-spanning domains and the first is a typical type II signal anchor with the positive charges adjacent the N-terminal side (see Fig. 4–10D). The protein initially inserts just as a type II protein with the N terminus in the cytosol and elongation proceeds until the second membrane-spanning domain reaches the translocon. It enters the pore and acts as a stop-transfer sequence, the translocon opens laterally so both membrane spanning domains are in the bilayer, and the protein has both its N- and C-terminal ends in the cytosol. If there were a third membrane-spanning domain in this protein, it would be equivalent to start-transfer anchor sequence in the sense that it would enter the translocon and serve as a membrane-spanning domain, although the ribosome does not detach from the RER and another SRP is not needed for synthesis to proceed. The orientation

of all subsequent membrane-spanning domains in type IV proteins is determined by the first one, with all others continuously threading through the membrane, each with an opposite orientation. Thus, if the first domain crosses the membrane with its N-terminal side in cytosol and the C-terminal side in the lumen, the next membrane-spanning domain will have its N-terminal side in the lumen and C-terminal side in the cytosol, and so on. Multipass proteins with even numbers of membrane-spanning domains end up with their N terminus and C terminus on the same side of the membrane, whereas proteins with an odd number of membrane-spanning domains have their two termini on opposite sides of the membrane.

Given their amino acid sequences, a reasonable topology for most membrane proteins can be deduced. This is especially important in medicine given that the human genome sequence is available, and that the orientation of a protein of unknown function can be predicted, offering clues to function. Deducing the topology is facilitated by using the primary amino acid sequence to prepare a hydropathy plot (Fig. 4–11).

In this plot, each amino acid is assigned a value proportional to its hydrophobicity. The values are summed over a window, usually 10 to 20 adjacent residues beginning with the amino terminus, and the sum is a point on the plot placed in the middle of the sequence window. The window is then moved by one residue toward the C-terminal end and the hydropathy value calculated again. By repeating the process with the window moving one residue each time until the C terminus is reached, a plot of the hydrophobicity of the residues in the window versus length is obtained. Spikes in hydrophobicity usually identify hydrophobic stretches that are either N-terminal signal peptides or membrane-spanning domains. Coupled with inspection of the charges on the N- or C-terminal side of membrane-spanning sequences, the topology of the protein can be suggested. For proteins with multiple membrane-spanning domains and complicated topologies, structural studies of the protein are necessary to confirm suggested topologies.

Some Membrane Proteins Are Anchored to the Bilayer by Covalently Attached Lipids

Several classes of proteins are covalently modified after synthesis to attach a hydrophobic lipid (either a fatty acid or a hydrophobic molecule such as geranyl or farnesyl groups) that inserts into the bilayer and serves as a membrane anchor. Many proteins in this class are synthesized in the cytosol, the lipid is covalently added, and protein inserts into the cytoplasmic side of a membrane. Proteins attached to membranes by this mechanism are highly relevant to medicine because many of them are protooncogenes that when mutated cause cancer. For example, mutations in the small GTP-binding protein Ras frequently cause cancer in humans, and active Ras has a covalently attached lipid that tethers it to the plasma membrane. Because the activity of Ras depends on the anchoring process, enzymes that catalyze the covalent addition of the lipid anchor have been studied intensely as potential targets for chemotherapy.

Another class of lipid-anchored membrane proteins contains the complex lipid glycosylphosphatidylinositol (GPI). This lipid anchor is added to certain integral membrane proteins that are already inserted into the

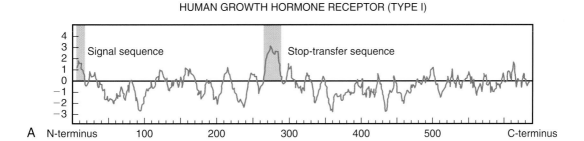

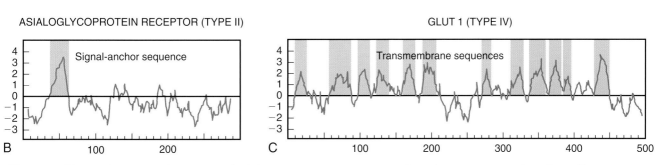

Figure 4–11. Hydropathy profiles can identify likely topogenic sequences in integral membrane proteins. Hydropathy profiles are generated by plotting the total hydrophobicity of each segment of 20 contiguous amino acids along the length of a protein. Positive values indicate relatively hydrophobic portions, and negative values relatively polar portions of the protein. Probable topogenic sequences are marked. The complex profiles for multipass (type IV) proteins, such as GLUT1 *(c)*, often must be supplemented with other analyses to determine the topology of the proteins. *(Adapted from Lodish H, Berk A, Matsudaira P, et al. Molecular Cell Biology, 5th ed. New York: W. H. Freeman and Company, 2004.)*

bilayer on cleavage of the original membrane-spanning anchor, replacing it by covalently attaching the GPI (Fig. 4–12). Because GPI-anchored proteins are on the luminal side of the ER, they appear on the extracellular face of the cell after transport to the plasma membrane.

Glycosylation of Secretory and Membrane Proteins Begins in the Endoplasmic Reticulum

Almost all secretory proteins and integral membrane proteins are cotranslationally glycosylated by the trans-

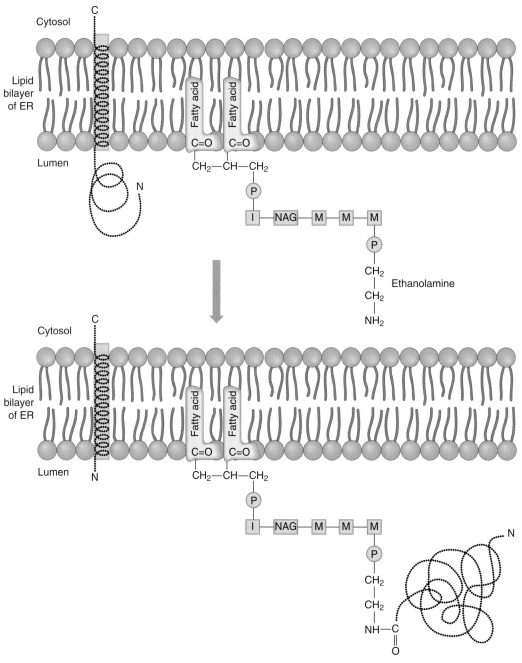

Figure 4–12. Formation of a glycosylphosphatidylinositol (GPI)-anchored membrane protein. An endoprotease within the lumen of the endoplasmic reticulum (ER) cleaves the proteins away from its C-terminal membrane-spanning domain, then attaches the new C terminus of the released protein to the amine within the ethanolamine moiety of the GPI anchor.

fer of a preassembled oligosaccharide unit to asparagine (Asn) residues of the proteins as they emerge from the translocon, catalyzed by the enzyme oligosaccharide protein transferase (Fig. 4–13).

This type of glycosylation is called *N-linked* because the linkage is via the amide nitrogen of Asn. All glycosylated Asn residues appear in the tripeptide sequence Asparagine-X-Serine (or Threonine) where X is any amino acid. This tripeptide sequence is necessary for glycosylation but is not sufficient, because not every appearance of the tripeptide in proteins contains an Asn that is glycosylated. The oligosaccharide contains two N-acetylglucosamine, nine mannose, and three glucose molecules with the branched sequence shown in Figure 4–13. The 14-residue oligosaccharide is preassembled by one-at-a-time addition of nucleotide-activated monosaccharides onto a large hydrophobic polyisoprenoid lipid termed a *dolichol*, with the first N-acetylglucosamine in a pyrophosphate link to the dolichol.

Three glucose residues and one mannose residue of the 14-residue oligosaccharide are trimmed in the RER.

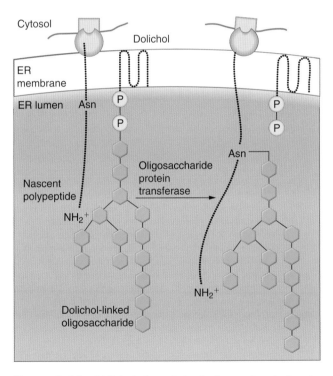

Figure 4–13. N-linked glycosylation in the rough endoplasmic reticulum (RER). In the early steps of N-linked glycosylation, the oligosaccharide is preassembled on a dolichol carrier, a large, aliphatic hydrocarbon terminating in a hydroxyl group that is phosphorylated. The oligosaccharide is assembled by one-at-a-time addition of monosaccharides. Interestingly, most of this assembly occurs on the cytosolic side of the membrane, and the oligosaccharide is then transferred through the membrane to the luminal side, all while attached to the dolichol (these early steps are not shown). Once an Asn residue in the correct sequence context emerges from the translocon, the enzyme oligosaccharide protein transferase attaches the oligosaccharide to the protein cotranslationally.

The N-linked oligosaccharide is further trimmed in the Golgi complex, and other sugars are added. N-linked oligosaccharides on proteins appear to have complex and sometimes overlapping functions. One function is to stabilize protein structure, related, in part, to the affinity of the sugars for water so that glycosylated regions of a protein are highly water soluble. N-linked oligosaccharides are frequently recognized by other proteins, providing a molecular handle for protein–protein interactions. For example, certain cell-adhesion proteins on the cell surface have sugar-binding domains that help one cell adhere to another. In addition, phosphorylation of mannose residues on the oligosaccharide in the Golgi complex provides a signal for targeting lysosomal enzymes to lysosomes (see later for a more detailed discussion). The 3 glucose residues at the end of the 14-residue oligosaccharide precursor have a special role in protein folding and quality control in the ER (see the next section).

Protein Folding and Quality Control in the Endoplasmic Reticulum

All the information for proteins to fold into their final three-dimensional structures is contained within their amino acid sequences, but the rate of folding is often too slow on a biological time scale. Hence, proteins have evolved that catalyze the folding of other proteins. One class of these folding catalysts, called *chaperones*, binds to nascent proteins as they emerge from free ribosomes in the cytoplasm to promote rapid folding. It is not surprising, therefore, that chaperones exist within the RER that cotranslationally assist in the folding of secretory and membrane proteins as they emerge from the translocon. Recent work has elucidated elegant quality-control machinery within the RER that facilitates protein folding, retains misfolded proteins within the RER to prevent their transport to the Golgi apparatus, and disposes of proteins that cannot fold properly. Protein misfolding in the ER (and elsewhere in the cell) is medically relevant because it impacts the pathology of numerous diseases.

Abundant chaperones exist within the lumen of the ER that promote protein folding in different ways. BiP, which stands for heavy chain binding protein, was first discovered as a binding partner for the heavy chain of immunoglobulins and is now understood to associate in an ATP-dependent process with hydrophobic peptide sequences of most proteins as they emerge from the translocon. The result is to shield hydrophobic residues and help them fold to the interior of the protein where they are protected from water. Secreted proteins, and many membrane proteins that are on the exterior surface of the plasma membrane, contain disulfide bonds between cysteine residues of the protein that help to stabilize proteins exposed to the variable environment outside the cell. (In contrast, cytosolic proteins exist in a stable and controlled environment and do not contain disulfide bonds.) To ensure that disulfide bonds are quickly and correctly formed by secreted proteins, another class of chaperones, called *protein disulfide isomerases*, promotes oxidation-reduction reactions that connect the sulfhydryl groups of cysteine residues into disulfide bonds.

Calreticulin and calnexin are another type of chaperone that binds to a glucose residue on the N-linked oligosaccharide of proteins to promote folding. Figure 4–14 provides a deeper look at calnexin action.

As proteins emerge from the translocon, BiP binds to assist folding and N-glycosylation also occurs. When synthesis is complete, the protein is released and a second quality-control process starts that involves calnexin and calreticulin. Calnexin, a single-pass membrane protein, and calreticulin, a soluble protein, are lectins (sugar-binding proteins) that bind to glycoproteins containing a single glucose residue. The action of calnexin is illustrated in Figure 4–14 (calreticulin is similar except that it is a soluble protein). The first two glucose residues on the N-linked oligosaccharide are removed by glucosidases I and II. Calnexin binds to the single glucose and forms a complex to assist folding in collaboration with ERp97, a disulfide isomerase that promotes disulfide bond formation. When calnexin and ERp97 are finished, the final glucose residue is cleaved by glucosidase II and the protein is released. If the protein is still not properly folded, it is a substrate

for a glycosyltransferase that re-adds another glucose residue, initiating another round of binding to calreticulin and folding assistance. When the protein is properly folded, glucose is not re-added, and the protein passes the quality-control standard to exit the RER via vesicles destined for the Golgi complex. This folding cycle illustrates that one function of N-linked glycosylation is to participate in chaperone action.

Two other pathways come into play when, despite the best efforts of chaperones, a protein cannot fold properly. One pathway is ER-associated degradation (ERAD). This system identifies improperly folded proteins and exports them to the cytosol where they are degraded by the proteasome. Many details of the pathway are not well understood, including what is the pore in the ER membrane through which the proteins are retrotranslocated to reach the cytosol. The pathway is known to involve addition of the peptide ubiquitin to proteins on the cytosolic side of the membrane that marks them as substrates for the degradation by the proteasome. If unfolded proteins become abundant in the lumen of RER, a second pathway induces the synthesis of additional chaperones and also increases the amount of proteins involved in ERAD to eliminate excess unfolded proteins. This pathway is called the unfolded protein response (UPR) and consists of a collection of mechanisms that up-regulate genes encoding proteins involved in chaperone function and in ERAD.

Why has the cell devoted so many resources to quality control in the RER? Improperly folded proteins invariably expose hydrophobic amino acid sequences to water,

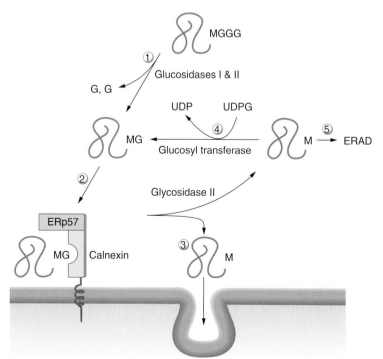

Figure 4–14. Calnexin and quality control in the endoplasmic reticulum (ER). N-linked oligosaccharide of N-glycosylated proteins has three terminal glucose residues. *(1)* The first of these is removed by glycosidase I and the second by glycosidase II, producing the monoglycosylated species. *(2)* Calnexin (and calreticulin, not shown) bind to N-linked oligosaccharides terminating in one glucose residue, in association with ERp57, a protein disulfide isomerase. Together, calnexin and ERp57 promote the folding of the glycoprotein. *(3)* Proteins that appear to be folded are released, and the last glucose residue is removed by glycosidase II. Properly folded proteins can then collect in the lumen of vesicles that carry cargo from the ER to the Golgi complex. *(4)* However, if a protein is still not folded, the protein glucosyltransferase senses the misfolds and reattaches a single glucose residue from UDP-glucose, once again creating the monoglucosylated species that can interact with calnexin for further folding help. *(5)* Proteins that are so badly misfolded that they cannot be recovered are shunted to the endoplasmic reticulum–associated degradation pathway (ERAD) whereby they are exported to the cytosol and degraded by the proteasome.

which causes unfolded proteins to aggregate, thereby shielding the hydrophobic surfaces from water. These aggregates can become large and could recruit other proteins into them by inducing folded proteins to unfold; thus, unfolded proteins could be catalysts that accelerate the unfolding of already folded proteins. If these aggregates left the ER, they could acutely interfere with the action of existing proteins at other locations in the cell. In addition, unfolded proteins present novel antigenic sites to the immune system that would elicit unwanted autoimmune reactions should the proteins be exposed on the cell exterior. The medical relevance of quality control in the RER is emphasized by the observation that there are more than 35 known diseases that are directly or indirectly related to improper protein folding in the RER. Among them is cystic fibrosis, where a single point mutation in a chloride transporter makes the protein difficult to fold and it accumulates in the RER. If the protein containing the mutation does manage to fold correctly, it functions normally in chloride transport. Observations such as this have launched major efforts to develop drugs that enhance protein folding in the RER.

Clinical Case 4-1

Peter Drysdale is a 13-year-old boy who was having a physical examination before changing from a public to a private high school. He was very thin, appeared younger than his stated age, and appeared chronically ill. The initial examination showed that he had scattered rales throughout his lungs, an area of coarse rhonchi in the right middle lobe, and a questionable area of atelectasis in the left lower lung field. Prompted by these findings, the school physician elicited a more detailed history from Peter and his parents.

Peter had been a full-term normal weight infant, but shortly after birth was returned to the hospital because he was not passing any stools. Fortunately, with 3 days of intravenous hydration, he began to have normal gastrointestinal function, and returned home. Though subsequently he grew slowly, his parents noted no particular problems until he began school at the age of 5. At that time he developed his first episode of pneumonia, which required antibiotic therapy for *hemophilus influenza*. Since then, he has had repeated pneumonias two or three times a year. These are usually heralded by fits of heavy coughing productive of thick green mucus. When cultured, his sputum shows variable organisms, and he has often responded to oral broad-spectrum antibiotics. Unfortunately, the last two cultures have shown predominantly *pseudomonas* variants.

In addition to his lung problems, Peter has recently developed right upper quadrant pain after eating, and an intravenous cholecystogram showed the presence of gallstones and probable cholecystitis. Furthermore, around the same time he noted that his stools, although formed, were larger and frequently floated in the toilet bowl. With this information in hand, the physician took a small sample of Peter's axillary sweat secretions.

Cell Biology, Diagnosis, and Treatment of Cystic Fibrosis

Peter has cystic fibrosis, which results from a mutation in a protein called the cystic fibrosis transmembrane conductance regulator (CFTR). The CFTR is a chloride channel that contains 12 membrane-spanning domains encoded by a gene on chromosome 7. More than two-thirds of cystic fibrosis patients contain a deletion of a single phenylalanine residue at position 508 of CFTR. Loss of this single residue destabilizes the protein so that it folds slowly in the RER with the result that it is retained and degraded in the RER by the ERAD pathway described in this chapter. Interestingly, reducing the temperature of cultured cells expressing the defect, or by the adding certain chemical chaperones to cells, increases the amount of the CFTR that is correctly folded, and the folded mutant protein functions normally in chloride transport. This has launched a major research effort to discover drugs that facilitate CFTR folding.

CFTR regulates the transport of chloride (and water) across epithelial cell membranes. A multitude of physiologic defects occur when this protein fails to reach its intended membranes. These include inadequate hydration of mucus secretions in the lung, which renders the patient particularly susceptible to bacterial infections; inadequate hydration of an infant's stool, which can produce meconium ileus at birth or pathologic constipation thereafter; thickening of cervical mucus that can limit female fertility; and inadequate water and chloride secretion in the accessory gastrointestinal tract, leading to gallstones and cholecystitis, as well as destructive desiccation of the exocrine pancreas. Perhaps of less importance physiologically, but most convenient diagnostically, it also leads to a marked increase in sweat chloride caused by the inability of the sweat gland epithelium to adequately resorb chloride.

Peter's sweat sample had a chloride of 110 mEq/L, which established the diagnosis. His prognosis is limited. Pulmonary toilet and specific antibiotic therapy will help his lungs temporarily. Oral lipase therapy may normalize his stools, and cholecystectomy may avert further biliary problems. Unfortunately, progressive, permanent multiorgan damage is likely.

LEAVING THE ENDOPLASMIC RETICULUM: A PARADIGM FOR VESICULAR TRAFFIC

The ER, the Golgi apparatus, lysosomes, the plasma membrane, and various endosomes are all connected by pathways of vesicular traffic whereby membrane from one organelle is transferred to another. A vesicle forms at the originating membrane and moves to and fuses with the acceptor membrane, thus transferring selected lipids, proteins, and the luminal contents of the vesicle to the destination organelle. All membranes involved in vesicle traffic must accurately select cargo to be loaded into the vesicle, bud the vesicle, target and fuse the vesicle to the correct membrane, and finally, return to the originating membrane those components dedicated to vesicle formation and fusion so they can be reused for another round of vesicular traffic. The molecular mechanisms of vesicular trafficking differ in detail for different membranes, but mechanistic similarities have been identified. An overview of features common to most pathways of vesicle traffic is presented next, followed by specific details of ER to Golgi vesicle traffic.

Overview of Vesicle Budding, Targeting, and Fusion

Cargo Selection Requires a Specific Coat

Cargo selection is initiated by the assembly of a protein scaffold on the cytoplasmic side of the membrane where the vesicle will bud. The scaffold is part of a structure called a *coat* that can be seen on membranes by electron microscopy, and different coat proteins exist for different organelles. The coat often contains two protein layers, an inner layer closest to the membrane often called the *adaptor* that interacts with specific integral membrane proteins. The integral proteins may be cargo themselves en route for function in another membrane, but they may also be transmembrane receptor proteins with binding sites on the luminal side of the membrane for binding to soluble proteins, thus concentrating the soluble proteins on the luminal side of the membrane where a vesicle will form. Cargo membrane proteins have structural features recognized by binding sites on the adaptor proteins of the coat. These structural features, called *sorting signals,* can be stretches of amino acids common to several cargo types, but they can also be more subtle, depending on tertiary protein structure. Membrane proteins that bind to the coat adaptor on the cytoplasmic side and to a soluble protein on the luminal side need an additional binding site for interacting with sorting signals on soluble proteins, and several of these signals are also known.

How is the coat recruited to the cytoplasmic side of an organelle membrane? A common theme is to first recruit a small GTP-binding protein to the cytoplasmic surface of the membrane. Different membranes use different GTP-binding proteins, but recruitment is initiated by the exchange of GDP for GTP, catalyzed by a guanine nucleotide exchange factor (GEF) that is already associated with the membrane, which triggers binding of the GTP-binding protein to the membrane. The GTP-binding proteins contain a hydrophobic helix that is buried inside the protein when it is soluble in the cytosol, but which is displayed on GTP binding so that it inserts into the membrane, anchoring the GTP-binding protein to the membrane. This anchored GTP-binding protein then interacts with coat components to start the coating process. Figure 4–15 illustrates a generic model of coat protein binding and cargo selection.

Vesicle Budding

To form a vesicle, the parent membrane must deform with a high radius of curvature. In general, protein components of the coat, or other proteins that bind to the coat, are believed to associate with the membrane and cooperate to shape the vesicle. Once the vesicle has separated from the parent membrane, the coat protein complexes must dissociate, providing access to proteins on the cytoplasmic surface of the vesicle membrane that are needed for targeting and fusion functions. In the best understood models, the hydrolysis of GTP by the GTP-binding protein that first recruited the coat is the signal for the coat to dissociate. Coat components and the GTP-binding protein in the GDP state are released to the cytosol, available for another round of vesicle formation.

Vesicle Targeting and Fusion

Once formed, the vesicle moves to the destination membrane and fuses with it, combining the vesicle membrane with the target membrane and mixing the soluble contents of the vesicle with the luminal contents of the destination organelle. Vesicle targeting and fusion must occur with great fidelity because delivery of the proteins and lipids to the wrong membrane could be lethal. There are three general levels to targeting and fusion: vesicle movement, vesicle tethering, and vesicle fusion. In small cells, such as yeast, vesicles are believed to move mainly by diffusion. However, in larger mammalian cells, vesicles often move on tracks of the cytoskeleton, such as microtubules. Proteins are known that connect vesicles to microtubules, and strong evidence has been found that motor proteins associated with different types of progression on microtubules provide for vesicle movement (see discussion in Chapter 3).

Once near the target membrane, the initial interaction of the vesicle with the target membrane is a tethering event. Tethering involves an assembly of proteins that form an extended structure long enough to tether the vesicle to the membrane. Vesicle transport among

Cytosol

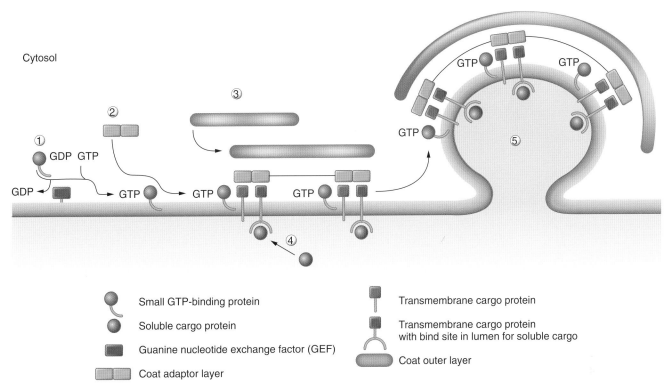

Legend:

- Small GTP-binding protein
- Soluble cargo protein
- Guanine nucleotide exchange factor (GEF)
- Coat adaptor layer
- Transmembrane cargo protein
- Transmembrane cargo protein with bind site in lumen for soluble cargo
- Coat outer layer

Figure 4–15. Generic model showing the initial steps in vesicle coating, cargo recruitment, and vesicle budding. *(1)* A small guanosine triphosphate (GTP)–binding protein is recruited to the membrane on the exchange of GDP for GTP catalyzed by a guanine nucleotide exchange factor (GEF) already located at the membrane. When activated, the small GTP-binding protein exposes a hydrophobic segment (sometimes with an attached lipid) that inserts into the membrane, anchoring the protein to the cytosolic side of the membrane. *(2, 3)* The activated GTP-binding protein interacts with soluble coat components, recruiting them to the membrane. There are usually two parts to the coat, the adaptor layer and the outer layer. *(4)* The adaptor layer binds to transmembrane proteins that present protein motifs containing amino acid sequences that are sorting signals. Some of the transmembrane proteins have domains on the luminal side of the membrane that bind to soluble luminal proteins, thus coupling the soluble proteins to sites on the membrane where a vesicle will form. *(5)* The coat layers accumulate and contribute to deforming the planar bilayer into a vesicle. The vesicle eventually buds off and the coat dissociates on hydrolysis of GTP to GDP by the small GTP-binding protein (not shown).

different membranes uses different tethering factors, but most appear to be regulated, at least in part, by a class of small GTP-binding proteins called *Rab proteins*. Rab proteins use a typical GTP/GDP cycle that is controlled by Rab GTPase activating proteins and GEFs. In the GDP state, Rab proteins are complexed with an escort protein in the cytosol, and when activated, they insert a covalently attached hydrophobic lipid anchor into their home membrane where they are included in vesicles when they bud. More than 60 Rab proteins are in mammalian cells, and a unique Rab protein appears to participate in each different vesicle targeting type, leading to the idea that Rab proteins contribute, at least in part, to the specificity of vesicle targeting. Distinct Rab proteins are also differentially expressed in different tissues. Several disease syndromes are known that result from defects in a Rab protein, a Rab escort protein, or a Rab regulatory protein. For example, the eye disease choroideremia is caused by a defective escort protein that normally binds to the GDP-bound form of Rab27a in the cytosol.

Vesicle fusion is mediated by integral membrane proteins called SNARES that are of two types, v-SNARES and t-SNARES ("v" stands for vesicle and "t" stands for target). Every vesicle when it buds incorporates v-SNARE proteins that have binding sites in their cytosolic domains for cognate t-SNARE proteins on the target membrane. Although there are variations on the theme, there is usually a single v-SNARE class in the vesicle and three different t-SNARES in the target membrane. When the SNARES interact, each contributes one amphiphilic α-helix to form an extremely stable coiled-coil structure where the four helices twist around one another, similar to the strands of a rope. The strength of this interaction is believed to bring the cytosolic face of the vesicle and the target membrane into close contact, facilitating fusion. A generic model of vesicle tethering and fusion is shown in Figure 4–16.

Evidence supporting this model of fusion includes studies where many different v- and t-SNARE combinations were incorporated into lipid vesicles and the ability of the vesicles to fuse was measured. Fusion activity was greatest for those v- and t-SNARE combinations that were known from other studies to be cognate pairs in living cells.

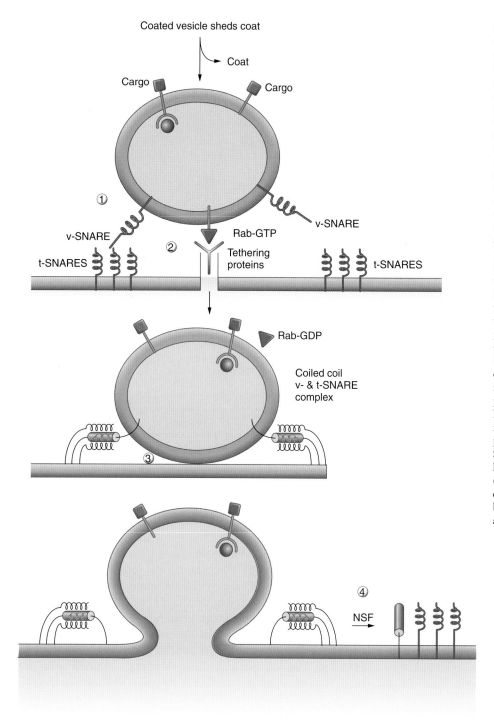

Coated vesicle sheds coat

Coat

Cargo

Cargo

①

v-SNARE

v-SNARE

②

Rab-GTP

Tethering proteins

t-SNARES

t-SNARES

Rab-GDP

Coiled coil v- & t-SNARE complex

③

④

NSF

Figure 4–16. Vesicle tethering and fusion. *(1)* A coated vesicle sheds its coat and moves to the target membrane either by diffusion or by directed movement attached to a cytoskeletal track. In addition to cargo, the vesicle contains a v-SNARE that was recruited at the time of vesicle formation. A Rab protein in the active guanosine triphosphate (GTP)–bound form is also recruited to the cytosolic face of the vesicle. *(2)* The vesicle docks with the target membrane in a tethering step that involves the Rab-GTP and a tethering protein complex on the target membrane. *(3)* The v-SNARE on the vesicle interacts strongly with three t-SNARES on the target membrane, bringing the cytosolic face of the two membranes close together so they can fuse. At some point in this process, the Rab protein hydrolyzes GTP to guanosine diphosphate (GDP) and dissociates into the cytosol in the inactive GDP-bound form complexed with a Rab escort protein. *(4)* After fusion is complete, NSF (*N*-ethylmaleimide sensitive factor), in association with other proteins, breaks apart the v- and t-SNARE complexes in a process that hydrolyzes adenosine triphosphate (ATP), thus freeing the v-SNARE to enter a return vesicle that takes it back to its home membrane for another round of vesicle transport.

Recycling v-SNARES

After membranes fuse, there must be a mechanism for dissociating the stable v- and t-SNARE complexes, enabling the v-SNARES to recycle to their original membrane and support another round of vesicle fusion. SNARE complexes are disassembled by a protein called NSF, an ATPase of the AAA family. NSF was originally discovered as a soluble protein necessary for vesicle fusion in assays that reconstituted vesicle transport among Golgi membranes. NSF was sensitive to inhibition by *N*-ethylmaleimide, a reagent that binds to free thiol groups of cysteine residues in proteins, hence the name *N*-ethylmaleimide sensitive factor. Although originally supposed to be necessary for the fusion reaction itself, it is now believed that inactivating NSF inhibited fusion because v- and t-SNARES became trapped in complexes, depleting the supply of free SNARES so that newly made vesicles lacked SNARES needed for fusion. Once the v- and t-SNARE complexes are dismantled by NSF, the v-SNARES are incorporated into vesicles and return to their original membranes by vesicle transport.

Endoplasmic Reticulum to Golgi Vesicle Transport and COPII-Coated Vesicles

Vesicles that carry cargo from the ER to the Golgi complex form at specialized regions of the RER called the transitional ER. Transitional ER sites can be identified by electron microscopy and are widely dispersed throughout the RER membrane as islands that lack bound ribosomes. The coat protein used in ER to Golgi transport is called COPII, where COP stands for coat protein. As described in the generic model of vesicle formation, a small GTP-binding protein called Sar1 is used to recruit COPII to membranes. Sar1 is bound to GDP in the cytosol and is activated by the exchange of GTP for GDP, catalyzed by the membrane-bound GEF protein Sec12, found on the transitional ER. A hydrophobic helix at the N terminus of Sar1 is exposed on binding GTP and inserts into the membrane as an anchor. Sar1-GTP binds to a dimeric protein complex containing proteins Sec23 and Sec24, the adaptor layer of the COPII coat protein complex. A second layer of the coat complex consisting of proteins Sec13/Sec31 subsequently binds to the Sar1/Sec23/Sec24 scaffold. Both Sar1 and Sec23/Sec24 may interact with sorting signals in the cytoplasmic domains of cargo membrane proteins. Sec24 contains two distinct cargo-binding sites that interact with specific amino acid sequences found in the cytoplasmic domains of cargo proteins. For example, one site on Sec24 interacts with proteins that contain the sequence Asp-X-Glu, termed the *diacidic signal* (where X is any amino acid), as well as other sorting sequences. The second site on Sec24 binds to cargo proteins that have different sorting signals. It is also likely that Sec23 and Sar1 contain additional cargo-binding domains.

ER sorting signals can be subtle and not easily decoded. For example, the binding of certain cargo proteins to Sec23/Sec24 depends on the oligomerization of the proteins that contain multiple subunits. In other instances, the cargo protein must bind to an escort protein to reveal the export signal. One class of escort protein is ERGIC 53, which is believed to interact with mannose portions of the N-linked oligosaccharide on exported proteins. There is an autosomal recessive human syndrome in which mutated ERGIC 53 results in the failure of clotting factors V and VIII to efficiently enter COPII vesicles, resulting in a bleeding disorder. One concept emerging from the study of ER export sorting signals is that their presentation can depend on whether a protein is correctly folded; that is, quality control is an important parameter of whether a protein will display the correct signal to be included in a vesicle that leaves the RER.

In the COPII vesicle system, evidence has been found that Sar1 itself actually initiates membrane deformation to bud the vesicle by inserting an amphiphilic helix into the membrane. The Sec23/Sec24 complex is then proposed to stabilize the bent shape and contribute to vesicle formation. Once the vesicle has separated from the parent membrane, the coat protein complexes dissociate, triggered by hydrolysis of GTP on Sar1-GTP, releasing the Sar1/Sec23/Sec24/Sec13/Sec31 COPII components.

In a step that is not well understood, uncoated COPII vesicles appear to aggregate together near ER transitional sites where they are formed and fuse together to make a larger compartment that morphologically contains intermixed vesicles and tubules. Fusion of vesicles to others of a like kind is called *homotypic fusion,* and it also occurs in other examples of membrane traffic. The tethering step of fusion uses the GTP-binding protein Rab1, which when activated by GTP binding recruits a coiled-coil protein called p115 as part of the tether. Fusion is completed by formation of SNARE complexes. Homotypic fusion requires that each vesicle contains both v-SNARES and t-SNARES, and there is evidence that this occurs. The compartment formed by homotypic fusion of COPII vesicles near the transitional ER is called the *vesicular tubular compartment* (VTC). To return SNARES to the transitional ER it would be efficient if retrograde recycling vesicles emerged from the VTC while in close proximity to the transitional RER, and strong evidence exists that recycling is mediated by another type of vesicle coated with a protein complex called COPI (COPI is discussed in more detail later in this chapter). After the VTC grows larger, it attaches to microtubules and moves en masse from the distributed transitional ER sites to the cis region of the Golgi apparatus.

The details of the COPII pathway have been (and continue to be) worked out by a combination of yeast genetics, *in vivo* studies, and *in vitro* reconstitution of vesicle formation with purified components. Many of the proteins involved in this pathway, and other pathways of membrane traffic, were first identified by mutations in yeast and subsequently found in mammalian cells, an example of how work in nonmammalian organisms has direct application to understanding human cell biology and disease.

THE GOLGI APPARATUS

The structure and morphology of the Golgi apparatus is illustrated in Figure 4–17. The forming face, where VTCs from the ER arrive, is called the *cis*-Golgi network (CGN). Secretory material arrives at the cis face and

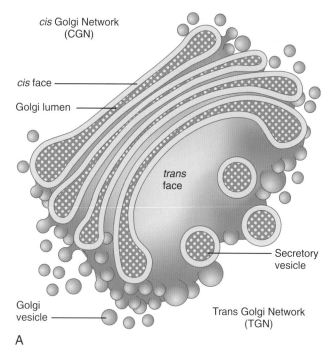

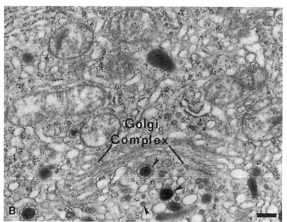

Figure 4-17. **Structure of the Golgi complex. A:** The Golgi contains stacks of flattened cisternae. These cisternae have dilated edges from which small vesicles bud. The forming or cis face of the Golgi is the *cis*-Golgi network (CGN) and arrives from the endoplasmic reticulum (ER). The exporting or trans face of the Golgi is the *trans*-Golgi network (TGN), where vesicles and cargo depart for other membranes. **B:** Electron microscopy of the Golgi apparatus of a mammalian luteal cell. Numerous secretory granules and vesicles of various sizes *(arrowheads)* can be observed budding from the trans face of the Golgi complex (large luteal cell from a rat corpus luteum 20 days pregnant). Scale bar = 0.27 μm. *(Electron micrograph provided by Dr. Phillip Fields.)*

progresses through the medial and trans stacks to reach the *trans*-Golgi network (TGN), where it is sorted and distributed to the next destination, either the plasma membrane or endosomes. Proteins passing through the Golgi complex undergo several types of covalent modification catalyzed by Golgi-resident enzymes. One main function of the Golgi apparatus is to modify N-linked oligosaccharides as they move through the Golgi stacks, to add a second type of carbohydrate to some proteins, and to add phosphate to

mannose residues that are the targeting signal for delivering lysosomal enzymes to lysosomes. In the TGN, some proteins are sulfated and certain proteins undergo controlled proteolytic events as part their maturation process. These modifications are described in the following section.

Glycosylation and Covalent Modification of Proteins in the Golgi Apparatus

The processing of N-linked oligosaccharides in the Golgi complex is summarized in Figure 4–18. In the RER, the N-linked oligosaccharide is trimmed to contain two N-acetylglucosamine and eight mannose residues. On arrival in the Golgi apparatus, further trimming and the addition of N-acetylglucosamine, galactose, fucose, and N-acetylneuraminic acid to various branches of the oligosaccharide tree happens in different Golgi subcompartments. Generally, the more terminal sugars are added in later Golgi compartments with the last sugar, N-acetylneuraminic acid (also called sialic acid), being added in the TGN. A special N-linked carbohydrate modification occurs for lysosomal enzymes in the CGN, the phosphorylation of one or more mannose residues on position 6. Mannose 6-phosphate is later recognized by a receptor, the mannose 6-phosphate receptor that selects lysosomal enzymes from all the other soluble secretory proteins and diverts them to lysosomes (see later for a more detailed description).

A second type of carbohydrate modification also occurs in the Golgi complex, the addition of O-linked oligosaccharide chains to proteins. O-linked carbohydrate is attached to proteins via the oxygen of the hydroxyl group of either serine or threonine, hence the term *O-linked*. O-linked oligosaccharides are built by one-at-a-time addition of UDP-activated monosaccharides, catalyzed by glycosyl transferase enzymes. Depending on the protein, there may be only a few or many O-linked carbohydrate residues added to make a glycoprotein whose mass is mainly carbohydrate. Taken together, there are hundreds of different glycosyl transferases distributed in Golgi subcompartments involved in N- and O-linked glycosylation. Variability is common in the exact oligosaccharide sequence even among different copies of the same protein.

Two other protein modifications occur in the TGN. One is sulfation of certain proteins on tyrosine residues. The second is the proteolytic cleavage of some secreted proteins in the TGN, or in secretory vesicles, that activate the protein. This is common for secreted polypeptide hormones (e.g., insulin) and is a way of ensuring that the hormone will not be in its active form until it is ready for secretion from the cell. An overview of protein modifications that occur in the Golgi apparatus is presented in Figure 4–19.

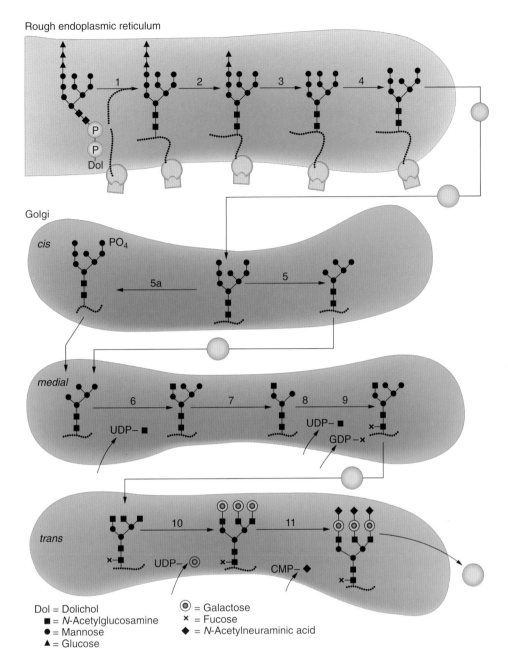

Rough endoplasmic reticulum

Golgi

cis   PO₄

5a   5

medial

UDP–■   6   7   8   9   UDP–■   GDP–×

trans   10   11   UDP–◉   CMP–◆

Dol = Dolichol
■ = N-Acetylglucosamine
● = Mannose
▲ = Glucose
◉ = Galactose
× = Fucose
◆ = N-Acetylneuraminic acid

Figure 4–18. Processing of glycoprotein N-linked oligosaccharides in the Golgi complex. The known stages occur in 11 enzymatic steps, beginning with the precursor N-linked oligosaccharide *(step 1).* First, three glucose residues are removed *(steps 2 and 3)* and four mannose residues are cleaved *(steps 4 and 5).* For lysosomal enzymes, one or more mannose residues are phosphorylated. One N-acetylglucosamine is added *(step 6),* two mannose residues are removed *(step 7),* one fucose and two N-acetylglucosamine residues are added *(steps 8 and 9),* three galactose residues are added *(step 10),* and then three N-acetylneuraminic acid residues are added *(step 11).*

Retrograde Transport through the Golgi Complex

In the face of abundant membrane traffic from the ER to the Golgi and farther, retrograde vesicular transport from Golgi membranes to the ER is robust. One function of this retrograde traffic is to return SNARES involved in anterograde traffic back to the RER.

Although perhaps not immediately obvious, another crucial reason for retrograde transport is to return resident ER proteins that have escaped back to the ER. This need arises because the volume of anterograde traffic from the RER to the Golgi can be so large that resident ER proteins are unintentionally included in COPII anterograde vesicles. These resident ER proteins would be lost unless captured by an efficient mechanism of

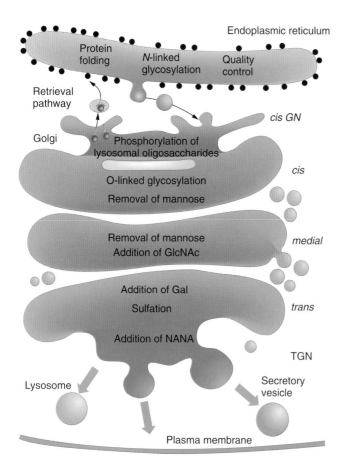

Figure 4–19. Overview of compartmentalized functions in the secretory pathway. Protein folding, N-linked glycosylation, and quality control begin in the rough endoplasmic reticulum (RER). Phosphorylation of mannose on lysosomal enzymes, O-linked glycosylation, and trimming of N-linked oligosaccharides begin in the *cis*-Golgi network (CGN). Additional carbohydrate trimming and addition of monosaccharide to N- and O-linked oligosaccharides occurs in the medial and *trans*-Golgi cisternae, as well as in the *trans*-Golgi network (TGN). Sulfation of proteins on tyrosine residues happens in the TGN.

not normally residents of the ER, with the result that the hybrid proteins became ER residents because they were so efficiently captured and returned whenever they left the ER.

The mechanism of retrograde vesicular transport conforms to the generic model of vesicle transport described earlier in this chapter (see Figs. 4–13 and 4–14). The coat used by retrograde vesicles is called COPI (coat protein I), present in the cytosol as a soluble complex of seven polypeptides termed *coatomer*. Coatomer units are recruited from the cytosol to Golgi membranes by a GTP-binding protein called Arf1. On activation by a GEF, Arf1 inserts into the Golgi membrane, initiating the binding of coatomer to form the COPI coat. Similar to COPII vesicles, there appear to be two layers to the COPI complex, an inner layer that interacts with sorting signals, and an outer layer that may form a scaffold to stabilize the coat. The inner layer has sites that bind cargo. One site directly binds the KKXX motif, concentrating proteins that display this motif in budding vesicles. For soluble proteins that contain the KDEL signal, there is a transmembrane protein, the KDEL receptor, which has a luminal domain that binds the KDEL signal and a cytoplasmic domain that binds to the coat, thus attaching luminal proteins that have the KDEL signal to sites where the vesicles bud. The coat dissociates after vesicle formation and the vesicles fuse with a target membrane using tethering factors and SNARES.

The two general target membranes for COPI-coated vesicles are the membrane of the preceding Golgi stack and the ER. Thus, COPI-coated vesicles may carry material in the retrograde direction from one stack to the next, trans to cis, and eventually from the cis stack of the Golgi to the ER. Understanding the directionality of COPI-coated vesicles has an interesting history. They were originally discovered with *in vitro* assays that reconstituted vesicle traffic between Golgi stacks, and it was thought that they were anterograde vesicles. Subsequent evidence strongly suggested that they carried retrograde cargo, but it remains a formal possibility that there are two types of COPI-coated vesicles, one for anterograde transport and the other for retrograde transport, differing in subtle ways not yet appreciated.

The study of trafficking in the Golgi apparatus is a vibrant field, often updated with new information. For example, recent work has discovered that the COPI-coated vesicle mechanism of retrograde transport is not the only process operating to transport material in the retrograde direction into the ER. There is at least one COPI-independent retrograde transport pathway, discovered because some medically important protein toxins (such as Shiga toxin made by pathogenic strains of *Escherichia coli*) use this pathway to damage cells. The details of this pathway are poorly understood to date. For another example of how the understanding of organelle membrane trafficking is constantly under revi-

retrograde transport to return them to the ER. Two types of proteins are returned by retrograde transport. Soluble ER-resident proteins that are included within the fluid volume of an anterograde COPII-coated vesicle when it buds, and ER-resident integral membrane proteins accidentally included in the membrane of the vesicle at budding. Soluble and membrane proteins contain different retrograde sorting signals that ensure they are recognized and included within retrograde vesicles. Soluble proteins contain at their C terminus the sequence KDEL (or a related sequence), whereas integral membrane proteins contain the sequence KKXX (or a related sequence) at their C terminus. These uncomplicated sorting signals were discovered by comparing the amino acid sequences of different soluble and membrane ER proteins. Their function as retrograde sorting signals was uncovered by grafting the signals at the DNA level onto other proteins, even proteins that were

sion, we turn next to anterograde traffic through the Golgi complex.

Anterograde Transport through the Golgi Complex

The two basic models for anterograde transport through the Golgi complex are the vesicle shuttle model and the cisternal progression model (compared in Fig. 4–20). The vesicle shuttle model supposes that each Golgi membrane stack is a permanent structure that receives vesicles carrying anterograde cargo from the adjacent cisternae on the cis side and then packages that cargo into new vesicles that go into the adjacent stack on the trans side. Retrograde transport operates in the reverse to recycle SNARES and other material to the preceding stack. In contrast, the cisternal progression model supposes that the Golgi stacks are transient structures whereby a given cisternae matures into the adjacent cisternae in the trans direction by remodeling its membrane. Membrane remodeling occurs by retrograde transport vesicles that transfer Golgi proteins, including glycosyl transferases, back to the preceding cisternae. In this model, each cisternae can be thought of as one large vesicle carrying material in the anterograde direction whereas at the same time sending membrane and enzymes back to the next cisternae in the production line. Each new CGN develops from VTCs coming in from the transitional ER and progresses through the stack to eventually become the TGN.

Until the late 1990s, the vesicle shuttle model was rarely challenged. However, several lines of evidence have reversed the preference of most cell biologists. One was the realization that COPI-coated vesicles carried retrograde cargo, and there was no clear substitute vesicle that could carry anterograde cargo, required by the vesicle shuttle model. Another line of evidence was that some cargo materials, such as fish scales and large collagen complexes, were too large to fit into small anterograde vesicles. Large complexes could, however, move through the Golgi complex within individual cisternae as the cisternae progressed by maturation. The study of cargo movement through the Golgi complex is an active area and new revelations could arise as the understanding of the transport process is refined.

Leaving the Golgi Complex

There are three general pathways of anterograde export from the TGN: constitutive secretory vesicles that constantly ferry material from the TGN to the plasma membrane; regulated secretory vesicles that collect material destined for the plasma membrane, but hold it until a signal for secretion triggers fusion with the plasma membrane; and vesicles that carry lysosomal enzymes to late endosomes or lysosomes. An overview of these pathways is presented in Figure 4–21.

Constitutive and Regulated Secretion

Vesicles constantly leave the TGN and fuse with cytosolic face of the plasma membrane, replenishing lipids and membrane proteins as a housekeeping function. It is not clear that there is a cargo selection step in this process, and indeed, one of the perplexing gaps in understanding membrane traffic at this level is that no coat protein has been identified for constitutive secretory vesicles. There is even evidence that vesicles may

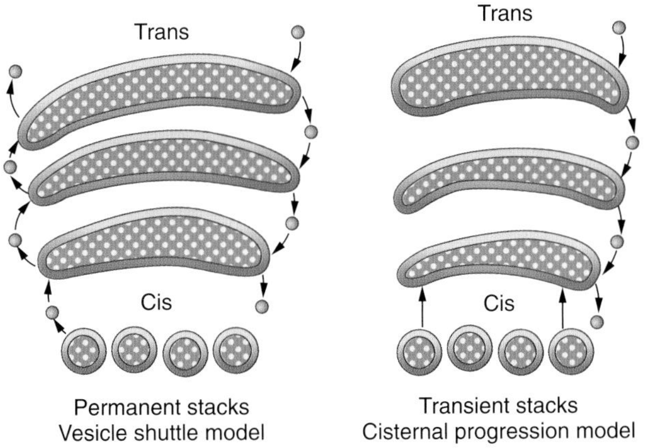

Figure 4–20. Models for transport through the Golgi complex. In the vesicle shuttle model (**left**), each Golgi cisternae is a permanent structure that receives vesicles from the preceding cisternae and sends vesicles to the next cisternae. In the cisternal progression model (**right**), the Golgi cisternae are transient structures and mature into the next cisternae on remodeling the membrane by sending material to the incoming stack on the cis side by retrograde transport vesicles.

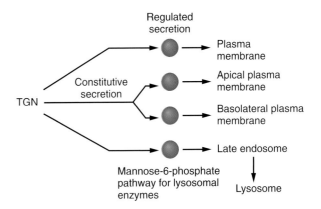

Figure 4–21. Overview of transport from the *trans*-Golgi network (TGN). There are three main pathways for material leaving the TGN: regulated secretion destined for the plasma membrane, constitutive secretion destined for the plasma membrane, and the mannose 6-phosphate pathway for delivery of lysosomal enzymes to late endosomes. Note that secretion to the plasma membrane in polarized cells involves targeting cargo and vesicles to either the apical or basolateral domains of the plasma membrane.

not be involved; rather, membrane tubules may extend out from the TGN, transiently contacting the plasma membrane to transfer material, but this mechanism is debated. There is an additional complication for constitutive secretion in polarized cells where there are two compositionally distinct plasma membranes, the apical and basolateral membranes: How is material destined for the apical membrane segregated from material destined for the basolateral membrane, and how is targeting to the different membranes achieved? Evidence exists that two different types of vesicles, each carrying its own distinct cargo and complement of SNARE proteins, bud from the same TGN and go either to the apical or basolateral membrane. Still, the coat proteins for these vesicles are unknown. Evidence also exists that some proteins, especially in hepatocytes, are first delivered to the basolateral membrane and then sorted into endocytic vesicles that carry selected material from the basolateral membrane to the apical membrane, a process called *transcytosis*.

Regulated secretory vesicles collect proteins that are to be stored in vesicles and released on demand. The secretion of insulin by pancreatic β-cells is an example of regulated secretion where the insulin is held in vesicles until blood glucose is elevated. The collection of proteins into regulated secretory vesicles appears to occur when the proteins selectively aggregate together at sites where the vesicles will form. The aggregates can contain several different types of proteins as passengers in the same vesicle. Thus, the sorting signal is part of the protein that encodes the ability of regulated secretory proteins to aggregate together. Proteins that do not join the aggregates are not segregated into the vesicles. A coat protein for regulated secretory vesicles remains to be identified.

Lysosomal Enzymes Are Targeted via a Mannose 6-Phosphate Signal

Lysosomes contain a wide variety of soluble hydrolytic enzymes capable of digesting most naturally occurring macromolecules. Lysosomal enzymes enter the lumen of the RER as typical secretory proteins containing a cleavable N-terminal signal peptide and receive typical N-linked oligosaccharides in the RER en route to folding and transport to the Golgi complex. To eventually reach lysosomes, a way must exist to select lysosomal enzymes from all the other secreted proteins and divert them to lysosomes. This process begins in the CGN with the phosphorylation of mannose residues on the N-linked oligosaccharides of lysosomal enzymes. The discovery of the pathway that lysosomal enzymes use to reach lysosomes is a great example of how basic and clinical sciences merge, and it begins with a disease called *inclusion body disease* (I-cell disease). Patients with I-cell disease secrete a wide variety of lysosomal enzymes,

rather than transporting them to lysosomes. Because so many different types of enzymes are affected, researchers realized that the origin of the disease was in a fundamental pathway used by cells to target an entire class of enzymes to their correct subcellular organelle. When the secreted lysosomal enzymes were collected from the medium of cultured cells from patients with I-cell disease and put in the medium surrounding normal cultured cells, the I-cell lysosomal enzymes were not taken up by the normal cells. However, when lysosomal enzymes from normal cells were put in the medium containing either normal or I cells, the normal enzymes were taken up by both cell types. This provided a clue that there was a structural difference between lysosomal enzymes from healthy individuals and from patients with I-cell disease. This difference was biochemically tracked to the presence of phosphate on mannose residues of normal enzymes, missing from the mannose residues of enzymes from patients with I-cell disease.

Subsequent studies identified the mannose 6-phosphate receptor, a transmembrane protein that binds to lysosomal enzymes that contain the mannose 6-phosphate sorting signal. Interestingly, the binding is strong at neutral pH but weak at acidic pH, providing an important clue to how the binding of lysosomal enzymes to the receptor is controlled by pH. The addition of phosphate to mannose residues occurs by a two-step process where N-acetylglucosamine phosphate is transferred from the UDP activate monosaccharide to the 6 position of mannose by the enzyme N-acetylglucosamine phosphotransferase, followed by cleavage of the N-acetylglucosamine, leaving the phosphate group on the mannose (Fig. 4–22).

In patients with I-cell disease, the phosphotransferase enzyme is defective; thus, lysosomal enzymes are not phosphorylated. The sorting signal for lysosomal enzymes appears to be different than other protein sorting signals discussed so far, such as signal peptides or the KDEL retrieval signal, because the signal is a phosphorylated carbohydrate. However, for the phosphotransferase to act, it must recognize some protein-based motif that is unique to lysosomal enzymes and not other secreted proteins. This signal is a common tertiary protein structure found only on correctly folded lysosomal enzymes. Thus, the sorting signal for lysosomal enzymes is ultimately encoded in the amino sequence of the proteins, just as with other sorting signals.

The pathway for transport of lysosomal enzymes from the TGN to lysosomes is illustrated in Figure 4–23. Lysosomal enzymes bearing the mannose 6-phosphate marker arrive together with other secretory proteins in the luminal fluid of the TGN. The pH in the TGN is slightly acidic but sufficient for the mannose 6-phosphate receptor to bind lysosomal enzymes and recruit them into vesicles that bud from the TGN. Vesicle budding uses a coat protein complex called *clathrin* that

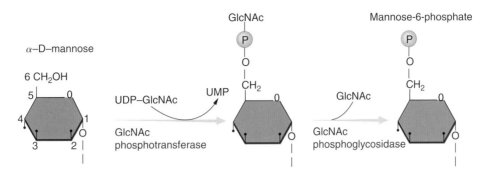

Figure 4–22. Biosynthesis of mannose 6-phosphate on lysosomal enzymes. The addition of mannose 6-phosphate occurs in two steps within the *cis*-Golgi network. First, N-acetylglucosamine-phosphotransferase uses UDP-GlcNAc to add GlcNAc phosphate in a phosphodiester linkage to the 6 position of mannose residues on N-linked oligosaccharides of a lysosomal enzyme. Next, a phosphoglycosidase removes the GlcNAc, leaving mannose phosphorylated on the 6 position.

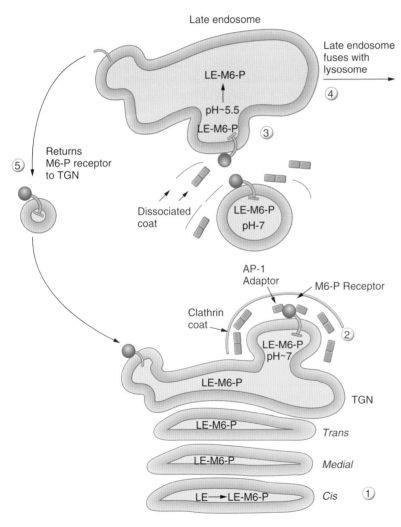

Figure 4–23. Mannose 6-phosphate (M6-P) pathway for delivery of lysosomal enzymes to lysosomes. *(1)* Lysosomal enzymes receive the M6-P marker in the *cis*-Golgi network (CGN) and migrate through the Golgi cisternae with other secretory enzymes to the *trans*-Golgi network (TGN). *(2)* The M6-P receptor in the TGN binds lysosomal enzymes via the M6-P marker, segregating lysosomal enzymes from other soluble secretory proteins. The clathrin/adaptor protein 1 (AP-1) adaptor coat is recruited by interactions with the small guanosine triphosphate (GTP)–binding protein ARF (not shown) and collects M6-P receptors with associated lysosomal enzyme cargo in the budding vesicles. *(3)* After vesicle formation and loss of the coat, the vesicles fuse with late endosomes where the low pH dissociates lysosomal enzymes from the M6-P receptor. *(4)* The lysosomal enzymes subsequently reach lysosomes when the late endosome fuses with existing lysosomes, or matures into a lysosome. *(5)* The M6-P receptor is returned to the TGN in vesicles that bud from the TGN, but the coat involved in this branch of the pathway is not yet known.

has the prototypical features of coat proteins discussed earlier. The initial interaction of the clathrin coat with the membrane relies on the small GTP-binding protein Arf1. Recall that Arf1 is also used in binding of COPI coats to Golgi membranes and, as we will see, to initiate binding of still other coats to membranes. How a single Arf1 protein functions in the binding of different coats to different membranes is unclear. The clathrin coat contains an inner layer, called *adapter protein 1* (AP-1), consisting of one copy of four different subunits. AP-1 binds to the cytosolic domains of cargo proteins, including the mannose 6-phosphate receptor, thus coupling lysosomal enzymes to the coat machinery. The outer layer of the coat contains a protein called *clathrin* composed of one heavy and one light chain that assemble into a trimer containing three legs called a *triskelion*. Triskelions associate to form the outer layer of the clathrin coat (see Fig. 4–23). Once the clathrin-coated vesicle buds, the coat dissociates and the vesicle fuses to late endosomes using Rab and SNARE proteins unique to this type of vesicle.

The pH within late endosomes is acidic (~5.0–5.5), established by ATP-dependent proton pumps in the endosomal membrane. The affinity of the mannose 6-phosphate receptor for mannose 6-phosphate is pH dependent, and at the low pH of the late endosome, the affinity is low, dissociating lysosomal enzymes from the receptor. This is an important driver for targeting lysosomal enzymes and underlies the asymmetric unloading of the enzymes in late endosomes, which releases the unoccupied receptor to recycle in a return vesicle to the TGN. The use of pH gradients to uncouple receptors and ligands is a theme seen again in the endocytic pathway. It is unclear what type of vesicle carries the unoccupied mannose 6-phosphate receptors back to the TGN. Under conditions where drugs block the acidification of late endosomes, lysosomal targeting to lysosomes is disrupted because all the receptors remain occupied with enzymes. The late endosome with its cargo of lysosomal enzymes eventually fuses with a lysosome and the phosphate groups on the mannose are cleaved off, thus removing the targeting signal.

At least two other types of coat protein complexes are known on the TGN, the GGA coat and the AP-3 complex. The GGA coat is an adapter layer that associates with clathrin as the outer layer and appears to also function in the transport of lysosomal enzymes from the TGN to lysosomes, although it may do other things as well. AP-3 is a third type of adaptor structurally related to AP-1 and may mediate direct delivery of some enzymes to lysosomes. Hermansky–Pudlak syndrome type 2, a disease associated with pigmentation defects, immunodeficiency, and blood disorders, is caused by a defect in one subunit of the AP-3 adaptor. Much unknown remains about leaving the TGN and lysosomal protein targeting; for example, lysosomal enzymes in hepatocytes from patients with I-cell disease correctly target the enzymes, perhaps because an alternate pathway of vesicle transport is used.

ENDOCYTOSIS, ENDOSOMES, AND LYSOSOMES

Endocytosis is the general term describing vesicles that form at the plasma membrane, carrying material into the cell. Endocytosis is unique because it is the portal by which many physiologically important molecules are brought into the cell, and it is often exploited by pathogens seeking access to the cell interior. There are different types of endocytosis, some found only in certain cell types, whereas others are found in almost all cells. Phagocytosis refers to the actin-dependent formation of large (≥1 micron) vesicles by macrophages and neutrophils that is frequently used to engulf large particles, such as invading bacteria. The large vesicles, called *phagosomes*, fuse with lysosomes, exposing ingested material to destructive lysosomal enzymes. Other types of endocytosis are divided into two general categories, clathrin-independent and clathrin-dependent endocytosis. Several types of clathrin-independent endocytosis exist. Macropinocytosis refers to a type of clathrin-independent endocytosis whereby extensions of the plasma membrane fuse to engulf a volume of extracellular fluid. A second type of clathrin-independent endocytosis forms vesicles that bud into the cytoplasm from sites on the membrane called *lipid rafts* that are enriched in cholesterol and sphingolipids. Yet a third type is the formation of small vesicles at lipid raft sites in association with a protein called caveolin, particularly abundant in endothelial cells, adipose cells, and fibroblasts. The mechanisms and functions of the various types of clathrin-independent endocytosis are not well understood. Clathrin-dependent endocytosis, as the name suggests, is the formation of endocytic vesicles using the coat protein clathrin. This form of endocytosis is active in almost all cell types and is detailed in the following section.

Clathrin-Dependent Endocytosis

A typical mammalian cell internalizes about half of its plasma membrane surface area per hour, mostly by formation of 50- to 100-nm clathrin-dependent vesicles. When a vesicle forms, extracellular material is brought into the cell in two ways, either as a soluble component in the fluid that happens to be engulfed by the vesicle or as a substance that has bound to a receptor in the membrane where the vesicle forms. Internalization in the fluid phase is classically called *pinocytosis* (cell drinking), whereas internalization of receptor-bound material is called *receptor-mediated endocytosis*. Receptor-mediated endocytosis is a much more efficient process than pinocytosis because the material taken up is concentrated on the membrane where vesicles bud. It is not surprising, therefore, that cells place many different types of specific receptors in the plasma to capture extracellular nutrients and signaling molecules by receptor-mediated endocytosis.

The general pathway of clathrin-dependent, receptor-mediated endocytosis conforms closely to the paradigm of vesicular traffic described in this chapter (see Figs. 4–15 and 4–16). The small GTP-binding protein that initiates coat binding to the cytosolic side of the plasma membrane is called Arf1. The coat consists of an inner layer of adaptor protein 2 (AP-2) and an outer clathrin layer. AP-2 has close structural similarities to the AP-1 adaptor used at the TGN. AP-2 has binding sites for the cytosolic domains of transmembrane receptors, ensuring that they will be efficiently included in the vesicles. About 2% of the cytoplasmic surface of many cells is covered at any time with clathrin/AP-2 coats, forming clathrin-coated pits that are visible by electron microscopy as electron-dense areas beneath the membrane. Some receptors constantly cluster at these sites via interactions with the AP-2, whereas other receptors do not expose AP-2 binding sites and do not associate with the pits until a ligand binds on the extracellular side. Different types of receptors may bind to the same clathrin-coated pits, hence entering the same vesicle.

After budding, the clathrin/AP-2 disassociates, and the vesicles either fuse with each other to form an early endosome or fuse with an existing early endosome, regulated by typical Rab proteins and SNARES. Early endosomes contain active proton-translocating ATPases that pump in protons from the cytosol, acidifying the endosome interior. As with the mannose 6-phosphate receptor and its ligand mannose 6-phosphate described at the TGN, the affinity of many (but not all) receptor-ligand pairs is regulated by pH, and at low pH, the receptor and ligands dissociate, releasing the ligand into the fluid within the endosome, whereas the unoccupied receptor remains in the membrane. This separation of ligand and receptor is a key event that allows the receptors to recycle back to the plasma membrane to initiate another round of endocytic uptake. Recycling begins from the early endosome and is believed to occur by the budding vesicles from endosomes that fuse with the plasma membrane, although coat proteins for this branch of the pathway have not been identified. Removal of membrane from the early endosome remodels the membrane, turning the early endosome into a late endosome. Recall that late endosomes are recipients of lysosomal enzymes from the TGN, thus bringing ligands released from receptors into the endosomal fluid together with digestive enzymes. The late endosomes then fuse with lysosomes, completing the delivery of endocytosed material together with lysosomal enzymes to lysosomes. Figure 4–24 provides an overview of the clathrin-dependent endocytosis pathway. To illustrate the endocytic pathways, we next look in more detail at the receptor-mediated uptake of two ligands, low-density lipoprotein (LDL) and transferrin.

Receptor-Mediated Endocytosis of Low-Density Lipoprotein and Transferrin

LDL is a lipoprotein particle in blood of 20 to 25 nm in diameter that contains an outer monolayer of phospholipids and a protein, apolipoprotein B-100 (apoB-100), that surrounds a core of cholesterol esters. LDL carries dietary cholesterol to cells and is taken up by receptor-mediated endocytosis. The LDL receptor is a single-pass transmembrane protein in the plasma membrane that has a binding domain on the cell exterior for apoB-100 and a cytosolic domain that binds the AP-2 adaptor. Studies of the human inherited disorder familial hypercholesterolemia identified mutations in the LDL receptor that contributed much to the current understanding of receptor-mediated endocytosis and cholesterol homeostasis. One class of LDL receptor mutants failed to cluster in coated pits. Analysis of the defective receptor demonstrated mutations that eventually defined a sequence, Asn-Pro-X-Tyr (NPXY in the single letter code, where X is any amino acid), that is a sorting signal for binding to the AP-2 adaptor. Interestingly, mutations in the subunit of the AP-2 adaptor that fail to bind the LDL receptor sorting signal, as well as other receptors that have this signal, have also been found.

Structural studies of the apoB-100 and the LDL receptor have demonstrated at the molecular level why LDL is released from the receptor at the low endosomal pH. There are histidine residues in a domain of the LDL receptor that when protonated interfere with apoB-100 binding, causing LDL release. LDL in the fluid of the vesicles is delivered to lysosomes, where the particle is degraded, releasing cholesterol for use by the cell. Studies of LDL uptake in patients afflicted with familial hypercholesterolemia also contributed to the understanding of important features of cholesterol regulation. Cells obtain cholesterol either from the diet, delivered by the LDL particles, or they make cholesterol *de novo*. When dietary cholesterol is available and delivered to cells, they sense the cholesterol level and a feedback loop suppresses cholesterol synthesis. However, in patients that have a defective LDL receptor, cells do not receive dietary cholesterol, even though it is in the blood, and make additional cholesterol to compensate. This inability to coordinate cholesterol synthesis with dietary availability greatly increases whole-body cholesterol and leads to premature atherosclerosis. Studies of familial hypercholesterolemia and the LDL system have made such important contributions to biology and medicine that the researchers most responsible for advances in the area, Drs. Joseph Goldstein and Michael Brown, received the Nobel Prize in Medicine and Physiology in 1985.

Cells obtain iron (Fe) by a variation in the receptor-mediated endocytosis theme. Transferrin is a blood protein that binds two atoms of Fe^{+3} and carries it to cells. Cells contain a transferrin receptor that typically clusters in coated pits via binding of the cytosolic domain to AP-2 adaptors, thus bringing the receptor and its cargo of iron-laden transferrin into cells. At the low pH in endosomes, however, the transferrin does not dissociate from the transferrin receptor; rather, the transferrin itself responds to the low pH with a conformational change that releases the cargo iron, which can then enter

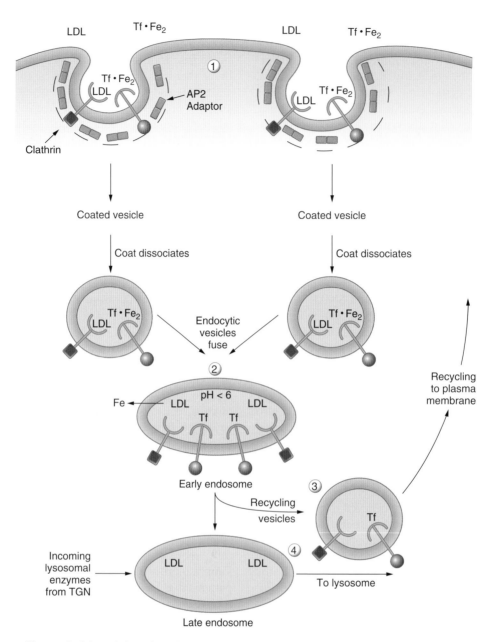

Figure 4–24. Clathrin-dependent endocytosis of low-density lipoprotein (LDL) and transferrin. *(1)* LDL and transferrin (Tf-Fe$_2$) each bind their respective receptors in the plasma membrane. The transmembrane receptors have peptide sequences in their cytosolic domains that bind to the adaptor protein 2 (AP-2) adaptor layer of the clathrin coat, and clathrin-coated vesicles bud from sites on the plasma membrane. *(2)* After the coat dissociates, the vesicles either fuse with themselves to form an early endosome or fuse with a preexisting early endosome. The pH within early endosomes is decreased by a proton-translocating ATPase, triggering the dissociation of LDL from its receptor and triggering the release of Fe from Tf-Fe$_2$ whereas the apolipoprotein (apo) Tf remains bound to the receptor. *(3)* Recycling vesicles then bud from the early endosome (or a related structure called a *recycling endosome,* not shown here) carrying the unoccupied LDL receptors and the apo Tf/Tf receptor complex back to the plasma membrane. *(4)* LDL in the lumen of the early endosomes goes with the fluid to the late endosome and eventually reaches the lysosome where it is degraded, releasing its cargo of cholesterol.

the cytoplasm. The transferrin without the bound iron, called *apotransferrin,* is recycled back to the plasma membrane with the receptor. At the neutral extracellular pH, the affinity of apotransferrin for the receptor is low, and the receptor is released into the extracellular fluid where it can bind more Fe^{+3}, acquiring a high

affinity again for transferrin receptors to begin another cycle of receptor-mediated endocytosis and iron delivery. Comparing the receptor-mediated pathways of LDL and transferrin shows how variations on a mechanistic theme are adapted to cellular needs. In the next section, yet another adaptation helps to explain how membrane

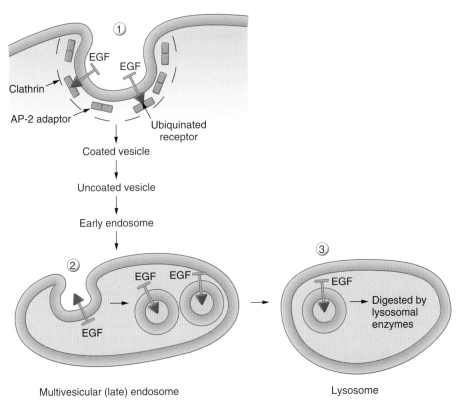

Figure 4-25. Multivesicular endosomes and epidermal growth factor (EGF) endocytosis. *(1)* EGF binds to the EGF receptor, a transmembrane protein that is ubiquinated on its cytosolic domain and that interacts with clathrin/adaptor protein 2 (AP-2) adaptors. After endocytosis and dissociation of the coat, the early endosome matures into a late endosome that can also become a multivesicular body. *(2)* The vesicles within multivesicular bodies are formed by invagination of the endosome membrane to form vesicles within a vesicle that are enriched in ubiquinated receptors, such as the EGF receptor. *(3)* The small vesicles within the multivesicular body are substrates for lysosomal enzymes when the endosome fuses with lysosomes, subjecting the vesicle membrane, and eventually the vesicle contents, to degradation. In this way, membrane-bound material originally brought into the cell by endocytic vesicles, which includes transmembrane receptors, can be digested.

components themselves are delivered in a digestible form to lysosomes.

Multivesicular Endosomes

Unlike the LDL and transferrin endocytic pathway, some ligands taken into the cell by receptor-mediated endocytosis are degraded in lysosomes, together with their receptors. For example, polypeptide hormones such as epidermal growth factor and its receptor are both degraded in lysosomes after receptor-mediated endocytosis. The destruction of both the hormone and its receptor transiently depress the sensitivity of cells to certain hormone signaling pathways because the number of hormone receptors on the cell surface goes down. Other membrane receptors are degraded in lysosomes as part of the natural turnover of cell components. This raises the question of how a segment of the endosomal membrane, together with bound ligands, could be delivered to lysosomes for digestion. Normally, the lysosomal membrane itself is protected from attack by lysosomal enzymes by a luminal layer of digestion-resistant glycoproteins. Electron micrographs of endosomal organelles showed more than 50 years ago the presence of small vesicles inside of larger endosomal vesicles, called *multivesicular bodies* or *multivesicular endosomes*. Recent advances have led to a better understanding that the vesicles inside endosomes are part of a regulated pathway to present membrane components to lysosomal enzymes for degradation.

Certain receptors in the plasma membrane, among them the epidermal growth factor receptor, are mono-ubiquinated in their cytosolic domains by a series of reactions that covalently attach the small protein ubiquitin to lysine residues. The ubiquitination serves as a signal for a variety of activities related to protein degradation, as described in the following. AP-2 complexes contain a cargo binding site for ubiquinated proteins to connect them to endocytic vesicles forming at the plasma membrane. After endocytosis, ubiquinated receptors are collected together in the endosomal membrane at sites where vesicles will bud into endosomes. Several protein complexes have been discovered that contribute to parts of the budding process, including ubiquitin binding proteins to select proteins to be included in the vesicles, and proteins that may help deform the membrane to bud vesicles into the endosome interior. Once within endosomes, the loose vesicles are delivered to lysosomes where the vesicles are digested (Fig. 4-25).

The space inside endosomes is equivalent to the extracellular space, and budding vesicles into the interior of an endosome is topologically similar to the process used by membrane-enveloped viruses, including HIV, to bud outside the cell. Many viruses do not encode all the proteins needed for extracellular budding and instead commandeer components used in making multivesicular endosomes for their own budding process. This understanding of the relation between multivesicular body formation and the life cycle of pathogenic viruses has fueled work in the area because it may be possible to devise interventions that prevent virus budding, thus blocking transmission.

LYSOSOMES

Lysosomes have a rich history. They were first discovered in biochemical assays as a collection of acid hydrolases that moved together as a large particle when cells were broken open and the contents analyzed by centrifugation. The large particles were biochemically characterized and predicted to be subcellular organelles before the development of techniques for electron microscopy of cell samples. Lysosomes were subsequently visualized in cells by electron microscopy, verifying the hypothesis that they were a membrane-bound subcellular organelle. For this work, Christian de Duve shared the 1974 Nobel Prize in Physiology and Medicine with Albert Claude and George Palade, a Nobel Prize that marks the founding of modern cell biology.

Lysosomes are membrane-limited vesicles, 0.2 to 0.6 μm in diameter, that contain a wide variety of hydrolytic enzymes that have an acidic pH optimum. The pH within lysosomes is ~4.5, generated by proton-translocating ATPases in the lysosomal membrane that are structurally related to the mitochondrial ATPase that uses proton gradients to power ATP synthesis. The hydrolases include proteases, nucleases, glycosidases, lipases, and more, sufficient to degrade almost any naturally occurring macromolecule to its monomeric constituents. We have already seen how many lysosomal enzymes are targeted to late endosomes via the mannose 6-phosphate marker. Before reaching lysosomes, most lysosomal enzymes are in an enzymatically inactive proenzyme form to control their hydrolytic capacity. When the proenzymes reach lysosomes, preexisting lysosomal proteases usually cleave one or more strategic peptide bonds that activate the proenzymes. The lysosomal membrane itself is resistant to action of its hydrolytic contents because of membrane glycoproteins that coat the luminal side of the lysosomal membrane with a protective polysaccharide layer.

Lysosomal digestion products, such as amino acids, sugars, or nucleotides, are transported across the lysosomal membrane to the cytosol where they join existing nutrient pools. Thus, one function of lysosomes is to obtain nutrients, illustrated earlier for the delivery of dietary cholesterol to cells via the LDL pathway. Lysosomes, however, have been adapted to a variety of other functions. One function is defense from microorganisms via phagocytosis, followed by digestion in the macrophage equivalent of a lysosome, the phagosome. Another function is the extracellular release of lysosomal enzymes by sperm during fertilization, which aids the sperm in penetrating the egg. A third example is the process of autophagy (self-eating) whereby the cell deliberately digests some of its own subcellular components. This occurs as part of the normal turnover process, but it is also enhanced in starved cells where cell components are recycled to provide nutrients. Autophagy begins when a cup-shaped membrane segment wraps around a subcellular structure to form a closed vacuole. Subcellular structures such as mitochondrial fragments have been seen within autophagic vacuoles by electron microscopy. The source of the enveloping membrane is uncertain, but it is probably either the ER or an endosome. The vacuole eventually fuses with lysosomes, exposing the enveloped material to digestive enzymes.

More than 40 known lysosomal storage diseases each result from a genetically defective enzyme. We have already seen one of these, I-cell disease, which was crucial in discovering the mannose 6-phosphate pathway for targeting lysosomal enzymes to lysosomes. Tay–Sachs disease is another well-known example, characterized by the accumulation of the glycosphingolipid ganglioside, especially in nervous tissue where gangliosides are prevalent. The cause is a defective α-hexosaminidase that normally removes a sugar residue required for further catabolism of the ganglioside. Children with severe Tay–Sachs disease rapidly develop mental retardation, paralysis, and die within 3 to 4 years after birth. Once it was realized that the lysosomal storage diseases were caused by single enzyme defects, the idea of enzyme replacement therapy with functional enzymes developed. Applying this simple idea is not trivial because it requires that replacement enzymes injected into the patient be correctly targeted to lysosomes in affected cells, and not targeted to unwanted sites. Early attempts at this therapy found that injected lysosomal enzymes were rapidly cleared from the blood by cells that destroyed the enzymes before therapeutic effects occurred. This clearance was traced to certain carbohydrate residues on the enzymes that were signals for uptake and destruction of the enzymes by some cells. Modifying the undesirable carbohydrates to avoid premature destruction of the enzymes aided the therapeutic value, and enzyme replacement therapy has been partially successful in ameliorating some symptoms of Gaucher's disease, where there is a defective glucocerebrosidase, and Fabry disease, caused by a defective α-galactosidase. Enzyme replacement therapy is another example of how basic cell biology and medicine together are used to treat disease.

The Ubiquitin-Proteasome System Is Responsible for Nonlysosomal Protein Degradation

In 2004, Aaron Ciechanover and Avram Hershko, from the Technion Israel Institute of Technology, and Irwin Rose, retired from the Fox Chase Cancer Center, shared the Nobel Prize in Chemistry for their landmark discovery of the ubiquitin-proteasome system. The ubiquitin-proteasome system is the major nonlysosomal pathway for ATP-dependent protein degradation. First, let us look at the mechanism of protein ubiquitination and then the function of the proteasome in protein degradation.

Ubiquitin is an 8.6-kDa highly conserved polypeptide found both free and covalently conjugated to other proteins in all eukaryotic cells. Protein ubiquitination is a posttranslational modification of target proteins, a dynamic process involving ubiquitin conjugating enzymes and deubiquitination enzymes. As shown in Figure 4–26, the conjugation of ubiquitin to target proteins is a multistep process.

A cascade of reactions catalyzed by several classes of enzymes is required to form an isopeptide bond between the C-terminal glycine of ubiquitin and a lysine ε-amino group of the target acceptor protein. Ubiquitin-activating enzyme (E1) uses ATP and forms a high-energy thioester bond between its active site cysteine and the C-terminal glycine of ubiquitin via formation of ubiquitin-adenylate, an activated ubiquitin intermediate. The activated ubiquitin is next transferred from E1 to an active site cysteine of the ubiquitin-conjugating enzyme (E2). The final step, formation of an isopeptide linkage between ubiquitin and the target protein, is sometimes catalyzed by an ubiquitin-protein ligase (E3). In general, E3 binds both E2 and the target protein, bringing them into close proximity. Ubiquitin is then transferred from E2 to the substrate either directly or via an E3-thioester intermediate. Two distinct E3 families, containing conserved protein domains, have now been identified. Members of the HECT (homologous to E6-AP carboxyl terminus) domain family, can form

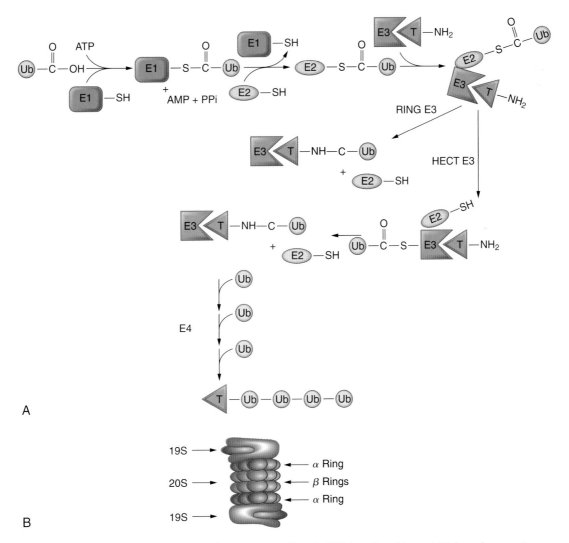

Figure 4–26. The ubiquitination pathway. **A:** Free ubiquitin (Ub) is activated in an ATP-dependent reaction with the formation of a thioester intermediate between E1 and the C terminus of ubiquitin. Ubiquitin is then transferred to an E2, again forming a thioester linkage. There are two classes of E3 enzymes. The RING domain E3s bind to the target (T) protein and directly transfer ubiquitin from the E2 to the target. HECT domain E3s form a thioester intermediate between ubiquitin and itself before transferring ubiquitin to the target. An E4 then adds a series of ubiquitins by linking through lysine 48 of one ubiquitin to the C terminus of the next. (**B**) Structure of the proteasome.

thioester intermediates with ubiquitin and conjugate it to substrate. Members of the other class, RING domain E3s, are now believed to mediate the direct transfer of ubiquitin to target proteins. Distinct E3s interact with a subset of E2s, suggesting that specific E2/E3 tandems might contribute to the specificity of protein ubiquitination. An additional conjugating factor, named E4, is required for polyubiquitination.

Ubiquitination is involved in many essential and diverse cellular processes such as cell-cycle progression, transcriptional activation, and apoptosis. For most known substrates, polyubiquitination leads to protein degradation by the 26S proteasomal complex. The poly-ubiquitin chain, linked by four or more ubiquitin molecules, is recognized by the proteasomes 19S regulatory complex, and the substrate protein is degraded by the 20S core complex. As shown in Figure 4–26B, the 20S complex has a barrel shape with four stacked rings. There are two identical outer α rings and two inner β rings. Each ring contains seven distinct subunits and the catalytic sites are associated with specific β subunits. Initially, the polyubiquitinated protein associates with the 19S complex via the ubiquitin chain. This binding widens the α ring, allowing entry of the protein into the barrel of the 20S core complex. The protein is degraded into small peptides, which are subsequently degraded into amino acids by amino and carboxypeptidases within the cytoplasm. The polyubiquitin chain is released and disassembled via the action of specific deubiquitinating enzymes.

Ubiquitin and ubiquitin-related proteins also regulate other cellular function not directly involving protein degradation, including translation, activation of transcription factors and kinases, activation of DNA repair, and regulation protein interactions. As discussed in Chapters 9 and 10, the dysfunction of the ubiquitin-proteasome system plays an important role in several cancers and neurologic disorders.

MITOCHONDRIA

One of the main functions of mitochondria is to provide the cell with ATP generated by oxidative phosphorylation (see Fig. 4–31). Mitochondria are also involved in many other metabolic functions, including heme biosynthesis, synthesis of iron/sulfur (Fe/S) clusters, steroid synthesis, metabolism of fatty acids, regulation of the cellular redox state, calcium homeostasis, amino acid metabolism, carbohydrate metabolism, and protein catabolism. In addition, mitochondria are the central regulators of programmed cell death (see Chapter 10). Some mitochondrial functions are performed only in specific cells. For example, mitochondria in liver cells contain enzymes that detoxify **ammonia,** a waste product of protein metabolism.

Electron microscopy of mitochondria indicates a roughly ellipsoid organelle with a length of 1 to 2 μm

and a width of 0.1 to 0.5 μm (Fig. 4–27). Mitochondria when viewed in living cells with a fluorescent dye, such as rhodamine 123, are highly dynamic organelles, changing shape, fusing, dividing, and moving. Mammalian cells typically contain hundreds or thousands of mitochondria; the number of mitochondria per cell appears to depend on the metabolic requirements of that cell.

Mitochondria are surrounded by two membranes: an outer and an inner membrane. This creates two compartments: an intermembrane space between the inner and outer membranes and the matrix, which is the compartment enclosed by the inner membrane (see Fig. 4–27). The outer membrane contains a major integral protein, called *porin.* Porins can form channels within the outer membrane through which molecules that are less than 5000 daltons can pass freely. The inner membrane contains proteins involved in the respiratory chain, ATP production, and the transport of small molecules and ions across the inner membrane. The inner membrane is invaginated to form **cristae,** which increase its surface area, thereby enhancing its ability to generate ATP. The inner mitochondrial membrane has a high concentration of the phospholipid, cardiolipin (diphosphatidylglycerol), which is believed to decrease the permeability of this bilayer to small ions. The matrix contains enzymes involved in the oxidation of pyruvate and fatty acids, as well as most of the enzymes of the citric acid cycle. In addition, the mitochondrial genome and ribosomes for protein synthesis are located in the matrix. The intermembrane space contains a number of proteins, including cytochrome *c,* which plays a critical role both in electron transport and in programmed cell death.

ATP Production by Oxidative Phosphorylation

One of the main functions of mitochondria is the generation of ATP via oxidative phosphorylation. In 1961, Peter Mitchell proposed that the production of ATP within mitochondria was powered by a mechanism that he called *chemiosmotic coupling.* To understand chemiosmotic coupling, we begin in the cytosol with the formation of pyruvate and fatty acyl-CoA. Glucose is converted to pyruvate in the cytosol by the glycolytic pathway (Fig. 4–28).

The net reaction is

$$\text{Glucose} + 2\ \text{NAD}^+ + 2\ \text{ADP} + 2\ \text{P}_i \rightarrow$$
$$2\ \text{Pyruvate} + \boxed{2\ \text{NADH}} + 2\ \text{ATP} + 2\ \text{H}^+ + 2\ \text{H}_2\text{O}$$

The two reduced nicotinamide adenine dinucleotide (NADH) molecules formed by glycolysis are reoxidized to NAD$^+$ by transfer of their electrons to complexes of the electron transport chain located in the mitochondria. Pyruvate also enters the mitochondrial matrix,

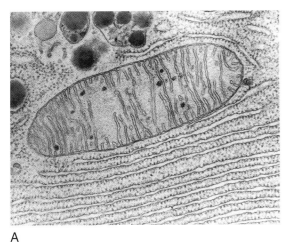

Figure 4-27. Structure of a mitochondrion. **A:** Transmission electron micrograph of a mitochondrion in a human pancreatic acinar cell. Mitochondria are surrounded by two membranes, an outer and an inner membrane. The inner membrane has many invaginations, called *cristae*. Several granules can be seen in the matrix. (Courtesy Keith R. Porter/Photo Researchers, Inc.) **B:** Diagrammatic presentation of mitochondrial compartments and membranes.

A

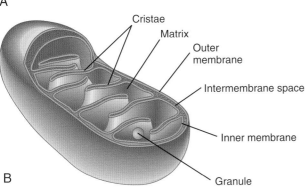

Cristae

Matrix

Outer membrane

Intermembrane space

Inner membrane

Granule

B

where it is converted to acetyl-CoA, the key substrate for the citric acid cycle. Another source of mitochondrial acetyl-CoA is the oxidation of fatty acids. Free fatty acids, which are stored primarily in adipose tissue as triacylglycerol, can be released into the bloodstream. Free fatty acids can cross the plasma membrane of cells and are converted in the cytosol to fatty acyl-CoA, which can be transported into the mitochondrion. Within the mitochondrial matrix, free fatty acyl-CoA is broken down by a cycle of reactions (Fig. 4–29) that removes two carboxyl carbons and produces one acetyl-CoA molecule per cycle.

The acetyl-CoA formed by the oxidation of pyruvate and fatty acyl-CoA fuels the citric acid cycle, also known as the *Krebs cycle* or *tricarboxylic acid cycle* (Fig. 4–30). During the citric acid cycle, acetyl-CoA is oxidized to two molecules of CO_2 (which are released from the cell), and the released electrons are transferred to NAD and flavin adenine dinucleotide (FAD). The net reaction for the citric acid cycle is

$$\text{Acetyl-CoA} + 3\ NAD^+ + FAD + GDP + P_i + 2\ H_2O \rightarrow$$
$$2\ CO_2 + \boxed{3\ NADH} + \boxed{FADH_2} + GTP + 2\ H^+ + \text{CoA-SH}$$

The 3 NADH + $FADH_2$ molecules formed by the citric acid cycle and the NADH molecules generated during glycolysis can now transfer electron pairs to acceptor molecules on the inner mitochondrial membrane, eventually leading to the reduction of oxygen and the formation of water.

This brings us to the steps of chemiosmotic coupling. When NADH (or $FADH_2$) is oxidized, it releases a hydride ion (H^-). The hydride ion is immediately converted to a proton (H^+) and two high-energy electrons ($2e^-$). It is these two high-energy electrons that reduce molecular oxygen to water during mitochondrial respiration. During this process, there is generation of approximately 53 kcal/mol of free energy, which would be lost as heat if the transfer was direct. Instead, the transfer of electrons from NADH (or $FADH_2$) to oxygen is catalyzed by a series of electron carriers associated with four multiprotein complexes located in the inner mitochondrial membrane. These multiprotein complexes are presented schematically in Figure 4–31. The two high-energy electrons from NADH are initially transferred to the NADH-ubiquinone oxidoreductase complex (Complex I). Complex I has a molecular weight of approximately 950,000 and consists of approximately 46 polypeptides. The high-energy electrons are transferred by a flavin and several iron/sulfur (Fe/S) prosthetic groups attached to the protein complex to

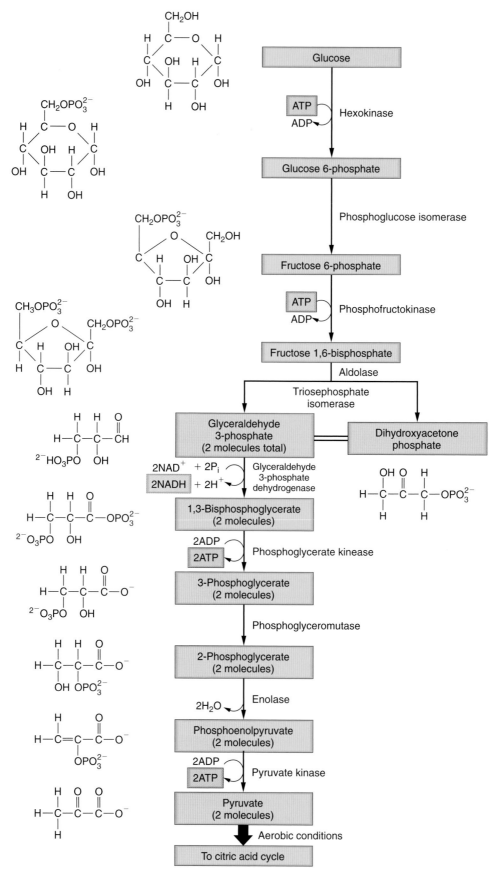

Figure 4–28. **Glycolytic pathway.** The breakdown of glucose into pyruvate yield two molecules of NADH and two molecules of ATP.

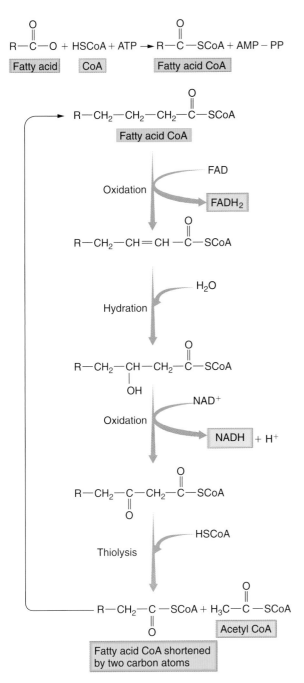

Figure 4–29. Pathway for the oxidation of fatty acids in mitochondria.

oxidoreductase complex or Complex III). The cytochrome b-c_1 complex is a dimer of identical monomers, each with 11 different subunits. There are three cytochromes and one Fe/S group per monomer. The cytochromes contain a bound heme group with an Fe^{3+} (ferric) iron atom that converts to an Fe^{2+} (ferrous) state when it accepts an electron. The cytochromes and Fe/S complex transfer the pair of electrons from ubiquinone to cytochrome c. Cytochrome c, which is attached to the inner mitochondrial membrane facing the intermembrane space, transfers the electrons to the cytochrome c oxidase complex (Complex IV). Cytochrome c oxidase contains 13 polypeptides that associate to form two attached proteins. Each protein monomer contains two cytochromes and two copper atom carriers that accept the electrons from cytochrome c and transfer them to oxygen.

As the high-energy electron pair is transferred to Complexes I, III, and IV, protons from the matrix are pumped into the intermembrane space (see Fig. 4–31). The inner membrane is impermeable to ions such as H^+, but because of the porin channels, once in the intermembrane space, the H^+ can freely pass into the cytosol. This movement of protons out of the matrix has two effects: the membrane potential of the inner membrane of the mitochondria is 160 mV and is negative on the matrix side, and the pH within the cytoplasm and intermembrane space (pH 7) is approximately 1 pH unit lower than in the matrix (pH 8). Hence, the first aspect of the chemiosmotic coupling theory is that the transfer of electrons from NADH (or $FADH_2$) down the electron transport chain causes protons to be pumped out of the matrix. This pumping of protons causes a proton motive force (PMF) to form, which is the additive force placed on a proton in the intermembrane space to move down its electrochemical gradient (proton concentration gradient + Δ membrane electric potential). In mathematical terms:

$$PMF = \psi - \left(\frac{2.3(RT)}{F} \times \Delta pH\right)$$

where

Ψ = membrane potential = −160 mV across inner membrane (negative inside matrix)
R = gas constant = 1.987 cal/degree · mol
T = temperature (°K)
F = Faraday constant = 23.062 cal/mV · mol

∴ at 37°C,

$$PMF = \Psi - (60 \times \Delta pH) = 160 - (60 \times (-1)) = 220 \text{ mV}$$

Therefore, protons have a PMF of approximately 220 mV at 37°C, pulling them across the inner membrane and back toward the matrix. How does this PMF drive ATP synthesis? This is accomplished by the ATP synthase (Complex V) of the oxidative phosphorylation system. The mitochondrial ATP synthase is a

coenzyme Q. Coenzyme Q, also known as ubiquinone, accepts the two electrons and is converted to ubiquinol. Electrons can also be transferred to coenzyme Q via the succinate-ubiquinone oxidoreductase complex (Complex II), which transfers high-energy electrons created during the conversion of succinate to fumarate from $FADH_2$ to an Fe/S center and then to ubiquinone. The hydrophobic ubiquinol can then move laterally in the membrane and transfer the electrons to the cytochrome b-c_1 complex (also called the ubiquinol-cytochrome c

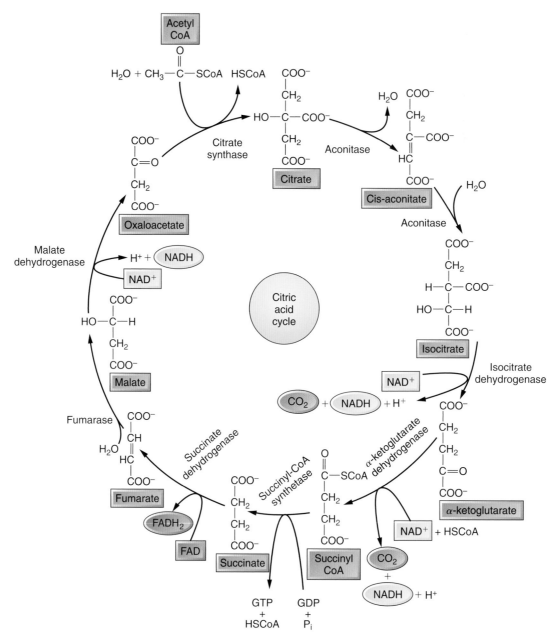

Figure 4–30.   Citric acid cycle.

membrane-bound multisubunit enzyme that couples the PMF across the inner mitochondrial membrane to the synthesis of ATP in the matrix. The ATP synthase is organized into two components, F_0 and F_1, which are linked together by a central and a peripheral stalk (Fig. 4–32). The F_1 region is the catalytic region and consists of five different subunits, with a stoichiometry of $\alpha_3\beta_3\gamma\delta\epsilon$. The F_0 region spans the inner membrane and contains a proton channel through which protons can flow back from the intermembrane space to the matrix. The PMF drives the passage of protons through the inner membrane. This passage of protons causes a ring of c subunits in the membrane region of F_0 to rotate. The c-ring

is attached to the central stalk (consisting of the γ, δ, and ϵ subunits), which also rotates. Rotation of the central stock within the $\alpha_3\beta_3$ subcomplex of F_1 causes the three catalytic β subunits to go through a series of conformational changes that leads to ATP synthesis. The ATP synthase, therefore, converts ADP + inorganic phosphate (P_i) to ATP and is powered by protons moving down their electrochemical gradient into the matrix, powered by the PMF. Therefore, the second point of the chemiosmotic coupling theory is that the PMF causes movement of protons down their electrochemical gradient by the ATP synthase, and this movement drives ATP synthesis.

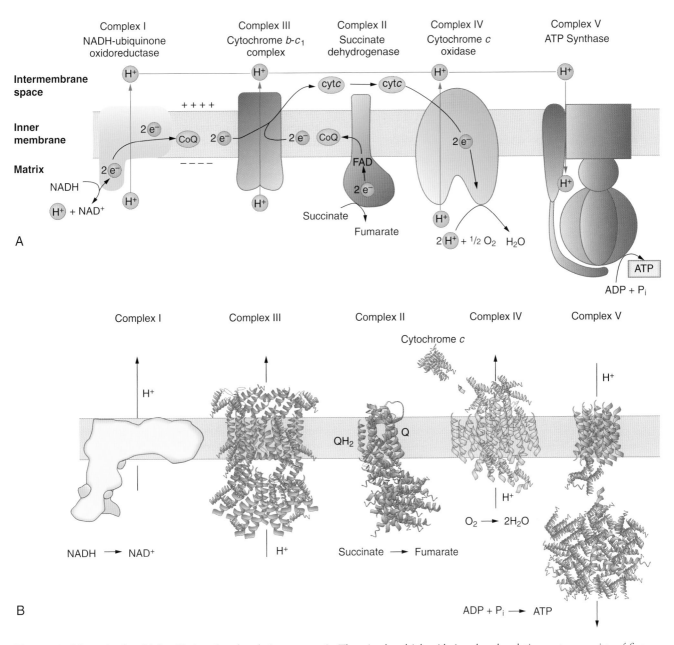

Figure 4–31. Mitochondrial oxidative phosphorylation system. **A:** The mitochondrial oxidative phosphorylation system consists of five complexes. Electrons from NADH enter the electron transport chain in Complex I. Electrons from succinate enter the electron transport chain via FADH₂ in Complex II. The electrons are then transferred from either Complex I or II to coenzyme Q (ubiquinone), which passes the electrons to Complex III. Electrons are then transferred to cytochrome *c*, then to complex IV, and then to molecular oxygen to form H₂O. Protons are pumped from the matrix to the intermembrane space in complexes I, III, and IV, which results in a proton gradient across the inner membrane. Movement of protons back to the matrix through Complex V is used to drive ATP synthesis. **B:** Structures of the protein complexes of the oxidative phosphorylation system (either mitochondrial or bacterial versions) are shown. The tertiary structure of Complex I has not yet been determined. (*From Brian E. Schultz and Sunny I. Chan, reprinted with permission from* The Annual Review of Biophysics Biomolecular Structure, *Volume 30 © 2001 Annual Reviews Inc.*)

The movement of protons down their electrochemical gradient is also coupled to the movement of small molecules and ions, such as P_i, ATP, ADP, Ca²⁺, and pyruvate, into and out of the mitochondrial matrix by transport mechanisms (see Chapter 2). The mechanism for getting the newly synthesized ATP out of the mitochondria and into the cytosol is accomplished by the adenine nucleotide translocator by which the movement of ATP down its electrochemical gradient (from the matrix to the intermembrane space and into the cytosol)

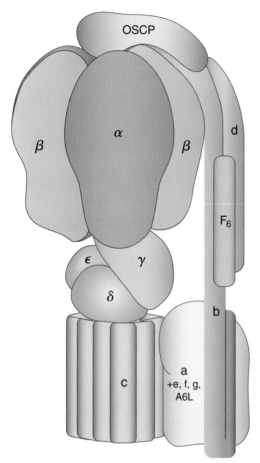

Figure 4–32. Structure of the mitochondrial ATP synthase. The mitochondrial ATP synthase (Complex V) is composed of two large oligomeric complexes, F_1 and F_0, which are linked together by a peripheral and a central stalk. The F_0 portion is membrane embedded forming a channel through which the protons can cross the membrane. The F_0 base consists of three different polypeptides, with a stoichiometry of ab_2c_{10-14}. The catalytic F_1 contains five different subunits: 3α, 3β, 1γ, 1δ, and 1ε. The α (red) and β (yellow) subunits are arranged alternatively within the F_1 head. The γ, δ, and ε subunits form a central stalk linking the $\alpha_3\beta_3$ subcomplex of F_1 to a ring of c subunits of the F_0 region. Movement of protons from the intermembrane space to the matrix, driven by the proton motive force, causes the ring of c-subunits and the associated central stalk to rotate. Rotation of the central stock within the $\alpha_3\beta_3$ subcomplex of F_1 causes the three catalytic nucleotide-binding sites to go through a series of conformational changes resulting in ATP synthesis. The peripheral stock connects subunit a (ATPase subunit 6) of the F_0 region with the F_1 head and acts to hold the $\alpha\beta$ subunits in a fixed position. The peripheral stock of the bovine mitochondrial ATP synthase contains subunits b, d, F_6, and OSCP. Thus, the ATP synthase complex catalyzes the conversion of ADP + inorganic phosphate (P_i) to ATP with energy supplied by protons moving down their electrochemical gradient from the intermembrane space to the matrix. (*Adapted from Walker J. E. and Dickson V. K. 2006. The peripheral stock of the mitochondrial ATP synthase. Biochim Biophys Acta 1757:286–296, with permission.*)

is coupled to the movement of ADP in the reverse direction. The synthesis of ATP within the mitochondrion also requires phosphate ions (P_i), so P_i must be imported from the cytosol. This is mediated by another membrane transport protein, the phosphate transporter, which imports one phosphate ion together with one H^+ into the mitochondrial matrix. Figure 4–33 summarizes this section.

Mitochondrial Genetic System

Mitochondria contain their own genetic system. The human mitochondrial genome is a circular double-stranded DNA molecule made up of two complementary strands, heavy and light (Fig. 4–34).

The human mitochondrial DNA (mtDNA) has been completely sequenced and is approximately 16,569 base pairs in length. It contains 37 genes, 28 of which are encoded by the heavy strand and 9 by the light strand. Human mtDNA encodes 2 rRNAs (12S and 16S rRNA) and 22 tRNAs. It also encodes 13 proteins involved in electron transport and oxidative phosphorylation: 7 subunits of the NADH-ubiquinone oxidoreductase complex (Complex I), 1 subunit (cytochrome *b*) of the cytochrome *b*-c_1 complex (Complex III), 3 subunits of the cytochrome *c* oxidase complex (Complex IV), and 2 subunits of the ATP synthase complex (Complex V). There are approximately 2 to 10 mitochondrial genomes per human mitochondrion and multiple mitochondria per cell.

There are a number of interesting features of the human mitochondrial genome. First, almost the entire genome is coding sequence. Mitochondrial protein synthesis requires only 22 mitochondrially encoded tRNAs, which leads to the second distinctive feature of the mitochondrial genome. Many of the mitochondrial tRNAs recognize any one of four nucleotides in the third position of codons. Therefore, there is far greater "wobble" in the third codon position within mitochondrial mRNAs, leading to "two-out-of-three" pairing of tRNA. Lastly, the human mitochondrial genetic code differs from the "universal" genetic code (Table 4–1). For example, UGA encodes a stop codon in the universal genetic code but codes for tryptophan in human mtDNA.

There are several important differences between the genetics of mtDNA and of nuclear genes. One key difference is that the human mitochondrial genome is maternally inherited. In addition, mtDNA is polyploid, and individual cells contain hundreds or thousands of copies of mtDNA. During cell division, mitochondria together with their genomes are distributed to daughter cells more or less randomly. In addition, the mitochondrial genome has a much greater mutation rate (~10-fold greater) than that of the nuclear genome.

Defects in Mitochondrial Function Can Cause Disease

As discussed earlier, mitochondria play a key role in energy production and also participate in a number of other biological functions. Defects in any of these mito-

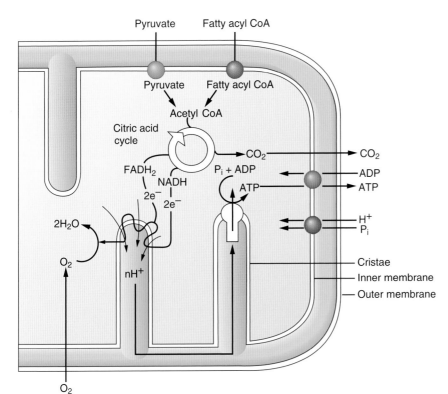

Figure 4–33. Summary of mitochondrial function. Pyruvate and fatty acyl-coenzyme A (CoA) are transported into mitochondria by specific transporter proteins and are metabolized to acetyl-CoA. Acetyl-CoA is then metabolized by the citric acid cycle. NADH and $FADH_2$ are produced by the citric acid cycle and electrons are transferred from NADH and $FADH_2$ to O_2 by a series of electron carriers in the inner mitochondrial membrane. A proton motive force is created by this electron transfer, and protons, moving back down their electrochemical gradient into the matrix, power the ATP synthase to produce ATP. ATP produced in the mitochondrial matrix is transported to the cytosol by the adenine nucleotide translocator (which exchanges ATP for ADP). Inorganic phosphate (P_i) is transported from the cytosol into the matrix of mitochondria by the phosphate transporter.

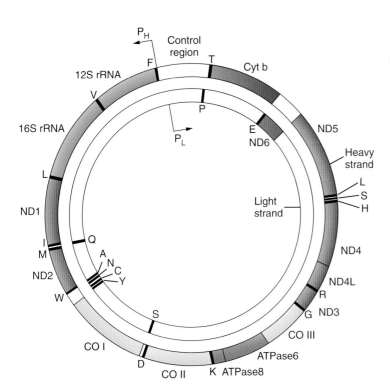

Figure 4–34. Organization of the human mitochondrial genome. Human mitochondrial DNA (mtDNA) is a circular double-stranded molecule and contains approximately 16,569 base pairs. Proteins and RNA encoded by each of the two strands (light and heavy) of human mtDNA are shown. The genes encoding the two ribosomal RNAs (rRNA; 12S and 16S rRNA) are shown in green. Genes encoding the 22 transfer RNAs (tRNAs) are denoted by the single letter code for their amino acids and are shown in black. *ND1, 2, 3, 4L, 4, 5,* and *6* indicate genes encoding subunits 1, 2, 3, 4L, 4, 5, and 6 of the NADH-ubiquinone oxidoreductase complex (Complex I) and are shown in red. *Cyt b* indicates the gene encoding cytochrome b of the cytochrome b-c_1 complex (Complex III) and is shown in purple. *COI, II,* and *III* represent genes encoding subunits 1, 2, and 3 of the cytochrome *c* oxidase complex (Complex IV) and are shown in yellow. ATPase6 and ATPase8 indicate genes encoding subunits 6 and 8 of the mitochondrial ATP synthase (Complex V) and are shown in blue. The control region contains the regulatory information for DNA replication and transcription: P_H, promoter heavy strand; P_L, promoter light strand.

chondrial processes can result in a mitochondrial disease. Because of their dual genetic control, mitochondrial diseases can be caused by mutations in either nuclear genes or mtDNA genes.

A wide range of human diseases have now been associated with mutations in the mitochondrial genome

(www.mitomap.org). Because mtDNA is inherited maternally, these diseases are usually inherited from the mother. Mitochondrial diseases caused by mtDNA mutations are clinically heterogenous; some mutations affect only a single tissue, whereas other mutations are multisystemic. Mutations in mtDNA can affect all of the

copies of the mtDNA in a cell (homoplasmic mutation) or only some copies (heteroplasmic mutation). With heteroplasmy, a minimum number of mutant mtDNAs must be present ("threshold effect") to cause mitochondrial dysfunction and clinical expression of the disease. In addition, because there is random segregation of mitochondria at mitosis, the proportion of mutant mtDNAs present in daughter cells can change. Thus, the clinical expression of a pathogenic mtDNA mutation depends on the specific mtDNA mutation, the proportion of mutant and normal mitochondria in a particular tissue, and the reliance of the tissue on mitochondrial ATP production. Tissues with high oxidative energy needs, such as the nervous system and muscle, tend to be the most affected by mutations in mtDNA.

Mutations in mtDNA include both large-scale mitochondrial DNA rearrangements and mtDNA point mutations. Several clinical syndromes are associated with mtDNA rearrangements, including Kearns–Sayre syndrome (KSS), progressive external ophthalmoplegia (PEO), and Pearson's marrow-pancreas syndrome. KSS, PEO, and Pearson's syndrome are the result of a large-scale mtDNA deletion that impairs mitochondrial protein synthesis. These mutations are heteroplasmic and usually arise spontaneously. Interestingly, PEO can also be caused by a mutation in one of several nuclear genes, including the *POLG* gene, which encodes the catalytic subunit of the mitochondrial DNA polymerase γ, and the *ANT1* gene, which encodes an adenine nucleotide translocator protein.

Mitochondrial diseases can also be caused by point mutations in mtDNA. mtDNA point mutations, in contrast with large-scale mtDNA rearrangements, are usually maternally inherited. Point mutations can occur in mitochondrial mRNA, tRNA, or rRNA genes. mtDNA point mutations can be either heteroplasmic or homoplasmic. Leber's hereditary optic neuropathy (LHON) is an example of a homoplasmic mtDNA point mutation. LHON is a rare, maternally inherited disease that results in blindness, usually in young adults, due to atrophy of the optic nerve. Approximately 90% of LHON patients carry one of three mtDNA missense mutations that alter a subunit of Complex I (NADH-ubiquinone oxidoreductase) of the oxidative phosphorylation system: 3460G>A *(ND1)*, 11778G>A *(ND4)*, and 14484T>C *(ND6)* (Fig. 4–35). LHON can also be caused by a missense mutation in the mtDNA gene encoding cytochrome *b*, a subunit of complex III of the oxidative phosphorylation system.

TABLE 4–1.   Differences between the Human Mitochondrial and the "Universal" Genetic Code

Codon	Universal Genetic Code	Mitochondrial Genetic Code
UGA	Stop	Trp
AUA	Ile	Met
AGA	Arg	Stop
AGG	Arg	Stop

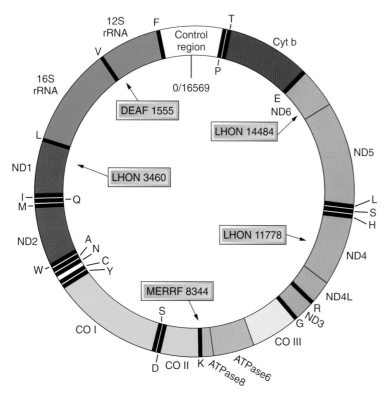

Figure 4–35. Human mitochondrial DNA (mtDNA) showing the location of several mutations that cause mitochondrial diseases. The human mitochondrial genome is shown. *Arrows* indicate the positions of several pathogenic mutations in mtDNA that result in a mitochondrial disease. *Numbers* indicate the nucleotide position of the mutation. DEAF, nonsyndromic deafness, LHON, Leber's hereditary optic neuropathy; MERRF, myoclonic epilepsy and ragged-red fiber disease.

Clinical Case 4-2

Peter Franklin is a 24-year-old first-year law student. Six months ago, after graduating from college, he celebrated and drank a bit too much with his roommates and actually passed out. However, he remembers no injuries and recovered fully after a "mega-hangover." Under considerable first-year student stress, he has recently found that one or two martinis help him to sleep after a long session with torts.

Three months ago, he noted some trouble reading his law texts. While watching Sunday football with his girlfriend, he determined that his vision problem was only in his left eye, because when she blocked his right eye view, he couldn't see the scrimmage line at all. His eye didn't hurt, and he didn't recall injuring it. However, because his older brother had complained of major vision problems before his death in a tragic auto accident 8 years earlier, he went to see an ophthalmologist at the student health center. The doctor found no evidence of conjunctivitis or damage to the surface of the eye, and corneal pressures were normal bilaterally. He told Peter that his retinas looked fine except for some minimal erythema of his left disk, with a few prominent capillaries. No signs of retinal hemorrhage visually or after an intravenous fluorescein test were evident. The right eye was completely normal, as was his blood thiamin level.

During the next 2 months, Peter's vision deteriorated progressively, until he had a complete loss of central vision on the left. He found that he was reading his law texts entirely with his right eye and, in fact, that he was more comfortable when he covered his left eye. He could only appreciate peripheral motion on the left, and occasionally found himself bumping into a door frame when he was entering from the left.

He sought treatment 6 months later with minimal, but similar, symptoms in his right eye. He no longer feels confident driving his car.

Cell Biology, Diagnosis, and Treatment of Leber's Hereditary Optic Neuropathy

LHON is a painless optic atrophy. It is caused by a point mutation in mitochondrial DNA. About 90% of LHON cases have a point mutation in either the *ND1*, *ND4*, or *ND6* genes, which encode subunits of the NADH-ubiquinone oxidoreductase, a critical enzyme of the mitochondrial oxidative phosphorylation system (see Fig. 4–35). Because the defect is in a mitochondrial gene, Peter's mother was the carrier. Though perforce both sons and daughters inherit the genetic defect from the mother, for reasons that are obscure, only about one-half of sons and one-tenth of daughters with the mutation develop vision loss.

Additional genetic or environmental factors must be involved. In retrospect, it seems likely that Peter's brother's visual problems were due to the same disease, and that they may have caused his automobile accident.

This disorder usually can be distinguished from toxic optic atrophy by family history, and in the case of the toxic optic neuropathy of alcoholism, by thiamin levels. It can be distinguished from optic neuritis by the absence of pain, especially on eye movement. Also, optic neuritis, which may be a precursor of multiple sclerosis, is primarily found in women. Leber's optic neuropathy usually develops in young adulthood and becomes bilateral. There is no treatment.

Point mutations in mitochondrial tRNA and rRNA genes can also result in mitochondrial diseases. For example, myoclonic epilepsy and ragged-red fiber disease (MERRF) is caused by a point mutation in the TΨCG loop of the mitochondrial tRNA lysine gene. In addition, nonsyndromic deafness (DEAF) can be caused by a point mutation in the 12S rRNA gene. Figure 4–35 illustrates several mitochondrial diseases that are caused by pathogenic point mutations in mtDNA.

Mutations in nuclear genes that encode proteins important for mitochondrial function can also result in mitochondrial diseases. In contrast with mutations in mtDNA, these mutations show Mendelian inheritance. A number of nuclear DNA mutations have now been identified that affect a wide variety of mitochondrial processes. Examples include mutations in genes that encode structural or assembly factors of the oxidative phosphorylation system, genes that encode proteins involved in nuclear-mitochondrial communication, genes that affect mitochondrial fusion or mobility, genes that affect mitochondrial protein import, genes that regulate the structure of the lipid membrane, and genes that regulate transport across the inner mitochondrial membrane. Interestingly, several neurodegenerative disorders, including Friedreich's ataxia, hereditary spastic paraplegia, and Wilson's disease, are caused by mutations in mitochondrial proteins.

There is increasing evidence that mitochondrial dysfunction plays an important role in a large number of degenerative diseases, including diabetes mellitus, cardiovascular disease, cancer, Alzheimer's disease, amyotrophic lateral sclerosis, Huntington's disease, and Parkinson's disease.

Mitochondria Import Most of Their Proteins from the Cytosol

As discussed earlier, human mtDNA encodes only 13 proteins. Analyses of the mitochondrial proteome indicate that mammalian mitochondria contain approximately 1500 unique proteins. Thus, most mitochondrial proteins are encoded by nuclear genes and imported from the cytosol. Most mitochondrial proteins are synthesized on soluble ribosomes as precursors and imported posttranslationally. Mitochondrial precursor proteins contain specific targeting signals that direct them to the organelle. These mitochondrial targeting signals are diverse in nature and are recognized by receptor proteins located in the mitochondrial outer membrane. Once imported, mitochondrial proteins need to be sorted to different suborganellar destinations—outer membrane, inner membrane, intermembrane space, and matrix.

We first examine how proteins are imported from the cytosol into the mitochondrial matrix. This import pathway is summarized in Figure 4–36. Most mitochondrial matrix proteins are synthesized as a larger precursor with a cleavable amino-terminal mitochondrial targeting sequence (called a *presequence*). These presequences are approximately 20 to 40 amino acids in length with multiple positively charged residues and have the characteristic of forming an amphipathic α-helix where one size of the helix is nonpolar and the other side contains positively charged amino acids. The precursor protein is maintained in an unfolded condition by the attachment of cytosolic chaperone proteins, such as members of the Hsp70 family of heat shock proteins. The presequence first binds to receptors in the outer membrane of mitochondria (Tom20 and Tom22). The precursor protein is then transferred to the general import pore (Tom40) of the TOM complex (translocase outer membrane), through which it can cross the outer membrane. For the protein to be translocated across the membrane, cytosolic Hsp70 must be released from the protein; this step requires ATP hydrolysis. The precursor protein is then transferred to the presequence translocase of the inner membrane (TIM23 complex), which forms a channel across the inner membrane. Insertion into the channel formed by the Tim23 protein requires the electrochemical gradient ($\Delta\Psi$) across the inner mitochondrial membrane. The ATP-driven presequence translocase-associated motor (PAM) is needed to complete the translocation of the precursor protein into the matrix. The central component of the PAM import motor is the mitochondrial Hsp70 (mtHsp70) chaperone. As the precursor protein emerges in the matrix, there is binding of a mtHsp70 protein, and ATP hydrolysis in the matrix is used to translocate the precursor protein into the matrix. Once inside the matrix, the presequence is cleaved by the matrix processing peptidase. Unfolded matrix proteins can be transferred to a mitochondrial chaperone of the Hsp60 family (mtHsp60), which, together with it cochaperonin (mtHsp10), assists in folding the protein into its final conformation.

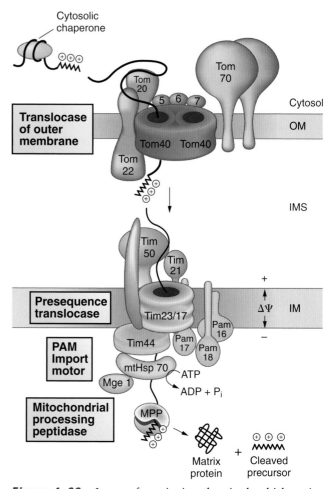

Figure 4–36. Import of proteins into the mitochondrial matrix. Proteins are targeted to the matrix of mitochondria by aminoterminal presequences that contain positively charged amino acids. Precursors are unfolded by cytosolic chaperones, such as heat shock protein 70 (Hsp70), before import into mitochondria. The presequence first binds to the receptors, Tom20 and Tom22, and is then transferred to the general import pore, Tom40, of the TOM complex (translocase of the outer membrane). After passage through the OM, the presequence binds the intermembrane space (IMS) domain of the Tom22 receptor protein. The presequence-containing precursor then binds the Tim50 protein and is transferred to the presequence translocase of the inner membrane (IM; TIM23 complex). The membrane potential ($\Delta\Psi$) across the inner membrane is required for insertion of the precursor protein into the channel formed by the Tim23 protein. The presequence translocase-associated motor (PAM), which is driven by ATP, is required for completion of protein translocation into the matrix. The central constituent of this import motor is the mitochondrial Hsp70 chaperone (mtHsp70). mtHsp70 and its nucleotide exchange factor, Mge1, are transiently recruited to the presequence translocase by Tim44. Pam16 and Pam18 act as cochaperones. Once the precursor emerges in the matrix, it is bound by mtHsp70, and multiple rounds of ATP hydrolysis are required to translocate the protein across the inner membrane. In the matrix, the presequence is cleaved by the mitochondrial processing peptidase (MPP). (*Adapted from Bohnert M, Pfanner N, and van der Laan M. 2007. A dynamic machinery for mitochondrial precursor proteins. FEBS Lett 581:2802–2810, with permission.*)

As noted earlier, some mitochondrial proteins are targeted to the outer membrane, inner membrane, or intermembrane space, rather than to the matrix. Mitochondrial precursor proteins are sorted to one of the submitochondrial compartments from the TOM complex. There are several modes for transport of precursor proteins to the inner mitochondrial membrane. Some proteins of the inner membrane are synthesized as a larger precursor with a presequence (similar to matrix proteins discussed earlier). These proteins use the presequence import pathway (TOM and TIM23 complexes). They are then inserted into the inner membrane from the TIM23 complex by a hydrophobic sorting signal located after the presequence and do not require the PAM import motor. Some proteins that are targeted to the inner membrane are hydrophobic proteins (such as the carrier proteins). These proteins do not contain presequences but have multiple internal mitochondrial import signals. These proteins are initially recognized by a different receptor, Tom70, in the mitochondrial outer membrane. They are translocated across the outer membrane through Tom40. These proteins are then recognized by small soluble Tim proteins (Tim9, Tim10 complex) of the intermembrane space. The proteins are then inserted into the inner mitochondrial membrane by the carrier translocase of the inner membrane (TIM22 complex). This insertion requires the membrane potential ($\Delta\Psi$) across the inner mitochondrial membrane. Other proteins targeted to the inner membrane are first transported across the inner membrane into the mitochondrial matrix. After removal of the presequence, a second sorting signal is uncovered. This sorting signal targets the proteins to another translocase of the inner membrane, Oxa1, where they are inserted into the inner mitochondrial membrane. Oxa1 is also the translocase for some inner membrane proteins that are encoded by mtDNA and synthesized on mitochondrial ribosomes.

The TOM complex is sufficient for insertion of a subset of outer membrane proteins, with a relatively simple topology, such as single-pass proteins. However, β-barrel proteins of the outer membrane, such as porin and Tom40, are first imported via TOM into the intermembrane space. These precursors are then transferred with the help of small Tim proteins of the intermembrane space to another complex of the outer membrane, the SAM complex (sorting and assembly machinery). These pathways for mitochondrial import are summarized in Figure 4–37.

Interestingly, defects in components of the mitochondrial import machinery can cause mitochondrial diseases. For example, the X-linked recessive neurodegenerative disorder Mohr–Tranebjaerg syndrome (deafness-dystonia syndrome) is due to a defect in a small Tim protein (Tim8a; also called the deafness-dystonia protein 1), which is required for protein import into mitochondria.

PEROXISOMES

Peroxisomes are small, ubiquitous organelles with a single membrane and a diameter ranging from 0.1 to 1 μm (Fig. 4–38). Peroxisomes participate in many different metabolic activities, including the oxidation of fatty acids, the breakdown of purines, the biosynthesis of cholesterol, the biosynthesis of bile acids, and ether-lipid biosynthesis.

Peroxisomes are a heterogeneous group of organelles and contain a wide variety of enzymes that can catalyze the O_2-dependent oxidation of substrates (labeled RH_2) with the production of hydrogen peroxide. The reaction is:

$$RH_2 + O_2 \rightarrow R + H_2O_2$$

Many different substrates are broken down by such oxidative reactions in peroxisomes, including uric acid, amino acids, purines, methanol, and fatty acids.

The hydrogen peroxide formed in this reaction is broken down by the enzyme, catalase, which is also present in peroxisomes, in one of two reactions. In one reaction, catalase uses the hydrogen peroxide produced to oxidize other substrates, such as alcohol, formaldehyde, nitrites, phenol, and formic acid. The general reaction is:

$$H_2O_2 + RH_2 \rightarrow R + 2\ H_2O$$

This reaction is an important reaction in liver and kidney cells where peroxisomes detoxify various toxic substances. For example, approximately one-fourth of consumed alcohol is detoxified by this mechanism. In a second reaction, catalase can convert hydrogen peroxide to H_2O.

$$2H_2O_2 \rightarrow 2\ H_2O + O_2$$

A major function of peroxisomes is the β-oxidation of long and very long chain fatty acids. In humans, fatty acids are oxidized both in peroxisomes and in mitochondria (see Fig. 4–29). Short-, medium-, and most long-chain fatty acids are oxidized in the mitochondria, whereas peroxisomes oxidize very long chain fatty acids and some long chain fatty acids.

In addition to oxidation reactions, peroxisomes are involved in the biosynthesis of some lipids. For example, in animal cells, dolichol and cholesterol are synthesized in peroxisomes, as well as in the ER. Peroxisomes also contain enzymes for the synthesis of plasmalogens, a family of ether phospholipids. Plasmalogens are abundant lipids in the myelin sheaths that surround the axons of neurons.

The peroxisome does not have its own genome or ribosomes; therefore, all of its proteins must be imported. Peroxisomes have two subcompartments, an internal matrix and an outer membrane. The majority of peroxisomal proteins are synthesized on free ribosomes and

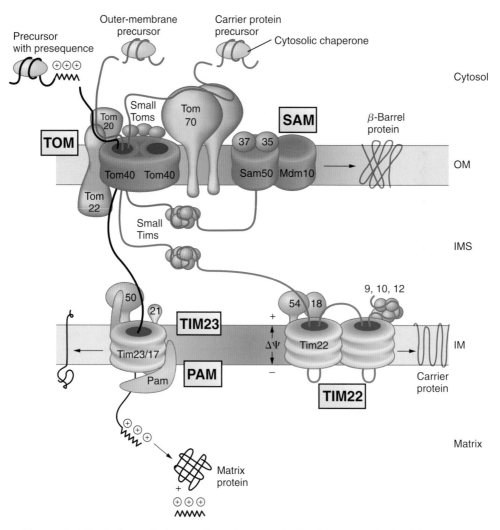

Figure 4–37. Pathways for import of proteins into mitochondria. Most mitochondrial precursor proteins are synthesized on soluble ribosomes in the cytosol. With the help of chaperones in the cytosol, these mitochondrial precursor proteins are transferred to receptors on the TOM complex (translocase of the outer membrane [OM]) and are then translocated across the outer membrane through the general import pore (Tom40). From the TOM complex, precursor proteins are sorted to one of the mitochondrial subcompartments. Precursors with amino-terminal presequences (such as most matrix proteins) are transported across the inner membrane (IM) by the presequence translocase of the inner membrane (TIM23 complex) and the ATP-dependent presequence translocase-associated motor (PAM complex). Translocation into the matrix requires both the membrane potential ($\Delta\Psi$) across the inner mitochondrial membrane and ATP hydrolysis in the matrix. Some inner membrane proteins with presequences are inserted from the TIM23 complex and do not require the PAM import motor. Hydrophobic carrier proteins of the inner membrane are transferred from the TOM complex to the Tim9-Tim10 chaperone complex of the intermembrane space (IMS). They are then inserted into the inner membrane by the carrier translocase (TIM22 complex). Membrane insertion of these precursors requires the membrane potential ($\Delta\Psi$) across the inner mitochondrial membrane. The precursors of the β-barrel outer membrane proteins are transferred from the TOM complex to the SAM complex (sorting and assembly machinery) of the outer membrane via small soluble Tim chaperones of the intermembrane space. (*Adapted from Bohnert M, Pfanner N, and van der Laan M. 2007. A dynamic machinery for import of mitochondrial precursor proteins. FEBS Lett 581:2802–2810, with permission.*)

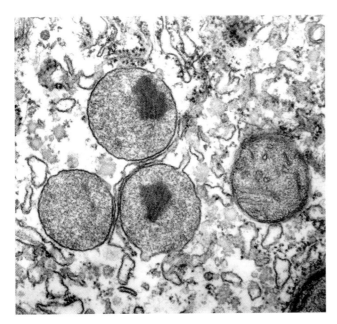

Figure 4–38. Electron micrograph of a cell showing peroxisomes. Transmission electron micrograph of a liver cell showing several peroxisomes. A urate oxidase crystal is evident within two of the peroxisomes. *(Courtesy Don W. Fawcett, Visuals Unlimited.)*

posttranslationally translocated across the peroxisomal membrane. Proteins are targeted to peroxisomes using specific peroxisomal targeting signals (PTS). Different targeting signals are used for peroxisomal matrix (PTS) versus peroxisomal membrane proteins (mPTS).

Two matrix PTS have been identified: one a tripeptide located at the carboxy terminus (PTS1) and one a 9-amino acid sequence close to the amino terminus (PTS2). The consensus sequence of the C-terminal PTS1 is S/A-K/R-L/M. Cytosolic soluble receptors recognize the PTS of peroxisomal matrix proteins and shuttle them to the peroxisome membrane. PTS1 is recognized by the receptor, Pex5p and PTS2 by Pex7p. Current evidence suggests that the peroxisomal matrix protein enters the peroxisome together with its receptor. Once inside the peroxisome, the receptor and matrix protein release, and the receptor recycles to the cytosol. Unlike targeting of proteins to the mitochondria or to the ER, the PTS is not removed upon import. Also, in contrast with import into mitochondria, proteins are imported into peroxisomes in a folded conformation.

Proteins are sorted to the peroxisomal membrane using a sorting mechanism that is distinct from the mechanism used by peroxisomal matrix proteins (although the molecular details have not been well worked out). At least one peroxisomal membrane protein, Pex3p, appears to be targeted to the peroxisome via the ER.

Most new peroxisomes are formed by the growth and division of preexisting peroxisomes. However, recent evidence suggests that there might also be a *de novo* pathway for peroxisome biogenesis.

Numerous genetic disorders of peroxisome function have been identified. These can be divided into two categories. The first category includes disorders resulting from a defect in a single peroxisomal enzyme, such as X-linked adrenoleukodystrophy (X-ALD). In X-ALD, there is an accumulation of very long chain fatty acids in the brain and adrenal cortex because of a defect in a membrane protein that transports these fatty acids into the peroxisomes. Excess long chain fatty acids in the brain destroy the myelin sheath surrounding nerve cells. The second category is a set of disorders that result from a deficiency in the biogenesis of the peroxisome and affects all of the metabolic pathways of the peroxisome. These disorders are known as peroxisomal biogenesis disorders and include Zellweger's syndrome. Zellweger's syndrome is a fatal genetic disorder where the defect lies in an inability to import proteins into the peroxisome. Mutations in at least 12 different *PEX* genes that encode proteins important for peroxisomal protein import have now been identified in Zellweger's syndrome.

SUMMARY

In addition to the plasma membrane surrounding the cell, eukaryotic cells contain a variety of membrane-limited subcellular organelles that can be placed in one of two categories: those that are connected by pathways of vesicle-mediated membrane traffic and those that are not. Subcellular organelles that are connected by pathways of membrane traffic include: (1) the ER, (2) the nucleus, (3) the Golgi apparatus, (4) various endosomes, and (5) lysosomes. The ER has the largest surface area of any subcellular organelle and is the site where most proteins and lipids are first inserted into a membrane. New lipids and proteins leave the ER for other sites by vesicle-mediated transport. The nucleus is bounded by two lipid bilayer membranes and contains dynamic pores through which DNA, RNA, and proteins pass between the nucleoplasm and the cytoplasm. The nucleus is also connected to the ER and can be considered a specialized extension of the ER. The Golgi apparatus consists of membrane stacks involved in modification and sorting of lipids and proteins. Membrane made in the ER passes through the Golgi apparatus en route to its final destination. Various endosomes are intermediates in the endocytic pathway that originates at the plasma membrane. Lysosomes contain enzymes capable of degrading most natural biopolymers. Two membrane-bound organelles that are not connected by pathways of vesicle-mediated transport are mitochondria and peroxisomes. Nuclear-encoded mitochondrial proteins and most peroxisomal proteins are synthesized on cytosolic ribosomes and imported posttranslationally. A major function of mitochondria is oxidative energy metabolism. Peroxisomes participate in a wide variety of biological reactions, including the detoxification of various compounds.

Suggested Readings

Aridora M, Hannan LA. Traffic jam: a compendium of human diseases that affect intracellular transport processes. *Traffic* 2000; 1:836–851.

Aridora M, Hannan LA. Traffic jams II: an update of diseases of intracellular transport. *Traffic* 2002;3:781–790.

Lodish H, Berk A, Matsudaira P, et al. *Molecular Cell Biology*, 5th ed. New York: W. H. Freeman and Company, 2004.

Scriver CR, Sly WS, Childs B, et al. In: Scriver CR, Sly WS, eds. *The Metabolic and Molecular Bases of Inherited Disease*, 8th ed. New York: McGraw-Hill Professional, 2000.

Mitochondria

Beal MF. Mitochondria take center stage in aging and neurodegeneration. *Ann Neurol* 2005;58:495–505.

Taylor RW, Turnbull DM. Mitochondrial DNA mutations in human disease. *Nat Rev Genet* 2005;6:389–402.

Wallace DC. A mitochondrial paradigm of metabolic and degenerative disease, aging and cancer: a dawn for evolutionary medicine. *Annu Rev Genet* 2005;39:359–407.

Wickner W, Schekman R. Protein translocation across biological membranes. *Science* 2005;310:1452–1456.

Wiedemann N, Frazier AE, Pfanner N. The protein import machinery of mitochondria. *J Biol Chem* 2004;279:14473–14476.

Peroxisomes

Michels PAM, Moyersoen J, Krazy H, et al. Peroxisomes, glyoxysomes and glycosomes. *Mol Membr Biol* 2005;22:133–145.

Wanders RJA, Waterham HR. Peroxisomal disorders I: biochemistry and genetics of peroxisome biogenesis disorders. *Clin Genet* 2004;67:107–133.

Chapter 5

Regulation of Gene Expression

CELL NUCLEUS

Nucleus Structure

The Nucleus Is Bounded by a Specialized Membrane Complex, the Nuclear Envelope

The most conspicuous organelle within a eukaryotic cell is the nucleus. It is the sequestering of nearly all the cellular DNA in the nucleus that marks the major difference between eukaryotic and prokaryotic cells. Nuclei are generally spherical and are bounded by a special membrane system, the **nuclear envelope**, which defines the nuclear compartment. The nuclear envelope is formed from two distinct lipid bilayers (Fig. 5–1).

The **inner nuclear membrane** is in close contact with a meshwork of intermediate filaments, the nuclear lamina, which provides support for this lipid bilayer. In addition, this membrane contains proteins that provide contact sites for chromosomes and nuclear ribonucleic acids (RNA), either directly or through proteins of the nuclear matrix. The **outer nuclear membrane** is contiguous with the membrane of the endoplasmic reticulum (ER). Ribosomes that are actively synthesizing transmembrane proteins are often observed associated with the outer nuclear membrane. The outer nuclear membrane is a specialized region of the ER. Proteins synthesized on ribosomes associated with the outer nuclear membrane are either destined for the inner or outer membranes or translocated across the membrane into

the region between the inner and outer nuclear membranes, termed the **perinuclear space**.

Nuclear Pores Allow Communication between the Nucleus and Cytosol

Inside the nucleus are all the components of the genetic apparatus. This includes **deoxyribonucleic acid (DNA)**, **ribonucleic acid (RNA)**, and nuclear proteins to organize and provide for nuclear function (Fig. 5–2).

The nuclear proteins, which include structural proteins of the matrix and lamina, RNA and DNA polymerases, and gene regulatory proteins, are synthesized in the cytoplasm and brought into the nucleus. Thus, nuclear proteins must pass the double-membrane barrier of the nuclear envelope. The transport of materials to and from the nucleus is facilitated by "holes" in the nuclear envelope called **nuclear pores**.

The interior of the nucleus and the cytoplasm of the cell maintain contact or communication through the nuclear pores. In electron micrographs, the pores appear as highly organized disklike structures surrounding a central hole or cavity. Nuclear pore structure is conferred by a set of protein granule subunits, which are arranged in an octagonal pattern and form the boundaries of the pore complex. These eight protein subunits consist of radial arm segments joined together by a set of proteins referred to as **spokes**, which traverse the pore membrane. Sophisticated electron microscopy and

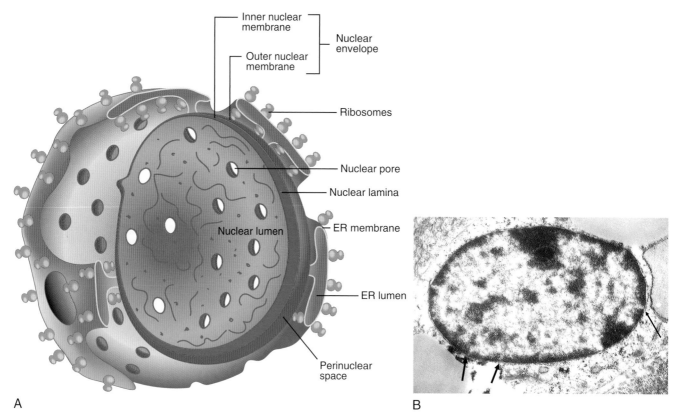

A   B

Figure 5–1.   Relation of the nuclear envelope with cellular structures. A: Diagram shows the double-membrane envelope that surrounds the nuclear compartment. The inner nuclear membrane is lined by the fibrous protein meshwork of the nuclear lamina. The outer nuclear membrane is contiguous with the membrane of the endoplasmic reticulum (ER). As illustrated, the outer nuclear membrane often has ribosomes associated with it that are actively synthesizing proteins that first enter the region between the inner and outer nuclear membranes, the perinuclear space, which is contiguous with the lumen of the ER. The double membrane of the nuclear envelope is perforated with holes or channels of the nuclear pores. B: Electron micrograph of a nucleus from a luteal cell. *Thick arrowheads* denote the inner and outer nuclear membranes of the nuclear envelope, which contains the nuclear pores *(thin arrow)*. *(Courtesy Dr. Wayne Sampson.)*

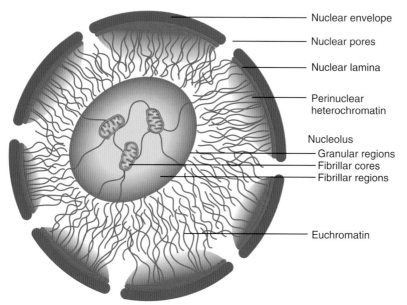

Figure 5–2.   Schematic model of a typical eukaryotic interphase nucleus. The organization of the internal nuclear compartment is shown. The inner nuclear membrane is in direct contact with the protein network of the nuclear lamina, which is often associated with a highly condensed DNA protein complex, referred to as the perinucleolar heterochromatin. Most of the nuclear compartment is filled with noncondensed DNA-protein complexes, the euchromatin. The most obvious structure within an interphase nucleus is the dark-staining nucleolus, which can be further divided into fibrillar and granular components.

genetic ablation studies have indicated the presence of a transporter subunit(s) that lines the central channel of the pore and, as detailed later, forms the barrier between the nucleus and cytoplasm. On the cytoplasmic and nucleoplasmic faces of the outer and inner nuclear membranes are ring structures formed from eight bipartite subunits that are in apparent contact with the spoke proteins and the membrane phospholipids. These are important to the nuclear pore complex, not only for overall structure, but because they anchor filaments that extend into the cytoplasm and nuclear compartments. These nuclear pore ring–attached filaments allow for direct coupling of the nuclear compartment with the cell cytoplasm through interaction with cytoskeletal and nucleoskeletal filaments. In addition, the nuclear pore ring filaments appear to participate in the recognition of molecules that need to be transported through the nuclear pore. Thus, the nuclear pore contains an intricate composition of proteins that form this important structure (Fig. 5–3).

The pore complex penetrates the double membrane of the nuclear envelope, bringing together the lipid bilayers of the inner and outer nuclear membranes at the boundaries of each pore. Although this would appear to allow exchange of components (i.e., proteins, phospholipids, and so forth) between these two membranes, evidence now indicates that these two membranes remain chemically distinct. Therefore, the protein components of the nuclear pore complex must provide a barrier preventing bulk exchange between these two membranes.

Measurements of nuclear pore complexes have demonstrated that they are highly organized structures with an outside diameter of approximately 100 nm and an internal channel 9 to 10 nm in diameter. The properties of transport through the nuclear pores have been addressed by injection of radiolabeled compounds into the cytosol and examination of the rate of their appearance into the nucleus. Such experiments have demonstrated that the nuclear pores are freely permeable to ions and small molecules, including proteins with a diameter smaller than 9 nm [equivalent to 60 kDa or less relative molecular mass (M_r)]. Nonnuclear proteins larger than 9 nm in diameter (greater than 60 kDa) are excluded from nuclear transit. However, nuclear resident proteins that are synthesized in the cytoplasm and are larger than 60 kDa are readily transported into the nucleus, indicating that there must be mechanisms for the selective transport of the molecules across the nuclear envelope.

Selective transport of large molecules and complexes across the nuclear envelope occurs through the nuclear pore by a receptor-mediated process. The key experiments illustrating this concept have made use of the protein nucleoplasmin, a 165-kDA M_r pentameric protein found in high concentrations in frog oocyte nuclei. When purified nucleoplasmin is injected into the cytosol, it accumulates into the nucleus at a rate greater than can be explained by simple diffusion, showing that the protein was concentrated into the nucleus by a

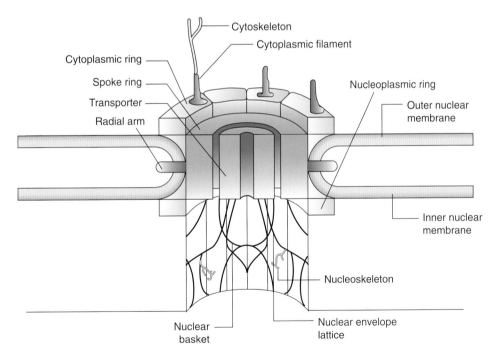

Figure 5–3. Diagram of nuclear pore complex that allows communication of the cell cytoplasm with the internal nuclear compartment. The diagram shows the proteins that comprise the ~100-nm, octagon disk-shaped nuclear pore complex. The pore complex is anchored into the nuclear envelope (outer nuclear membrane and inner nuclear membrane) by the radial arms and spoke rings. Subunits that make up the transporter are just inside the spoke ring and form the aqueous channel of the nuclear pore complex. Adjacent to these structures are the cytoplasmic and nucleoplasmic rings that anchor the connections of cytoplasmic and nucleoplasmic filaments (Basket). It is through these filament structures that potential connections to the cytoskeleton *(outside)* and nuclear matrix *(inside)* are made, adding a physical connection between the nuclear compartment and the rest of the cell. *(Adapted from Goldberg M, Allen T. Curr Opin Cell Biol 1995;7:301–307, by permission.)*

selective uptake mechanism. Furthermore, by separating the nucleoplasmin protein into fragments, it was shown that the ability to selectively import the entire 165-kDA protein was conferred by a small domain of the protein. That is, the domain contained amino acids forming a signal that marked the nucleoplasmin as a nuclear resident protein. Such signals are called **nuclear localization sequences/signals (NLS)**.

In addition, the nucleoplasmin nuclear localization signal could be linked to other proteins and thus mark them for import into the nuclear compartment; this is true even for proteins that are normally never found in the nucleus. The selective transport of material into the nucleus occurs only when energy-generating molecules (guanosine triphosphate [GTP]) are present. These early studies conclusively showed that nucleoplasmin contains a domain that functions as a signal sequence for nuclear localization of the protein, and this transport is an energy-dependent, specific transport through the nuclear pores.

The mechanism of nuclear localization signal-mediated protein import appears to encompass a variety of proteins, both soluble in the cytoplasm and located on the nuclear pore complex, that work in a multistep pathway. First, a protein that has a nuclear transport signal, or NLS, binds a receptor complex. This is a multisubunit receptor, soluble within the cell cytoplasm, which functions to dock the protein to be transported with filaments extending from the cytoplasmic ring of the nuclear pore complex. After the docking, more proteins associate with the complex; most important are GTPases and their activating proteins, which provide the energy of the protein translocation by the hydrolysis of GTP. Not all of the proteins that form the translocation complex go through the pore to the nucleoplasm, and those factors that do are apparently recycled by some mechanism to the cell cytoplasm for further rounds of nuclear transport. A model for nuclear import, based on current data, is shown in Figure 5–4.

One mechanism for recycling factors is to use them to export materials out of the nucleus into the cytoplasm. The export cycle (see Fig. 5–4) is similar to import, except that cargo destined for export would contain a signal marking it for transport to the cytoplasm. Such cargo exported out of the nucleus would be messenger RNA (mRNA), ribosomal RNA and ribosomes, and perhaps proteins that shuttle to the cytoplasm. These nucleic acid and proteins associate with factors containing an export signal, generally a short domain of leucine-rich, hydrophobic amino acids. Although many factors that participate in this process have been examined, there are likely many more that have not yet been identified.

Nuclear transport signal sequences have been identified from a variety of nuclear-targeted proteins. Table 5–1 lists nuclear resident proteins for which an import signal sequence has been identified.

Each identified sequence is a small region of the total protein (about four to eight amino acids in length), and most are basic; however, no apparent consensus of sequences exists, either in primary structure or in any location within the different proteins. The same nuclear pore complexes are responsible for the transport of RNA out of the nucleus to the cytosol. Transport out of the nucleus is selective and requires a receptor in the nuclear pore complex that binds with the exported protein-RNA complex. It is now clear that the complexes to be exported contain an export signal sequence. Thus, it would appear likely that the nuclear pore complex contains multiple receptor sites that recognize a variety of signal sequences on protein complexes to be transported across the nuclear envelope. Whether this is a property of a single- or multiple-subunit protein within the pore complex remains to be elucidated.

In summary, the nuclear pore is an important channel of communication between the interior of the nuclear compartment and the cytoplasm of the cell. It displays properties of a molecular filter, in that ions and small molecules are freely permeable through the aqueous channel, whereas larger protein complexes are selectively transported through the pore. This selective transport is highly specific because of the presence of amino acid signal sequences within the protein complexes that cause them to bind with receptor-like proteins on the pore complexes. Then an energy-dependent process is responsible for the translocation across the nuclear envelope.

The Structure of the Nucleus Is Determined by Proteins of the Nuclear Lamina and the Nuclear Matrix

Lining the inner surface of the nuclear envelope in interphase cells is a protein meshwork, the nuclear lamina. This meshwork forms an electron-dense layer 30 to 100 nm in thickness that provides connections between the inner nuclear membrane and perinuclear chromatin. When examined by electron microscopy, the lamina appears as a square latticework built from filaments that are about 10 nm in diameter. These filaments are classified as intermediate filaments and are composed of three extrinsic membrane proteins, called **lamins A, B, and C**, which have M_r of 60 to 70 kDA. The mRNA for lamins A and C are formed from alternately spliced transcripts of the same gene and encode identical proteins, except that lamin A contains a COOH-terminal extension of 133 amino acids. Lamin B is encoded from an mRNA that is synthesized by a gene distinct from the lamin A/C gene.

Isolated lamins have a rodlike structure approximately 52 nm in length and a globular head domain. Similar to other intermediate filament proteins, the

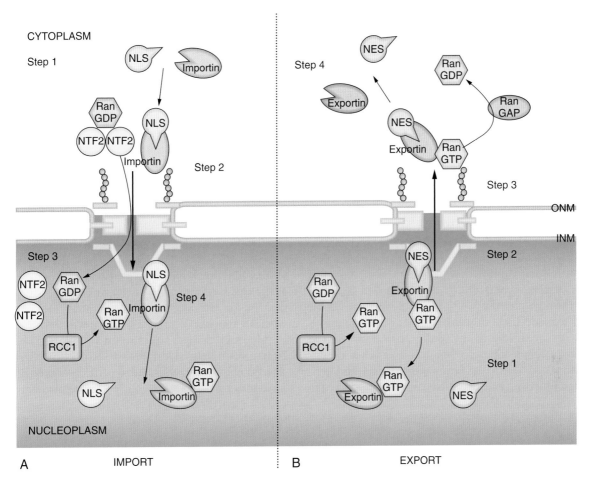

Figure 5–4. Diagram of the steps necessary for import and export through the nuclear pore complex. **A:** Proteins to be transported to the nucleus demonstrate its nuclear localization sequence (NLS), which is recognized by a cytoplasmic receptor molecule (Importin) forming a soluble complex in the cytoplasm, called *step 1*. This binding/recognition leads to a docking of the complex with fibers emanating from the cytoplasmic ring of the pore complex (step 2). This docking recruits accessory proteins, the Ran-GTPases and NTF2, to promote interaction between the transported complex and the control transporter channel and the Ran-GTPases that catalyze the hydrolysis of GTP to provide for transport (step 3). The hydrolysis of energy, GTP to GDP, allows for movement through the pore (step 4). **B:** Export from the nucleus follows a similar plan. The exported cargo demonstrates a nuclear export signal recognized by a receptor called *exportin* (step 1), after which the complex becomes associated with the pore complex (step 2). This docking recruits the energy-generating Ran-GTPases to provide the energy (step 3) to export the cargo into the cytoplasm of the cell (step 4). Thus, import and export through the nuclear pores follows similar pathways.

TABLE 5–1. Nuclear Import Signal Sequences Derived from Various Nuclear Resident Proteins

Protein	Location of Signal Sequence	Amino Acids of Signal Sequence
SV40-large T antigen	Internal: residues 126–132	Pro-Lys-Lys-Lys-Arg-Lys-Val
Influenza virus nucleoprotein	COOH terminus: residues 336–345	Ala-Ala-Phe-Glu-Asp-Leu-Arg-Val-Leu-Ser
Yeast mat α 2	NH₂ terminus: residues 3–7	Lys-Ile-Pro-Ile-Lys
Yeast ribosomal protein L3	NH₂ terminus: residues 18–24	Pro-Arg-Lys-Arg

Summarized from Dingwall C, Laskey R. *Annu Rev Cell Biol* 1986;2:367–390; Lyons RH, Ferguson B, Rosenberg M. *Mol Cell Biol* 1987;7:2451–2456; Newmeyer DD, Forbes DJ. *Cell* 1988;52:641–653; Christophe D, et al. *Cell Signal* 2000;12:337–341.

formation of the long filaments is mediated by the globular head domain. Lamin B is different from lamins A and C in that it is posttranslationally modified by the addition of an isoprenyl group, which allows membrane lipid attachment. The inner nuclear membrane contains a receptor molecule of about 58,000 M_r that binds specifically to lamin B. Multiple receptor proteins are on the inner nuclear membrane, such as the nesprin protein family, which facilitates binding of the nuclear lamina to the membrane. The nesprin proteins also contain actin-binding domains, and thus may be transducers of signals from the cytoskeleton to the nuclear skeleton. Lamins A and C then interact with lamin B, which mediates interactions with the lamina and chromatin. Therefore, in interphase cells, all three lamin proteins are found adjacent to the inner nuclear membrane, forming the nuclear lamina complex.

As the chromatin condenses in the prophase stage of mitosis, there is an apparent disappearance of the nuclear membranes and the nuclear lamina. Figure 5–5 summarizes a model to explain the participation of the nuclear lamins in the breakdown and re-formation of the nuclear envelope during the cell cycle.

Examination of cells by electron microscopy shows that during prophase the nuclear membrane fragments into smaller vesicles that remain associated with the ER. Lamin B is found tightly coupled with these vesicles, whereas lamins A and C are depolymerized and are found throughout the cell. This depolymerization and subsequent breakdown of nuclear membrane is thought to be mediated by phosphorylation of the lamins by a lamin kinase, p34/cdc2, and other downstream kinases. When the lamins are phosphorylated, they depolymerize, and the nuclear membrane breaks down into small vesicles, whereas the chromosomes condense. During telophase of the cell cycle, the nuclear membrane and associated structures reassemble around the separated daughter chromosomes. This reassembly of the nuclear envelope appears to be mediated by the lamins and is coincident with the removal of phosphates from these proteins. The re-formation of the nuclear membrane closely follows the decondensation of the daughter chromosomes. As the chromatin becomes dispersed, it apparently induces the dephosphorylation of the lamins, allowing them to polymerize, which in turn causes the small vesicles associated with lamin B to fuse and form a normal interphase nuclear membrane. Although the interactions that lead to reassembly of the nuclear membrane are not yet clearly defined, it is thought that the phosphatase responsible for removal of phosphates from the lamins is tightly associated with the chromatin, possibly a component of the internal nuclear structure called the **nuclear matrix.**

Although defined structurally and biochemically, the function of the nuclear matrix is unclear. Perhaps its most obvious role would be to provide organization and structure to the internal nuclear compartment. Newly

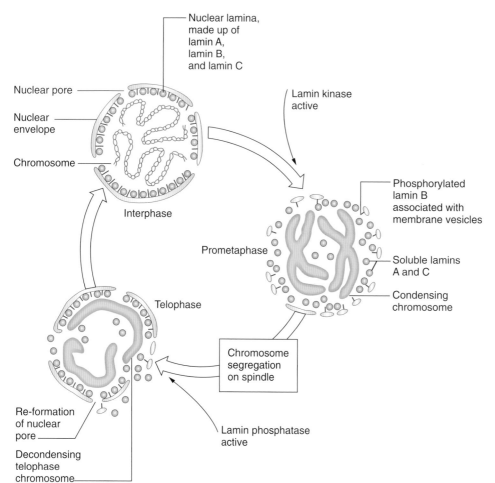

Figure 5–5. Correlation of lamin protein phosphorylation and nuclear envelope structure during mitosis. As cells proceed from interphase to prophase of the cell cycle, there is a condensation of the chromosomes and a breakdown of the nuclear envelope. Nuclear envelope breakdown is concomitant with the activation of lamin kinase, which phosphorylates the lamin proteins (A, B, and C) and causes the depolymerization of the nuclear lamina matrix. Lamin B remains associated with remnant membrane vesicles when phosphorylated, whereas lamins A and C are dispersed within the prometaphase cell. Coincident with the decondensation of chromosomes in daughter cells *(telophase)*, the phosphates are removed from the lamin proteins by an activated lamin phosphatase. This allows the polymerization of the nuclear lamina, using the membrane-bound lamin B as a nucleation site and the formation of a nuclear envelope.

replicated DNA and the enzymatic components necessary for DNA synthesis are associated with the matrix, suggesting a role in the organization of the DNA replication machinery. Recent evidence indicates that actively transcribed genes, and the products of their transcription (e.g., heterogeneous nuclear RNA [hnRNA]), are enriched in nuclear matrix preparations. Localization of RNA transcripts in the nucleus using fluorescent-labeled nucleic acid probes has shown that the RNA follows tracks in the nuclear compartment, with a more intense fluorescent signal seen near the nuclear borders. Thus, after transcription, RNA does not diffuse within the nucleoplasm, but is possibly bound to nuclear matrix fibers as they are spliced to form mature RNA bound for the cytoplasm. These experiments support the concept that the nuclear matrix plays an important role in the organization of the nuclear compartment. In addition, many components of the actin cytoskeleton, including actin itself, have been recently observed in the nuclear compartment. This emerging view of the nuclear compartment is one of a highly organized compartment, and this organization is fundamental to efficient operation of nuclear function. We must also remember that receptors on the nuclear envelope make contact with the cellular cytoskeleton, the nuclear lamina, and likely the nuclear matrix. Thus, cellular organization extends inside important cellular organelles/compartments and aids in the efficient functioning of the cell, such as providing vital links from outside the cell through the cytoplasm and to the nuclear compartment. This allows the cell to adapt quickly to a changing environment and survive many of the challenges it must confront.

Nuclear Function

The Genome of the Cell Is Sequestered in the Nuclear Compartment

The nucleus contains almost all of the genetic information of the cell in the form of DNA. DNA is composed of four nucleotides: two are purines that have a double-ring structure (adenine and guanine), and two are pyrimidines that have a single-ring structure (thymine and cytosine). The basic structure of DNA, derived in 1953 by Watson and Crick, is that of two polynucleotide chains that are held together by hydrogen bonds between adenine and thymine (A-T base pairing) and guanosine and cytosine (G-C base pairing). The two chains are antiparallel or complementary and are coiled into a double helix approximately 2 nm in diameter. The nucleotides are arranged in a nonrandom fashion, such that the genetic information is contained in the specific linear arrangement of bases. The information is stored in "words" consisting of three nucleotides, termed the **codons**, of the genetic code. A model of DNA as genetic information and its relation to chromosomes

stored in a eukaryotic nucleus is presented in Figure 5–6.

In eukaryotic cells, each DNA molecule is packaged into linearly arranged units, termed the **chromosomes,** and the total genetic information stored within the chromosomes is referred to as the **genome** of the organism. The human genome contains about 3×10^9 nucleotide pairs that are packaged into 24 separate chromosomes (22 autosomes and 2 different sex-determinant chromosomes). In diploid somatic cells, there are two copies of each chromosome present—one inherited from the mother and one from the father, except for the sex chromosomes in male individuals, in which the Y chromosome is from the father and X chromosome from the mother. Thus, a diploid human cell contains 46 chromosomes and approximately 6×10^9 nucleotide pairs of DNA. For these chromosomes to remain as discrete functional units, they must possess the ability to replicate, separate into daughter cells at mitosis, and maintain their integrity between cell generations. Experiments examining chromosomal architecture and function have defined three domains or elements necessary for the maintenance and propagation of individual chromosomal units (Fig. 5–7).

To replicate, the DNA contains specific regions that function as the focal point for the initiation of DNA synthesis, termed the **DNA replication origin.** Each chromosome contains many replication origins dispersed throughout its length, which become activated in an asynchronous fashion during the S phase of the cell cycle. These are specific nucleotide sequences at which DNA synthesis begins, although not all of the origins are active at the same time. This suggests that there must be heterogeneity among these sequences, allowing the ordered replication of the chromosome. DNA replication and repair are important processes for cellular maintenance and in disease (see later for more detailed discussion).

A second sequence element that is responsible for attachment of the chromosome to the mitotic spindle during M phase of the cell cycle is termed the **centromere.** Each chromosome contains one centromere region in which the DNA interacts with a complex set of proteins, forming a structure called the **kinetochore.** This structure is responsible for the segregation of the chromosome into daughter cells on cell division.

The third sequence element that is required for maintenance of chromosomal structure is located at termini of the linear chromosome and is called the **telomere.** This is a specialized sequence that defines the ends of the linear chromosome and is built from sequence repeats enriched in guanosine and cytosine bases. These sequences are replicated by telomerase, which folds the guanosine-rich DNA strand to form a special structure that protects the end of the chromosome. When the telomere is not replicated correctly, there can be a shortening of the chromosome. This shortening has been

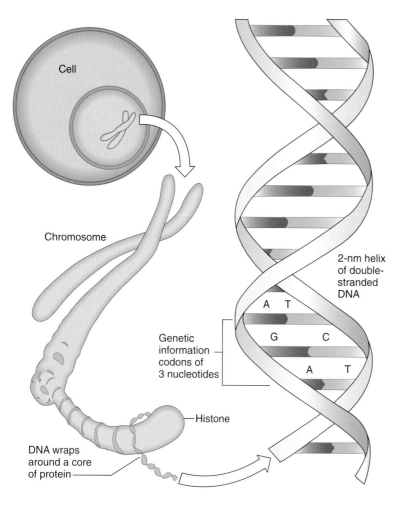

Cell

Chromosome

2-nm helix
of double-
stranded
DNA

A   T

Genetic
information
codons of
3 nucleotides

G   C

A   T

Histone

DNA wraps
around a core
of protein

Figure 5-6. Relation of molecular details of the genetic code stored in DNA and chromosomes within the nucleus of the cell. At the right is a model of DNA in which two antiparallel strands (one is 5′ to 3′ top to bottom and the other is the reverse) are held together by pairing of nucleotide bases. It is the strict arrangements of the nucleotides, in codons of three bases, that are the stores of genetic information. As indicated, the 2-nm helix of DNA is associated with protein, forming the chromatin fibers of individual chromosomes that are housed in the nucleus of eukaryotic cells.

correlated with diseases such as cancer. Moreover, chromosomal shortening is also found in the aged. Thus, researchers are making a concerted effort to find compounds to enhance telomere maintenance.

DNA REPLICATION AND REPAIR ARE CRITICAL NUCLEAR FUNCTIONS

Replication of DNA Occurs during the Synthetic (S) Phase of the Cell Cycle

The ability of cells to divide and multiply is critical for normal growth and development of the organism. In addition, there are tissues in adult organisms that undergo constant, rapid cell renewal, such as blood and digestive tract epithelium, which indicates that cell multiplication is a vital process. Cellular growth occurs through a highly regulated process called the **cell cycle**, which results in the creation of two equal "daughter cells." During the cell cycle, each component of the cell must be replicated so that the resultant cells can provide their function to the organism. As discussed earlier, one of the important processes is the replication of the cell's

genome and maintenance of the chromosome. This chapter discusses the mechanisms of DNA replication and repair; the regulation of the cell cycle is examined in Chapter 9.

DNA replication occurs in the part of the cell cycle called the S phase, so named because it is the synthesis of DNA that is ongoing during this time. DNA synthesis begins at specific locations along the chromosome called the **replication origins**. These are distinct sequences; however, there does not appear to be a singular origin sequence in higher eukaryotes. In contrast, bacteria, viruses, and yeast have very well-defined nucleotide sequences that form their replication origins. Replication origins in human cells are sequence enriched in A-T base pairs, likely because it takes less energy to pull apart A-T pairs. Thus, the separating of the DNA strands so they can be copied would be facilitated at origins containing A-T bases.

The replication origins are recognized by a group of protein factors that form the **initiation complex.** It is interesting that not all replication origins are initiated at one time; indeed, there are groups of replication origins that initiate DNA synthesis at different times during S phase. In general, genes that are actively transcribed are replicated during the first part of the S phase

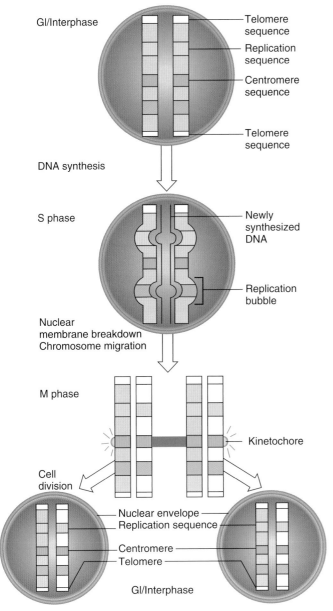

Figure 5-7. Sequence elements of chromosomes that are necessary for maintenance of structure and propagation. Three sequence elements are needed to maintain chromosomes as individual units in nuclei of eukaryotic cells. First is the telomere sequence, which caps the ends of the chromosomes, keeping degradative enzymes from attacking the units during interphase or G phase of the cell cycle. For duplication, chromosomes contain numerous origins of replication that serve as points of initiation for DNA synthesis during the synthetic (S) phase of the cell cycle. After S phase, the nuclear envelope breaks down and the chromosomes are segregated into the daughter cells by use of the kinetochore, which forms at the region of chromosomal constriction called the centromere DNA element.

replication initiation factors that help govern chromosomal replication. Heterogeneous complexes exist for initiating replication and for recognizing damaged DNA for repair, and mutations found in these factors form the basis of disease processes.

Once the origin has been identified by the initiation complex factors, the parental DNA strands are separated and each is used as a template for the synthesis of new DNA. This synthesis is catalyzed by the enzyme DNA polymerase. In eukaryotes, four DNA polymerases have been recognized, termed polymerases α, β, γ, and Δ. DNA polymerase α is important for replication, whereas DNA polymerase β primarily functions in DNA repair. Polymerase γ is enriched in mitochondria and is involved with replication and maintaining the mitochondrial genome. DNA polymerase Δ plays a role in both replication and repair, primarily elongating DNA strands.

All DNA polymerases copy the DNA template from 3′ to 5′ and produce a newly synthesized strand in the 5′ to 3′ direction. Because polymerases all work in a singular direction, the replication of DNA must involve the actions of many proteins. DNA replication occurs from the origin in a bidirectional fashion. This means it begins at a specific location (as indicated in Fig. 5–8); then the process moves simultaneously in both directions.

Because there are many origins on the eukaryotic chromosome, the bidirectional movement ensures replication of the entire chromosome is accomplished in a relatively short timeframe. As shown in Figure 5–8, once synthesized, each daughter molecule contains one intact parental strand and one newly synthesized strand, which are joined by appropriate base pairing interactions. Thus, the resultant chromosomes are replicated by what is called a **semiconservative process.** The "bubble" of DNA formed at each origin is termed the **replication fork.** The replication fork is where active synthesis is ongoing and requires the coordinated efforts of multiple proteins. A class of protein called helicases unwinds the parental DNA strands, which are then kept separated by the actions of single-strand binding proteins and a second group of proteins called **topoisomerases.** With these protein complexes the DNA strands are maintained separately, and each is then copied by the DNA polymerases.

DNA is synthesized in both directions from the replication fork, and the simplest mechanism to accomplish this would be binding a polymerase to each strand and synthesizing the DNA strands. This would require synthesis of DNA in two directions: 5′ to 3′ and 3′ to 5′. However, DNA polymerases synthesize DNA only in the 5′ to 3′ direction; thus, mechanisms to synthesize the genome must take the strandedness of DNA synthesis into account. Using radioactive DNA precursors, researchers showed that synthesis of the new DNA strands near the origin was asymmetric and not equal.

and DNA that is not actively transcribed is replicated at a later time. This suggests two concepts that are actively being pursued experimentally: (1) Transcription and replication may share factors that link the two processes; and (2) because no two cell types express the exact same genes, there may be heterogeneity in the

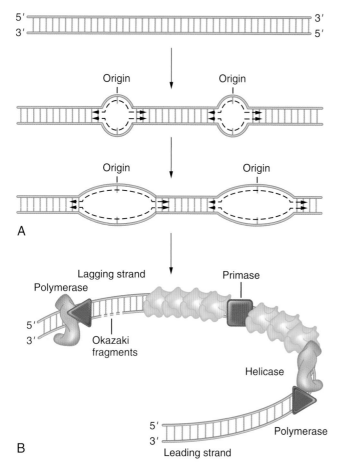

A

B

Figure 5–8. Schematic of events leading to chromosome replication. **A:** The major events of DNA replication are shown, beginning with two separate origins of replication. After recognition of the replication origins, the synthesis of DNA proceeds in both directions from the initiation complexes and proceeds down the DNA molecule. Once synthesis complexes meet, the DNA strands are ligated or joined together, forming a daughter DNA strand composed of one parental strand and one newly synthesized strand. (B) Some of the proteins that are responsible for DNA synthesis from a replication origin. Because DNA polymerases synthesize DNA only from 5′ to 3′, there are two sets of reactions occurring as synthesis occurs away from a replicating origin. On one side, the leading strand, there is synthesis, 5′ to 3′, using a singular polymerase complex. On the other strand, there is the participation of many proteins. This includes single-strand binding proteins keeping the DNA strands apart, a primase activity to "prime" DNA synthesis using an RNA strand, and the synthesis from these RNA strands creating the smaller DNA fragments called the *Okazaki fragments*. Once synthesized, these smaller fragments are ligated together, forming the newly synthesized DNA strand.

Moreover, these experiments showed that near the replication fork there were short pieces of DNA about 1 to 200 bases in length. The asymmetry of DNA synthesis is created by the continuous synthesis 5′ to 3′ of a template strand, called the **leading strand**, and the synthesis of short fragments, called **Okazaki fragments**, on the other strand, which is referred to as the lagging strand. Because the Okazaki fragments are also synthesized 5′ to 3′, the direction of nucleotide addition is opposite to

the overall growth of the DNA strand. After the synthesis of the short Okazaki fragments, they are joined or ligated together to form a continuous strand of new DNA.

Clinical Case 5-1

Nils Ericson is the 9-year-old son of the postmaster of Ogunquit, Maine. His mother took him to a dermatologist, claiming that "his sunburn has not gone away." On physical examination, Nils is a bright, alert 9-year-old, who has a curious gaze because his corneae seem to be "milky." His mother says this has been present for years, but that he seems to see perfectly well. He does well in school.

Examination of Nil's skin shows that in places it is covered with nearly confluent freckles, and in these areas the skin is dry, thickened, and leathery. Interestingly, these regions are confined to his face, ears, and wrists. In addition to the freckles, there is a deep rosy erythema in these areas, and there are a few small raised white warty regions in among the freckles. Also, he complains of a sore on the tip of his tongue, which has not healed all summer long.

On questioning, the mother stated that "Nils has always been overly sensitive to sunburn," and because of that she has been scrupulously careful to keep him covered with shirt and long pants and to slather on sunscreen whenever he is out on the beach playing in the sand dunes. She also reports that Nils's brother and sister are "fine out in the sun," but that Mr. Ericson is very sun-sensitive and has had "skin cancers" removed repeatedly in the past several years.

Despite the boy's protests, the dermatologist took three small punch biopsies of his wrist lesions and a thin slice biopsy from the tip of his tongue.

Cell Biology, Diagnosis, and Treatment of Xeroderma Pigmentosa

Nils has xeroderma pigmentosa. This autosomal recessive hereditary disorder usually develops in early childhood and is associated with a plethora of sun-induced premalignant and malignant changes in the skin. Actinic keratoses and basal cell carcinomas are common. Squamous cell carcinomas occur less frequently, but regularly, and there is a 5% lifetime risk for melanoma for patients with xeroderma. Corneal clouding occurs in many, and probably represents the same epithelial damage in the cornea as in the skin.

Patients with xeroderma pigmentosa are exquisitely sensitive to the sun because of inherited inadequate DNA repair mechanisms. In particular, they lack various endonucleases that usually remove undesirable light-induced DNA adducts. Exposure to ultraviolet light, especially of the UVB type, which penetrates the epidermis, can produce reactive cyclobutanol dimers, and other 6,4-photo products that react to form dysfunctional adducts with pyrimidines in skin cell DNA. Normally, such adducts are removed by the repair endonucleases that mitigate the damage, but in the absence of these enzymes, patients with xeroderma pigmentosa do not adequately repair the biochemical lesion. If the damaged DNA segments involve and inhibit tumor suppression genes such as p53, unbridled proliferation and tumor genesis may occur in the affected cells.

Two of Nils's skin biopsies showed actinic keratoses, but the third showed a basal cell carcinoma that was fully contained within the margins of the biopsy. The tongue sample showed an invasive squamous cell carcinoma with local invasion beyond the biopsy site. This required subsequent surgical excision.

In addition to strict avoidance of the sun, the dermatologist prescribed a cream containing a liposome-encapsulated preparation of the T4N5 endonuclease, which he hoped would avert future carcinogenesis.

DNA Repair Is a Critical Process of Cell Survival

DNA replication occurs with few errors. This is likely enhanced by having polymerases that synthesize DNA in a singular direction and by having cellular mechanisms that efficiently detect and repair errors. In addition to errors of synthesis, cellular DNA is constantly being challenged by outside forces, such as ultraviolet light, chemicals, and products of cellular reactions such as oxygen radicals. In general, the repair of damaged DNA follows three steps: (1) recognition of the damaged or altered DNA, (2) assembling the proteins needed to repair the damage, and (3) repairing the DNA (Fig. 5–9).

It is critical that damaged DNA is repaired accurately, because the propagation of the errors will lead to detrimental effects for the organism. Because there are many ways that DNA can be damaged, several proteins can accurately detect the damage and signal to assemble appropriate repair complexes. It is therefore not surprising that mutants of these detection systems can be a root cause for diseases. Table 5–2 shows a partial list of diseases whose primary cause can be linked to mutants in DNA surveillance complexes.

After recognition of the DNA lesion, damaged DNA is then repaired by one of two mechanisms. One pathway, called **base excision repair**, relies on the lesioned base to be removed by an enzyme called **DNA glycosylase**, followed by excision of damaged DNA. The second repair pathway, called **nucleotide excision**, depends on a small patch of DNA surrounding the damage being removed as a unit. Subsequent to removing the damage, both pathways then resynthesize the DNA and ligate the repaired DNA into place, completing the repair. Efficient repair systems allow the stability of DNA, ensuring the fidelity of stored information and enhancing the survival of the organism.

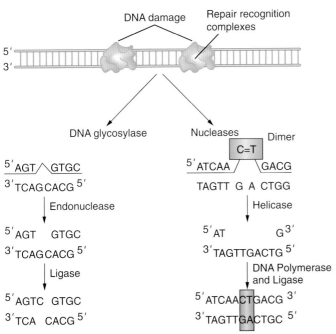

Step 1: Recognition

Step 2: Assembling complexes

Step 3: Repair damage

Figure 5–9. Major events of DNA repair processes in eukaryotic cells. The major steps in the DNA repair pathway are: (1) recognition of the DNA to be repaired, (2) assembling of the appropriate repair complexes, and (3) repair of the DNA. The recognition step requires constant surveillance of the DNA by complexes that can identify mutant or damaged DNA. This means that there are many components involved, which is likely why so many diseases involve mutations in these protein complexes. Recruiting the appropriate repair complexes requires the decision to use either the base excision or nucleotide excision repair pathway. Finally, the actual repair requires the polymerases and ligases to "patch" the double-stranded DNA molecule.

TABLE 5–2.   Diseases Associated with DNA Repair Defects

Disease	Defect/Name
Skin cancer, ultraviolet light sensitivity, neurologic defects	Nucleotide excision repair/xeroderma pigmentosa (XP)
Colon cancer	Mismatch repair
Leukemia, lymphoma, genome instability	ATM protein kinase/ataxia-telangiectasa (AT)
Breast cancer	Homologous recombination repair network/BRCA-2
Premature aging, genome instability	DNA helicase/Werner syndrome
Stunted growth, genome instability	DNA helicase/Bloom syndrome
Congenital abnormalities, leukemia, genome instability, cancer at various sites	DNA intrastrand repair/Fanconi anemia

Summarized from Cleaver JC, Mitchell DL. *Cancer Medicine*, Vol. 1. 2003; Mitchell JR, et al. *Can Opin Cell Biol* 2003;15:232–240; Yang Y, et al. *J Neurosci* 2005;25:2522–2529; Proietti DSL, et al. *DNA Repair* 2002;1:209–223; Wood RD, et al. *Science* 2001;291:1284–1289; Sancur A, et al. *Annu Rev Biochem* 2004;73:39–85.

Clinical Case 5-2

Nathan Rubenstein is a 16-year-old boy who is referred to a pediatric hematologist because he is sufficiently anemic to be symptomatically dyspneic with minimal effort. The referring doctor found a hemoglobin level of 7.2 grams, a white count of 2200, and a platelet count of 76,000. He has had no bleeding or infectious episodes. Nathan's mother is alarmed because one of his older brothers died after a rapid course of acute myelogenous leukemia several years ago. Three other brothers and a sister are in excellent health.

On examination, Nathan is a pale, frail "funny-looking kid" with short forearms and thumbs, who seems younger and shorter than his stated age. He is afebrile, but his pulse is elevated at 108, as are his respirations, at 23. He is in mild respiratory distress after climbing one flight of stairs to the office. The physician notes that Nathan has several "café au lait" spots on his shoulders and back. The rest of the examination is unremarkable, including the absence of any palpable organs or masses in his abdomen.

The hematologist draws some blood for laboratory studies and obtains a bone marrow core biopsy and aspiration. He also orders a skeletal bone series of radiographs. In addition, he asks the mother to bring in Nathan's four remaining healthy siblings for tissue-typing studies and to return for laboratory results in a week. He tells Mrs. Rubenstein that Nathan is profoundly anemic and suggests that he avoid any

significant exercise for the next week. He tells her, however, that he does not want to transfuse him at this time because he does not want to sensitize him to different blood groups.

Nathan's blood smear shows red cell macrocytosis, reduced white cells with 30% polys, and reduced, but large-sized, platelets. The marrow examination results show substantial hypoplasia of all three formed elements, with an increase in marrow fat to 85%. The red cell series shows early megaloblastic changes; the white cell series shows a shift to the left, with increased promyelocytes and occasional "Pelger Huet–like" peculiarly lobed anomalous nuclei in the more mature forms. The megakaryocytes show a small number of binucleate forms. The hematologist reads this picture as marrow hyperplasia with substantial trilineage dysplasia. He also asks the laboratory to perform an assay of the DNA stability on the lymphocytes from Nathan's blood sample. After incubation with a mild DNA cross-linking reagent, they find a 36-fold increase in the incidence of obvious chromosome breaks in comparison with a concurrent control. The bone X-ray films show only vestigial radii and short-thumb metacarpals bilaterally.

Cell Biology, Diagnosis, and Treatment of Fanconi's Anemia

On the strength of these findings, the hematologist makes a diagnosis of Fanconi's hypoplastic anemia. On the return visit, to discuss the patient's laboratory findings and the tissue matching of his siblings, the physician explains that untreated, Nathan has a poor prognosis, with a high likelihood of death from marrow failure, leukemia, or some form of gastrointestinal cancer before the age of 25. He believes, however, that he can avert much of this risk with a peripheral blood stem-cell bone marrow transplant.

Fortunately, one of Nathan's older brothers is a perfect tissue-type match, and Nathan undergoes a successful marrow transplant using mobilized stem cells from him 1 month later. After the return of adequate marrow function in 4 months, Nathan is no longer dyspneic, his hemoglobin level is 12.5 grams, and his white cell and platelet counts are within normal limits. Nathan understands that he has avoided many of the serious effects of his genetic disorder, but that he must remain vigilant for early signs of cancer for the rest of his life.

DNA Is Packaged by Nuclear Proteins to Form the Nuclear Chromatin

The DNA within the cell nucleus is associated with a variety of nuclear proteins, and the DNA protein complex is referred to as **chromatin.** Proteins associated with the DNA can be divided into two general categories: the histone and the nonhistone chromosomal proteins. Nonhistone proteins are a heterogeneous class of polypeptides that includes structural proteins (the high-mobility group of proteins [HMG]), regulatory proteins (those that appear to have a direct role in gene regulation; e.g., *Fos, Myc*), and enzymes needed for nuclear function (RNA polymerases, DNA polymerases). The histones are found only in eukaryotic cells and are by far the most abundant proteins present in the nucleus. Histones are relatively small proteins that are rich in positively charged amino acids (arginine and lysine), which gives them an overall strongly positive charge (basic) that enables them to bind tightly with the negatively charged (acidic) DNA molecules.

There are five types of histones, designated H1, H2A, H2B, H3, and H4. Four of the histones, H2A, H2B, H3, and H4, are termed the **nucleosomal histones,** because they are responsible for formation of the inner core of a DNA-protein complex called the **nucleosome.** The nucleosome is the basic unit of chromatin fiber and gives chromatin the beads-on-a-string appearance in electron micrographs. Examination of the structure of histone-DNA chromatin complexes has relied on digestion of the chromatin with nonspecific nucleases (Fig. 5–10).

These studies have shown that the basic structure of chromatin can be resolved into a repeating unit, called the **nucleosomal bead.** Each nucleosome bead is formed from an octamer of proteins containing two copies of each of the H2A, H2B, H3, and H4 histones, around which is wrapped about 150 nucleotide pairs of DNA. This is the amount of DNA that will make two complete turns around the octamer core of nucleosomal histones, forming a chromatin fiber that is approximately 11 nm in diameter.

Because it contains the simplest arrangement of DNA and protein, the 11-nm chromatin fiber is considered the basic unit of chromatin packaging in the nucleus. However, only a small portion of the DNA is found packaged as an 11-nm fiber in an interphase cell and is probably limited to those regions of DNA that are actively transcribing gene sequences. When nuclei are treated gently and examined by electron microscopy, most of the chromatin is found in a fiber that measures 30 nm in diameter. This 30-nm chromatin fiber is thought to represent the packaging of the nucleosomes by the remaining histone, H1. One model that accounts for the formation of the 30-nm chromatin fiber is the cooperative binding of H1 molecules to nucleosomal DNA. Each histone H1 molecule binds through the central region of the molecule to a unique site on the nucleosome and extends to contact sites on adjacent nucleosomes (Fig. 5–11). This cooperative binding would compact the nucleosomes such that they are pulled together into regularly repeating arrays, forming the 30-nm chromatin fiber.

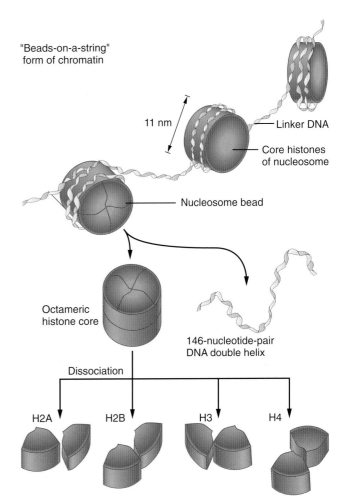

Figure 5–10. Outline of experiment examining chromatin repeating structure. The digestion of chromatin with nonspecific nucleases, such as DNase I, results in the release of repetitive units termed the *nucleosome*. This represents the "beads" of the chromatin fiber. Analysis of the components of the nucleosome demonstrates that the bead is made from DNA, a repeating size of 146 base pairs, wrapped around a core of protein. The protein core is made from two molecules each of the core or nucleosomal histones: H2A, H2B, H3, and H4.

Chromatin inside an interphase eukaryotic cell nucleus in interphase has been divided into two classes, based on its state of condensation. Chromatin that is highly condensed and considered to be transcriptionally inactive is referred to as **heterochromatin.** In electron micrographs of interphase nuclei, the heterochromatin is generally concentrated in a band around the periphery of the nucleus and around the nucleolus. The amount of heterochromatin present in the nucleus is correlated with the transcriptional activity of the cell. That is, little heterochromatin is present in transcriptionally active cells, whereas nuclei of mature spermatozoa, a transcriptionally inactive cell, contains predominantly highly condensed chromatin. In a typical eukaryotic cell, about 90% of the chromatin is thought to be transcriptionally inactive. This amount of inactive chromatin is much more than can be accounted for as the highly condensed

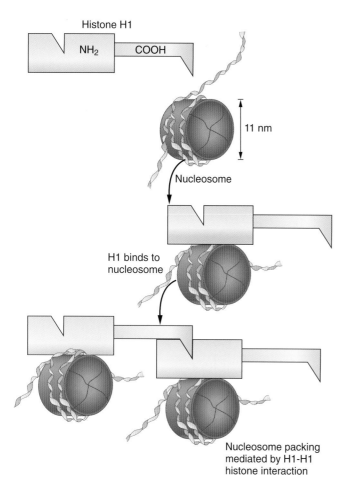

Figure 5–11. Model for histone H1 packaging of chromatin into a 30-nm fiber. A histone H1 molecule contains two distinct domains: a globular NH₂-terminal domain and a COOH-terminal "arm" segment. In the presence of nucleosomes, the H1 molecule binds to a specific region of the nucleosome through its globular domain with the COOH-arm segment able to reach out to subsequent H1-containing nucleosomes. The COOH-terminal domain of histone H1 is then able to interact with specific sites on the adjoining H1 nucleosome by cooperative H1-H1 protein interactions.

heterochromatin. Therefore, heterochromatin is thought to be a special class of inert chromatin that may have specialized functions. For example, the DNA near the centromere region is composed of repetitive DNA, and these sequences appear to constitute a major portion of the heterochromatin DNA. The remaining 10% of chromatin that is transcriptionally active is found in a more extended, dispersed conformation and is called **euchromatin.** Euchromatin is responsible for providing the RNA molecules that exit the nucleus and encode the proteins of the particular cell type.

Chromosomes are visible as distinct units in the light microscope when the chromatin is extensively condensed at mitosis. As a 30-nm fiber, the chromatin could not account for the degree of DNA condensation in metaphase chromosomes. Consequently, higher order packaging units are required to achieve this state. From studies examining the appearance of specialized chro-

mosomes, such as the lampbrush chromosome found in frog oocytes, it is thought that regions of the chromosome are present as extended loops of the 30-nm chromatin fiber held together at the base of the loop by a specific protein-DNA complex. The model of chromatin condensation presented in Figure 5–12 shows that to account for the size of the typical human chromosome (~1.4 μm) in its most condensed state, the extended loop structures of the chromatin must be condensed again, possibly by drawing in the loop domains to form a tightly wound helical formation.

Thus, to achieve the compaction necessary to fit the ~10⁷ base pairs of DNA in the individual chromosomes into the ~1.4-μm chromosome seen at metaphase, there must be at least four orders of packaging in the DNA above the 2-nm double-helical chain of the DNA molecule.

The human genome contains 46 chromosomes in each diploid cell—1 pair of the sex-determining chromosomes and 22 pairs of autosomal chromosomes. Cytologic methods, involving staining of fully condensed metaphase chromosomes with various stains or dyes, have been useful for the identification of individual chromosomes. For example, staining metaphase chromosomes with the Giemsa reagent results in a characteristic pattern of bands on each chromosomal unit, termed **G banding.** Once the chromosomes have been stained, they can be examined under the microscope. The display of the chromosomes prepared in such a manner is referred to as the karyotype of the organism. Figure 5–13 shows an example of karyotype analysis, which is a Giemsa-staining pattern of the metaphase chromosomes of a normal human female (46,XX karyotype).

With the aid of these methods, it has been possible to correlate a variety of human syndromes with abnormalities in chromosome number. An example of Down syndrome, which is characterized by the presence of an additional chromosome 21, is shown in Figure 5–13. Other abnormalities that result in the loss or movement of a particular region of an individual chromosome in the genome (e.g., **cri du chat** syndrome, which results from the loss of a portion of the small arm of chromosome 5) can also be identified by these technologies. Thus, karyotype examination provides a powerful tool for the recognition of chromosomal abnormalities associated with particular genetic diseases, and it is a particularly useful technique for prenatal diagnosis of such disorders.

The Nucleolus Is a Dense Nuclear Organelle That Specializes in the Formation of Ribosomal RNA

When interphase cells are examined under the microscope, the most prominent feature observed in the nucleus is a dense structure, termed the **nucleolus.** The

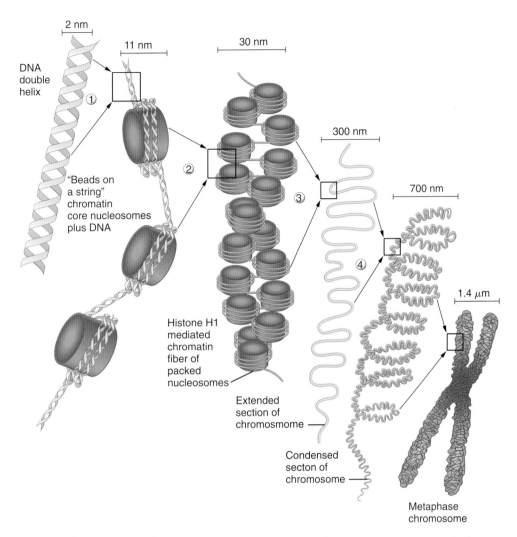

Figure 5–12. Model of chromatin condensation needed to achieve the packaging observed of metaphase chromosomes. The first-order packaging involves the formation of the 11-nm chromatin fiber by association of the DNA helix with the core nucleosome proteins. To form the 30-nm fiber, there is the cooperative binding of histone H1, molecules pulling the nucleosomes into close apposition. The 30-nm fiber is representative of a looped section of the chromosome, and this folding results in a 10-fold packing unit of ~300 nm. The looped domains are thought to be arranged in a secondary loop, folding the chromatin into a 700-nm structure; however, the interactions that result in this packaging are not well-defined.

size and shape of the nucleolus is dependent on its activity. In cells that are actively synthesizing large amounts of proteins, the nucleolus may occupy up to 25% of the total nuclear volume, whereas in dormant cells, it may be hardly visible. Examinations of cells from different physiologic states have shown that the observed differences in nucleolar size are primarily due to differences in cellular status. Cells that are active in protein synthesis contain more of the maturing ribosomal precursor particles in their nucleus. This increased nucleolar size probably reflects the time necessary to assemble the ribosomal RNA (rRNA) with proteins of the ribosomal subunits, because electron micrographs of cells containing large nucleoli demonstrate an increase in the number of active ribosomal genes and an increase in the apparent rate that each gene is transcribed. The size and shape or number of visible nucleolar centers in the

nucleus can be used in pathologic examinations as a determinant of the particular cell.

In general, the nucleolus is visible only in interphase cells. Concomitant with the condensation of chromosomes as the cell approaches mitosis, the nucleolus is observed to decrease in size, then disappears as RNA synthesis stops. In humans, the rRNA genes represent clusters of DNA segments located near the tip of five different chromosomes (chromosomes 13, 14, 15, 21, and 22); thus, there are 10 different ribosomal gene loci in diploid somatic cells. Following mitosis, rRNA synthesis is restarted initially on small nucleoli located at the 10 ribosomal gene loci, which are often referred to as **nucleolar-organizing regions (NORs)**. These small NORs usually are not observed separately because of the rapid infusion of the NORs that forms the larger characteristic interphase nucleolus.

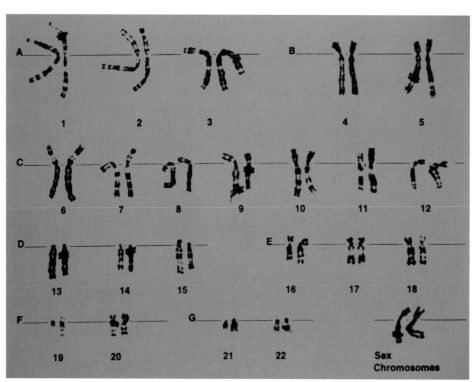

A

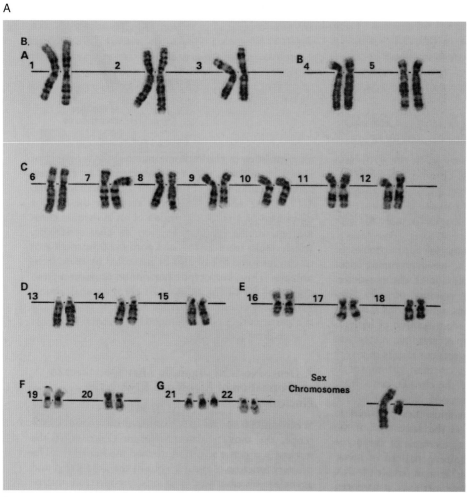

B

Figure 5–13. Giemsa-staining pattern of human metaphase chromosomes. **A:** An example of G-binding karyotype analysis. Shown are the aligned metaphase chromosomes of a normal female human, as illustrated by the 46,XX karyotype. **B:** Example of a karyotype from an individual with Down syndrome is shown with an extra chromosome 21. *(Courtesy Dr. Cathy Tuck-Muller, Department of Medical Genetics, University of South Alabama, College of Medicine.)*

In interphase cells, the nucleolus is responsible for synthesis of ribosomal RNA and the production of mature ribosomal subunits by complexing the rRNA with appropriate proteins. The completed subunits are transported to the cytoplasm through the nuclear pores, where they provide the machinery for translation.

REGULATION OF GENE EXPRESSION

Genomics and Proteomics

The Study of Gene Expression Has Been Facilitated by Recombinant DNA Technology

A major focus of modern cell biology is to understand how a cell works in molecular detail. Although classic biochemical approaches have made possible the purification and examination of many cellular components, until recently, the only way to investigate the informational content of the cellular genome was the examination of the phenotype of mutant organisms to deduce gene function. This approach remains an important investigative and diagnostic tool (karyotyping analyses); however, investigators now possess the ability to directly examine a specific gene and how it functions in normal and pathologic circumstances through the application of techniques, referred to as **recombinant DNA technology**. Although new technical advances occur rapidly in this field of study, the key techniques that constitute the basis of recombinant DNA technology are:

1. The cleavage of DNA at specific locations by restriction endonuclease, which facilitates the identification and manipulation of individual gene sequences.
2. The propagation of eukaryotic DNA fragments in bacterial cells by gene cloning, which allows the isolation of large quantities of a specific DNA.
3. The determination of the order of nucleotides contained in a purified DNA by DNA sequencing, which allows the examination of gene structure and the amino acid sequence it encodes.
4. The direct amplification of a DNA sequence by the polymerase chain reaction (PCR), which enhances the ability to quickly examine specific regions of the genome for genetic defects.

Because recombinant DNA techniques are becoming increasingly important as clinical diagnostic tools, we devote an initial section to explaining these techniques, after which we examine the concepts of gene expression, leading to a discussion of the importance of these concepts toward establishing viable genetic therapies.

Restriction Nucleases: Enzymes That Cleave DNA at Specific Nucleotide Sequences

One of the most important developments of recombinant DNA technology was the discovery of enzymes that catalyze the double-stranded cleavage of DNA at specific nucleotide sequences, the restriction endonucleases. This discovery came from an understanding of the defense mechanism used by bacteria to protect themselves from foreign DNA molecules carried into the cell. The genome of a bacteria contains a host-specific pattern of DNA methylation, and when a DNA that does not contain this pattern is encountered by the cell (e.g., carried in by bacteriophage), it is degraded, thereby protecting the bacteria from the foreign DNA. The first of these enzymes was purified from *Escherichia coli*. Remarkably, this enzyme catalyzed the double-stranded cleavage of DNA within a specific short nucleotide sequence. Hundreds of enzymes capable of cleaving DNA at specific nucleotide sequences have been isolated from different species of bacteria, providing powerful tools for the characterization of DNA molecules.

With these enzymes, the DNA isolated from a particular cell can be cleaved into a series of discrete fragments called **restriction fragments.** The size of DNA fragments produced by an enzyme digest can be analyzed by resolving the cleaved DNA on an electrophoretic gel. Thus, by examining the sizes of restriction fragments produced from a particular gene region after treatment with combinations of different restriction endonucleases, a map of that region can be drawn that shows the location of each restriction site relative to adjacent restriction sites (Fig. 5–14).

Because restriction endonucleases cleave DNA at positions of specific nucleotide sequences, a restriction map reflects the arrangement of these sequences within a given fragment of DNA. This is useful to characterize similarities and differences between isolated, homogenous DNA fragments (e.g., cloned DNA).

The DNA fragments containing a specific sequence can be identified within the thousands of fragments produced when a population of DNA molecules, such as the total genome of an organism, is cleaved with restriction enzymes by using a technique referred to as **Southern hybridization** (Fig. 5–15).

Southern hybridization is a powerful way to examine the organization of specific genetic loci among individual members of a family or population. A difference in restriction maps between two individuals is called a **restriction fragment length polymorphism (RFLP)**. This analysis has become an important technique to identify loci that are close to, or contain, a defective gene associated with a genetic disease (Fig. 5–16). For example, the molecular basis for Duchenne's muscular dystrophy (the dystrophin gene) and cystic fibrosis (the cystic fibrosis transporter gene) was elucidated using RFLP technology.

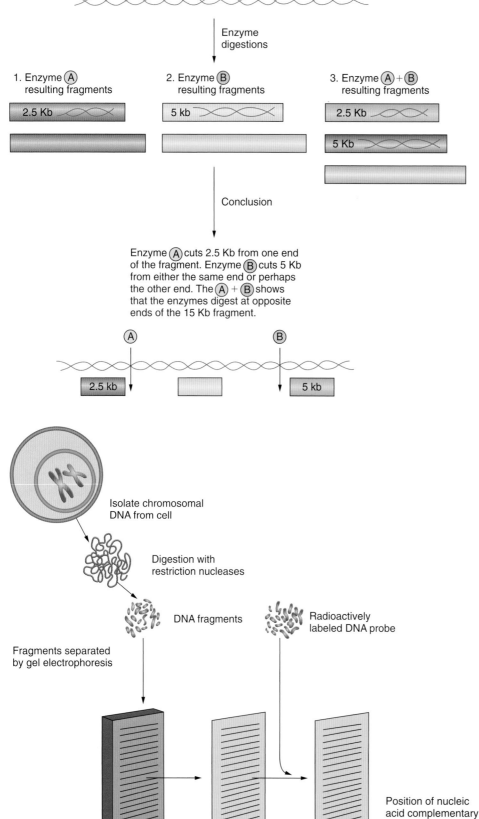

15 Kb DNA Fragment

Enzyme digestions

1. Enzyme (A) resulting fragments

2.5 Kb

2. Enzyme (B) resulting fragments

5 kb

3. Enzyme (A) + (B) resulting fragments

2.5 Kb

5 Kb

Conclusion

Enzyme (A) cuts 2.5 Kb from one end of the fragment. Enzyme (B) cuts 5 Kb from either the same end or perhaps the other end. The (A) + (B) shows that the enzymes digest at opposite ends of the 15 Kb fragment.

(A)   (B)

2.5 kb   5 kb

Figure 5–14. Example of restriction mapping. Restriction enzymes are useful tools to characterize segments of DNA. Shown is an experiment demonstrating how the cleavage sites for different restriction endonucleases are positioned relative to each other to create a restriction map of a DNA fragment. *Kb* is an abbreviation that means 1000 nucleotides or 1000 nucleotide pairs.

Isolate chromosomal DNA from cell

Digestion with restriction nucleases

DNA fragments

Radioactively labeled DNA probe

Fragments separated by gel electrophoresis

Transfer DNA fragments to special paper

Incubate, wash

Position of nucleic acid complementary to labeled DNA probe is detected by autoradiography

Figure 5–15. Southern blotting analysis. The size or migration of specific DNA fragments within a mixture of DNA can be examined by the Southern hybridization technique. After treatment with restriction endonucleases, the DNA fragments are resolved on electrophoretic gels, and the fragments contained within the gel are then transferred to a nitrocellulose paper by blotting. The paper is incubated with a radioactive DNA fragment under conditions that permit this DNA probe to bind with complementary molecules on the paper sheet (hybridization). After hybridization, the sheet is washed free of nonspecific probe binding, and the immobilized molecules complementary to the probe are visualized as radioactive bands on X-ray films placed next to the nitrocellulose paper.

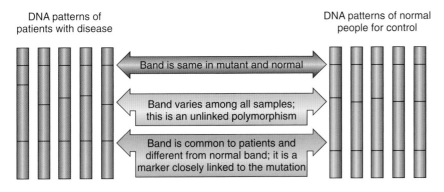

DNA patterns of patients with disease

DNA patterns of normal people for control

Band is same in mutant and normal

Band varies among all samples; this is an unlinked polymorphism

Band is common to patients and different from normal band; it is a marker closely linked to the mutation

Figure 5–16. An example of restriction fragment length polymorphism analysis. Southern blotting examination of the chromosomal DNA from patients who have a disease trait and comparison with the DNA from unaffected individuals has been instrumental in elucidating the molecular basis for many genetic disorders. *(Modified from Lewin B. Genes, 4th ed. New York: Oxford University Press, 1991, by permission.)*

Gene Cloning Can Produce Large Quantities of Any DNA Sequence

In 1973, Boyer and Cohen recognized that DNA molecules from any source are capable of being joined covalently by DNA ligase, an enzyme that links DNA molecules together. In their pioneering experiments, Cohen and Boyer linked a fragment of eukaryotic DNA to a DNA molecule isolated from bacteria that was capable of directing its own replication in bacterial cells. This hybrid DNA formed in vitro is called a **recombinant DNA molecule** because it is formed from the end-to-end joining of two different DNA. Moreover, because the bacterial DNA fragment replicates when the bacterium grows, the eukaryotic DNA fragment of the recombinant molecule is also replicated.

This experiment is an example of **DNA cloning.** Successful cloning requires two basic elements. The first is a suitable host bacterial strain. Most, if not all, bacteria contain a restriction-modification system. However, many bacterial strains have been modified or engineered by classic genetic selection such that they no longer contain this defense mechanism. Because these bacterial strains can undergo genetic transformation (the uptake of DNA) with a foreign DNA molecule with the DNA becoming resident within the recipient cell, they provide an excellent host for cloning experiments. The second element is the segment of bacterial DNA capable of directing replication in these modified bacterial cells. These DNA are commonly referred to as DNA vectors and can be used to "carry" the foreign DNA linked to them in the bacterial cell (Fig. 5–17).

Cloning has been helpful in isolating specific eukaryotic DNA for analysis. Two basic types of clones have been used: complementary DNA (cDNA) clones, which are copies of a specific mRNA, and genomic clones, which are pieces of the genome spliced with a DNA segment capable of replication in bacteria. Because the bacterial segment contains an origin of replication that directs replication of the recombinant DNA molecule (e.g., the bacterial vector and the foreign DNA linked to it) as the bacteria grow, there is an unlimited supply of the target DNA for further analyses.

The Primary Structure of a Gene Can Be Rapidly Determined by DNA Sequencing

Many human diseases are the consequence of single-base changes in genes, causing abnormal proteins to be formed, or perhaps abnormal function of the gene. Sickle-cell anemia, for example, is the result of a single-base change that ultimately replaces a glutamic acid at residue 6 with a valine in the β-chain of the hemoglobin molecule. Other diseases of hemoglobin caused by single-nucleotide changes exist; however, the changed base is not within the protein-coding sequence of the gene. For example, a particular class of disease in which hemoglobin is not produced (the β-thalassemias) is caused by single-nucleotide change of the β-globin gene that creates improper processing of the pre-mRNA to the mature mRNA capable of directing the synthesis of the protein. The basis of these thalassemic diseases remained a mystery until the genetic material from affected individuals was cloned and the primary structure of the genes was determined by the DNA sequencing.

DNA sequencing is simply the determination of the order of the nucleotides in a particular DNA fragment. There are two principal methods for determining DNA sequence. The first method is based on the chemistry of the DNA molecule. The second depends on an enzyme that synthesizes DNA from a single-stranded template and is referred to as the dideoxynucleotide chain termination method. Both methods are reliable and have been used by individual research laboratories to determine sequence and are generally displayed on special electrophoretic gels that allow separation of DNA fragments that differ by a single DNA nucleotide (Fig. 5–18).

It is now more common to determine the sequence of a protein by deducing the amino acids from a nucleotide sequence; there are only four bases in nucleic acids and 20 different amino acids, making DNA a chemically simple molecule. Moreover, the importance of knowing and understanding the primary structure of a gene as related to human disease can be inferred from the recent

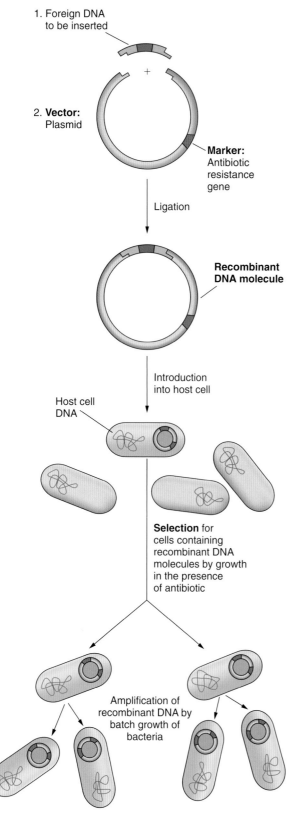

Figure 5-17. **Cloning DNA fragments.** Two DNA fragments can be covalently joined together, forming a recombinant DNA molecule. If one of the DNAs contains a bacterial origin of replication and a vector (such as a plasmid, as shown here), then the products of the in vitro ligation reaction can be placed into a host bacteria. Plasmid vector molecules also contain genes that confer resistance to antibiotics, allowing the selection of bacteria that take up the recombinant molecule.

effort leading to the determination of the complete sequence of the 10^9 nucleotides of the human genome. Indeed, by applying the same techniques, genomes from a variety of species have been sequenced in their entirety, which has resulted in an emerging technology termed *bioinformatics*. This is essentially a way to analyze and/or use the volumes of data generated by today's automated technologies. Bioinformatics will be important in the future to help determine why certain individuals tolerate certain drugs better than others. As discussed later, the newer technologies of genomics and proteomics will eventually lead to "individualized or personalized medicine."

Specific Regions of the Genome Can Be Amplified with the Polymerase Chain Reaction

The techniques discussed in this chapter have revolutionized our understanding of how cells work. However, they are somewhat time-consuming. In 1987, a novel technology was introduced that allows amplification of a nucleic acid sequence without the need to clone it. This technique, called *PCR*, does require knowledge of the sequences that surround the region to be amplified. Synthetic oligonucleotides complementary to these sequences are used as primers in a series of reactions that use a special thermostable DNA polymerase isolated from a bacterial species that lives at high temperatures (Fig. 5–19).

The reactions involve a cycle of steps that first denatures the DNA duplex at high temperature, then allows binding of the oligonucleotides by cooling to a lower temperature and extension of the oligonucleotide primers by DNA polymerase. Because the reagents are not inactivated by high temperatures, they can be added to a single tube, and the cycles of denaturation, annealing, and extension can be repeated multiple times. Each cycle increases the concentration of the duplex DNA bound by the oligonucleotide primers such that microgram quantities of DNA are isolated from the nucleus of a single cell (a specific sequence is amplified exponentially; e.g., 30 cycles would represent 228, or 27 million-fold amplification). Moreover, this can be accomplished in hours, compared with weeks or months needed for conventional cloning techniques.

Although PCR is a relatively new technique, it is rapidly becoming an important diagnostic tool. Because of its speed and sensitivity, PCR is becoming a vital technique for the clinical diagnosis of infectious diseases such as acquired immune deficiency syndrome (AIDS). The PCR-based technologies can detect the presence of the viral genome much earlier than can antibody-based tests that require months (and even years) of infection to detect the presence of a viral protein. In addition, PCR tests have been developed for prenatal diagnosis

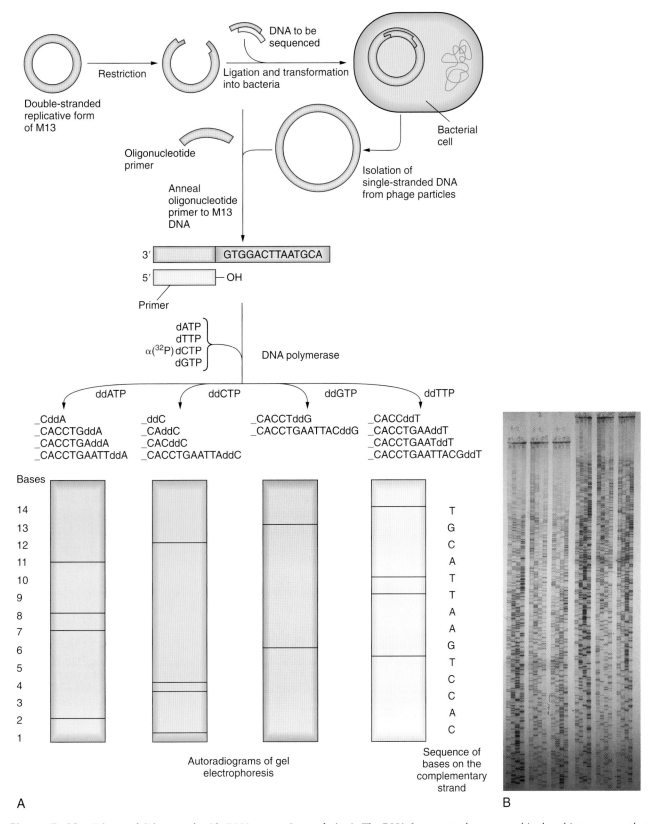

Figure 5–18. Scheme of dideoxynucleotide DNA sequencing analysis. **A:** The DNA fragment to be sequenced is cloned into a vector that is capable of synthesizing a single-stranded version of the recombinant DNA molecule. The single-stranded DNA is then used as a template for in vitro DNA synthesis in the presence of a dideoxynucleotide. The resultant DNA fragments are resolved on an electrophoretic gel, which allows determination of the DNA sequence. **B:** Example autoradiogram from a dideoxynucleotide chain termination sequence experiment. *(Courtesy Dr. Adrienne Kovacs.)*

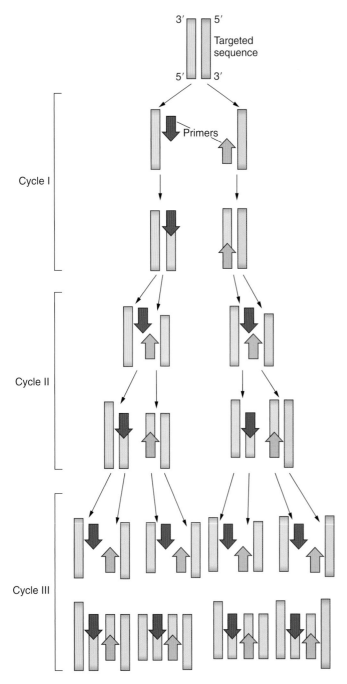

Figure 5–19. Polymerase chain reaction (PCR). Amplification of a DNA fragment through three cycles of a PCR experiment. In each cycle, the DNA strands are denatured, permitting annealing of synthetic oligonucleotides and the synthesis of the complementary DNA strand. Notice that there is an exponential increase in the amount of the target DNA at the end of each cycle, resulting in an amplification of the DNA segment.

of a variety of genetic diseases. Finally, because PCR is such a sensitive technique, it is possible to analyze DNA from a tissue sample as small as a human hair. Thus, PCR is an ideal technology for use in forensic medicine.

Bioinformatics: Genomics and Proteomics Offer Potential for Personalized Medicine

Comparative Genomics Allow the Generation of New Therapies

With the automation of DNA sequencing techniques has come the need to organize and evaluate vast amounts of data. This has opened a new area of science called **bioinformatics**. Essentially, bioinformatics is the application of computer science to give an understanding of increasing loads of data generated from automated experiments. Indeed, complete sequences are now available for a large number of species, from bacteria to humans, and the emphasis is to use these data. Direct comparisons of genomes have shown that many human genes are similar to genes found in other mammals (e.g., 98% similarity human/chimpanzee and 90% similarity human/mouse), and this similarity extends to other organisms. Thus, by identifying gene regions of high similarity, scientists can begin to predict features of genes that are critical for the function of a protein. Extending these findings, these studies will give greater information to the understanding of human gene structure and function, thereby allowing the development of better strategies to combat common diseases. In addition to gene segments, the DNA surrounding genes has been found to be similar among species. These analyses will promote finding genomic features conserved among disparate species, and thus provide important information concerning signals that regulate gene expression. This will lead to new, innovative therapies for human genetic diseases. For example, certain primates share ~98% genomic identity with humans, yet they are not susceptible to certain human syndromes such as malaria. Thus, from an understanding of why primates do not suffer from these diseases, scientists will be able to generate better therapies for certain human disease states.

Another technique that has developed from the human genome sequencing project is the ability to examine the expression of every gene or potential gene in cells or tissues. This technique is called **microarray analysis** (Fig. 5–20), and it involves isolating RNA from cells and testing the whole mixture by hybridization with special slides or chips that contain small spots of nucleic acid, each spot representing a specific gene. Two components that allow this technique to be useful are the ability to make short nucleotide sequences representing each gene as derived from the complete genome sequence information and the ability to make Pico liter spots of these DNA sequences on special glass plates. By hybridizing the total RNA content of a cell to the glass slides or gene chip, scientists can accurately determine which genes of the entire genome are active in the cells of interest and which genes are not active in these cells. In addition, newer technologies are available that

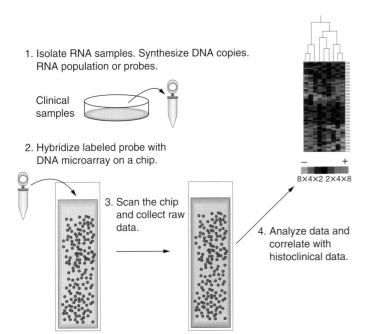

1. Isolate RNA samples. Synthesize DNA copies. RNA population or probes.

Clinical samples

2. Hybridize labeled probe with DNA microarray on a chip.

3. Scan the chip and collect raw data.

4. Analyze data and correlate with histoclinical data.

8×4×2 2×4×8

Figure 5–20. Outline of a microarray analysis. An outline of expression profiling of RNA samples from cells, called *microarray analysis*. The initial step *(1)* requires that RNA samples are isolate from the cells or tissues to be examined. These may be cell populations that are derived from a tumor and normal cells, from cells before and after a drug therapy, or from other clinical samples. The RNA populations are then made into DNA copies, or cDNA, at which time fluorescent dyes are incorporated into the copied DNA. This creates a DNA probe for hybridization to the gene chip *(2)*. This is a special slide that contains small spots of nucleic acids representing every gene in the human genome. Once the hybridization is performed, the slide is scanned to collect the expression data *(3)*, and the data can then be analyzed *(4)* by bioinformatics computer systems. The expressed RNA can determine whether whole pathways (e.g., gluconeogenic pathway or signaling cascades) are activated by treatment of the cells, and this information can be correlated with clinical data to affect a therapeutic course for the patient.

can link or label RNA populations with different fluorescent tags and use them in microarray analyses. Thus, one can isolate RNA from cell populations, such as from a specific tumor and the normal cells, label them with different fluorescent dyes, then mix the labeled RNA to hybridize to the same microarray gene chip. By incorporating sensitive color scanning and detection equipment, it is now possible to examine whole-genome expression patterns *and* simultaneously know how the patterns change in cancer (tumor) cells from the same individual patient. This has revolutionized medicine because physicians can order microarray analysis of a patient's tumor cells, or cell populations affected by specific drug treatments, and quickly determine whether specific metabolic pathways are altered. Thus, although currently experimental, the gene chip microarray holds great promise to move the practice of medicine to include individualized therapeutics.

A direct application of the bioinformatic analyses of genomic sequences has come in the ability to predict how an individual patient will respond to certain drugs. It is well known that although most drugs can be generally effective (i.e., therapeutic), a particular patient may not respond to the therapy. Often, the nonresponsive patient can be grouped with other nonresponsive patients by some criteria, suggesting that there may be a genetic component to the ability of an individual to respond to drug therapies. An example can be seen in patients with clinical depression, where 20% to 40% of the patients do not respond to drug therapy. Moreover, a significant fraction of these patients exhibit resistance to asthmatic, ulcer, and antihypertensive therapies. Comparing the genetics of these individuals has demonstrated variations in the enzyme systems that are responsible for drug metabolism, the cytochrome P450 system.

The study of heredity and responsiveness to drug therapy is referred to as **pharmacogenetics**, and the application of bioinformatics to these analyses is called **pharmacogenomics**. As shown in Table 5–3, numerous variations of the cytochrome P450 enzyme system lead to an individual's ability to respond to specific drug therapies.

Similar pharmacogenetic analyses are in process for other classes of metabolic enzymes. Thus, in the future, medicine will be individualized based on the genetic makeup of the patient, which will allow for more efficient delivery of drug therapies.

The Next Level of Biological Complexity Is the Study of Protein Structure and Function or Proteomics

Now that the entire human genome has been sequenced, a central challenge is to understand the structure and function for all the proteins encoded by the DNA. Proteins perform most of the biological functions within the cell; thus, to complete an understanding of how the cell works, you must know which proteins are expressed in the cell and how these interact with each other to allow cellular function. The entire protein complement expressed by a cell is referred to as the proteome of the cell. Proteomics encompasses a broader definition, which includes protein sequence, posttranslational modifications, protein copy number, protein network formation, protein structure and function, and the localization of proteins within the cell confines. Proteomics includes techniques discussed in Chapter 1, namely, protein purification, protein separation, and gel electrophoresis; however, with the recent application of sophisticated

TABLE 5–3. Pharmacogenetics of the Cytochrome P450 System: Examples of Clinical Importance

Disease	Affected Enzyme	% of Natural Dose[a]		Examples
		Ultrarapid Metabolism	Poor Metabolism	
Depression	cyp2C9	—	—	Bipolar disorder and valproate
	cyp2C19	—	40	Selective serotonin reuptake inhibitors
	cpy2D6	200	30	Side effects of tricyclic antidepressants
Psychosis	cpy2D6	160	30	Haloperidol and Parkinsonian side effects, oversedation and thioridazine
Ulcers	cyp2C19	—	20	Proton pump inhibitors and pH/gastrin
Cancer	cyp2B6	—	—	Cyclophosphamide metabolism
	cyp2D6	250	60	Nonresponse to antiemetic drugs
Cardiovascular	cyp2C9	—	30	Warfarin dosing and blood pressure response
Pain	cyp2D6	—	—	Codeine dosing and nonresponders
Epilepsy	cyp2C9	—	—	Phenytoin dosing and unwanted side effects

[a]Represents the percentage change of dose for normal population.
cyp = cytochrome P450.
(Katzung T. *Pharmacology*. New York: McGraw-Hill, 2002).

mass spectroscopy, a new level of information concerning the cellular proteome is now possible.

Two basic mass spectrometric techniques are used in proteomic analyses, as well as variations of these methods. Both methods examine proteins derived from a cell or tissue that (as illustrated in Fig. 5–21) has been initially digested with proteases to liberate manageable, smaller peptide fragments. Peptides are placed into mass spectrometers that can ionize the peptides into a gas phase and then detect the mass-to-charge ratios of the individual peptides that are created. In the technique referred to as MALDI (matrix-assisted laser desorption/ionization), the peptides are mixed with a small organic molecule with a chromophore that absorbs light of a specific wavelength. The sample and matrix is then dried, and the crystal lattice heated with a beam of light from a laser. The matrix absorbs photons and transfers energy to the peptides that are released as ions. A time-of-flight instrument then measures the individual peptide mass-to-charge ratio. This mass fingerprinting analysis gives a highly individualized result that can provide specific identification of the peptide when compared with databases of peptide mass information. In electron spray ionization, the peptides are in an acidic solution and passed through a needle at high voltage. The resulting positively charged ions enter the first component of a tandem mass spectrometer yielding a primary spectrum (see Fig. 5–21). The mass ion can be isolated and fractured at peptide bonds via energetic collision with particular gasses, and the resultant fragments analyzed by a tandem mass spectrometer, a process called **MS/MS analysis**. The fragmentation pattern is diagnostic of the peptide composition, and examining the fragmentation pattern against databases can lead to the sequence of the original peptide fragment. In addition, because the MS/MS is extremely sensitive, alterations to a particular residue of the peptide can be determined. That is, posttranslational events are also easily determined using the MS/MS technique. Thus, the peptide fragments give information of the "proteome" or proteins expressed in particular cellular populations and can also derive information of the status of a specific protein within the analyzed cellular lysate. This can be important for determining the status of cells undergoing changes in response to disease or drug treatments and can provide information regarding molecular control of cellular processes.

The application of automated data handling (bioinformatics) to compare the calculated peptide mass against databases can determine the origin of a peptide, then identify a protein by its "peptide footprint." The combination of increasingly automated techniques and the improvements in data handling have made proteomics and MS/MS a powerful technique for drug discovery and diagnostics in the clinical setting. For example, a number of companies are now in trials to develop diagnostic tools for identifying cancer stages using proteomic scanning of body fluids (blood and urine). In addition, it is possible to monitor the efficacy of a treatment by examining markers in body fluid, using high-throughput proteomic techniques. Comparing the proteome in a disease state versus control, or in a disease state plus or minus a specific treatment, is called protein profiling. Therefore, treatments will become more individualized with the utilization of genomics and proteomics, thus making strides toward the promise of personalized medicine to improve health care and change the landscape of health care delivery.

Transgenic Mice Offer Unique Models of Genetic Diseases

Medical science has prospered dramatically when there is the ability to model a disease process in animals. The

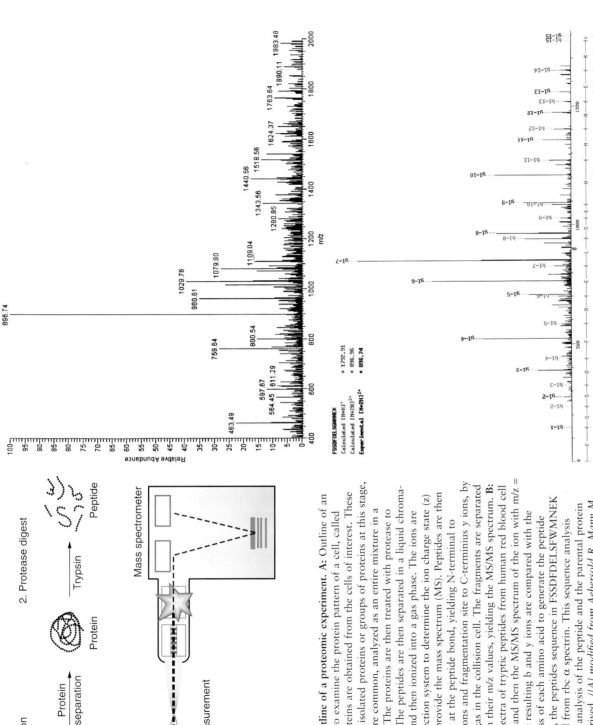

Figure 5–21. Outline of a proteomic experiment. **A:** Outline of an experiment designed to examine the protein pattern of a cell, called *proteomics*. First, proteins are obtained from the cells of interest. These can be separated into isolated proteins or groups of proteins at this stage, or as is becoming more common, analyzed as an entire mixture in a "shotgun" approach. The proteins are then treated with protease to liberate peptides (2). The peptides are then separated in a liquid chromatography apparatus and then ionized into a gas phase. The ions are passed through a detection system to determine the ion charge state (z) and the mass (m) to provide the mass spectrum (MS). Peptides are then fragmented, primarily at the peptide bond, yielding N-terminal to fragmentation site *b* ions and fragmentation site to C-terminius *y* ions, by collision with argon gas in the collision cell. The fragments are separated and detected based on their m/z values, yielding the MS/MS spectrum. **B:** A typical full mass spectra of tryptic peptides from human red blood cell (rbc) α spectrin (*top*) and then the MS/MS spectrum of the ion with m/z = 896.74 (*bottom*). The resulting *b* and *y* ions are compared with the known molecular mass of each amino acid to generate the peptide sequence. In this case, the peptides sequence in FSSDFDELSFWMNEK can only be generated from rbc α spectrin. This sequence analysis provides unequivocal analysis of the peptide and the parental protein from which it was derived. (*[A] modified from Aebersold R, Mann M. Mass spectrometry-based proteomics. Nature 2003;422:198–207, and [B] MS and MS spectra provided by S. Goodman.*)

mouse has presented an excellent model for much of this work, in large part because of the tremendous knowledge base of mouse genetics. Many experiments mapping human genes and diseases have relied on somatic cell techniques using mouse cells. The finding of rodent genes that can complement the function of the human counterparts, and the similarities observed for many genes at the primary structure (DNA sequences) and chromosome locals between mice and humans lend support for the use of mice as human disease models.

It has been demonstrated in a variety of eukaryotic systems that linear DNA fragments introduced to a cell will be ligated together to form tandem arrays (i.e., DNA molecules joined end to end) and can be integrated into the genome at random positions. This provides a technique to test the function of a gene within the confines of the recipient cell. If the gene-modified chromosome is allowed to enter the germ-line cells, it can then be passed on to progeny. These animals, if maintained, contain permanently altered genomes and are said to be transgenic animals with the foreign DNA referred to as **transgenes.**

An example of a gene-targeting mouse experiment is shown in Figure 5–22. In this particular mouse, deleting a particular segment from the mouse genome using homologous recombination turned off a gene encoding a factor important in the regulation of genes in visceral smooth muscle cells. The gene, called *Nkx3.2,* indeed

was found to be important for proper development of gastrointestinal tract smooth muscle because these mice demonstrate altered stomach and intestinal morphologies. Another interesting and unexpected result from this experiment (see Fig. 5–22) is that Nkx3.2 also exerts effects on certain bone structures such that Nkx3.2-deleted mice have altered limb growth and "kinky" tails. This occurs from an altered regulation of growth and development of only certain bone structures, such as the forelimbs shown in the Nkx3.2-deficient mouse. This result indicated that multiple mesenchyme-derived structures could be linked through a single regulatory molecule and has provided a clue into an understanding of human birth defects that exhibited anomalies in seemingly unrelated organ systems.

Many transgenic mice have been formed by the introduction of a human gene that houses altered information leading to disease, allowing the function of the human gene to be examined. Furthermore, the recent ability to add DNA that integrates at a specific location as a gene within the mouse genome (e.g., gene targeting) has allowed the ability to generate transgenic mice with altered expression of the mouse gene or to generate a mouse that expresses an altered gene product—that is, altered at the information storage level of the genome. These types of experiments have provided powerful knowledge of human disease states, from the ability to model the disease, and have enhanced our understanding of genetics so that gene-based therapies for human diseases are now within reach.

Figure 5–22. *Nkx3.2-deficient mice have altered limb growth.* Two-month-old mouse that does not express the regulatory molecule Nkx3.2. These mice do not have normal gastrointestinal development, and as illustrated, they exhibit altered growth of the forelimbs. This altered growth was not predicted, but the phenotyping of this mouse strain has shown that multiple organ systems can be affected through a single molecule. Furthermore, in annals of pediatric genetics and development, cases describe linking birth defects in certain bone growth and gastrointestinal problems. Thus, this mutant mouse has helped to describe the molecular basis for defects in seemingly unrelated organ systems and provides an important model of human disease. *(Courtesy of Dr. Monique Stanfel.)*

Gene Expression: The Transfer of Information from DNA to Protein

According to standard definitions, the word **gene** is defined as "a complex molecule associated with the chromosomes and acting as a unit or in various biochemically determined combinations in the transmission of specific hereditary characteristics." The word **expression** is defined as an "outward indication or manifestation of some feeling, condition, or quality." Thus, the phrase **gene expression** is an outward indication or manifestation of the complex molecules associated with the chromosomes.

What is a measure of this outward manifestation? The most obvious answer is that the appearance of a functional protein within a given cell would be the outward indication of the expression of a gene. The aggregate functional capabilities of the collective set of proteins expressed in a particular cell are what specify the biochemical and phenotypical properties of the cell. For example, skeletal muscle is composed of elongated, multinucleated cells that are able to contract when stimulated by neuronal input because of the collective set of proteins expressed in these cells. This does not mean that all of the proteins found in a skeletal muscle cell are expressed exclusively in this cell type; in fact, many of these proteins are expressed in a variety of cells. However, there are proteins unique to skeletal muscle, implying that there must be mechanisms that stringently govern their appearance. The appearance of proteins in a cell occurs by a transfer of the information housed within the genetic material to the machinery responsible for protein synthesis. Thus, the mechanisms that govern the expression of proteins must, in some way, act on this information transfer.

The Basic Steps of Information Transfer Are Transcription and Translation

The process of information transfer from DNA to protein involves two major steps: transcription and translation (Fig. 5–23). Transcription occurs in the nucleus and is the synthesis of a single-stranded RNA

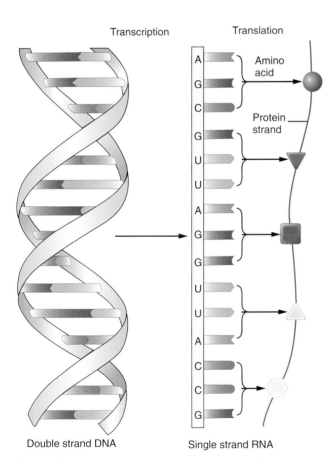

Figure 5–23. Two major steps are involved in information transfer from DNA to protein. The first step involves the synthesis of a single-stranded RNA molecule from a double-stranded DNA template. This process, catalyzed by the enzyme RNA polymerase, is called *transcription.* The product of transcription by eukaryotic RNA polymerase II is an RNA that directs the synthesis of a protein molecule by the translation machinery, the *ribosome.* The genetic code of three nucleotide units, the codons, stored in the double-stranded DNA, are passed to the single-stranded RNA molecules (the messenger RNA), which are the "words" that specify a particular amino acid building block of the protein sequence.

copy of the information stored in the double-stranded DNA molecule. Synthesis of the RNA is catalyzed by the enzyme RNA polymerase, and the synthesis of RNA from DNA is asymmetric; that is, only one strand of the DNA is transcribed to make the RNA copy. In eukaryotes, three types of RNA molecules are transcribed: transfer RNA (tRNA), ribosomal RNA (rRNA), and mRNA, each of which is synthesized by a different RNA polymerase. Ribosomal RNA (transcribed by RNA polymerase I) and tRNA (transcribed by RNA polymerase II) are often considered to be structural RNA because they make up the integral components of the translational machinery, the ribosome. The RNA molecules that are synthesized by RNA polymerase II have a special role; they carry the information needed to code a protein sequence from the DNA to the ribosome. The information is stored in this RNA as discrete "words" made up of three nucleotides, referred to as **codons.** Because this RNA provides the link between the genome and the protein synthesis machinery, it is called **messenger RNA (mRNA).**

Regardless of the RNA that is synthesized, the process of transcription begins with binding of RNA polymerase to specific sequences of DNA in or near the gene, called **promoter DNA elements.** Synthesis of the RNA then occurs by the progressive addition of ribonucleotides to form an RNA chain with the polarity of synthesis being 5′ to 3′. Therefore, transcription is a multistep process, and the interruption or enhancement at any of the steps necessary for transcription would be important factors for the regulation of expression. Transcriptional controls are thought to be paramount regulatory mechanisms because the synthesis of RNA from the DNA template constitutes the primary step of gene expression.

The second basic step of information transfer is the translation of codons encrypted in the nucleic acid of the mRNA into a chain of amino acids. Protein synthesis is conducted by the ribosomes, and in eukaryotic cells, this synthesis occurs in the cytoplasmic compartment of the cell. Ribosomes are large complexes of RNA and protein composed of one large subunit and one small subunit that come together to form the protein synthesis machinery. The basic function of the small subunit is binding the mRNA and tRNA, whereas the large subunit is required for catalyzing the peptide bonds of the growing protein chain. Thus, the synthesis of a protein requires all three of the RNA classes synthesized in the nucleus working in concert in the cytoplasm.

Translation occurs through a series of distinct steps (Fig. 5–24). The first step is binding of the small ribosomal subunit with an mRNA molecule. The small ribosomal subunit is prepared for binding of mRNA by its association with proteins, referred to as **initiation factors,** and its binding of special tRNA, called the **initiator tRNA,** which contains a methionine amino acid residue. The binding of the small ribosomal subunit

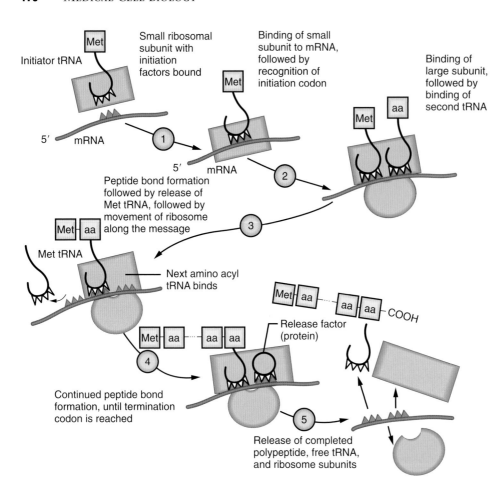

Initiator tRNA

Small ribosomal subunit with initiation factors bound

Binding of small subunit to mRNA, followed by recognition of initiation codon

Binding of large subunit, followed by binding of second tRNA

5′ mRNA

5′ mRNA

Peptide bond formation followed by release of Met tRNA, followed by movement of ribosome along the message

Met tRNA

Next amino acyl tRNA binds

Release factor (protein)

Continued peptide bond formation, until termination codon is reached

Release of completed polypeptide, free tRNA, and ribosome subunits

Figure 5-24. Scheme of events for protein translation from a messenger RNA (mRNA) molecule. The process of translation begins with binding of activated small ribosomal subunits with the mRNA. On identification of an initiation codon (AUG), the large ribosomal subunit binds to form the translation machinery that recruits tRNA specified by the three nucleotide codons of the mRNA. This enables the ribosome to join the amino acids of the polypeptide chain. When a termination codon is encountered, the protein chain is released, and the translation machinery breaks up, which can then re-form a translation complex on reactivation of the small ribosomal subunit. *(Modified from Widnell C, Pfenninger K. Essential Cell Biology. Baltimore: Williams & Wilkins, 1990, by permission.)*

occurs at or near the 5′ end of the mRNA, after which the activated ribosomal subunit scans the messenger for an AUG initiation codon for methionine. This results in the formation of an initiation complex.

After the formation of the initiation complex, the large ribosomal subunit binds to the small-subunit mRNA structure, permitting progressive "reading" of subsequent codons. The reading of mRNA involves the recognition of the three-nucleotide codon, with recruitment and binding of the appropriate tRNA-amino acid complex to the ribosome. Once the tRNA is bound, the amino acid it is carrying is linked to the growing protein chain by peptide bond formation, catalyzed by the large ribosomal subunit. The formation of the peptide bond causes a release of the tRNA from the previous cycle, a shift of the tRNA polypeptide chain within the ribosomal complex, and the subsequent reading of the next codon sequence. This cycling continues until the ribosome encounters a codon that specifies the end of the protein, called the **termination** or **stop codon,** after which the mature protein is released from the ribosome. The ribosome-mRNA complex dissociates into separate components, which, after the appropriate charging of the small ribosomal subunit, can reassemble to make more protein.

Because mRNA is composed of 4 nucleotides (adenosine, cytosine, guanosine, and uridine), there are 64 possible combinations available to form the 3 nucleotide codon sequences. Three of these nucleotide combinations, UAA, UAG, and UGA, do not specify an amino acid, but rather instruct the ribosomal machinery to end the protein synthesis; they are referred to as **stop codons**. This leaves 61 possible combinations to encode only 20 amino acids; the amino acids are specified by binding of complementary nucleotides carried in specific tRNA molecules. Two amino acids, methionine and tryptophan, are encoded by a single codon, and the other 18 amino acids are encoded by multiple codons. Thus, the genetic code is said to be degenerate, meaning that most amino acids are specified by more than one triplet sequence.

In prokaryotic cells, mRNA is available to the translational machinery as soon as it is transcribed. Translation is often observed to begin before the transcription process is finished. This is not true in eukaryotes, in which there is a separation of the transcriptional and translational machinery. In eukaryotic cells, transcription occurs in the nucleus, and the transcribed product, called the **primary RNA transcript**, is often larger than the final, mature mRNA. The primary RNA transcript is modified or processed to form the mature RNA, which is then transported from the nucleus to the cytoplasm to participate in protein synthesis. Thus, gene expression in eukaryotes is more complex than that of prokaryotes; it requires additional steps, and each step in the process provides a potential point of regulation.

Each Cell Type of Multicellular Organisms Contains a Complete Complement of Genes

Two central observations indicate that the entire blueprint, or genetic plan, for an organism is contained within the nuclei of all different cell types. First, examination of the DNA content of different cell types within an organism demonstrated that all somatic cells have approximately equal amounts of DNA, and this DNA content was twice that found in the gametic cells. For example, physical, chemical, and kinetic experiments showed that a liver cell contained the same DNA content as a brain cell. This is referred to as the **constancy of DNA.** However, it is difficult to imagine that cells that differ so dramatically in morphologic structure and function do not suffer some type of irreversible change in genetic material (e.g., loss of nonessential genes) during the progressive specialization of the cells during development. Because only a fraction of the total genetic material is expressed in a given cell, it was reasoned that a loss of genetic material might be so minimal that it had escaped detection when examining gross DNA content. This was shown to be incorrect by an elegant

set of experiments by Gurdon and colleagues, who examined differentiation in frogs. The seminal experiment from these studies is diagrammed in Figure 5–25.

When a nucleus of a fully differentiated frog cell (e.g., intestinal epithelium cells) was injected into a frog egg from which nuclei had been removed, the injected donor nucleus was capable of programming the recipient egg to produce normal, viable tadpoles. Because a tadpole contains the full range of differentiated cell types found in adult frogs, it was concluded that the nucleus of the original differentiated donor cell contained all the necessary information to specify the frog's many different cell types. No irreversible loss of important DNA sequences was evident in the differentiated cell nucleus; the development of an organism requires the expression of a specific set of sequences at the appropriate time. These seminal findings have been extended with recent experiments cloning a number of mammalian species, including sheep, cows, cats, and mice. These experiments will ultimately expand medical therapeutics as better techniques become available to replace diseased organs with newly constructed systems, perhaps grown in culture systems, derived from individual genomes. Such therapies, called **regenerative medicine,** have the potential to cure or treat a vast number of diseases; however, a convergence of science and ethics is necessary to determine how this potential is used.

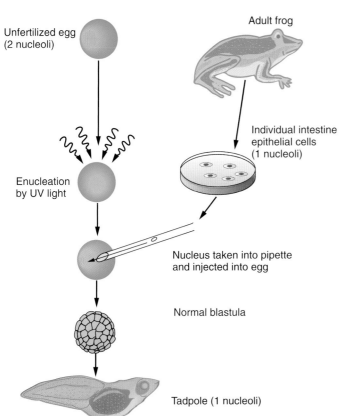

Unfertilized egg
(2 nucleoli)

Adult frog

Enucleation
by UV light

Individual intestine
epithelial cells
(1 nucleoli)

Nucleus taken into pipette
and injected into egg

Normal blastula

Tadpole (1 nucleoli)

Figure 5–25. Nuclear transplantation experiments to examine the constancy of DNA in eukaryotic cells. Diagram outlines an experiment that examined the capacity of nuclei from an adult differentiated tissue to express specific gene sequences. Nuclei from the intestinal epithelium of an adult frog were injected into an oocyte that had had its genetic material inactivated by ultraviolet (UV) light. Genetic markers for the two kinds of nuclei were derived from the difference in the number of nucleoli expressed in the cells; the donor nucleus had one nucleolus, and the acceptor oocyte had two nucleoli. The injected nuclei have the capacity to express all the genes necessary to make a tadpole and subsequently, an adult frog, demonstrating that differentiated cells contain an entire complement of genes to specify the many differentiated cells of an adult organism. (*Modified from Gurdon JB. Sci Am 1968;219:24–35, by permission.*)

The Molecular Definition of a Gene as a "Unit" of Information

The primary function of the genome is to produce RNA molecules, and only specific regions of the DNA sequence are copied or transcribed into a corresponding RNA nucleotide sequence that can function to encode a protein (mRNA) or as structural molecules (tRNA and rRNA). Therefore, every segment of the DNA molecule that produces a functional RNA would constitute a gene.

Until recently, genes were defined primarily by their abilities to confer a biochemical or phenotypic trait on a cell. The application of molecular-cloning technologies has allowed a refinement of the definition of a gene based on the structural organization of the DNA nucleotide sequence. One of the most notable findings from these analyses is that many of the DNA sequences transcribed by RNA polymerase II produce functional mRNA that represent more nucleotides than are found in the cytoplasmic messenger.

One of the first eukaryotic genes in which extra DNA sequences were identified was the chicken ovalbumin gene (Fig. 5–26). The extra length of nucleotides consists of long stretches of noncoding DNA that interrupt the segments of informational DNA. The informational or coding sequences are called **exons,** and the interrupting stretches of noncoding DNA are referred to as **introns.** Thus, the noninformational segments of the RNA molecule as synthesized from the DNA (called the **RNA primary transcript**) must be removed and segments and informational content joined together to form the mature mRNA. These events occur in the nucleus and are termed **RNA splicing.** Therefore, in eukaryotes, a gene does not necessarily reflect, nucleotide for nucleotide, the functional RNA that it encodes.

A second refinement of the definition of a gene has evolved from a correlation of structural organization of the DNA sequence with the ability to visualize regions of chromatin supporting active transcription. Electron micrographs of transcribed DNA segments (genes) demonstrate the transcribed segment as an expanded, bead-on-a-string region of chromatin (e.g., ~11-nm chromatin fiber) on which RNA polymerase molecules appear as globular particles, with a single RNA molecule trailing the polymerase particle. Active RNA polymerase II molecules are often observed to be single units, indicating an infrequent transcription of many gene segments.

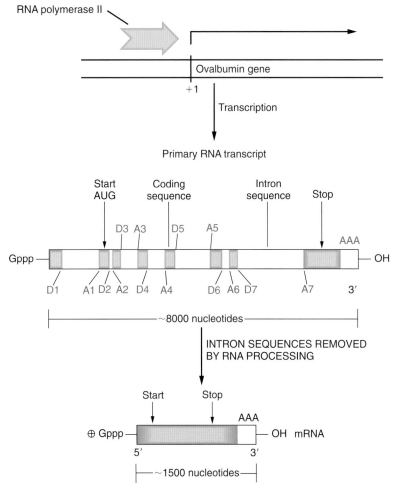

Figure 5–26. The chicken ovalbumin gene makes RNA that is larger than the translated messenger RNA (mRNA). Transcription of the ovalbumin gene leads to the formation of RNA that is larger than the mRNA found in the cytoplasm for translation. This larger RNA, called the *primary RNA transcript,* contains segments of noninformational content—introns—that are found in the double-stranded DNA of the gene. The intron segments are removed, and the segments containing informational content are joined together at specific sites, called *donor* (D) and *acceptor* (A) sequences, to form the mature mRNA by a mechanism of RNA processing.

However, on occasion, many polymerase particles and associated transcripts are observed on a single gene, indicative of high-frequency transcription. In that situation, the length of the associated RNA molecules is observed to increase progressively in the direction of transcription. Micrographs of such gene regions demonstrate a characteristic "Christmas tree" pattern. More importantly, these experiments demonstrated that transcription of a DNA segment begins and ends at discrete sites, defining a transcription unit (Fig. 5–27).

Because transcription begins with the binding of a polymerase molecule to specific DNA segments, an expanded definition of a gene as a unit of transcription must then take into account the segments of DNA that are associated with the transcribed segments. These associated DNA regions are said to direct or promote transcription. Comparisons of DNA sequences from several genes show that certain of these elements are conserved in sequence and position relative to the transcribed sequences.

Because nuclei of somatic cells from multicellular organisms contain a complete complement of genetic material, the drastic physiologic and biochemical differences among differentiated cell types must arise by the specific activation of individual genes, allowing their expression within a given cell. This activation process includes sequences that are copied into a RNA molecule, as well as adjacent DNA sequences that are required for appropriate transcription, causing transcription units. Moreover, the primary transcripts synthesized from eukaryotic genes contain informational segments (exons) and noninformation segments (introns) and must be significantly modified to form a functional RNA sequence.

Accordingly, the pathway of information transfer from DNA to protein in eukaryotic cells involves a complex set of steps; altering the pathway at any of these steps can be a point of regulatory control for gene expression. As shown in Figure 5–28, a cell may exert control of the proteins it expresses by the following steps:

1. Transcription control specifying when and how often a gene sequence is copied into RNA
2. Processing of the primary transcript altering the modifications of the synthesized RNA molecule to form a functional mRNA
3. RNA transport selection of which mRNA is exported from the nucleus to the cytoplasm
4. mRNA stability selectively degrading the mRNA molecules in the cytoplasm
5. Translation control selection of which mRNA in the cytoplasm is translated
6. Protein posttranslational control activating, inactivating, or compartmentalizing specific polypeptide chains after they have been translated.

Although there is evidence to show that each of these steps may function as points of gene expression control, the expression of any given gene is governed by the collective set of interactions along the pathway of information transfer.

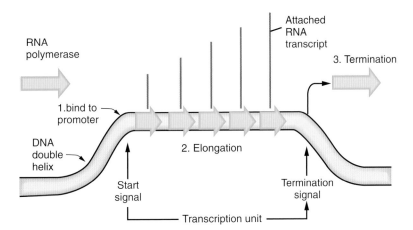

Figure 5–27. Model of a transcription unit, derived from electron microscopy experiments, visualizing transcribed DNA. A transcribed segment of DNA appears as an unfolded 11-nm chromatin fiber that is thought to be looped from the chromosome. Transcription begins with the binding of an RNA polymerase to a promoter DNA element that specifies the starting place for polymerization of the RNA molecule. The growing RNA chain remains linked to the polymerase as transcription proceeds, forming the characteristic "Christmas tree" pattern observed in the electron microscope. Transcription terminates by the release of the primary RNA transcript, which undergoes RNA-processing reactions. Thus, the transcription unit is defined by discrete start and termination sites of the transcription process. *(Modified from Alberts B, Bray D, Lewis J, et al.* Molecular Biology of the Cell, *2nd ed. New York: Garland Publishing, 1989, by permission.)*

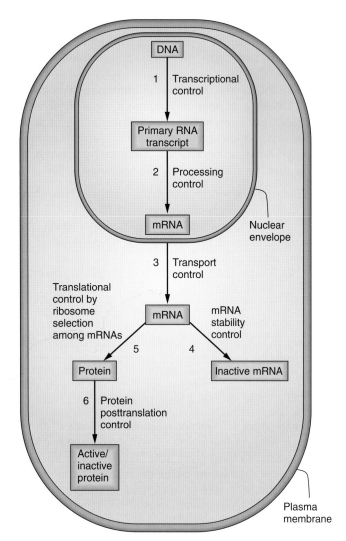

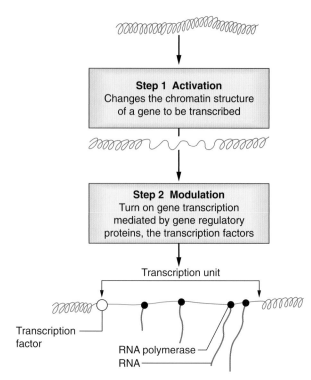

Figure 5–29. Steps involved in the transcriptional control of gene sequences. The transcriptional control of gene expression occurs in two discrete stages. The first step, *activation*, involves changes in the structure of chromatin containing the DNA sequences to be transcribed. The molecular interactions that cause this change in chromatin structure are unknown, but they result in the formation of a region of DNA-protein complex that is biochemically distinct from that of inactive (condensed) chromatin. The second step of transcriptional control, *modulation*, involves the binding of gene regulatory proteins to the activated DNA sequences that operate to fine-tune the transcription of a specific gene.

Figure 5–28. Six steps of information transfer in eukaryotes that constitute potential regulatory points of gene expression. Three steps of potential regulation occur in the nucleus of the cell: *(1)* transcription, *(2)* RNA processing, and *(3)* transport of the messenger RNA (mRNA) from the nucleus to the cytoplasm. Once in the cytoplasm, RNA is subjected to degradation: *(4)* mRNA stability control or selective translation and *(5)* translation control by the pool of ribosomes in the cytoplasm. The translated protein may be modified *(6)* to form an active protein, to inactivate the protein, or to compartmentalize the protein by posttranslational controls.

Transcriptional Control Requires Two Basic Steps: Activation and Modulation of Gene Sequences

The control of gene transcription occurs at two levels: activation, the conversion of compacted chromatin to an extended structure; and modulation, the fine-tuning of transcription mediated by DNA-binding proteins called **transcription factors** (Fig. 5–29).

When chromatin is in its most compacted state, such as at metaphase of the cell cycle, there is little or no RNA transcription. This inhibition of transcription occurs because the DNA is packaged so tightly to form the metaphase chromosomes that it becomes inaccessi-

ble to the nonhistone proteins responsible for transcription (RNA polymerases, transcription factors, and others). To be transcribed, a gene sequence must first be made available to the RNA polymerases and other proteins (activated state) so that the regulatory proteins can provide their functions of influencing the rate that a gene is transcribed (modulation).

In electron micrographs of transcription units (a DNA segment that is being transcribed), the chromatin appears as an 11-nm fiber. Nucleosomes are present; however, the transcribed segment of DNA is in an extended fiber arrangement of chromatin. These observations correlate well with experiments showing that transcriptionally active gene segments are arranged in a chromatin structure that is biochemically distinct from inactive genes. For example, if a gene is expressed in skeletal muscle but not in liver cells, the gene sequence is more accessible to probes, such as nucleases, in nuclei isolated from skeletal muscle than in those isolated from liver cells. Chromatin in this accessible state is referred to as **active chromatin,** and its nucleosomes are thought to be altered in such a way that their packing is less condensed.

The mechanisms that form active chromatin are not understood. It is thought that active chromatin acquires an extended loop structure that emerges from the surrounding highly condensed chromatin. This model implies that there must be some way to recognize the gene sequences that are to be expressed, such as the possibility that they are converted by a specific protein-DNA interaction to an active, less-condensed form. Evidence now indicates that the conversion of chromatin from an inactive to an active form requires DNA synthesis, and that the DNA near active gene sequence contains fewer modified bases, in particular, methylated cytosine residues. In addition, there is less histone H1 and more nonhistone proteins associated with active chromatin. It is unknown whether these changes cause the activation of gene sequences, or whether they occur subsequent to the conversion of the chromatin to an active state.

Eukaryotic cells contain a variety of sequence-specific DNA-binding proteins that function to modulate gene expression by turning transcription on or off. Collectively, these proteins are known as **gene regulatory proteins.** These proteins contain structural domains that can "read" the DNA, allowing their binding to specific sequences. DNA-binding domains are conserved, allowing the broad classification of regulatory proteins as helix-loop-helix, homeodomain, zinc finger, or leucine zipper proteins. Gene regulatory proteins are generally present in few copies in the individual cells and perform their function by binding to a specific DNA nucleotide sequence. The DNA sequences recognized by these proteins can be classified into two broad categories: the core or basal promoter sequences and the enhancer sequences (Fig. 5–30).

Core promoter sequences are generally located close to the transcribed portion of DNA and function to specify the exact point of RNA chain initiation. These sequences are often enriched in adenine and thymidine bases located approximately 20 to 30 (called **TATA sequence**) and 70 to 80 bases (called **CAAT sequence**) to the 5′ side of the transcribed DNA segment. Because they appear to function in all cell types, these sequences are thought to promote basal transcriptional activity. The second class of DNA elements, the enhancer sequences, are regulatory DNA sequences that activate or enhance transcription from a core promoter, with RNA synthesis beginning at the site specified by the core promoter, and appear to function regardless of their sequence polarity and location relative to the transcribed DNA. The enhancer sequences are variable and capable of providing their function even when located at relatively long distances away from a transcribed gene. In some cases, enhancer sequences have been found in DNA that follows (to the 3′ side) the transcribed gene. Each of the different enhancer elements appears to bind a specific, distinct protein factor, allowing the specificity

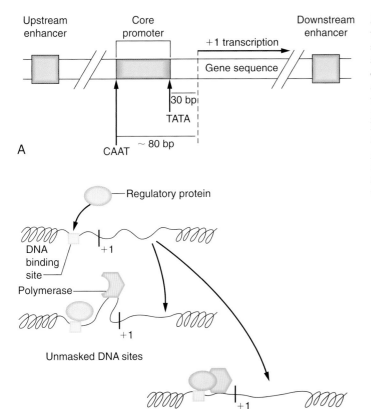

Figure 5–30. DNA segments that can modulate transcription by binding gene regulatory proteins. Several DNA sequences have the potential to bind gene regulatory proteins and are important in the modulation of transcription. **A:** These can be divided into two categories: core promoter elements that are conserved A–T-rich DNA sequences required for basal transcription activity, and enhancer DNA elements that may be placed 5′ (upstream) or 3′ (downstream) relative to the gene. **B:** Two general mechanisms can be evoked to explain the action of DNA-binding sites: The binding of a gene regulatory protein may change the surrounding DNA, such that it is more favorable for binding the transcriptional machinery (polymerases), or the bound regulatory protein may directly interact with the transcriptional machinery.

of its function to be based on the appearance or absence of the protein factor within the nucleus of a given cell.

RNA polymerase catalyzes the synthesis of RNA from a DNA template at a rate of about 30 to 40 nucleotides per second. Because the rate of synthesis is constant, the absolute rate that a gene is transcribed is effectively regulated by the number of polymerases that are synthesizing RNA from the gene sequence. Thus, the principal role of the gene regulatory proteins is to modulate the number of polymerase molecules actively synthesizing RNA from a given segment of DNA. That is, an increase in transcription requires an increase in active RNA polymerase molecules, and reduction in transcription is accomplished by fewer active RNA polymerase molecules. There are two ways that binding of a protein to a specific DNA sequence can affect gene transcription (see Fig. 5–30).

First, binding of a regulatory protein may affect the conformation of surrounding DNA. The binding may unmask sequences, allowing an optimized presentation of polymerase-binding sequences to increase transcription, or conversely, the binding may present a block to transcription by tightening DNA conformation or by occupation of polymerase-binding elements, or both.

Second, a regulatory factor may interact directly with the polymerase. The binding of a protein to a specific sequence near the gene could create a complex that attracts polymerase molecules by an optimal binding complex formed between the polymerase and the regulatory factor. Recent studies have shown that some proteins do not bind DNA directly, but rather influence transcription through binding to a regulatory factor, and perhaps facilitating interactions with RNA polymerase. Experiments have shown that both mechanisms are used (often in combination) and support the conclusion that the major function of regulatory factors is the recruitment of RNA polymerase molecules to a particular gene locus.

It would appear that a way to have tissue- and cell-specific transcription of a gene would be simply to have a single regulatory region for each gene. Transcription could then be governed by a single protein-DNA interaction. However, most eukaryotic genes contain multiple DNA elements that are able to bind to different regulatory factors. The synthetic rate of a gene is then governed by the sum-total effects derived from multiple protein-DNA interactions. The best example of this regulatory scheme is the expression of the β-globin gene. The DNA surrounding the chicken β-globin gene contains 13 distinct sites (seven that are 5′ and six that are 3′ to the gene) capable of binding eight different protein factors (Fig. 5–31).

In early development, the chromatin-containing β-globin gene is converted to an active state, and nine of the binding sites for regulatory factors are occupied, even though the gene is not transcribed. Later, all but one of the sites have bound their regulatory factors

allowing transcription, after which the change in binding of proteins at the 5′ region of the gene again inhibits the transcriptional capacity of the β-globin gene. Even though this represents a singular example, most eukaryotic genes appear to be subject to transcriptional regulation derived from the sum of activities of multiple regulatory proteins. The cellular specificity of gene transcription is thereby governed by a delicate balance of regulatory protein activities.

A class of regulatory factors has been defined that is able to regulate multiple genes specifying a particular cell type (Fig. 5–32). MyoD1, a nuclear protein found in skeletal muscle, is an example of this class of master regulatory proteins. The introduction of DNA containing the *MyoD1* gene into cultured fibroblasts (cells that never express muscle-specific gene products) can convert the fibroblast to muscle cells, phenotypically as well as biochemically. The exact process by which MyoD1 is able to convert cells to a skeletal muscle phenotype is unknown; however, this protein binds to the regulatory region of several (but not all) skeletal muscle–specific genes. That this protein does not bind to the regulatory regions of all skeletal muscle genes implies that it must trigger other regulatory events required for the expression of the muscle phenotype. Furthermore, the regulation of specific gene sequences must be governed by a combination of regulatory-binding activity. Regardless of the mode of action, the transcriptional regulation leading to cellular specialization is subject to a coordinate activation of gene sequences, with the subsequent modulation by the binding of sequence-specific regulatory protein factors.

Primary Transcripts Are Modified to Form Mature Messenger RNA

The primary transcripts of RNA synthesized by RNA polymerase II (mRNA) are modified in the nucleus by three distinct reactions: the addition of a 5′ cap, the addition of a polyadenylic acid (poly-A) tail, and the excision of the noninformational intron segments. These modifications are required to form a mature RNA capable of supporting the translation of a protein, and the entire set of events is called **RNA processing.**

The 5′ end of the mRNA (the end that is synthesized first during transcription) is capped by the addition of a methylated guanosine nucleotide. The addition of the 5′ cap is the first modification of the mRNA primary transcript and occurs almost immediately with the onset of transcription (Fig. 5–33).

The formation of the cap involves the condensation of the triphosphate moiety of a GTP molecule with the diphosphate group of the nucleotide at the 5′ end of the initial transcript. The enzymes responsible for the capping reaction(s) are thought to reside within the subunit structure of RNA polymerase II. The addition

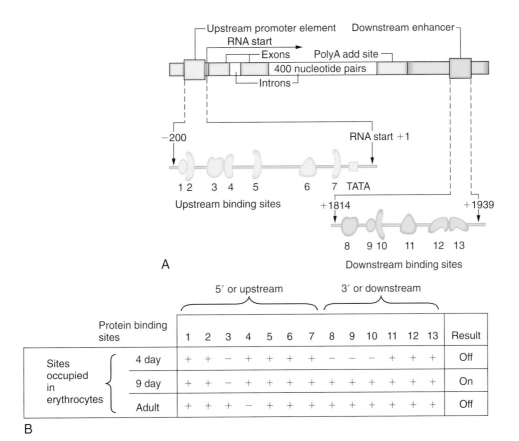

Figure 5–31. Site of gene regulatory protein binding that controls the expression of the chicken β-globin during development. **A:** Diagram of the chicken β-globin gene with the known binding sites of the 13 different gene regulatory proteins. Notice that some of the regulatory proteins (shown by the different shapes) have two binding sites (e.g., sites 3 and 8), and that some binding sites are close to each other—sites 1 and 2, 3 and 4, 9 and 10, 12 and 13—to promote protein–protein interactions between the two regulatory factors. **B:** The occupation of regulatory protein-binding sites during development is shown by the *plus* (+) occupied and *minus* (−) unoccupied sites. As indicated by this chart, the difference in the α-globin gene being on (9 days) and off (4 days and adult) is due to a balance of the activities of multiple gene regulatory proteins that may exhibit binding to different DNA sequences near the gene. (*Modified from Alberts B, Bray D, Lewis J, et al.* Molecular Biology of the Cell, *2nd ed. New York: Garland Publishing, 1989, by permission.*)

of the 5′ cap structure is critical for mRNA to be translated in the cytoplasm and appears to be needed to protect the growing RNA chain from degradation in the nucleus.

The second modification of an mRNA transcript occurs at its most 3′ end, the addition of a poly-A tail. The 3′ end of most polymerase II transcripts is not defined by the termination of transcription, but by the specific cleavage of the RNA molecule and the addition of adenosine residues to the cleaved molecule by a separate polymerase, poly-A polymerase. The signal for cleavage is the appearance of the sequence AAUAAA in the growing RNA chain, with the actual cleavage occurring about 10 to 30 nucleotides away from this signal sequence. Immediately on cleavage, poly-A polymerase adds 100 to 500 residues of adenylic acid to the 3′ end of the cleaved RNA molecule. The RNA polymerase II appears to continue transcription well beyond the cleavage site, with the subsequent RNA being rapidly

degraded, presumably because they lack 5′ cap structure. The exact functions of the poly-A tail are not well-defined; however, experimental evidence suggests that it plays an important role in the export of mature mRNA from the nucleus to the cytoplasm. In addition, it may serve a regulatory function, in that some genes contain multiple sites for poly-A addition.

After modifications of the 5′ and 3′ ends of the primary transcript, the noninformational intron segments are removed, and the coding exon sequences are joined together by RNA splicing. The specificity of exon joining is conferred by the presence of signal sequences marking the beginning (called the 5′ *donor site*) and the end (called the 3′ *acceptor site*) of the intron segment (Fig. 5–34).

These signal sequences are highly conserved (they are approximately the same in all known intron segments), and as might be predicted, alterations in these sequences lead to aberrant mRNA molecules. For example, a

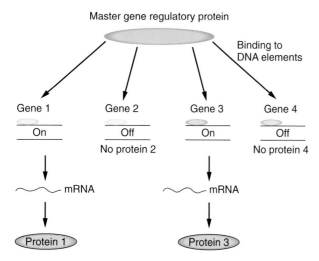

Figure 5–32. Scheme of the activity of a master gene regulatory protein. The expression of a protein, such as MyoD1, that has the ability to regulate the expression of multiple genes—some positive, some negative—could lead to cellular specialization. The activities of proteins that are subject to regulation by the master gene protein can be a biochemical trait for a differentiated cell (actin and myosin in skeletal muscle) or can be a protein that is required in the nucleus to regulate specific genes of the differentiated cell.

group of genetic diseases, collectively called the β-**thalassemia syndromes** (characterized by the abnormally low expression of hemoglobin), are directly attributable to single-base changes in the genome at splice junctions of the β-globin gene that disrupt the appropriate joining of exon segments. Therefore, splicing reactions must occur with exquisite precision to ensure that a functional RNA molecule is formed.

The excision of an intron segment from RNA is conducted by a ribonucleoprotein complex called the **spliceosome.** The spliceosome is formed from a set of undefined proteins complexed with a series of small RNA molecules referred to as *U1* through *U12*. The splicing reaction occurs in steps that include: (1) recognition of consensus 5′ donor and 3′ acceptor sequences, (2) cleavage of the 5′ splice site and formation of a looped RNA structure termed the **lariat,** and (3) the cleavage of the 3′ place site and subsequent ligation of the RNA molecule. The exact role of individual components of a spliceosome is currently being studied; however, it is known that the excision of introns requires the energy of ATP hydrolysis. The excised intron is degraded almost immediately after its release from the primary RNA transcript.

Although it would be logical that the splicing of RNA proceeds by the removal of the most 5′ intron to the last or 3′ intron, experimental evidence has demonstrated that the removal of introns from any given transcript follows a preferred path, often beginning with introns internal to the transcript. This appears to be an inefficient mechanism, and initially, it was viewed with skepticism; however, the recent discovery that a single gene

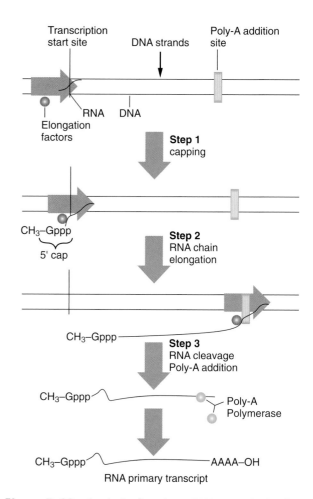

Figure 5–33. Synthesis of a primary RNA transcript involves two modifications of the RNA strand. Almost immediately after RNA synthesis is initiated, the 5′ end of the RNA is capped by a guanosine residue *(step 1)*, which protects it from degradation during the elongation of the RNA chain *(step 2)*. On reaching the signal sequence for the addition of the 3′ polyadenylic acid (poly-A) tail, the RNA is cleaved *(step 3)*, allowing the poly-A polymerase to add multiple adenosines to the 3′ end of the RNA. This RNA is the primary transcript and is ready for splicing of intron segments to form the mRNA. Although the steps involved in transcriptional termination are not well-defined, one model is that the polymerase is altered in its activity and continues to synthesize RNA, but this synthesis is not productive because the RNA is degraded.

may express multiple different proteins by the selected joining of exon sequences to form different mRNA has shed some new light on the pathways of intron removal. For example, a single gene encoding the protein troponin T can produce at least 10 distinct forms of the molecule by simply joining different combinations of encoded exon segments. This variability can be influenced by cell type or by factors extrinsic to a cell, enabling the expression of protein isoforms needed to compensate for alterations in cellular metabolism. The ability of certain genes to form multiple proteins by joining different exon segments in the primary transcript is called **alternate splicing** and has caused a reexamination of the concept of "one gene, one protein."

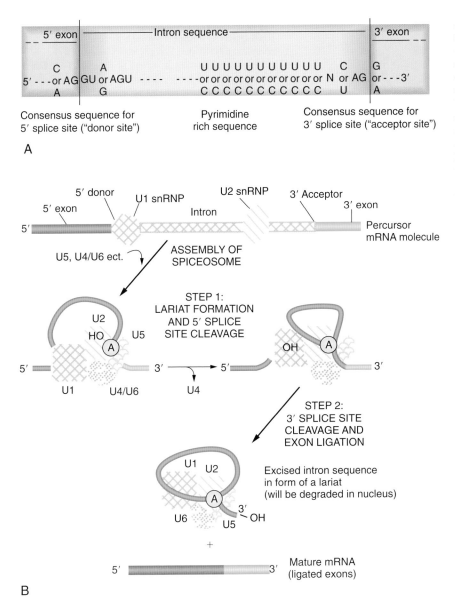

Figure 5–34. Mechanism of RNA splicing to form mature messenger RNA (mRNA) molecules. **A:** RNA splicing occurs at discrete locations that are marked by conserved sequences. The consensus sequences for RNA splicing, listed here, have been determined by comparison of many eukaryotic polymerase II gene sequences. The most conserved nucleotides *(shaded regions)* mark the boundaries of the intron sequence. **B:** The mechanics of RNA splicing involve the recognition of signal sequences by the U1 (5′ donor) and U2 (polypyrimidine sequence), which leads to the formation of the spliceosome (a combination of many small nuclear ribonucleoprotein [snRNP] molecules). Once the spliceosome is formed, the 5′ donor is cleaved by the formation of an RNA lariat, the 5′ donor is then ligated with the 3′ acceptor, and the spliced intron is degraded into the nucleus. *(Modified from Alberts B, Bray D, Lewis J, et al. Molecular Biology of the Cell, 2nd ed. New York: Garland Publishing, 1989, by permission.)*

RNA Transport to the Cytoplasm Occurs via the Nuclear Pore Complex

After the modifications of the primary transcript, the functional mRNA must transverse the nuclear envelope to the cytoplasm to direct the synthesis of a protein. Although this step is important, it is probably the least understood stop of the gene expression pathway. It is clear that transport of mRNA to the cytoplasm requires that RNA passes the nuclear envelope through the nuclear pores, presumably by an active transport mechanism that requires the recognition of the RNA, or a protein bound to it, by a receptor molecule either within the nucleoplasm or directly associated with the pore complex (Fig. 5–35).

Experiments have shown that primary transcripts that cannot be appropriately processed are retained in the nucleus and will be degraded; they are not allowed transport into the cytoplasm until all processing steps are completed. Therefore, there must be components within the nucleus that cause selective retention of RNA molecules. This selective retention might be operative on a broader scale, allowing the possibility of nuclear components serving as a filtering mechanism in the determination of which RNA is transported. Currently, the mechanisms that allow the transport of RNA are largely unclear, although without this transport, an RNA cannot complete the path of information transfer. It is clear that this involves receptors and an energy-dependent transport system through the pores, which is similar to protein import processes (see Fig. 5–4). Thus, it is a necessary step that has potential regulatory control of gene expression.

RNA in the Cytoplasm Is Subject to Degradation

Once RNA reaches the cell's cytoplasm, it is subject to degradation by nuclease components resident in the

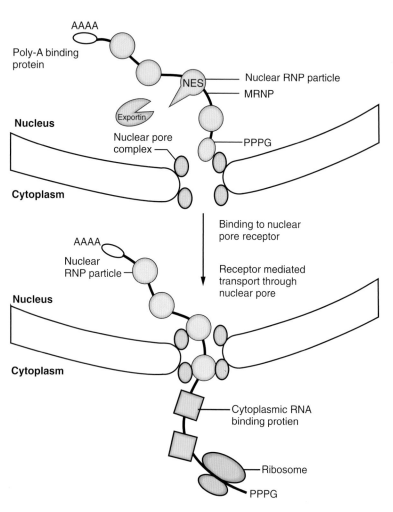

AAAA

Poly-A binding
protein

Nuclear RNP particle

NES

MRNP

Exportin

Nucleus

Nuclear pore
complex

PPPG

Cytoplasm

Binding to nuclear
pore receptor

AAAA

Nuclear
RNP particle

Receptor mediated
transport through
nuclear pore

Nucleus

Cytoplasm

Cytoplasmic RNA
binding protien

Ribosome

PPPG

Figure 5–35. Potential mechanism of messenger RNA (mRNA) transport through the nuclear pore complex. mRNA that is ready for transport to the cytoplasm is bound by a variety of proteins, including a polyadenylic acid (poly-A)–binding protein, ribonuclear protein (RNP) particles, and perhaps others. These proteins protect the mRNA from degradation and perhaps bind with a receptor molecule on the nuclear pore complex. Once bound, the nuclear pore receptor may facilitate the transport of the mRNA to the cytoplasm, similar to the mechanism of protein import to the nuclear compartment.

cytoplasm; that is, RNA, as well as other cellular components, are continuously being replenished by a balance of their synthesis and degradation. In eukaryotic cells, mRNA is degraded at different selective rates. A measure of degradative rate for a particular mRNA is called its **half-life** (the period it takes to degrade an RNA population to half its initial concentration), and RNA with longer half-life measurements are said to be more stable. For example, β-globin mRNA has a half-life longer than 10 hours, whereas RNA encoding the growth factors called *fos* and *myc* have measured half-lives in the same cells of approximately 30 minutes. Therefore, the β-globin RNA is more stable than are *fos* and *myc*, and these experiments demonstrate the ability of selective mRNA degradation within the cell. In addition, this example is indicative of how selective degradation of mRNA might control expression. Simply stated, a β-globin mRNA, by virtue of its longer resident time in the cytoplasm, has the ability to direct the synthesis of more protein because it is in contact with the synthetic machinery (the ribosomes) for a longer period than are *fos* and *myc* RNA.

The stability of mRNA can be influenced by extracellular signals. The primary response of cells to steroid hormones is an increased transcription rate of selective genes. However, the hormone also can influence the expression of these gene products by increasing the stability of their mRNA in the cytoplasm of the cell. Certain signals may cause the selective degradation of RNA, leading to less protein being expressed. For example, the addition of iron to cells decreases the stability of the mRNA that encodes the ion-scavenging transferrin receptor (Fig. 5–36).

The altered stability of the mRNA is mediated by a specific nucleotide sequence within the 3'-nontranslated region of the molecule. This region of the transferring receptor mRNA is bound by an iron-sensitive receptor protein that, when resident, protects the RNA from degradation. In the presence of excess iron, the receptor is dislodged from the 3'-nontranslated binding site, and the RNA is rapidly degraded, preventing the synthesis of the transferring receptor protein.

As suggested by experiments examining transferrin mRNA stability, the selective degradation of much mRNA is controlled, at least in part, by specific nucleotide sequences within the 3'-nontranslated region of the RNA molecule. This concept was demonstrated by genetic engineering, mixing, and matching specific regions from RNA displaying different stabilities, such as the experiments shown in Figure 5–29. When the 3'

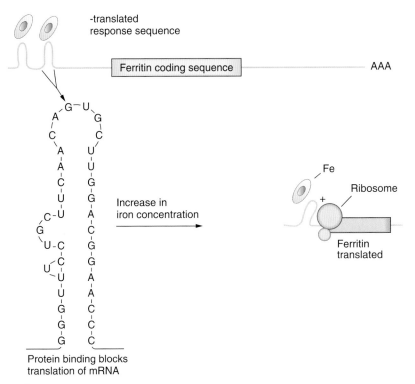

A

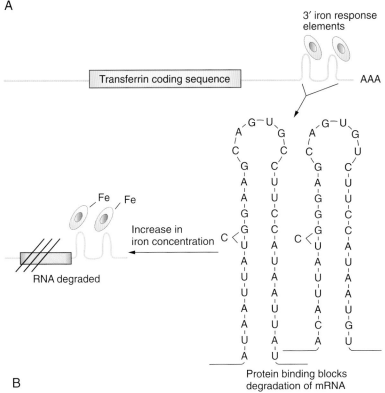

B

Figure 5–36. Iron metabolism in cells is regulated by modulation of ferritin translation and destabilization of transferring messenger RNA (**mRNA**). An iron-sensitive receptor protein is able to bind to specific sequences of the ferritin and transferring mRNA. **A:** The ferritin sequence is located in the 5′-nontranslated region of the mRNA, and the bound protein blocks the translation of this mRNA. **B:** The transferring mRNA contains similar binding sequences, and the binding of the same iron-sensitive receptor protein to these sequences stabilizes this mRNA, allowing more protein to be translated. An increase in iron concentration in cells is sensed by the receptor, and on binding the excess iron, the conformation of the protein changes so that it no longer is bound to the mRNA. The result is a release of the translational block of ferritin mRNA and a destabilization of the transferring mRNA. Thus, the regulation of intracellular iron concentration is accomplished quickly by posttranscriptional regulation of gene expression.

noncoding segment of a stable mRNA, such as globin, is substituted for the analogous region of a nonstable growth factor mRNA (e.g., *fos*), the resultant growth factor mRNA displays a stability similar to that of the globin RNA. That is, the engineered growth factor RNA becomes more stable solely because of its new 3′-nontranslated segment. Similarly, when the 3′ terminus of

histone RNA, an RNA that shows selective stabilization during the DNA S phase of the cell cycle, is placed onto a globin mRNA, the globin mRNA then acquires the cell cycle–dependent degradation characteristics of the histone mRNA. The conclusion drawn from these mix-and-match experiments is that the specific ability of RNA is governed, in part, by sequences resident within the 3′ noncoding portion of the RNA molecule. However, the cellular components responsible for this selective degradative process are not well-established.

Gene Expression Can Be Controlled by Selective Translation of Messenger RNA

The second basic step of gene expression is the translation of mRNA into protein. In eukaryotes, the translation of proteins occurs in the cytoplasm; however, not all mRNA is translated on arrival to the cytoplasm. The RNA molecules in the cytoplasm, as in the nucleus, are constantly associated with proteins, some of which may function to regulate translation. Most of the defined mechanisms that regulate translation operate to repress protein synthesis (negative translational controls), although some evidence has been reported (derived from studies examining viral RNA) that positive or enhanced translation of certain RNA may be operative. Translational controls are important in many fertilized eggs, which must rapidly switch from making the proteins required for maintenance of a quiescent oocyte to making proteins required for cell division and growth. These eggs have stored as RNA-protein complexes maternal mRNA that is not translated until the egg is fertilized.

Another important negative translational control has been demonstrated for the expression of iron storage protein ferritin. Ferritin mRNA in the cytoplasm shifts from an inactive RNA-protein complex to a translationally active polyribosome on increase of intracellular iron concentration. The block of ferritin mRNA translation is mediated by a 30- to 40-nucleotide segment of the RNA at the 5′ leader (5′-nontranslated region) segment of the molecule (see Fig. 5–36). This segment of the RNA binds a repressor protein that blocks the ability of ribosomes to form active complexes on the mRNA. This repressor, called the **iron response molecule,** is the same protein that is bound to the 3′-nontranslated segment of the transferring receptor mRNA discussed in the previous section. Thus, this iron response protein allows exquisite control of intracellular iron metabolism by increasing the degradation of the mRNA encoding an iron-salvaging protein, the transferring receptor, and simultaneously releasing the translational block of the iron-binding protein, ferritin. This provides rapid, sensitive controls of gene expression without affecting the synthetic rates (transcription) of these mRNA.

Modification of Protein Posttranslationally Can Affect the Expression of an Active, Functional Molecule

Once a protein has been synthesized by the ribosomal complexes in the cytoplasm, the functional capabilities of the protein often are not realized until the protein has been modified. Although these mechanisms, collectively referred to as *posttranslational modifications,* are not often thought of as gene expression controls, it is recognized that the manifestation of gene expression is not complete until a protein is performing its function within the cell. An example of posttranslational modifications occurs in the formation of a functional insulin protein. Insulin is a secreted polypeptide hormone that is synthesized on ribosomes associated with the ER. It is synthesized as a single polypeptide referred to as a *preproinsulin.* The prefix *pre-* refers to a signal peptide sequence that directs the translocation of the proinsulin molecule across the ER membrane. This presequence is immediately removed by protease cleavage, the first posttranslational modification. The proinsulin molecule then folds such that the NH_2- and COOH-terminal ends are held in close proximity by sulfhydryl bonds. The proinsulin is inactive until a second cleavage event removes the connective, or C peptide, leaving two chains formed from the NH_2- and COOH-terminal domains of the proinsulin peptide. Thus, multiple posttranslational events are required to form an active insulin hormone. Moreover, the sequestering of the hormone into secretory vesicles can be viewed as a regulatory event, in that some proteins are required at specific cellular locations for them to exert the functional properties. Any number of events or modifications alter the activity of a protein (e.g., phosphorylation, methylation, glycosylation, and so forth), or may play a role in the selective degradation of proteins (the addition of a ubiquitin molecule) that can constitute posttranslational controls of gene expression.

Structural RNA (Transfer and Ribosomal RNA) Are Subjected to Regulatory Mechanisms

The preceding sections have limited discussions to RNA synthesized by polymerase II, the mRNA. However, it is important to point out that the structural class of RNA—rRNA and tRNA—is also subject to regulation. RNA polymerase III transcribes the tRNA and a class of small RNA, referred to as 5S RNA, each of which is made as precursor molecules that are subsequently modified to form functional molecules. Moreover, the transcription of these RNA is facilitated by the specific binding of proteins to the promoters of these genes, one of which has been purified and is called *transcription factor (TF) IIIC.* Similarly, RNA polymerase I transcribes ribosomal RNA, a process facilitated by the binding of factors. For example, there is a protein called TFID that regulates the transcription of the 45S rRNA precursor, which is modified in the nucleolus to form 18S (small) and 28S (large) rRNA. Because each of the RNA (mRNA, rRNA, tRNA) must work in concert to provide the cell with functional proteins, it is clear that regulatory events for each RNA class are important for the overall transfer of information from DNA to protein.

GENETIC THERAPY

There Are Many Obstacles to the Development of Effective Gene Therapies

With the development of recombinant DNA technology has come the promise, or potential, of it dramatically improving the practice of medicine. Indeed, as advances have been realized, with various aspects of DNA technologies, there are parallel applications of new methods in the clinical management of patients. RFLP analysis is commonly used to aid in the diagnosis of specific inherited diseases. The sensitivity of the PCR technique makes it useful not only for the diagnosis of inherited diseases and for latent viral diseases (e.g., AIDS), but also for forensic medicine. An ability to understand the roles of a specific gene product in the pathogenesis of human disease has allowed precise and effective clinical intervention. Moreover, recombinant DNA technologies have led to the development of new therapeutic products made possible by the ability to engineer the overexpression of genetic material in bacteria and eukaryotic cells. Although recombinant DNA technologies have made significant inroads toward diagnosis and management of human disease, the treatment of human disease through transfer of genetic material to a patient (e.g., gene therapy) is not yet commonplace in medical practice.

In general, a variety of questions must be addressed as a prelude to the implementation of an effective gene-based therapy. First, the gene that is the root cause of the genetic disease must be well-studied. This would include its identification with cloning and sequencing, as well as an understanding of its regulation and the function of the expressed gene product within the cell. Second, the gene must be delivered to the appropriate cell(s) and maintained stably expressed at an appropriate level within the cell. For example, the targeting to and expression of a normal β-globin molecule in a neuron of the brain would not provide an effective therapy for sickle cell anemia. Finally, the expression of the gene product must be able to correct or reverse the disease process. This is important, particularly for somatic cell therapies (discussed later), when the expression of an appropriate gene product will be able to effect a curative response; that is, it will reverse the disease phenotype. Few inherited diseases are understood at the level of complexity outlined earlier. However, the daily contribution to our knowledge of the molecular bases underlying inherited diseases will soon bring genetic therapies to the forefront. Two considerations for successful gene-based therapies are a suitable model system for study and strategy for gene replacement.

Many Strategies Are Available for Gene-Based Therapies

The essence of a successful genetic therapy is the availability of a strategy for gene replacement. There are basically two types of gene replacement therapies: altering of germ line cells and altering of somatic cells. Whereas the fixing of germ line cells could potentially cure the disease, many ethical issues must be addressed before this is a viable alternative. The best opportunity for gene therapy is the ability to alter somatic cell expression of the diseased gene. As stated earlier, the disease must be studied to the greatest of molecular detail before a strategy can be devised. Perhaps the best examples to date, and those for which strategies have been devised, are sickle cell anemia, adenosine deaminase deficiency, cystic fibrosis, Duchenne's muscular dystrophy, and familial hypercholesterol anemia. In each case, the gene deficiency is known, its expression pattern studied, and the function of the expressed product defined.

To effect a successful gene-based therapy, one must be able to target the gene to the appropriate cell. This can be done in two ways, ex vivo (the cells are outside organisms) or in vivo (the cells remain within the organism). Many protocols rely on ex vivo techniques because of the ability to control the cellular environment during the delivery of the genetic molecule. A potential ex vivo strategy for the genetic intervention of familial hypercholesterol anemia is outlined in Figure 5–37.

The potential therapeutic DNA must be delivered to the cell in a way that it becomes part of the recipient cell's genome. There are essentially two methods for accomplishing this: physically adding the DNA and allowing it to integrate, or using a virus to facilitate DNA uptake. The viral approach is most used because of the ability to have DNA integrate in higher numbers of cells; however, the virus (retrovirus or DNA virus are both used) can cause immunologic problems later. In either case, the cells containing the altered gene must be replaced into the organism.

There are as many strategies as there are genetic diseases, and no one approach will be applicable to all diseases. For example, it is possible to stably express protein for longer periods in skeletal muscle cells. Thus, if the gene product is used outside the cell (e.g., α1-antitrypsin is a secreted molecule made in the liver, but its absence leads to lung disease), it could be possible to use the skeletal muscle as a synthetic protein "factory." As the efficiency of genetic-based therapies becomes greater and our knowledge base expands, using better molecular techniques and more expansion model systems, there is great promise in the ability to treat human disease with a specifically designed DNA molecule.

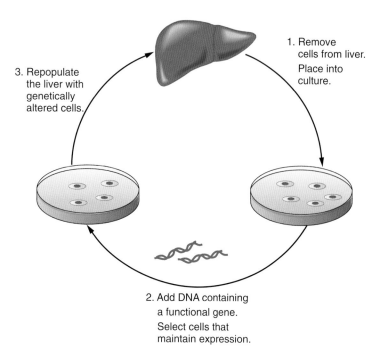

1. Remove cells from liver. Place into culture.

2. Add DNA containing a functional gene. Select cells that maintain expression.

3. Repopulate the liver with genetically altered cells.

Figure 5–37. A potential ex vivo method for genetic-based therapy of a liver disease. Hepatocytes are first isolated from the liver and placed into culture dishes with an appropriate media, such that they remain as hepatocytes. The DNA carrying an altered gene is added to the cells—either by physical means or facilitated by virus vectors—and the cells that received the gene are selected for the culture. The cultured cells containing the altered gene are then added back to the original liver tissue. It is necessary for the cells to have either a selected advantage or the ability to repopulate the liver tissue, so that the genetically altered cells will be effective in their therapeutic nature.

SUMMARY

The central regulator of the function of a cell is the nucleus. The nucleus houses most of the DNA within the cell, and it has the ability to use this stored information to allow the cell to develop and survive. The genome of the organism is acted on by proteins within the nucleus that allow the cell to identify the specific gene segments needed and to express these genes in an appropriate fashion. There are several steps to this process, and regulation of gene expression can occur at any or all of these steps. Misexpression of genes can be the root cause of disease processes. Thus, the cell has efficient ways to detect and fix genomic segments when they become altered. Mutants in specific enzyme systems can affect how the organism responds to outside influences. For example, mutations in the cytochrome P450 genes expressed in liver can cause an individual to be a poor metabolizer of a specific drug, leading to complications if the drug is prescribed. With the advent of whole-genome sequencing, and new technologies of proteomics, medicine is now beginning a new path toward individualized therapeutics. This means that by understanding the genetic makeup of an individual, the medical team can give therapeutics an individualized, optimized format. This will lead to major changes in medical delivery because unwanted side effects of therapeutics can be somewhat eliminated. Thus, a continued understanding of genetics and how genetics works will facilitate medicine in the future.

Suggested Readings

Asbersold R, Mann M. Mass spectrometry-based proteomics. *Nature* 2003;422:198–207.

Dean DA, Strong DD, Zimmer WE. Nuclear entry of nonviral vectors. *Gene Therapy* 2005;11:881–890.

Dewar JC, Hall IP. Personalized prescribing for asthma: is pharmacogenetics the answer? *J Pharm Pharmacol* 2003;55:279–289.

Dietel M, Sers C. Personalized medicine and development of target therapies: the upcoming challenge for diagnostic molecular pathology. A review. *Virchows Arch* 2006;6:744–755.

Hunter CV, Tiley LS, Sang HM. Developments in transgenic technology: applications for medicine. *Trends Mol Med* 2005;11: 293–298.

Ingelman-Sundberg M. Pharmacogenetics of cytochrome P450 and its application in drug therapy: the past, present, future. *Trends Pharmacol Sci* 2004;16:337–342.

Kozarova A, Petrinac S, Ali A, Hudson JW. Array of informatics: applications in modern research. *J Proteome Res* 2006;5:1051.

Lewin B, Ed. *Genes,* 7th ed. New York: Oxford University Press, 1999.

Tsongalis GJ, Silverman LM. Molecular diagnostics: a historical perspective. *Clin Chim Acta* 2006;2:350–355.

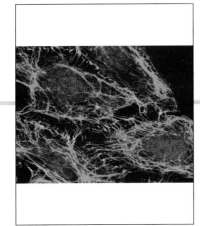

Chapter 6

Cell Adhesion and the Extracellular Matrix

Multicellular organisms exist because cells are able to bind to each other. The cellular property that mediates the binding is called *cell adhesion*. Cells may adhere either directly to each other in **cell-cell adhesion** or to extracellular components that provide a structural framework for cell binding. These extracellular components are collectively called the *extracellular matrix* (ECM). The binding of cells to the ECM is termed **cell-matrix** or **cell-substratum adhesion**.

Cell adhesion and the ECM are, together, crucial for the development and maintenance of tissue structure and function. It follows that abnormalities of the matrix or of adhesion compromise tissue function and cause human diseases. These abnormalities may be specific such as a mutation in the gene encoding a molecule involved in leukocyte adhesion. Alternatively, they may be part of an accumulation of changes that cause cancer, where loss of adhesion promotes metastasis or spread of cancer cells to other tissues.

Cell adhesion is not just a structural element that binds things together, it is also a highly dynamic process. For example, the cells of the epidermis, called *keratinocytes*, have to be tightly bound together to provide the barrier properties of the skin that protect humans against fluid loss, infection, and the wear and tear of everyday activities. However, epidermal keratinocytes are constantly being lost from the outer surface and replaced from the basal layer of the multilayered epidermis such that all cells are renewed approximately once every 28 days depending on body site. Therefore, there is a constant upward migration of keratinocytes, and this involves changes in both cell-cell and cell-substratum adhesion. Other cells such as blood platelets must circulate freely in the blood and thus must be nonadhesive. However, when wounding occurs, they must become adhesive rapidly to participate in hemostasis, that is, the coagulation of the blood that prevents bleeding. Clearly, cells must possess mechanisms for modulating their adhesiveness; that is, signals must exist that arise intracellularly that modify the stickiness of the cell surface. Moreover, signals associated with cell adhesion are by no means all one way. In addition to the "inside-out" signaling referred to earlier, cell adhesion generates "outside-in" signals that provide cells with information about their environment. Such signals are important regulators of cellular functions such as gene activity and differentiation, cell proliferation, and programmed cell death or apoptosis. A dynamic relation also exists between cells and the ECM. Cells produce the matrix and also the enzymes that degrade it, thus regulating its composition and turnover. A prime function of ECM is to provide tissues with shape, strength, and elasticity, as well as to act as a substratum for cell adhesion. However, the matrix also regulates cellular activity. It is an important reservoir of growth factors, which bind to it, and many of the "outside-in" cell adhesion signals are generated by specific interaction between cells and the matrix.

To understand cell adhesion, one must know about the molecules and structures that mediate it. From structural considerations one can move on to adhesion dynamics and the adhesive signals that regulate cell behavior and function. Similarly, knowledge of the components of the ECM and their structure must precede an examination of how cells interact with the matrix.

Molecules that mediate cell adhesion are called **cell adhesion molecules** (CAMs). They are also sometimes called *cell adhesion receptors* to indicate their specific binding to ligands that may be matrix proteins or other CAMs.

CELL ADHESION

Most Cell Adhesion Molecules Belong to One of Four Gene Families

The majority, but not all, of CAMs belong to one of four gene families: cadherins, the immunoglobulin (Ig) family, the selectins, and the integrins (Fig. 6–1). In human tissues, most of these are single-pass transmembrane proteins. Thus, they have an extracellular domain that participates in adhesion, a transmembrane domain that anchors the protein in the cell membrane, and a cytoplasmic domain that mediates attachment to the cytoskeleton. Almost invariably the NH_2 terminus is extracellular and the COOH terminus is intracellular.

The cytoplasmic domain is also the region that interacts with signaling molecules in the cell, and thus is involved in signal transduction and adhesive regulation. Adhesive binding by adhesion molecules may be either **homophilic,** meaning that the molecule binds to another of the same type, or **heterophilic,** meaning that binding is to a molecule of different type (Fig. 6–2). Cell-cell adhesion is commonly, but not always, homophilic, and cell-matrix adhesion is always heterophilic.

Cadherins Are Calcium-Dependent Cell-Cell Adhesion Molecules

The cadherins are a large family of calcium-dependent cell-cell adhesion molecules (see Fig. 6–1). They are exemplified by **E-cadherin** (epithelial-cadherin), which is widely expressed in epithelial cells and was discovered principally because of its role in early development of the mammalian embryo. Typically for cadherins, the extracellular (EC) domain consists of five similar subdomains that are barrel-shaped and held in an extended configuration by bound Ca^{2+} ions. Removal of calcium causes the EC domain to collapse and adhesion to be lost. Adhesive binding by most cadherins, including E-cadherin, is preferentially homophilic, involving interaction between the terminal subdomains of molecules on opposite cell surfaces (see Fig. 6–2).

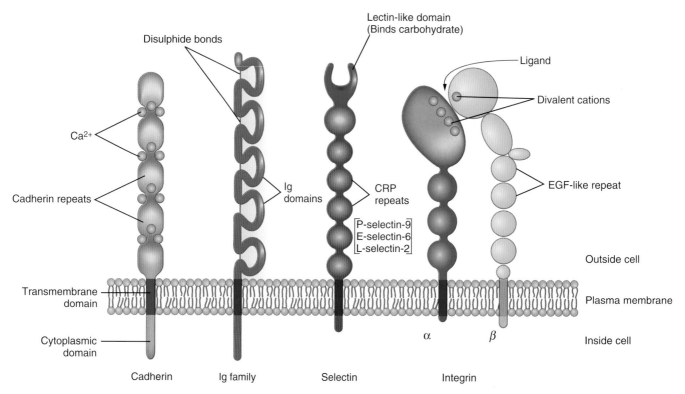

Figure 6–1. Most cell adhesion molecules belong to one of four major gene families: the cadherins, the immunoglobulin family, the selectins, and the integrins. The molecular structure of each is shown diagrammatically. *(Modified from Garrod DR. Cell to cell and cell to matrix adhesion. In: Latchman D, ed. Basic Molecular & Cell Biology, 3rd ed. Oxford, United Kingdom: Blackwell BMJ Books, 1997:80–91, by permission.)*

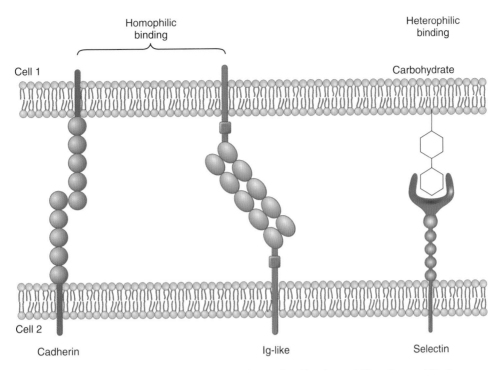

Figure 6–2. Binding by cell adhesion molecules can be either homophilic or heterophilic. In homophilic adhesion, a molecule on one cell surface binds to another identical molecule on the opposite cell surface. In heterophilic adhesion, the molecule binds to a molecule of a different type on the other cell.

The cadherin cytoplasmic domain is important for adhesive function. It binds molecules called α-catenin, β-catenin, and p-120 catenin (see Fig. 6–9). (*Note:* The nomenclature is strange here. β-catenin and p-120 catenin are distantly related members of the same family, but α-catenin belongs to a different family.) Experimental mutation of the cytoplasmic domain of E-cadherin to prevent catenin binding abolishes adhesion. It is thought that E-cadherin, β-catenin, and α-catenin may form a complex that binds to the actin cytoskeleton, thus anchoring the adhesion molecule in the cell. However, recent evidence suggests that α-catenin can form a complex with actin, or with E-cadherin and β-catenin, but that all four molecules do not form a complex simultaneously.

Homophilic binding by cadherins is believed to contribute to tissue segregation during embryonic development. For example, when the neural tube, the precursor of the central nervous system, is first formed it expresses N-cadherin, whereas the ectoderm from which it is derived expresses E-cadherin. E-cadherin is mutated and/or down-regulated in some cancers, for example, gastric and breast cancer, possibly facilitating the spread of cancer.

The Immunoglobulin Family Contains Many Important Cell Adhesion Molecules

The extracellular domains of CAMs are characterized by the presence of at least one but usually multiple subdomains that resemble the structure of the basic subdomain of antibody molecules or immunoglobulins (see Fig. 6–1). The structure of these Ig subdomains is stabilized by disulfide bonds rather than by Ca^{2+} ions so these molecules participate in calcium-independent adhesion. Adhesion may be homophilic (see Fig. 6–2) or heterophilic and is usually cell-cell, although some cell-matrix interactions are known. Several adhesion molecules of the nervous system are within this group, including the **neural cell adhesion molecule (NCAM)**, L1, and TAG, which are involved in neuronal guidance and fasciculation or bundling during development and regeneration of the nervous system. In epithelial cells, the Ig family molecule nectin is closely associated with E-cadherin. Ig family members are also important in the adhesion of leukocytes. Thus, heterophilic binding between two Ig-like molecules, lymphocyte function–related antigen 2 (LFA-2) and LFA-3, contributes to cell-cell adhesion of T cells. Leukocyte adhesion to endothelial cells, an important part of the inflammatory response, is mediated by heterophilic binding between the so-called **intercellular adhesion molecule (ICAM)** or the **vascular cell adhesion molecule (VCAM)** on endothelial cells and integrins (see later) on the leukocytes.

The cytoplasmic domains of this diverse group of adhesion molecules are involved in interaction with the actin cytoskeleton. For example, NCAM has several alternative COOH termini that arise through differential **splicing of messenger RNA** derived from a single gene. This generates a series of different molecules, a soluble form with no membrane anchor and three membrane-anchored forms of different sizes, 120, 140, and 180 kDa (Fig. 6–3).

The two largest forms of molecules have transmembrane and cytoplasmic domains, but the 120-kDa form has neither, instead being anchored to the cell membrane by a glycosylphosphotidylinositol linkage. This variation in structure offers the opportunity for alternative functional regulation. The cytoplasmic domains of the 140-kDa form bind to the cytoskeletal protein α-actinin and the 180-kDa form to α-actinin, actin, and spectrin. Another family of actin-binding protein, ezrin, radixin, and moesin (the ERM proteins) interact with the cytoplasmic domain of L1. The cytoplasmic domain of nectin binds to afidin that links the adhesion molecule to the actin cytoskeleton. ICAM also interacts with the actin cytoskeleton and transduces outside-in signals that mediate cytoskeletal reorganization.

The Ig family is large and diverse probably because the basic structure of the Ig subdomain is versatile and readily adaptable to a variety of different binding functions. However, only the T-cell receptor and the immunoglobulins themselves have somatically variable domains necessary for antigen recognition; the majority of Ig family subdomains are of constant structure. In this context, it is interesting that the involvement of Ig family members in adhesion preceded the immune system in evolution. The requirement for all cell adhesion to mediate multicellularity is much older than the development of complex immune responses.

Selectins Are Carbohydrate-Binding Adhesion Receptors

The selectins participate exclusively in heterophilic cell-cell binding because they have a **lectin**-like domain at their NH₂ terminus that binds to specific carbohydrate residues on the opposite cell surface (see Figs. 6–1 and 6–2). An epidermal growth factor–like domain followed by a series of complement regulatory protein repeats comprise the remainder of the extracellular domain, and they are followed by a transmembrane domain and a short cytoplasmic domain that interact with the cytoskeleton.

The selectin family is composed of three members: L- (leukocyte), E- (endothelial), and P- (platelet) selectin. Adhesive binding is calcium dependent. Ligands for selectin include so-called sialyl-Lewis X saccharides, and the best characterized counterreceptor is a mucin-like glycoprotein (GP), P-selectin GP ligand-1, present on leukocytes. Selectins have a major physiologic role in initiating the adhesion of leukocytes and platelets during the inflammatory and hemostatic responses. L-selectin is also a "homing receptor" that mediates lymphocyte binding to endothelium in peripheral lymph nodes.

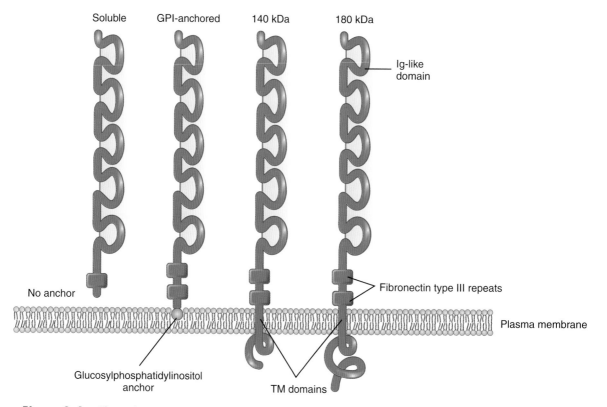

Figure 6–3. Alternative messenger RNA splicing in neural cell adhesion molecule (NCAM). This gives rise to alternative structure of the NCAM COOH terminus and absence or different mechanisms of anchorage to the plasma membrane. TM, transmembrane.

Integrins Are Dimeric Receptors for Cell-Cell and Cell-Matrix Adhesion

Unlike other adhesion receptors, integrins are **heterodimers** that consist of one α subunit and one β subunit (see Fig. 6–1). They are an important family of cell-matrix and cell-cell adhesion molecules that participate exclusively in heterophilic interactions. Eighteen different α subunits and eight different β subunits are known in vertebrate tissues. These subunits may associate in various binary combinations, and integrins may be classified into different subfamilies depending on which β subunit is involved. Thus, β_1 integrin may associate with one of nine different α subunits to generate a series of predominantly matrix receptors of differing ligand specificity. In contrast, the β_2 integrins are a group of cell-cell adhesion receptors of lymphoid cells with three alternative α subunits. Furthermore, some α subunits can associate with different β subunits (for example, $\alpha_6\beta_1$ and $\alpha_6\beta_4$).

Some integrins have quite specific binding properties, whereas others are promiscuous. Thus, $\alpha_5\beta_1$ binds to the **arginine-glycine-aspartic acid** (**RGD** in single-letter amino acid code) tripeptide sequence of the ECM protein fibronectin, but $\alpha_v\beta_3$ binds to several matrix components including vitronectin, fibronectin, fibrinogen, von Willebrand factor (vWF), thrombospondin, and osteopontin. An interesting example is $\alpha_4\beta_1$, which can bind to a specific domain of fibronectin but also to the Ig family adhesion receptor VCAM on endothelial cells. A further complication is that individual cell types usually express multiple integrins. For example, blood platelets express predominantly $\alpha_{IIb}\beta_3$ (GPIIb/IIIa), which binds to fibrinogen, fibronectin, vWF and vitronectin, and lesser amounts of $\alpha_v\beta_3$, $\alpha_5\beta_1$, $\alpha_2\beta_1$ (collagen), and $\alpha_6\beta_1$ (laminin) (see Fig. 6–22).

Integrin subunits have large extracellular domains, single transmembrane domains, and short cytoplasmic domains (except β_4, which has a large cytoplasmic domain). Both α and β subunits are involved in ligand binding. The extracellular domains can switch between a low-affinity state that does not bind ligand and a high-affinity, ligand-binding state (Fig. 6–4). Switching can be triggered by signals arising within the cell, thus enabling cells to change from nonadhesive to adhesive. This mechanism is important in regulating cell behavior (see later) and is called **integrin activation.** The high-affinity state is dependent on the divalent cations Mn^{2+} or Mg^{2+}, and the low-affinity state is stabilized by Ca^{2+}.

Integrin cytoplasmic domains are involved in the transmission of "inside-out" signals that modulate adhesive binding. They also mediate interaction of integrins with the actin cytoskeleton and with a series of signaling molecules that are involved in transduction of "outside-in" signals (see later).

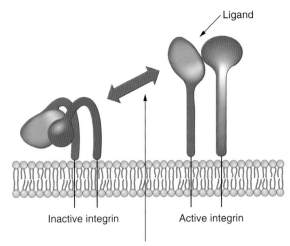

Figure 6–4. Signals from within cells regulate the adhesive binding activity of integrins. The collapsed form is inactive and unable to bind ligand, whereas the extended form is active in ligand binding.

INTERCELLULAR JUNCTIONS

CAMs may be widely distributed on the cell surface. However, they are often collected together in cell-surface structures or organelles called *junctions*. In addition to those junctions primarily involved in cell-cell or cell-substratum adhesion, there are others that function to seal the spaces between cells and regulate cell polarity or to participate in direct cell-cell communication.

Cell junctions were identified and much studied by electron microscopy long before anything was known about their molecular composition. Intercellular junctions are exemplified in the ultrastructure of the **junctional complex,** an association of three junction types at the apicolateral borders of simple epithelial cells such as those of the intestinal mucosa (Fig. 6–5). From apical to basal the junctional complex consists of a tight junction or zonula occludens (ZO; plural, zonulae occludentes), an adherens junction or zonula adherens (plural, zonulae adherentes), and a desmosome or macula adherens (plural, maculae adherentes). The apicobasal order in which these junctions are arranged is of great importance in cell and tissue function, as will be seen when they are considered in more detail later in this chapter. The fourth type of intercellular junction is the gap junction or nexus, which has a major role in intercellular communication.

At the cell-substratum interface of various epithelial cells, adhesion to the matrix is maintained by hemidesmosomes, structures that resemble half desmosomes but differ from desmosomes in molecular composition. Another type of cell substratum junction is the focal contact or focal adhesion. This has been studied mainly in cultured cells, and much has been learned about cell substratum interactions. The structure and function of

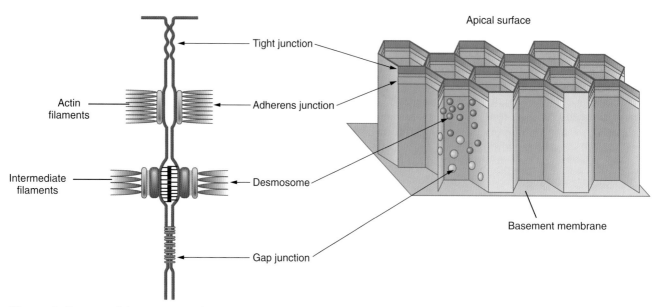

Figure 6–5. Intercellular junctions and the junctional complex. Simple epithelial cells have a junctional complex at their apicolateral borders. The components are the tight junction, the adherens junction, and the desmosome. The tight and adherens junctions are zonular, extending right around the cells, whereas desmosomes are punctate. Desmosomes are also present beneath the junctional complex, as is another punctate junction, the gap junction.

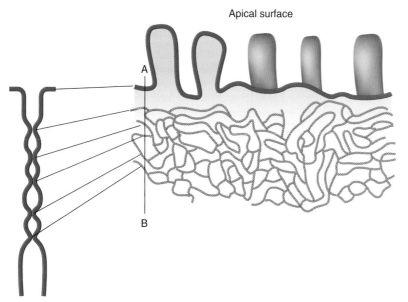

Figure 6–6. Ultrastructure of the tight junction. Drawing of the type of anastomosing network of strands of membrane particles seen by freeze fracture and electron microscopy at the apicolateral margins of intestinal epithelial cells. A section in direction *A-B* would appear as in the diagram (left). The "membrane kisses" of the tight junction correspond to the particle strands. The zonular nature of the junction is apparent as such a network extends completely around each cell.

these junctions are considered in more detail in the following section.

Tight Junctions Regulate Paracellular Permeability and Cell Polarity

Electron microscopy shows tight junctions as regions where the outer leaflets of the plasma membrane of adjacent cells touch, so-called membrane kisses, so that the width of the intercellular space is reduced to zero. A technique called *freeze-fracture* that splits cell membranes between the inner and outer leaflets enabled electron microscopists to show that tight junctions consist of a network of membrane particles that form a zone or belt around the apicolateral margin of the cell (Fig. 6–6). Addition of electron opaque substances to the upper surfaces of epithelia demonstrated that these could not penetrate through tight junctions to the basolateral intercellular space. These observations justify the name zonula occludens: "zonula" because the tight junction is a zone that extends around the cell, and "occludens" because it occludes or blocks the so-called paracellular channels, the spaces between cells.

One of the major roles of tight junctions, called the **"gate" function,** is to occlude the spaces between cells or, more precisely, to regulate the permeability of these spaces selectively (Fig. 6–7, A). This function is essential to the **barrier properties of epithelia,** such as the cell

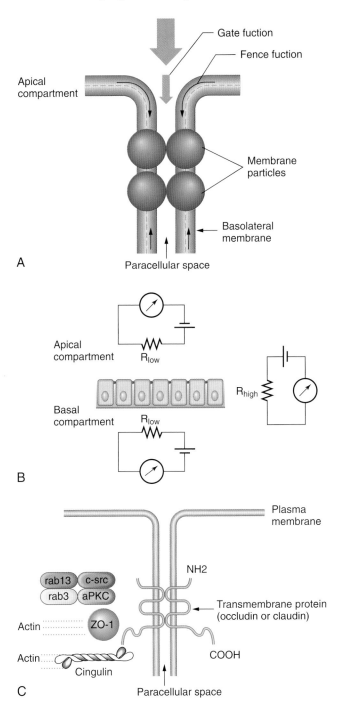

A

B

C

Figure 6–7. Structure and function of tight junctions. A: The "gate" function of the tight junction refers to its property of regulating the permeability of the paracellular channels, and the "fence" function maintains the separation of molecules in the apical and basolateral cell membranes. B: The low permeability of tight junctions causes high electrical resistance across the epithelial cell layer between the apical and basal compartments. C: Some of the molecules that contribute to the structure and function of the tight junction are shown.

layers that line the intestine, the kidney tubules and the airways, and the endothelia that line blood vessels. It means that an epithelial cell layer can act as a barrier separating biological compartments, and that the living elements of that barrier, the cells, can regulate what passes across from one compartment to another. If the electrical resistance between the apical and basal components is measured, it is found to be much greater than that of the solutions or media on either side. This arises because the membranes of the cells, being composed of tightly packed lipid layers, have a high electrical resistance called the **transepithelial resistance** and because the intercellular spaces are occluded by tight junctions (see Fig. 6–7, B). Ions that carry electric current thus can move much more easily within the apical or basal media than they can across the epithelium. If tight junctions are disrupted experimentally, the high electrical resistance of the epithelium is lost and both ions and large molecules such as proteins can pass readily between the apical and basal compartments. Reduction in epithelial barrier function has important consequences for human diseases such as inflammatory bowel disease and Crohn's disease in the intestine and allergic rhinitis and asthma in the airway.

A second vital role of tight junctions is called the **"fence" function** (see Fig. 6–7, A). Epithelial cells are polarized; they present different faces with differing functions to the compartments above and below, and this **polarity** is crucial for their function. Key to this function is the presence of different molecules in the apical and basolateral cell membranes. Unless they are restrained by linkage to the cytoskeleton, lateral diffusion of molecules in cell membranes is extremely rapid; that is, molecules can move around and quickly change places with each other in the plane of the membrane. Thus, if there was nothing to prevent them, the molecules on the apical and basolateral surfaces of the cell would soon get mixed up. Tight junctions act as "fences" to separate the molecules in the apical and basolateral membranes, thus helping to maintain cell polarity. Loss of cell polarity is a common feature of epithelial cancer cells (carcinoma cells) that have become poorly differentiated.

Because tight junctions are not considered as primarily "adhesive" junctions, their molecular components have not been considered among the CAMs described earlier. However, mutual binding by tight junction molecules on adjacent cells is clearly vital for the occluding function. The membrane components of tight junctions are the proteins **occludin** and **claudin** (see Fig. 6–7, C). Occludin is the product of a single gene and is ubiquitous in tight junctions. Claudins are a family of some 20 proteins that contribute differentially to the composition and function of tight junctions in different epithelia. For example, claudin 5 is particularly important in the tight junctions of endothelial cells that form the blood–brain barrier, and claudin 16 is important in magnesium resorption in the kidney tubule. A genetic disease called *familial hypomagnesia* results from mutation of the claudin 16 gene.

Although unrelated to each other, occludin and claudin are both tetraspanin proteins. Thus, they have four transmembrane domains and, because their NH_2 and COOH termini are both cytoplasmic, two extracellular and one cytoplasmic loop (see Fig. 6–7, C). The extracellular loops are responsible for homophilic binding between adjacent cells, and the COOH-terminal tail is primarily responsible for interaction with cytoplasmic components. Occludin and claudins are the components of the membrane particles characteristic of freeze-fracture pictures of tight junctions.

Many cytoplasmic components of tight junctions have been identified. These components include the cytoplasmic adaptor proteins **ZO-1**, ZO-2, ZO-3, and cingulin, which interact with both membrane proteins and the actin cytoskeleton, a variety of signaling molecules including **protein kinases** and **phosphatases, small GTPases** (guanosine triphosphatases) and **G proteins**, and some **transcription factors** that shuttle between the tight junction and the nucleus (see Fig. 6–7, C). The presence of these molecules and experimental finding relating to them suggest that tight junctions contribute to the regulation of many cellular processes including proliferation and differentiation.

Adherens Junctions Are Important for Cell-Cell Adhesion

Situated immediately beneath the tight junction in the junctional complex, the adherens junction is characterized by parallel cell membranes separated by an intercellular space approximately 20 nm in width, a cytoplasmic plaque of low electron density and association with microfilaments of the **actin cytoskeleton** (see Fig. 6–5).

In simple epithelial cells, the adherens junction is zonular (hence the name zonula adherens), extending around the entire circumference of the cells accompanied by a ring of actin filaments. In other cell types, adherens junctions take various forms such as the extensive **fasciae adherentes** of cardiac muscle and the smaller, punctate junctions formed by fibroblasts and other migratory cells.

The primary function of the adherens junction is cell-cell adhesion. In epithelia where adherens junctions and desmosomes are present, both contribute to the cohesiveness of the tissue. In simple epithelia where desmosomes are relatively few, adherens junctions probably make the major adhesive contribution, whereas in stratified epithelia where desmosomes are abundant, adherens junctions probably make the minor contribution. Where they occur together, adherens junctions and desmosomes are interdependent. Initial contacts between cells are made by fine processes called *filopodia* (singu-

lar, filopodium) that interdigitate and adhere by means of small adherens junctions forming an "adhesion zipper" (Fig. 6–8, A). Contacts are stabilized by the formation of desmosomes. Further maturation of adherens junctions is dependent on stabilization by desmosomes. When desmosomal adhesion is lost experimentally in mice or through inherited mutations in humans, adherens junctions do not stabilize and the epidermis falls apart through loss of keratinocyte adhesion. Contraction of the actin ring that underlies the zonula adherens can cause morphogenetic shape change of the cell sheet, for example, in rolling of the **neural plate** to form the neural tube, the precursor of the central nervous system, during embryonic development. Such contractility is probably of wider importance but has not been thoroughly investigated (see Fig. 6–8, B). The zonula adherens probably contributed to the stabilization of tight junctions in simple epithelia.

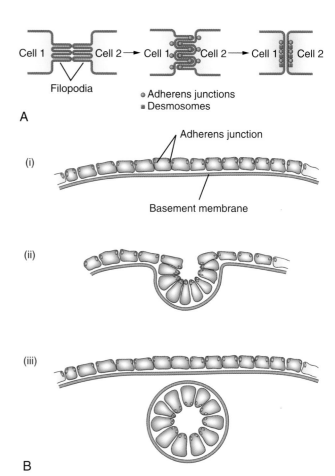

Figure 6–8. Functions of the adherens junction. **A:** Initial cell contact is made by filopodia. These interdigitate and punctate adherens junctions are formed to constitute an adhesion zipper. The adhesion is then expanded and stabilized by the formation of desmosomes. **B:** Bending of epithelial cell sheets during embryonic development is accomplished by contraction of the actin filament ring underlying the apicolateral adherens junctions. This acts like a purse string to narrow the apices of the cells. This process can result in tube formation as in generation of the neural tube.

In epithelia, the principal adhesion molecule of adherens junction is E-cadherin, though this is replaced by other cadherins in other tissues (Fig. 6–9). The Ig family molecule **nectin** (actually a small subfamily) is also a component of epithelial and other adherens junctions. The respective roles of these two CAMs in the adherens junction are unclear. However, because the formation and stability of adherens junctions is calcium dependent under experimental conditions, the role of E-cadherin appears to be crucial.

As discussed earlier, the cytoplasmic partners of E-cadherin are α- and β-catenin, and that of nectin is **afidin.** In view of recent doubt cast on the ability of the catenins and E-cadherin to form a complex with actin, it may be that the nectin-afidin complex is more important in associating the adherens junction with the cytoskeleton. The cytoplasmic plaque of the adherens junction has been shown to contain other actin-binding proteins such as **vinculin** and α-**actinin,** but it is unclear how these are involved.

In addition to being an adherens junction component, β-catenin is an important signaling molecule (Fig. 6–10). Any β-catenin that is free in the cell cytoplasm is normally proteolytically degraded. Activation of the **Wnt signaling pathway** by Wnt molecules binding to their cell-surface receptor blocks β-catenin degradation causing it to be stabilized in the cytoplasm. It can then enter the nucleus, forming a complex with members of the **TCF/LEF** (T-cell factor/leukemia enhancer factor) family and activating transcription of a number of genes that affect cell proliferation and transformation. The Wnt signaling pathway is important in embryonic development, in normal tissue maintenance, and when abnormally activated, in cancer.

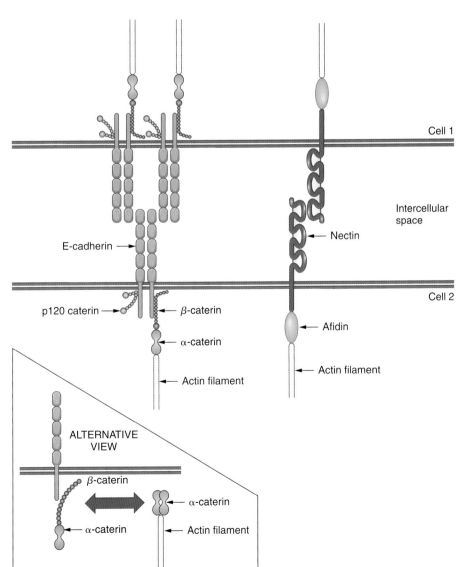

Figure 6–9. Molecular structure of the adherens junction. Two types of adhesion molecule are involved, E-cadherin and the immunoglobulin (Ig) family protein nectin. The cytoplasmic domain of E-cadherin binds β-catenin, which, in turn, binds α-catenin. This complex may link E-cadherin to the cytoskeleton, but an alternative view (inset) suggests that the whole complex cannot form simultaneously. Nectin is linked to actin by afidin.

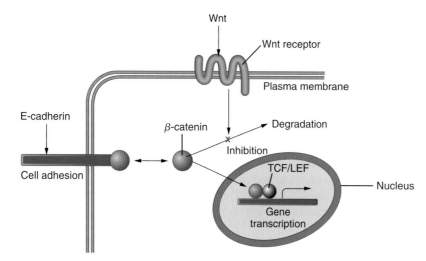

Figure 6–10. The Wnt signaling pathway. TCF/LEF, T-cell factor/leukemia enhancer factor. For further details see text on p 199.

Desmosomes Maintain Tissue Integrity

The third and most basal component of the junctional complex is the desmosome (see Fig. 6–5). Unlike the zonula occludens and the zonula adherens, the desmosome is a spotlike or punctate junction that occupies a roughly circular region of 0.5 μm or less in diameter on the cell surface. However, desmosomes are not confined to the junctional complex and commonly have a wider distribution at cell-cell interfaces. Its principal function is to maintain tissue integrity by providing strong intercellular adhesion and acting as a link between the cytoskeletons of adjacent cells (Fig. 6–11). Many tissues have such structural scaffolding extending through their cells. Cytoskeletal intermediate filaments (e.g., cytokeratin) provide the scaffolding poles, and desmosomes provide the couplings between them. The **desmosome-intermediate filament complex** is particularly well developed in tissues such as epidermis that are subject to constant sheer stress and abrasion, and is fundamentally important in maintaining tissue structure.

Electron microscopy shows that the cytoplasmic face of the desmosome consists of a dense plaque that is joined to a bundle of intermediate filaments (Fig. 6–12). The intercellular space is more than 30 nm wide and is characterized by the presence of a midline with branches extending between it and the plasma membranes of the adhering cells. This structure probably represents a highly organized arrangement of the adhesive material, and this organization may, in turn, explain why desmosomes are so strongly adhesive.

Desmosomes have two types of adhesion molecules, **desmocollin** and **desmoglein**, representatives of the cadherin family and known as the desmosomal cadherins (Fig. 6–13). Their extracellular domains form the midline structure, and their cytoplasmic domains lie in the dense plaques where they bind three other molecules, **plakoglobin, plakophilin, and desmoplakin**, that provide the link to the cytoskeleton. Plakoglobin and plakophilin

are related to β-catenin, one of the cytoplasmic components of the adherens junction, whereas desmoplakin belongs to the plakin family of cytoskeletal linker proteins.

Human diseases involving desmosomes are rare but include both autoimmune and inherited conditions. Such conditions can result in abnormalities of skin or cardiomyopathy. An autoimmune disease that affects desmosomes, pemphigus, is considered in Clinical Case 6–1. Although they have an essential structural role in tissues, desmosomal adhesion has to be dynamic and reversible. It has to permit the upward movement of cells (stratification) required for renewal of epithelia such as epidermis, and it has to release cells for movement when this is required, for example, during the closure or re-epithelialization of wounds.

Clinical Case 6-1

Phyllis Jacobson is a 56-year-old white married synagogue secretary who is now consulting the University Dental School for persistent sores in her mouth. For more than 2 months, she has been troubled by what she thought were "cold sores" inside her cheeks. Initially, she treated two isolated lesions with an over-the-counter camphor preparation that had worked for her in the past. However, there was no response in 2 weeks, and she developed two or three additional sores on her tongue. She then saw her local dentist, who believed the lesions were "viral" and treated them with a touch of a silver nitrate swab. This only made the condition worse and seemed to induce local ulceration and infection. Two weeks later, when she returned to her dentist's office and showed no improvement, the dentist said he saw vesicles and prescribed her a course of oral acyclovir. On re-examination 3 weeks later, the dentist saw several additional lesions, thought her voice was hoarse and that her larynx was involved, and now

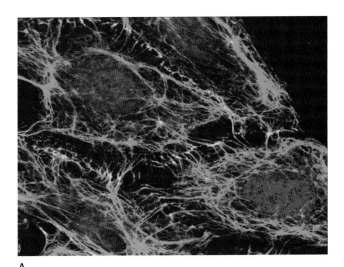

A

Figure 6–11. The desmosome-intermediate filament complex strengthens epithelia, particularly the epidermis. **A:** Fluorescence micrograph showing desmosome-intermediate filament complex in cultured epithelial cells. Desmosomes (red); intermediate filaments (green); nuclei (blue). **B:** Diagram showing the desmosome-intermediate-hemidesmosome complex in the basal epidermis. *(A: From Garrod DR. Retinoids and Lipid-Soluble Vitamins in Clinical Practice 2002;18:115–118, by permission; B: Modified from Ellison JE, Garrod DR. J Cell Sci 1984;72:163–172, with permission.)*

Epidermis

Desmosome

Intermediate filaments

Hemidesmosome

Basement membrane

Dermis

B

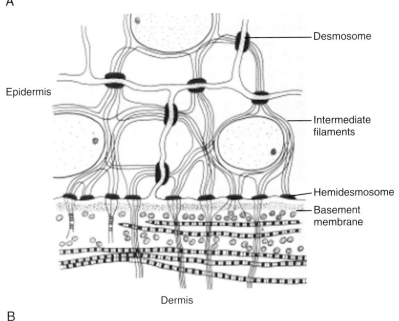

Figure 6–12. Electron micrograph of a desmosome from human epidermis. IDP, inner dense plaque; IF, intermediate filaments; ML, midline; ODP, outer dense plaque. The desmosome is about 0.5 μm wide. *(From Garrod DR. Retinoids and Lipid-Soluble Vitamins in Clinical Practice 2002;18:115–118, by permission.)*

thought that she had oral candidiasis. After a combined blood count, which was normal, he stopped the acyclovir and started her on nystatin mouthwashes. In response to her anxiety, he assured her that she did not have "mouth cancer" because the lesions were on multiple surfaces.

After 10 days of mouthwashes, Mrs. Jacobson was having difficulty chewing and swallowing and had lost confidence in her dentist. She sought help at the University Dental Clinic. Before he even looked in her mouth, the third-year dental student noted that she had several small, flaccid blisters on the midline of her forehead, just at the scalp juncture. After he saw multiple, now ulcerated, lesions in her mouth, he made a provisional diagnosis and called his senior attending. The attending performed a simple manipulation of her forehead skin and confirmed the student's opinion.

Continued

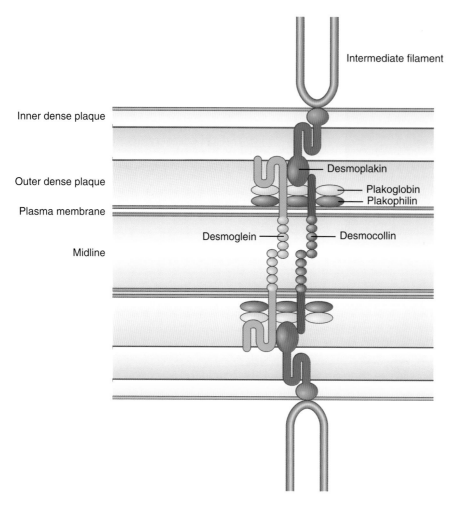

Intermediate filament

Inner dense plaque

Outer dense plaque

Plasma membrane

Midline

Desmoplakin

Plakoglobin
Plakophilin

Desmoglein

Desmocollin

Figure 6-13. Molecular composition of the desmosome.

Cell Biology, Diagnosis, and Treatment of Pemphigus Vulgaris

Mrs. Jacobson had experienced development of pemphigus vulgaris, an acquired autoimmune disorder often found in middle-aged Ashkenazi Jews. The simple test that the senior dentist performed was to apply firm lateral sliding pressure on the normal-appearing forehead skin adjacent to her blisters. This immediately induced a painless local separation of the basal layer and the first suprabasal layers of the epidermis—the so-called Nikolsky sign—which is pathognomonic for pemphigus. The clinical history was typical: Initial lesions in the oral cavity are frequently and repeatedly misdiagnosed as aphthous stomatitis, or viral or fungal infection. Also, eventual involvement of the larynx or nasal mucosa and progression to involve nonmucosal skin is common without treatment.

The pathogenesis of this formerly often lethal disorder is based on acquired autoantibodies directed against keratinocyte transmembrane adhesive proteins called *desmogleins*. Desmogleins are the adhesive molecules of the desmosomes that normally bind the acanthocytes together. The autoantibodies target the membrane surface epitopes of these cadherin-type molecules, and either directly block their adhesive function or induce their proteolytic dissolution. Whatever the intimate mechanism, the antibodies functionally interfere with their targets' essential adhesive role in binding keratinocytes together to form the remarkably stable, elastic, deformable, and complex organ known as skin. The loss of these crucial intercellular bonds permits stress cleavage between the keratinocyte layers. This acantholysis induces the blisters, as well as the structural weakness, which can be demonstrated by the Nikolsky sign.

A skin biopsy was obtained adjacent to one of Mrs. Jacobson's forehead lesions. Immunofluorescent staining demonstrated fixed, so-called intercellular IgG autoantibodies throughout her epidermis. This finding, together with the parallel demonstration of circulating intercellular autoantibodies, unequivocally confirmed the student's clinical diagnosis. After a negative tuberculin test, a normal glucose-tolerance test, and a brief clinical evaluation to rule out other major contraindications to steroid therapy, Mrs. Jacobson was prescribed 60 mg prednisone daily. She had a substantial and prompt regression of her signs and symptoms in less than 3 weeks. It is anticipated that with tapering doses she will be able to be maintained on low or minimum steroids and experience few permanent sequelae.

More recalcitrant patients may require more intense anti-immune therapy, including intravenous IgG, plasmapheresis, dapsone, and even antimetabolites or alkylating agents. However, with this armamentarium, most patients can now be successfully managed, and the diagnosis no longer carries its formerly lethal implications.

Gap Junctions Are Channels for Cell-Cell Communication

A fourth type of cell junction that is not part of the junctional complex but has a more general distribution on intercellular membranes is the gap junction. So called because electron microscopy demonstrates a regular intercellular space or gap of 2 nm in width, the principal function of the gap function is **cell-cell communication.** Gap junctions isolated from tissues and seen in plan view rather than in section by electron microscopy have the appearance of rafts of circular particles each about 7 nm in diameter and each with a dot at its center. Each particle is called a **connexon,** and each dot represents the end of a channel that passes through the middle of the connexon (Fig. 6–14). It is this channel that provides the pathway for intercellular communication.

Gap junctional communication is an important integrator of cell function in both excitable tissues such as nerves and cardiac muscle and nonexcitable excitable tissues such as epithelia. In cardiac muscle, gap junctions provide the route though which electrical impulses are propagated between the muscle fibers and thus are important in coordinating the heartbeat. The muscle fibers are said to be electrically coupled. In epithelia, they function in **metabolic co-operation** in which small metabolites and signaling molecules pass between cells. The latter function is vital during embryonic development, as well as in adult tissues.

Communication is possible because the central channels of the connexons provide an aqueous link between the cytoplasm of adjacent cells through which small soluble molecules can pass. The size restriction of the channels is about 1000 daltons so that molecules such as inorganic ions and small sugars and peptides can penetrate, but proteins and nucleic acids cannot.

Connexons are composed of a single protein **connexin** (now known to be a family of 20 genes; see Fig. 6–14, A). Connexin has four transmembrane domains, similar to but not related to the tight junction proteins

occludin and claudin. NH_2- and COOH-termini of connexin are intracellular, so the transmembrane domains are joined by one intracellular and two extracellular loops. Six connexin molecules form a hexamer that constitutes the connexon or hemichannel (see Fig. 6–14, B). By docking of the connexin extracellular loops with those of a connexon on the surface of another cell, gap junctional communication is established (see Fig. 6–14, C). The connexon channel can be regulated by intracellular signals to adopt an open or closed configuration so that intercellular communications can, in turn, be regulated. Intercellular communication through gap junctions can be demonstrated by passing an electric current between cells through intracellular microelectrodes (see Fig. 6–14, D) or by injecting a low-molecular-weight fluorescent dye into one cell and detecting its spread to adjacent cells by fluorescence light microscopy.

Connexin gene mutations are associated with a variety of human diseases including cardiovascular anomalies and cataract. For example, mutation of one family member, connexin $43\alpha_1$, is linked to a syndrome called *oculodentodigital dysplasia* that may involve developmental abnormalities of the face, eyes, limbs, and teeth.

Hemidesmosomes Maintain Cell-Matrix Adhesion

Cell-matrix adhesion of certain epithelia, especially epidermis, is mediated by specialized junctions called *hemidesmosomes*. So named because they resemble half desmosomes by electron microscopy, hemidesmosomes are responsible for strong binding between the basal surface of the epithelial cells and the underlying basement membrane and, intracellularly, for providing a link to the intermediate filament cytoskeleton.

Hemidesmosomes have dense cytoplasmic plaques that interact with the cytoskeleton (Fig. 6–15). Within the basement membrane, fine filaments called *anchoring filaments* appear to link to the outer surface of the plasma membrane opposite to the plaque. The anchoring filaments, in turn, connect to anchoring fibrils that extend from the basement membrane into the underlying collagenous matrix. Thus, hemidesmosomes appear to provide the link in a contiguous series of filaments that extends from the cell cytoplasm, through the basement membrane, and into the matrix beneath.

Any ultrastructural resemblance to half desmosomes disappears when the molecular composition of hemidesmosomes is analyzed (Fig. 6–16). The major adhesion molecule of hemidesmosomes is $\alpha_6\beta_4$ **integrin.** Also present is a type II membrane protein (NH_2-terminal cytoplasmic, COOH-terminal extracellular) called **BP180.** (BP represents bullous pemphigoid, an autoimmune blistering disorder in which the autoantibodies

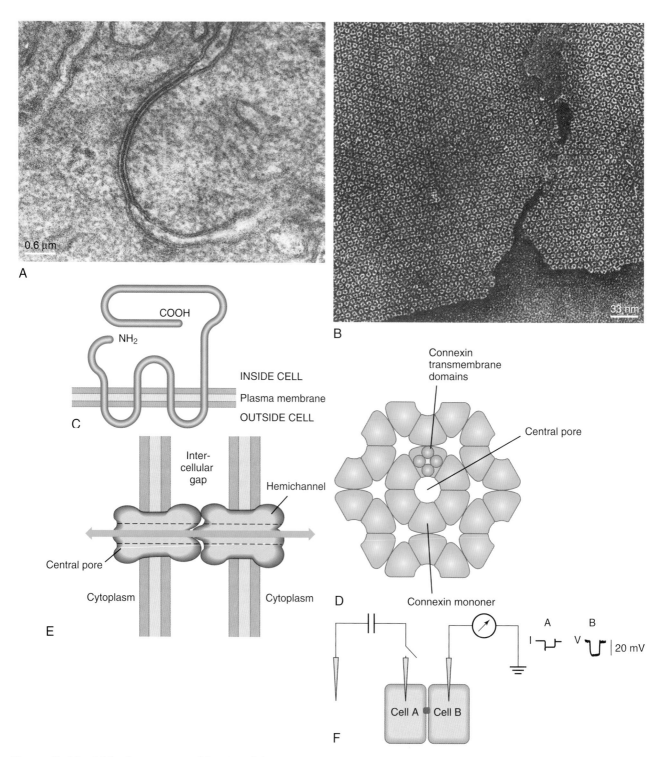

Figure 6–14. Molecular structure and function of the gap junctions. **A:** Electron micrograph of a gap junction showing the close (2 nm) approach of the cell membranes. Scale bar = 0.6 µm. **B:** Electron micrograph of surface view of an isolated gap junction showing connexons with central pores. Scale bar = 33 nm. **C:** Connexin has four transmembrane domains. **D:** Six connexin molecules form the channel of the connexon. **E:** Docking of two hemichannels between adjacent cells establishes intercellular communication. **F:** Adjacent cells are coupled electrically by gap junctions. *(A, B: From Gilula NB. Gap junctional contact between cells. In: Edelman GM, Thiery J-P, eds. The Cell in Contact: Adhesions and Junctions as Morphogenetic Determinants. New York: John Wiley & Sons, 1985:395–405, by permission.)*

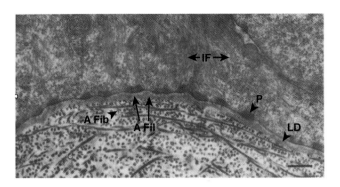

Figure 6-15. Structure of the hemidesmosome as observed by electron microscopy. A Fib, anchoring fibril; A Fil, anchoring filaments; IF, intermediate filaments; LD, lamina densa of basement membrane; P, plaque. Scale bar = 0.4 μm. *(From Ellison JE, Garrod DR. J Cell Sci 1984;72:163–172, by permission.)*

target this 180-kDa protein.) Within the plaque, two molecules are involved in linking to cytokeratin, **BP230** and **plectin,** both members of the plakin family and related to desmoplakin. Outside the membrane, the anchoring filaments appear to be composed of a member of the laminin family of ECM proteins, laminin 5, and form the substrate for $\alpha_6\beta_4$ integrin binding. The anchoring fibrils are composed of collagen type VII, a specialized member of the collagen family.

The continuity of structure formed by cytokeratin filaments, hemidesmosomes, **anchoring filaments,** and **anchoring fibrils** literally anchors the epidermis to the dermis. A variety of genetic diseases affect this **dermal-epidermal junction.** Mutations in the genes for specific **cytokeratins,** $\alpha_6\beta_4$ **integrin, laminin 5,** or **collagen VII** give rise to various forms of **epidermolysis bullosa (EB),**

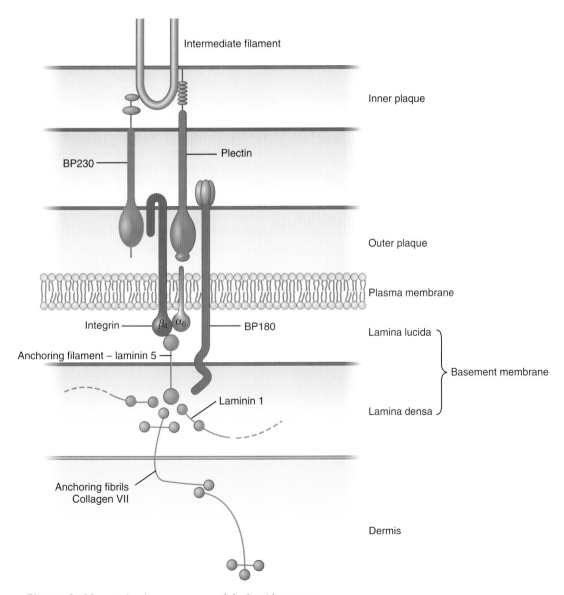

Figure 6-16. Molecular components of the hemidesmosome.

a group of blistering diseases that result in differing degrees of epidermal detachment that vary in severity from extreme debilitation to neonatal lethality. (EB simplex involves keratin filaments and causes epidermal blistering; EB junctional form involves $\alpha_6\beta_4$ integrin or laminin 5 and is lethal in early infancy; EB dystrophica involves anchoring fibers and causes epidermal blistering that leads to syndactyly.) Like desmosomes, hemidesmosomes need to provide strong adhesion but also to relinquish this when required. For example, replenishment of cells in the epidermis occurs in the basal layer. As necessary, cells need to move up from the basal layers to join the upper epidermal cell layers. To do this cells must lose their hemidesmosomal adhesions to relinquish contact with the basement membrane, and this must be done in a regulated fashion. Similarly, during cell migration to close epidermal wounds, cells must lose their hemidesmosomal adhesions and reacquire them when wound closure is complete.

Focal Contacts Are Adhesions Formed with the Substratum by Cultured Cells

When cells are plated on glass or plastic in tissue culture, they adhere to the surface and spread over it, often resembling the shape of a fried egg with thin edges and a bulky nucleus, the yolk, somewhere near the center. They do not actually adhere directly to the glass or plastic, but rather to a thin layer of ECM molecules that adsorb to the surface. Principal among these is fibronectin, which is abundant in the serum component of tissue culture medium and is usually also secreted by the cells (Fig. 6–17).

Adhesion to this adsorbed layer is mediated principally by integrins, with $\alpha_5\beta_1$ **integrin** being the principal fibronectin receptor. To function effectively, integrins

must cluster together on the cell surface. At the edges of spreading cells, integrins cluster to form small structures (<1 μm) called **focal complexes.** As cell spreading progresses, these complexes evolve into slightly larger, elongated structures call **focal adhesions** or focal contacts (Fig. 6–18); these are effectively adhesive junctions formed between the cell and the substratum. The underside of the cell is not flat. Focal contacts are the regions of closest association (about 15 nm) between the cell membrane and the substratum. Cytoplasmically, focal contacts are associated with the actin cytoskeleton. In well-spread cells, actin filaments are bundled into **stress fibers** that originate toward the center of the cell and terminate at focal contacts.

As well as being points of adhesion, focal contacts are important sites of signal transduction. Signals from outside the cell regulate cell function, whereas signals originating within the cell regulate cell adhesion. The cytoplasmic domains of the integrin subunits may recruit up to 50 different types of structural and signaling molecules to participate in adhesive linker and signal transduction functions.

The earliest formed focal complexes recruit cytoskeletal linker proteins such as vinculin and paxillin to the cytoplasmic face, accompanied by a signaling molecule, **focal adhesion kinase (FAK)** (Fig. 6–19). FAK is a **tyrosine kinase,** an enzyme that adds a phosphate group to specific tyrosine residues in protein substrates. **Phosphorylation** by protein kinases and dephosphorylation by protein phosphatases are important regulators of protein activity and function. Phosphorylation of components of the early focal complex causes recruitment of other structural components, the linker proteins **talin** and **tensin,** some with signaling potential such as **zyxin,** and another tyrosine kinase, **src.** Further changes can result in either maturation or turnover of the complex.

Regulation of the actin cytoskeleton also determines the formation of focal contacts and stress fibers. Key signaling molecules involved in this process are the Rho family of **small GTPases.** These molecules are active when they have GTP bound to them but inactive when

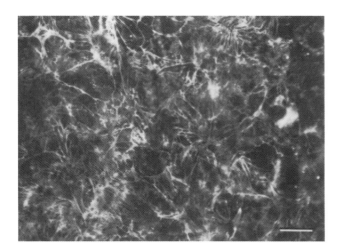

Figure 6–17. Fluorescence micrograph showing fibronectin produced by cells in culture. *(From Mattey DL, Garrod DR. J Cell Sci 1984;67:171–188, by permission.)*

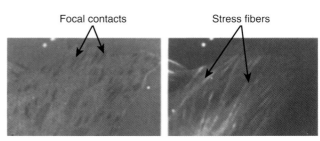

Focal contacts   Stress fibers

Figure 6–18. Focal contacts and stress fibers of cells in culture. **A:** Micrograph of cells under interference reflection microscopy showing focal contacts. **B:** Fluorescence micrograph showing actin stress fibers. Note how each stress fiber terminates at a focal contact. *(From Morgan J, Garrod DR. J Cell Sci 1984;66:133–145, by permission.)*

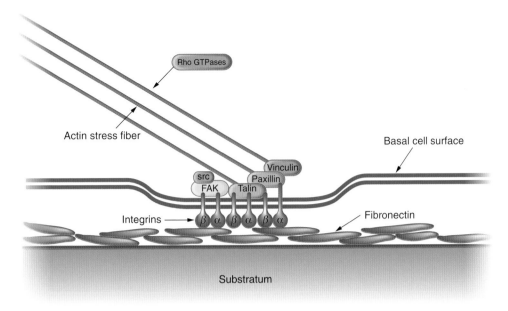

Figure 6–19. Molecular composition of the focal contact.

the bound GTP is cleaved to guanosine diphosphate (GDP). Important members of this family are ***cdc42, Rac,*** and ***Rho.*** *Rho* promotes the formation of focal contacts and stress fibers, whereas *Rac* causes the formation of large, thin **lamellipodia** with small adhesions localized to the extreme edge. *Cdc42* promotes the formation of **filopodia,** narrow processes that extend from the cell surface that are supported internally by actin filament bundles and make long adhesions with the substratum.

Much is known about focal contacts and their regulation because they are relatively easy to study in culture. They give some idea of the dynamic nature of cell adhesion and the complexity of its regulation. Such regulation is believed to be important for controlling cell motility *in vivo.* Some cells such as leukocytes migrate substantially as part of their normal function. To do this, they must dynamically regulate their adhesions and cytoskeleton. Other cells such as epithelial cells are normally less motile, but when they develop into tumor cells, they invade the surrounding matrix and spread or **metastasize** to other parts of the body. These processes involve loss of cell-cell adhesion at the primary site and dynamic regulation of adhesion during migration. Thus, it is important to investigate these processes to understand both normal and abnormal cell behavior.

CELL ADHESION HAS MANY IMPORTANT ROLES IN TISSUE FUNCTION

To this point the chapter has dealt with the nuts and bolts of cell adhesion—the adhesion molecules and intercellular junctions. It is now appropriate to consider how cell adhesion is involved in tissue function and cell behavior.

Junctions Maintain Epithelial Barrier Function and Polarity

More than 200 different cell types exist in the human body, and perhaps surprisingly, about 65% of these are epithelial; that is, they are the components of the cell sheets that line body surfaces and cavities. **Epithelia** provide functional and physical separation between biologic compartments within the body and often also have a protective or barrier role. Thus, the epidermis protects the human body from water loss and from the entry of environmental pathogens and toxins, and resists the minor abrasion and shear stress to which it is constantly subjected. The airway epithelium separates inhaled air from tissue fluid, providing an absorptive surface but also maintaining the cleanliness of the airway and providing a protective barrier against the entry of airborne pathogens and allergens. The intestinal mucosa separates the gut contents from tissue fluid, performing a digestive function, selective absorption of digested products, and preventing the entrance of allergens and bacteria.

To perform these functions, all epithelia must be polarized; that is, their apical surfaces are structurally and functionally different from their basal surfaces (Fig. 6–20). In stratified epithelia such as the epidermis, the basal layer is concerned with attachment to the underlying matrix and producing new cells from its population of **stem cells** to replenish those lost from the outer surface. By contrast, the outer layers are dead or dying, but in the process of progressing upward have

A STRATIFIED EPITHELIUM

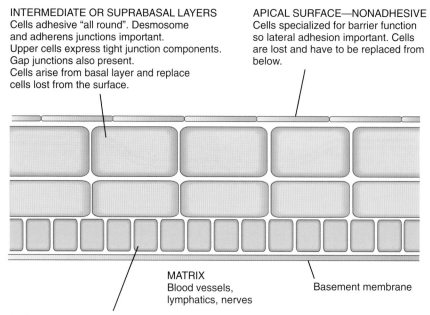

INTERMEDIATE OR SUPRABASAL LAYERS
Cells adhesive "all round". Desmosome
and adherens junctions important.
Upper cells express tight junction components.
Gap junctions also present.
Cells arise from basal layer and replace
cells lost from the surface.

APICAL SURFACE—NONADHESIVE
Cells specialized for barrier function
so lateral adhesion important. Cells
are lost and have to be replaced from
below.

MATRIX
Blood vessels,
lymphatics, nerves

Basement membrane

BASAL LAYER
Lateral adhesion to other cells and apical
adhesions to suprabasal cells involve
desmosomes and adherens junctions.

Adhesion to basement membrane by integrins
and hemidesmosomes. Stem cells present to divide
and replace cells lost from the surface. Upward movement
of cells from basal layer known as stratification.

B SIMPLE EPITHELIUM

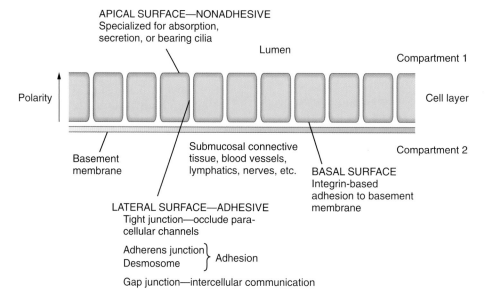

APICAL SURFACE—NONADHESIVE
Specialized for absorption,
secretion, or bearing cilia

Lumen

Compartment 1

Polarity

Cell layer

Basement
membrane

Submucosal connective
tissue, blood vessels,
lymphatics, nerves, etc.

Compartment 2

BASAL SURFACE
Integrin-based
adhesion to basement
membrane

LATERAL SURFACE—ADHESIVE
Tight junction—occlude paracellular channels

Adherens junction ⎫
Desmosome   ⎬ Adhesion

Gap junction—intercellular communication

Figure 6–20.   **Cell adhesion in epithelia. A:** Stratified and (**B**) simple epithelium showing
distribution of adhesive properties.

developed tough, impenetrable properties essential for barrier function. In simple epithelia such as the intestinal mucosa, polarity of structure and function lies within each cell. Cells of the small intestine have an apical surface specialized for absorption and a basolateral surface that has distinct properties for transferring absorbed molecules to the tissues.

Cell adhesion is central to epithelial function because it maintains the integrity of the cell layers and enables the polarity to be maintained. In a simple epithelium, the cell-cell adhesion junctions are on the lateral surface, and integrin-based adhesions, located on the basal surface, mediate attachment to the basement membrane. It is just as important that the apical surface be nonadhesive; otherwise, the opposite faces of the intestines would stick together and the lumen would be occluded.

Leukocytes Must Adhere and Migrate to Combat Infection and Injury

In contrast with epithelial cells that need to be in a constant adhesive state, other cell types need to be constitutively nonadhesive, but to increase their adhesiveness when this is functionally required. Leukocytes are one major example (Fig. 6–21). Mostly they circulate freely in the blood, showing no tendency to attach either to each other, to other blood cells, or to the endothelial cells that line blood vessels. However, when the need

arises because of tissue damage or infection, they must leave the blood and congregate at the appropriate site to combat the problem. This is the so-called **inflammatory response.** It involves changes in adhesive properties both for leukocytes and the **endothelial cells** lining the small blood vessels near to the site of injury.

The inflammatory response is initiated by release of diffusible molecules, inflammatory mediators, from the injured tissue, or by complement activation. These cause the rapid expression of P-selectin, stored inside the cells in vacuoles called **Weibel–Palade bodies,** on the surfaces of endothelial cells and a slower increase in the surface expression of E-selectin. The newly exposed selectins bind to carbohydrates on the surfaces of circulating leukocytes, causing them to become loosely attached to the endothelial cells. This initial attachment is such that the leukocytes roll along the endothelial cell surface under the force of blood flow.

Initial adhesion and inflammatory mediators trigger a response from the leukocytes that results in firm adhesion to the endothelial cells. This involves an "inside-out" signal that activates the normally nonfunctional integrin dimers on the leukocyte surface, enabling them to adhere firmly to the endothelial cells by binding to the Ig family adhesion molecule ICAM. Although this adhesion is firm, it permits cell migration for the next phase of the process, extravasation. Here, leukocytes migrate between the endothelial cells into the matrix of the tissue. Extravasation involves loosening of the junctional contacts of the endothelial cells in response to

LEUKOCYTE ADHESION, EXTRAVASATION AND MIGRATION

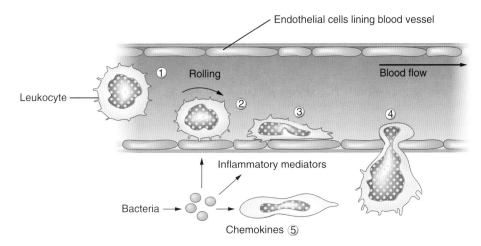

1. Leukocyte free in circulation—nonadhesive to endothelial cells
2. Leukocyte tethered to endothelium and rolling under force of blood flow—selectins—seconds
3. Leukocyte firmly bound to endothelium and migrating—integrins—ICAM
4. Extravasation of leukocyte from blood vessel—JAM, PECAM—minutes
5. Leukocyte migrates to source of infection or injury—integrins

Figure 6–21. Leukocyte adhesion and migration. (*Modified from Garrod DR. Cell to cell and cell to matrix adhesion. In: Latchman D, ed.* Basic Molecular & Cell Biology, *3rd ed. Oxford, United Kingdom: Blackwell BMJ Books, 1997:80–91, by permission.*)

inflammatory mediators and poorly understood adhesive interactions between the leukocytes and the endothelial junctional adhesion molecules. After extravasation, the leukocytes migrate to the site of injury or infection within the tissue guided by **chemotaxis** toward tissue-released diffusible molecules called **chemokines.** The mechanism of migration is not well understood, but probably involves the type of dynamic regulation of adhesion and the cytoskeleton that is referred to earlier in the discussion of focal contacts.

The inflammatory response is a defensive mechanism aimed at returning homeostasis after injury or localized infection. It is immensely important when it operates in a regulated fashion. However, it can be overactive, in which case it results in the tissue damage of inflammatory diseases such as arthritis. Conversely, rare human mutations in β_2 integrin gene cause a disease called **leukocyte adhesion deficiency.** Patients who have this disease cannot make **pus,** an accumulation of white blood cells, and are susceptible to death from overwhelming infection.

Platelets Adhere to Form Blood Clots

Blood platelets—small, anucleate cells that normally circulate freely in the blood—need to become adhesive rapidly to assist in the formation of **blood clots** at sites of blood vessel damage. They are present in enormous numbers, ~ 1.5–4.0×10^{11} per liter of blood. Once activated, platelets can adhere to matrix components, including collagen that is exposed by blood vessel damage; vWF, a matrix protein that is released from Weibel–Palade bodies of endothelial cells and platelet α **granules;** and fibrin, a protein that is generated from plasma to form the blood clot (Fig. 6–22). Platelets can also aggregate together. Platelet activation can be triggered by a variety of soluble agonists including **thrombin,** adenosine diphosphate, and collagen or peptides

A ADHESIVE COMPONENTS IN BLOOD PLASMA, PLATELETS AND
ENDOTHELIAL CELL BASEMENT MEMBRANE

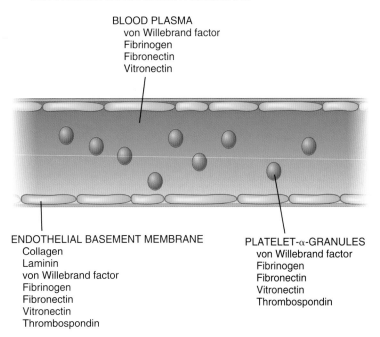

BLOOD PLASMA
von Willebrand factor
Fibrinogen
Fibronectin
Vitronectin

ENDOTHELIAL BASEMENT MEMBRANE
Collagen
Laminin
von Willebrand factor
Fibrinogen
Fibronectin
Vitronectin
Thrombospondin

PLATELET-α-GRANULES
von Willebrand factor
Fibrinogen
Fibronectin
Vitronectin
Thrombospondin

Figure 6–22. Adhesion of blood platelets. **A:** The adhesive environment. Diagram summarizing the molecular adhesive components in blood plasma, platelets, and the endothelial cell basement membrane. **B:** The adhesion receptors of platelets and their ligands. PSGL-1, P-selectin glycoprotein ligand-1.

B PLATELET ADHESION RECEPTORS

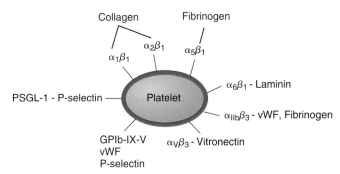

derived from it. Platelet adhesion involves a number of cell-surface adhesion receptors that have a complex nomenclature that originates from a time before integrins and other adhesion molecules were recognized. The three major platelet adhesion receptors are **GPIIb-IIIa** ($\alpha_{IIb}\beta_3$ integrin), **GPIa-IIa** ($\alpha_2\beta_1$ integrin), and **GPIb-IX-V,** a nonintegrin adhesion complex consisting of four gene products. GPIIb-IIIa is a promiscuous integrin that exhibits binding to the matrix molecules fibrinogen, vWF, fibronectin, vitronectin, and thrombospondin. GPIa-IIa is the major receptor for collagen and GPIb-IX-V for insoluble vWF. Platelet activation leads to activation of these receptors, release of vWF, and consequent adhesive binding.

Platelet adhesion is subject to a number of human genetic diseases that involve mutations in the genes for the adhesion proteins, including von Willebrand disease (see Clinical Case 6-2), **Bernard–Soulier syndrome** (involving GPIb-IX-V), and **Glanzmann thrombasthenia** (involving GPIIb-IIIa). Platelet adhesion can also be triggered inappropriately so that platelets adhere to endothelial cells in much the same way as leukocytes during the inflammatory response. This causes the development of **atherosclerotic lesions** (see Fig. 6–37) and **atherothrombosis,** which are enormously important health problems.

Clinical Case 6-2

Hjordis was a little blond girl who was born on a small island between Finland and Sweden in the Baltic Sea. She had a large immediate family, several dozen cousins, and young aunts and uncles. Many other islanders also had the same last name as Hjordis.

When she was 5 years old, Hjordis had her first serious nosebleed, which did not stop for 3 days. The island country doctor who treated it with cotton packing was alarmed, but not particularly surprised by this, because both of the child's parents and three of her surviving siblings also had recurrent nosebleeds and severe bleeding—with minor trauma for the boy, and with their periods for the girls. Remarkably, of a total of 10 brothers and sisters, 3 had died of uncontrolled gastrointestinal bleeding, another from oral bleeding after a tooth extraction, and a fifth from sudden hematemesis. Many of her extended family and the other islanders had similar histories.

In view of the severity of the nosebleed, the island doctor requested a consultation from a visiting professor from the mainland. From the family history the professor immediately recognized that he was

dealing with a hereditary bleeding disorder. When he examined Hjordis, he found no physical abnormalities (including her knees and elbows). However, when he made a small (3-mm) incision in her earlobe, she continued to bleed for 28 minutes before he stopped the slow drip with a pressure bandage. He also obtained a simple, glass-activated clotting time and found that it was normal at 5 minutes. Finally, he made a blood smear, which showed only the stigmata of iron-deficiency anemia and normal numbers of platelets. He defined her disease from these observations.

Though her initial epistaxis eventually abated, her bleeding difficulties continued as she grew up. When she was 7, she had another severe nosebleed; at age 8, she had an episode of gastrointestinal bleeding with melena; and at age 11, after a large celebratory fish dinner, she suffered a substantial episode of hematemesis, which responded to antacids, and it was thought that she might have a bleeding duodenal ulcer.

Unfortunately, when Hjordis started to menstruate at age 12 she had very heavy flow for the first three cycles. On the fourth cycle, the bleeding was even more extensive and uncontrollable with tamponade. Adequate blood transfusions were not available, and she died of exsanguination in 3 days.

Cell Biology, Diagnosis, and Treatment of von Willebrand Disease

This unfortunate young girl had von Willebrand disease. Our clinical vignette is loosely taken from the history of the actual proband, Hjordis, whom Dr. Erik von Willebrand studied to define his eponymic disease, which he published in 1926.

The essence of the diagnosis was then, and remains, an abnormal bleeding time with a normal clotting time and normal platelet numbers. (A ristocetin platelet aggregation study can also be used for more sensitive determinations in some patients.) Modern blood transfusion was not even in its infancy in 1926; thus, no useful therapy beyond physical methods was available for Hjordis.

Because the disease he was observing was found in both sexes, Dr. von Willebrand recognized that it was a new disease distinct from classic (male individuals only) hemophilia, and he termed it "Hereditary Pseudo-Hemophilia." However, it rapidly became known by his name. Ironically, a full century earlier in 1826, Francis Minot,* an alert Boston physician, had described 46 cases of serious hemorrhage from the umbilicus in mixed-sex newborns. These cases almost certainly included many patients

---

*Coincidentally, Francis Minot was the grand uncle of George Minot, who made the momentous discovery of the cure for pernicious anemia—also in 1926!

Continued

with von Willebrand disease, and for some time the disorder was also called Minot–von Willebrand disease.

The pathophysiologic basis of von Willebrand disease is either a quantitative or a qualitative defect in the vWF. This large GP, primarily synthesized by the endothelium, is of fundamental physiologic importance. It has an essential role in the crucial initial adhesion of platelets to wound collagen and their subsequent aggregation. This adhesive event is mandatory to promote the activation and secretion of those platelets. which then begins the evolution of, and reinforces the development of, normal hemostasis and coagulation.

Quantitative defects in vWF (type I disorders) are the most common (perhaps 1% of the population!). They are usually inherited dominantly and are comparatively benign with mostly superficial bruising and mucosal bleeding. (This is in contrast with the hemophilias, which are notorious for crippling large joint bleeds.) They can, however, cause significant problems after surgical trauma.

The qualitative defects (various type II and III disorders) are usually much more severe and are frequently associated with spontaneous and visceral bleeding. They can be inherited in both recessive and dominant patterns, though the genetic presentations can be confusing because of compound heterozygosity for the recessives and because of "decreased penetrance" for the dominants. In these qualitative disorders, several complex biochemic defects in the structure of the von Willebrand protein, or its proteolysis, prevent the formation of its high-molecular-weight multimer forms, which are required for its full adhesive function. This results in both severe platelet adhesion defects and protein-binding defects. The latter can include inadequate binding of vWF to the plasma procoagulant protein, factor VIII. This, in turn, influences the stability and activity of that crucial element in the clotting cascade. Hjordis certainly had one of these more serious variants.

Clinically, the most important element in the common type I disorder is its recognition so that proper preparation for surgical procedures can be made, and so that patients can be warned to avoid aspirin and other agents that inhibit platelet activity. For the other, more serious types, replacement of the vWF is frequently needed in the form of plasma cryoprecipitates or other concentrates that contain vWF complexed to factor VIII. Finally, stimulation of the endogenous synthesis or release of vWF by the patients' own endothelium with a synthetic vasopressin is useful in many patients of both types when they are subjected to bleeding stress.

Embryonic Development Involves Many Adhesion-Dependent Events

Cell adhesion has a crucial function throughout embryonic development. The first morphogenetic event in mammalian development is **compaction,** in which the loosely attached cells or blastomeres of the eight-cell embryo "zip up" their adhesions to become tightly bound together (Fig. 6–23, A [i, ii]). This event involves the adhesion molecule E-cadherin and other adherens junction components. Also at this stage tight junctions are beginning to form between the cells. By the time a hollow ball of cells, the **blastocyst,** is formed, the first epithelium, the **trophectoderm** that will generate the **placenta,** has a full complement of tight junctions, adherens junctions, and desmosomes, and this is still before implantation (see Fig. 6–23, B). Shortly after implantation, the process known as **gastrulation** is initiated. This is the event (some say the most important event in our lives!) that generates the three-layered organization of the embryo, with **ectoderm** on the outside, **endoderm** on the inside, and **mesoderm** in between (see Fig. 6–23C). (Ectoderm forms the epidermis and nervous system, mesoderm the muscles and bones, and endoderm the gut and some related organs.) Gastrulation involves a massive (in embryonic terms) amount of cell movement. This generates the correct embryonic shape and positions the cell layers with respect to each other. These events are crucially dependent on cell adhesion, with molecules such as E-cadherin again playing a key role.

Formation of the neural tube, the forerunner of the central neurons system, involves expression of different cadherin molecules between it and the ectoderm from which it segregates (see Fig. 6–23, D). From the junction between the neural tube and the ectoderm arises a migrating cell population called the **neural crest** or ectomesenchyme (Fig. 6–24). This forms mainly nerves and bones in the head and parts of the autonomic and peripheral nervous systems in the trunk. Migration and precise positioning of neural crest cells is regulated by a series of changes in the expression of adhesion molecules, especially cadherins, fibronectin, and integrins. These initiate and guide migration, then cause the cells to stop and aggregate when they have reached the correct position.

Development of the nervous system involves multiple examples of directed cell migration and the extension of nerve fibers. During each of these events, the migrating cells must reach the correct target. For example, motor neurons must reach and form synapses called *motor end plates* with the appropriate skeletal muscles. Similarly, sensory nerve fibers must form a precisely mapped series of connections with the appropriate part of the brain; for example, those from the eye with the contralateral **optic tectum** in chicks and frogs or the **visual cortex** in mammals. Specific adhesions involving

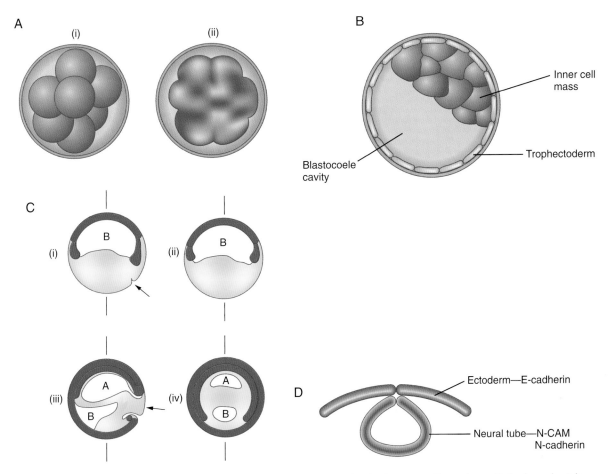

Figure 6–23. Cell adhesion in embryonic development. **A:** Compaction in the mammalian embryo. *(i)* At the early eight-cell stage, the embryonic cells, the blastomeres, are loosely attached. *(ii)* Without further cell division they "zip up" their adhesive contacts. **B:** At the blastocyst stage, the first epithelium, the trophectoderm, is formed. This contains the inner cell mass and the fluid-filled blastocoele cavity. **C:** Gastrulation is shown in the amphibian embryo because it is easier to visualize than in the mammal where gastrulation is more complex because of the presence of extraembryonic tissue that forms the placenta. *(i)* Longitudinal and *(ii)* transverse sections of the early gastrula. Cell invagination is just beginning with the formation of the blastopore lip *(arrow)*. *(iii, iv)* Comparable sections of the late gastrula where invagination is almost complete though the blastopore *(arrow)* is not quite closed. Ectoderm (blue); mesoderm (red); endoderm (yellow). A, archenteron, the future gut cavity; B, blastocoele. *Vertical lines* indicate the relationship between the sections. **D:** Formation of the neural tube, the future central nervous system, involves expression of different adhesion molecules.

Ig and cadherin family members play important roles in guiding this complex wiring process.

CELL ADHESION RECEPTORS TRANSMIT SIGNALS THAT REGULATE CELL BEHAVIOR

Transduction of signals by or involving CAMs has been alluded to several times earlier. It is now appropriate to consider this important topic in more detail. To form an integral component of a tissue, cells need to monitor their environment and to respond accordingly to the signals they receive. They have a variety of surface receptors that receive signals from diffusible molecules such as **growth factors** and chemokines. However, they also receive signals from insoluble components of their

environment, that is, from the ECM and from other cells. Transduction of these signals is an essential second function of adhesion molecules or adhesion receptors.

An important example of adhesion signaling that was studied before any adhesion molecules had been characterized is known as **contact inhibition** (Fig. 6–25). Thus, when a cell such as a fibroblast moving over the surface of a substratum in tissue culture encounters another cell, its leading lamellipodium first forms an adhesive contact with the surface of the other cell and ceases to extend in the direction of movement. Contact has inhibited movement in the initial direction, and eventually the cell moves off in another direction. The net result of this behavior is that cells do not move over each other in culture but instead remain as a monolayer on the substratum. In general, cells in confluent mono-layers, that is, continuous layers of cells in contact with

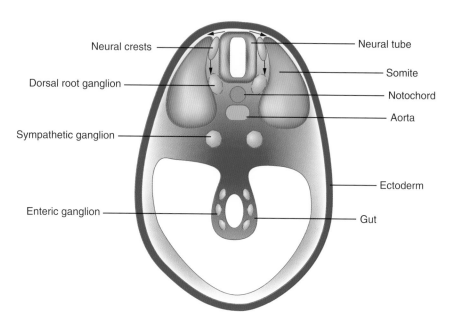

Neural crests

Dorsal root ganglion

Sympathetic ganglion

Enteric ganglion

Neural tube

Somite

Notochord

Aorta

Ectoderm

Gut

Figure 6-24. Neural crest and its derivatives in the trunk. *Arrows* indicate paths of migration of neural crest cells. The ventral pathway gives rise to ganglia and their associated nerves. The dorsal pathway gives rise to pigment cells of the skin. Cell adhesion plays a crucial role in guiding the migration and final positioning.

each other, though not completely static, do not move around very much. However, if the monolayer is experimentally wounded to produce a free edge, cells at that edge begin moving into the wound. Early excitement about the phenomenon of contact inhibition arose because many types of transformed cells were found not to be subject to it. Instead, they were found to be able to move freely over the surfaces of other cells, thus apparently mimicking the **invasive behavior** of tumor cells. Contact inhibition of cell movement should not be confused with **density-dependent inhibition** of cell growth. The latter describes the property of untransformed cells that causes them to slow down and eventually cease cell division once confluence is reached. This cessation of growth is caused principally by depletion of growth factors and not by formation of intercellular contacts. It is frequently and wrongly referred to as "contact inhibition of growth." In fact, many cell types continue to divide after confluence is reached, that is, after they have come into all-round contact with other cells. Cell division then gradually slows down. It is patently obvious that contact inhibition of growth does not generally occur *in vivo;* otherwise, epithelial cells that always remain in continuous sheets (unless wounded) would be unable to divide to replace lost cells.

Modern research has reported many examples of signaling by specific adhesion molecules. Some important examples are considered in the following sections.

Cell Growth and Cell Survival Are Adhesion Dependent

Cell-substratum adhesion is an important regulator of cell division. To proliferate, cells need to attach and

spread on the substratum (Fig. 6–26). Cells that are well spread proliferate faster than those for which spreading is restricted. This aspect of cell regulation is referred to as **anchorage dependence.** This is a feature of normal or untransformed cells. By contrast, transformed cells (i.e., cells that are able to produce tumors) are commonly anchorage independent. Thus, they can proliferate with little or no contact with the culture substratum and also when suspended in soft agar. Anchorage-independent growth in culture mimics the ability of tumor cells *in vivo* to grow in abnormal situations such as when they have become detached from the basement membrane or even when in suspension in ascites fluid in the peritoneal cavity. A signaling pathway involved in regulation of cell proliferation by substratum adhesion involves integrins and a cytoplasmic protein kinase called Erk. When cells are substratum attached, Erk can enter the cell nucleus to regulate proliferation, but if cells are held in suspension, Erk remains cytoplasmic and proliferation does not occur. Regulation of Erk activity appears to depend on the actin cytoskeleton and the ability of integrin adhesion to regulate assembly into stress fibers.

Conversely, cells that are released from substratum contact tend to undergo **programmed cell death** or **apoptosis** (see Fig. 6–26). This particular form of cell death that is triggered by release from the substratum has been named *anoikis*. Another property of many transformed or tumor cells is the ability to survive in abnormal situations, for example, not to be susceptible to **anoikis.** Normal cells require survival signals from cell-substratum adhesion to avoid anoikis; many tumor cells do not require such signals.

An example of the regulation of cell survival and cell death in normal tissue function comes from the **mammary gland** (Fig. 6–27). During pregnancy, the

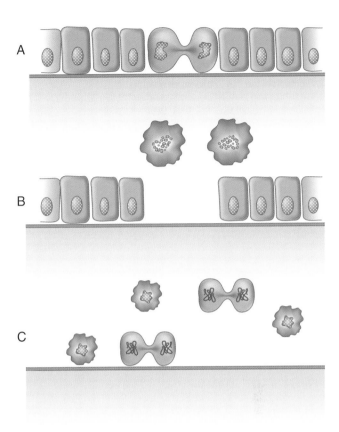

Figure 6–26. Cell attachment, division and death. **A:** Normal cells are anchorage-dependent requiring attachment to the substratum to proliferate. **B:** Cells that become detached from the substratum undergo anoikis, a form of programmed cell death or apoptosis. **C:** Many tumor cells both survive and proliferate in suspension.

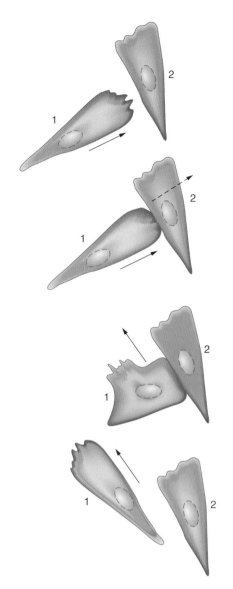

Figure 6–25. Contact inhibition of cell movement. Cell 1 moving in the direction of the *arrow* makes contact with and adheres to cell 2 (shown completely stationary for the purpose of illustration). Movement in the initial direction is inhibited. Cell 1 forms a new leading lamella or lamellipodium and moves off in a different direction.

mammary gland enlarges because of an elaboration of ECM and growth of the mammary epithelium so that the latter can produce milk. When suckling is over, the process is reversed. Enzymes called **matrix metalloproteinases (MMPs)** degrade the matrix. As a consequence, many epithelial cells have no matrix to which to adhere. They therefore lose their integrin-mediated **survival signals** and undergo apoptosis.

Cell Adhesion Regulates Cell Differentiation

The mammary gland also provides a well-characterized example of the regulation of cell differentiation by cell

adhesion (see Fig. 6–27). The primary function of the mammary gland is to produce milk during **lactation.** It has been shown that activation of the genes that code for milk proteins is dependent on a combination of two signals, one a diffusible signal from the hormone **prolactin** and the other an adhesive signal mediated by β integrins. Thus, if epithelial cells from lactating mammary glands are cultured on collagen in the presence of prolactin, they survive but do not produce milk proteins; however, if laminin is the culture substratum, milk proteins are produced. Many other examples also exist of the regulation of gene expression and cell differentiation by signals originating through cell contact and adhesion.

EXTRACELLULAR MATRIX

All tissues consist of two components, a cellular component and an extracellular component. The latter comprises a variety of specialized structures that constitute the ECM. Molecular components of this matrix are secreted and to some extent assembled by the cells of the tissue.

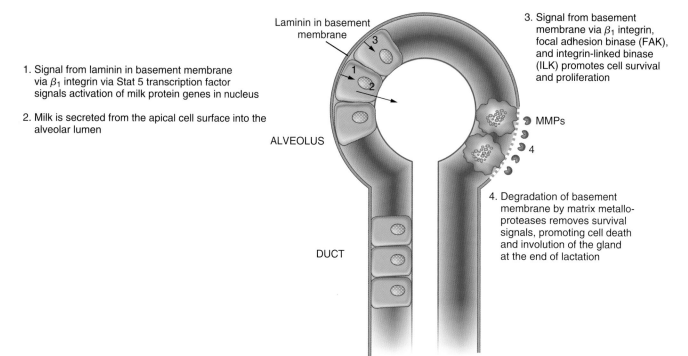

1. Signal from laminin in basement membrane via β_1 integrin via Stat 5 transcription factor signals activation of milk protein genes in nucleus

2. Milk is secreted from the apical cell surface into the alveolar lumen

Laminin in basement membrane

ALVEOLUS

DUCT

3. Signal from basement membrane via β_1 integrin, focal adhesion binase (FAK), and integrin-linked binase (ILK) promotes cell survival and proliferation

MMPs

4. Degradation of basement membrane by matrix metallo-proteases removes survival signals, promoting cell death and involution of the gland at the end of lactation

Figure 6–27. **Mammary gland.** A good example of the regulation of cell survival and differentiation by cell adhesion signaling.

The amount of ECM varies enormously between different tissues. Thus, in **bone** and **cartilage** and in the **dermis** of the skin, the bulk of the tissue is composed of matrix. In contrast, in epithelia and muscles, most of the tissue is cellular, the matrix being confined to a basement membrane or basal lamina that surrounds or underlies the cellular component. Composition and amount of ECM differ according to the function of the tissue. Bone is calcified for strength and consists largely of ECM, enabling it to fulfill its functions of providing strength and support for soft tissues and of carrying muscle attachments to facilitate its lever function in movement. Cartilage also consists mainly of ECM, but it has very different properties from bone because it needs to provide articulation in joints, while at the same time needing to resist compression and provide a cushioning effect between hard bones. The dermis connects the epidermis to the underlying tissues and needs to provide great strength and elasticity to dissipate the stresses impinging on the skin. The basement membrane is essentially a thin supporting layer for cell attachment, but it has other specialized functions in tissues such as the kidney.

In adult organisms, the majority of extracellular matrices exhibit slow turnover; that is, they are permanent or semipermanent in nature. They do, however, need to retain the capacity to respond to changes such as injury, for example, in the healing of fractures or wounds. Another type of matrix, the blood clot, needs to form rapidly and in the correct location in response to injury, but then needs to disperse as the injury is repaired. Modulation of the ECM is also important in angiogenesis, the generation of new blood vessels in response to injury or tumor growth. The role of the matrix is not exclusively structural; it provides the basis for signals transmitted to cells by adhesion receptors that bind to its components, and it acts as a reservoir for growth factors that also bind reversibly to its constituents. The major components of the ECM are considered in more detail in the following section.

Collagen Is the Most Abundant Protein in the Extracellular Matrix

Rather than a single protein, collagen comprises a family of 26 genetically distinct proteins. Collagens are the principal structural elements of all connective tissues. They are characterized by the presence of a repeated sequence of three amino acids—a **tripeptide, glycine-X-Y**, where X and Y are commonly proline or **hydroxyproline.** (Hydroxyproline is proline that has been posttranslationally modified by addition of a hydroxyl group.) All collagens are trimers in which at least some and often most of the protein chains are involved in forming a triple helix. The Gly-X-Y tripeptide plays a key role in the triple-helix structure.

The different family members can be divided into groups according to the structures that they form. These groups are fibril-forming or **fibrillar collagens,** fibril-associated collagens, network-forming collagens, anchoring fibrils, transmembrane collagens, **basement-membrane collagens,** and others. By far the most abundant group is the fibrillar collagens, which constitute about 90% of total collagen. Of these, fibrils composed of collagen types I and V form the structural framework of bone, and collagen types II and XI contribute to the fibrillar matrix of articular cartilage. The structure of these collagens gives them great tensile strength and torsional stability, which are essential properties for these tissues. The flexible triple helices of type IV collagen form a meshwork in basement membranes. Types IX, XII, and XIV are fibril-associated collagens that associate as simple molecules with collagen fibrils formed by other collagens. Type XVII collagen, which is discussed earlier as BP180, an adhesion molecule of hemidesmosomes, has its collagenous domain within the basement membrane and is a transmembrane protein with its noncollagenous domain extending into the hemidesmosomal plaque.

Different collagens have different structures and properties because they are composed of different trimeric combinations of protein chains (Fig. 6–28). They may be either homotrimers composed of three identical chains or heterotrimers composed of two or three different chains. Type II and III collagens are examples of homotrimers and types I and IV of heterotrimers. The chains are called α chains; the formula or molecular composition for type II collagen is $[\alpha 1 \, (II)]_3$ and for type I collagen $[\alpha 1 \, (I)]_2 \, \alpha 2 \, (I)$. Each different α chain is encoded by a different gene.

Characteristic of all collagens is the presence of a **triple helix** formed by mutual coiling of the three α chains. The so-called **collagenous domains** that form this triple helical substructure possess the $(Gly-X-Y)_n$ repeat amino acid sequence, which is prerequisite for α-helix formation. The collagenous domain may embrace the majority of the molecule as in the fibrillar collagens such as type I, or it may be limited to part of the molecule as in specialized collagens such as type XVII. The presence of hydroxyproline at either the X or Y position in many of the tripeptide repeats is essential for the stability of the helix by enabling the formation of intermolecular hydrogen bonds. In fibril-forming collagens, the triple helical domain is 300 nm (about 1000 amino acids) in length (see Fig. 6–28). Many of these triple-helical molecules (rather confusingly called monomers) assemble into fibrils through interaction of their nonhelical end domains and the side chains of amino acids that are exposed on the surfaces of the helices. In the fibril, the ends of the helical monomers are separated by a distance of 40 nm and a continuous row of monomers is staggered by about 27 nm with respect to the adjacent row, causing a periodicity (called the D-period) of about 67 nm. This staggered arrangement accounts for the characteristic **banded appearance of fibrillar collagens** viewed by light and electron microscopy.

It would be inappropriate here to enter into more detail about the variety of collagen substructure. Type IV collagen is considered further later in this chapter in the discussion of basement membranes.

The structures formed by collagen fibrils are truly remarkable. For example, **tendons** consist of large-diameter collagen fibrils of indeterminate length that are strictly parallel to each other and that enable the tendons to withstand repeated application of tension. Formation of these long parallel fibril bundles depends partly on self-assembly of the collagen molecules and partly on the activity of the cells that produce them.

Synthesis of collagen protein occurs in the **endoplasmic reticulum** of fibroblasts (Fig. 6–29). Here posttranslational modifications such as the hydroxylation of proline residues and the addition of carbohydrate chains occur. Here also the protein chains are assembled into a triple helix called **procollagen.** In procollagen the protein chains are longer at the NH_2 and COOH termini than those of mature collagen. These extra bits are called N- and C-propeptides. The C-propeptide initiates the formation of the triple helix, which then progresses toward the N terminus. Procollagen then proceeds

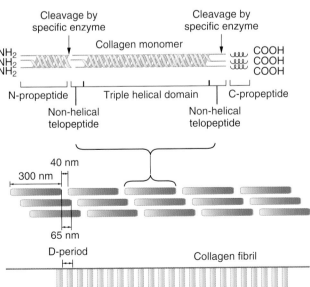

Figure 6–28. Molecular structure of fibrillar collagen. **A:** Electron micrograph showing the banded appearance of a type I collagen fiber from tendon. (Courtesy Drs. Helen Graham and Karl Kadler) **B:** Diagram showing the fiber assembly. The N- and C-propeptides of procollagen are cleaved to give a collagen monomer that is triple-helical with nonhelical telopeptides at each end. Monomers assemble in a regular manner to cause the banded collagen fibril.

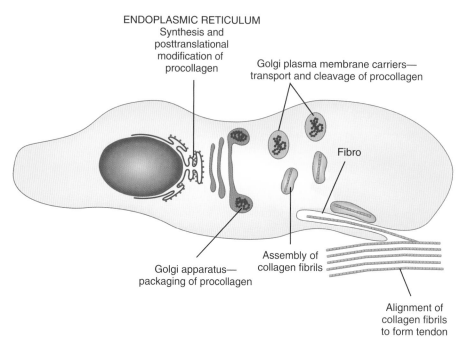

ENDOPLASMIC RETICULUM
Synthesis and
posttranslational
modification of
procollagen

Golgi plasma membrane carriers—
transport and cleavage of procollagen

Fibro

Golgi apparatus—
packaging of procollagen

Assembly of
collagen fibrils

Alignment of
collagen fibrils
to form tendon

Figure 6–29. How cells synthesize and assemble collagen fibrils into the parallel array that constitutes a tendon.

through the **Golgi apparatus** where it is packaged in the *trans*-**Golgi network** (TGN) into vesicles called **Golgi to plasma membrane carriers** (GPCs). During the formation of GPCs, specific enzymes cleave the N- and C-propeptides, the essential step that initiates **self-assembly** of banded collagen fibrils. The nascent fibrils then grow in length and number within the GPCs. A GPC containing several fibrils then joins the cell surface forming a structure called a **fibropositor** that extrudes the fibrils from the cell where it joins in a regular parallel array with other fibrils to form a bundle. The parallel alignment of fibrils in the tendon is therefore generated by the tendon fibroblasts. This process occurs only during fetal development. Thereafter tendons grow by addition of further fibrils to the parallel array.

Mutations in collagen genes result in a variety of human diseases including **chondroplasia, osteogenesis imperfecta,** Alport syndrome, Ehler–Danlos syndrome, and dystrophic EB, and other collagen abnormalities contribute to osteoarthritis and osteoporosis. During wound healing, remodeling of collagen needs to occur. Certain members of the matrix metalloproteinase family of enzymes are involved in the necessary degradation of collagen required for this process. These enzymes are produced by a variety of cell types including fibroblasts, inflammatory cells such as granulocytes, as well as hypertrophic chondrocytes, osteoblasts, and osteoclasts that are involved in the remodeling of cartilage and bone.

Glycosaminoglycans and Proteoglycans Absorb Water and Resist Compression

Other major bulk constituents of the ECM are long carbohydrate chains called **glycosaminoglycans (GAGs).** GAGs are usually linked to proteins to form proteogly-

cans. GAGs consist of repeating disaccharides linked into long, unbranched chains. One of the sugars in the repeating unit is an **amino sugar,** *N*-acetylglucosamine, and the other is a **uronic acid,** either glucuronic or iduronic acid (Fig. 6–30). GAGs are strongly negatively charged because most of the sugars bear **carboxylic acid groups,** and in **chondroitin sulphate, dermatan sulphate, heparan sulphate,** and **keratan sulphate** the amino sugars are commonly **sulphated.** These long carbohydrate chains have two important properties that underlie their major role in tissue structure and function. First, unlike protein chains, carbohydrate chains do not fold into compact units. Second, their negative charge attracts cations such as Na^+ that are **osmotically active** and thus attract large amounts of **water.** These properties mean that GAGs fill large volumes of space and are able to resist compressive forces such as the huge pressures that are exerted on cartilage in joints.

Hyaluronic acid (HA), a nonsulphated GAG, consists of up to 25,000 disaccharide units and is widely distributed in tissues. HA molecules can reach molecular weights of several million daltons, and a single molecule, swollen with water, can occupy a space of 10^7 nm^3. HA is a lubricant in **joints** and facilitates cell migration in embryonic development and wound healing.

Proteoglycans consist of sulphated GAGs covalently linked to a polypeptide chain, the **core protein** (Fig. 6–31). They vary enormously in size and carbohydrate composition. The largest can consist of up to 95% carbohydrate by weight, and **aggrecan,** a major constituent of cartilage, has a molecular weight of about 3 million Da. At the other extreme, **decorin** has a molecular weight of 40 kDa and a single carbohydrate chain. Already an enormous molecule in its own right, aggrecans in cartilage form giant complexes that have molecular weights of the order of 100 million Da and occupy

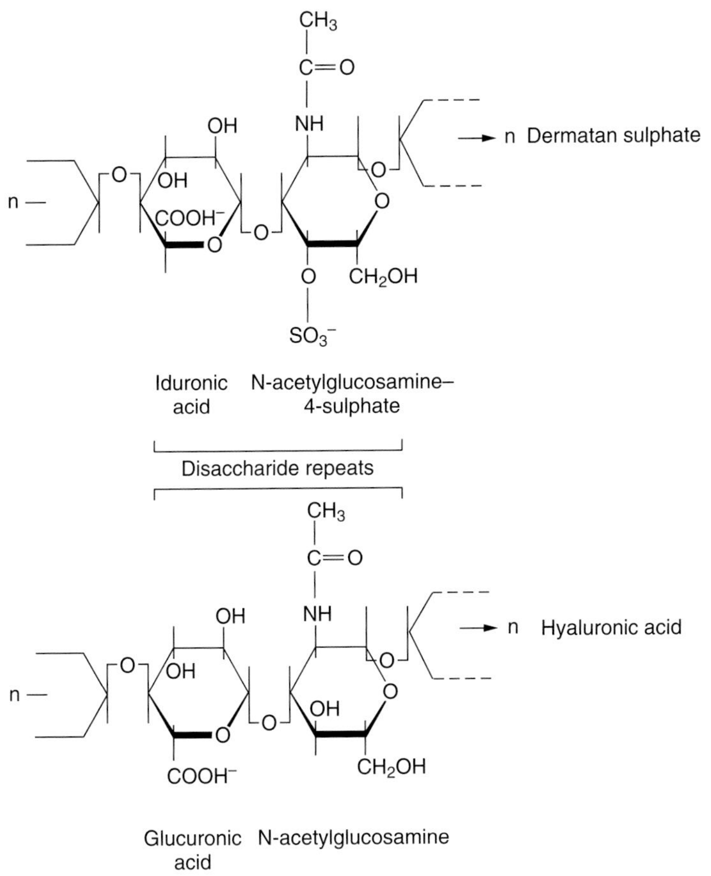

Figure 6-30. The repeating carbohydrate units that constitute sulphated and nonsulphated glycosaminoglycans.

Iduronic acid   N-acetylglucosamine– 4-sulphate

Disaccharide repeats

n Dermatan sulphate

n Hyaluronic acid

Glucuronic acid   N-acetylglucosamine

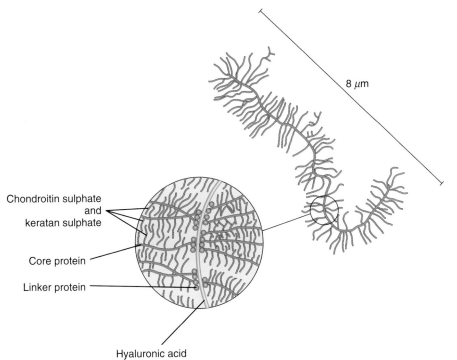

Figure 6-31. Structure of a large proteoglycan, aggrecan, from cartilage.

8 μm

Chondroitin sulphate and keratan sulphate

Core protein

Linker protein

Hyaluronic acid

a space of 5×10^{16} nm^3. An aggrecan aggregate consists of a central core HA molecule with many aggrecan molecules joined to it laterally by means of linker proteins; the entire substructure resembles a bottle brush under the electron microscope. Proteoglycans have several additional functions apart from their space-filling and mechanical properties. For example, by binding to growth factors such as **fibroblast growth factors** (FGFs) or **transforming growth factor-α** and chemokines, they can regulate the activity/availability of these diffusible signaling molecules. Decorin is an example of a proteoglycan that has such regulatory powers and is also involved in the formation of collagen fibers through its ability to bind to collagen. Another proteoglycan called *perlecan* is a critical component of the basement membrane of the kidney where its properties contribute to the filtration of plasma. Some proteoglycans are transmembrane proteins rather than components of the ECM. For example, **syndecans** are integral membrane proteoglycans that contribute to the adhesive properties of focal contacts.

Elastin and Fibrillin Provide Tissue Elasticity

In addition to resisting tensile, torsional, and comprehensive forces, tissues require considerable elasticity, the ability to return to normal shape after being disturbed. This property is particularly important in the skin, lungs, and blood vessels. **Tissue elasticity** resides largely in a network of elastic fibers that are interwoven with collagen fibers. The principal components of **elastic fibers** are the proteins elastin and fibrillin (Fig. 6–32). Elastin is the major component and constitutes up to 50% by weight of large arteries. It contains a series of **hydrophobic domains** that are responsible for its elastic properties and α-helical linker segments, rich in lysine, that are involved in forming cross-links to adjacent molecules. The resulting ECM complex consists of a network that confers on the fibrils five times the extensibility of an elastic band of the same size.

Elastic fibers are covered with a sheath of 10-nm diameter **microfibrils** composed of the protein fibrillin. These are important for fiber assembly. Mutations in the fibrillin gene cause a human hereditary disease called **Marfan syndrome** in which the integrity of elastic fibers is compromised leading to a rupture of the **aorta** in severe cases. Mutations in the elastin gene cause narrowing of major arteries.

Fibronectin Is Important for Cell Adhesion

Fibronectin is the best studied of many noncollagenous ECM proteins that play a role in regulating cell adhesion and cell behavior. There was considerable excitement when it was first discovered because it was found to be substantially less abundant in cultures of certain tumor cells than those of normal cells, suggesting that it might contribute to the lower adhesiveness and metastatic properties of tumors.

Fibronectin is important in embryonic development where it provides a substratum for guiding gastrulation movements and the migration of neural crest cells. As well as being a component of the ECM, a soluble form of fibronectin is abundant in blood plasma, where it is believed to contribute to blood clotting, wound healing, and **phagocytosis.**

Fibronectin is a dimer that consists of two similar or identical protein chains, each about 200 kDa molecular weight, that are linked together near their COOH termini by two disulfide bonds (Fig. 6–33). The major structural element of these chains is the barrel-like **fibronectin type III** repeat. Distributed along the chain are various sites for interaction with other molecules including domains for heparin, collagen, cell binding, and self-association. The major cell-binding site consists of a tripeptide sequence (Arg-Gly-Asp, or RGD in single-letter amino acid code) that is present on an exposed loop extending from one of the type III repeats. This is the site for binding $\alpha_5\beta_1$ integrin, the principal cellular fibronectin receptor. RGD sequences have subsequently been discovered in other matrix protein, for example, the blood clot protein fibrinogen. Snakes produce an RGD-containing protein, disintegrin, in their venom to prevent blood clotting, and drugs based on RGD peptides have been developed as anticlotting agents.

Laminin Is a Key Component of Basement Membranes

An important component of basement membranes is the trimeric protein laminin. Laminin consists of three different protein chains—α, β, and γ, the products of different genes. Each is actually a family of genes, and the various α, β, and γ chains and five different α chains,

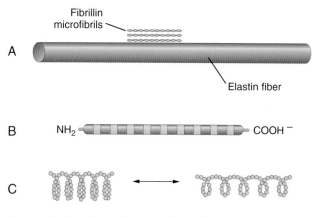

Figure 6–32. Elastic fibers. **A:** Elastic fibers are composed of a central core of elastin surrounded by microfibrils of fibrillin. The whole is cross-linked by γ-glutamyl-lysine bonds. **B:** Elastin molecules consist of tandem repeats of hydrophilic (purple) and hydrophobic (pink) domains. **C:** The hydrophobic domains are responsible for elasticity.

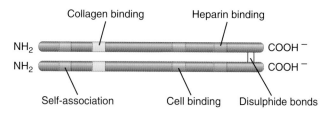

Figure 6–33. Fibronectin dimer showing various binding sites.

three β chains, and three γ chains can be combined to give a variety of laminins.

The classical laminin molecule is **laminin-1.** Laminin-1 consists of an α chain of about 400 kDa and β and γ chains of about 200 kDa each (Fig. 6–34). These chains together form a cross-shaped molecule. Globular domains are at the NH₂-terminal regions of all three chains and at the COOH terminus of the α chain. The COOH regions of the β and γ chains form a coiled-coil α-helical domain that associates with a rodlike region of the α chain. The NH₂-terminal globular domains of the crosslike structure contain domains for self-association enabling laminin to form a network that is the basis of basement membrane structure. The COOH-terminal globular domain contains a site for cell binding, for example, through $\alpha_6\beta_1$ integrin. Other forms of laminin, such as laminin-5, are composed of three NH₂-terminally truncated chains that do not form a cross but that nevertheless bind to other basement membrane components to form a substratum for hemidesmosomal adhesion.

Basement Membranes Are Thin Matrix Layers Specialized for Cell Attachment

Basement membranes are thin (50–100 nm), continuous layers of ECM that underlie epithelial and endothelial cell sheets and surround muscle cells, fat cells, and Schwann cells. They form a substratum for cell attachment and a link to the underlying connective tissue. The electron microscope shows two components to basement membranes, a clear or electron lucent layer next to the basal surfaces of the cells and a dark or electron-dense layer beneath this. These layers are called the **lamina lucida** and the **lamina densa,** respectively (see Fig. 6–15).

Basement membranes are highly cross-linked complexes of several proteins and proteoglycans (Fig. 6–35). Constitutive components are **type IV collagen,** laminin, a protein called **nidogen/entactin,** and heparan sulphate proteoglycan. In total, about 50 basement membrane proteins have been identified, of which collagens, espe-

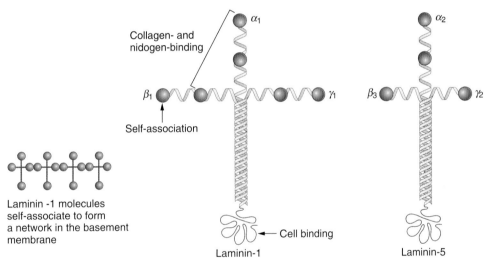

Figure 6–34. Structure of laminin.

Figure 6–35. Molecular composition of the basement membrane.

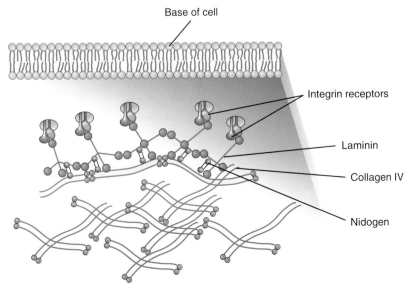

cially type IV, constitute 50% of all basement membranes. The basement membranes of different tissues have specific properties in addition to the general requirement for cell attachment. Specificity is conferred by different isoforms of type IV collagen, laminin, and heparan sulphate proteoglycan. Thus, 7 different type IV collagens and 12 different laminins are known. Such specificity is important for regulating the varied functions of different tissues and organs. Of the major basement membrane components, laminin and type IV collagen have the ability to self-assemble into sheetlike structures, whereas the other components do not. Studies on basement membrane assembly by cultured cells indicate that laminin first forms a network that associates with the cell surfaces through integrin adhesion receptors (especially β_1) and dystroglycan, a transmembrane proteoglycan. Type IV collagen forms an independent but associated network, the interaction between the two being facilitated by nidogen/entactin. This complex then forms a scaffolding for the binding of other basement membrane constituents.

Basement membranes are the targets of a number of human diseases. Mutations in the gene for the α_5 chain of type IV collagen are associated with **Alport syndrome,** a disease that involves nephritis and deafness. Junctional EB, a severe blistering disease of the skin that results in early infantile death, is associated with mutations in the genes that encode the laminin-5 chains, whereas dystrophic EB, a severely debilitating blistering disease that results in syndactyly, is due to mutations in collagen VII genes and consequent absence of anchoring fibrils. Autoantibodies to type IV collagen α_3 chain, which is present in the glomerular basement membrane of the kidney, are associated with **Goodpasture syndrome.**

A key component of **tumor growth** is **angiogenesis,** the elaboration of new blood vessels. Growing tumors cannot exceed a few millimeters in diameter without acquiring a new blood supply, or they will die of anoxia. Tumor cells produce growth factors that promote angiogenesis. The basement membrane inhibits the proliferation and migration of endothelial cells, thus preventing them from branching out to produce new vessels. During tumor growth, inflammatory and stromal cells within the matrix near the tumor produce matrix metalloproteinases that degrade the vascular basement membrane, thus enabling the endothelial cells to proliferate, migrate, and form new blood vessels to supply the tumor. Research on this process offers hope that inhibition of angiogenesis may be used to prevent tumor growth.

Fibrin Forms the Matrix of Blood Clots and Assembles Rapidly When Needed

A major ECM component of blood clots is a protein called *fibrin* that forms an **elastic network** to which cells and other ECM components bind. Polymerization of fibrin to form the network occurs when its precursor molecule **fibrinogen,** present in substantial quantities (2–4 g/L) in blood plasma, is cleaved by the enzyme thrombin. Fibrinogen molecules are elongated structures 45 nm in length that consist of two sets of Aα, Bβ, and γ chains linked by disulfide bonds (Fig. 6–36). Each molecule consists of two outer D domains linked to a central E domain by a coiled-coil segment. Thrombin cleaves a small fragment called *fibrinopeptide A* from the Aα chains to initiate polymerization. This involves formation of double-standard fibrils by end-to-middle interaction of the D and E domains, as well as lateral and branching fibril associations to form the network of the clot. The network is stabilized by covalent cross-linking through intermolecular E-(γ-glutamyl) lysine

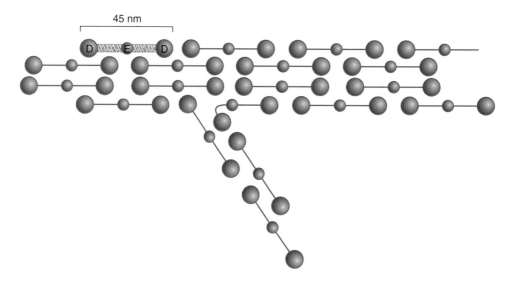

Figure 6–36. Structure of the fibrin molecule and the branching fibers that it forms in blood clots.

bonds by the action of an enzyme called *factor XIII* or plasma **transglutaminase.** The cross-linked clot has substantial elasticity and can recover its original form after extension by up to 1.8 times its length. Inhibiting mechanisms exist for both thrombin generation and factor XIII activity so that the clotting process can be regulated.

Fibrin has binding interactions with a variety of extracellular components including the ECM components fibronectin and heparin, the growth factors FGF-2 and vascular endothelial growth factor, and the cytokine **interleukin-1.** It also has binding sites for a number of CAMs including **vascular endothelial (VE)-cadherin,** the platelet integrin $\alpha_{IIb}\beta_3$, and the leukocyte integrin $\alpha_m\beta_2$ (Mac-1). These are important for promoting angiogenesis, incorporating platelets into the developing **thrombus,** and recruiting **monocytes** and **neutrophils,** respectively.

In addition to needing to form quickly when required, clots need to disperse when their function is no longer required, or if they form inappropriately. The dispersal process called *fibrinolysis* is mediated by an enzyme called **plasmin** that cleaves fibrin. Plasmin is activated by cleavage of a precursor protein **plasminogen,** through the action of another enzyme, **tissue plasminogen activator,** which binds to fibrin.

von Willebrand Factor in Normal and Abnormal Blood Clotting

vWF plays a key role in the major response of platelets to vascular injury by mediating the initiation and progression of thrombus formation (Fig. 6–37). Blood flow produced substantial **shear forces** at the blood vessel wall, and these forces oppose cell adhesion. vWF forms a bridge between collagen in the vessel wall and blood platelets sufficient to enable cell adhesion to develop.

Mature vWF is a multimeric protein that consists of a variable number of identical subunits linked together by disulfide bonds. Each precursor subunit is a protein chain of 2050 amino acids with a substantial number of carbohydrate chains bound to it. These subunits become linked together by disulfide bonds between their COOH termini to form dimers of about 500 kDa molecular weight. Further assembly requires cleavage of a propeptide from the NH$_2$ terminus by an enzyme,

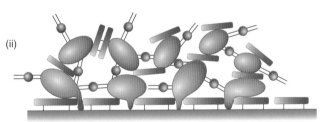

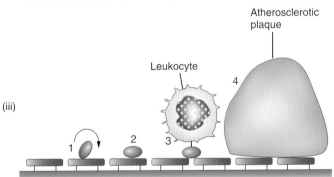

Figure 6–37. Normal and abnormal adhesion of blood platelets. *(i)* von Willebrand factor (vWF) (blue) attaches to exposed endothelial basement membrane collagen (green), and platelets adhere loosely, rolling under the force of blood flow. *(ii)* Platelets form stable adhesions to collagen and more plates attach by aggregation mediated by vWF and fibrinogen (brown). The thrombus then continues to grow. *(iii)* Platelets may adhere to the surfaces of endothelial cells initiating the formation of a thrombus by recruitment of other plates and leukocytes and, eventually, development of atherosclerotic plaque.

1. Rolling adhesion of platelet to endothelium – GPIb – P-selectin/vWF
   – PGSL-1 - P-selectin
2. Firm adhesion to endothelium – α^{IIb}integrin-fibrinogen
3. Recruitment of other platelets and leukocytes – PGSL-1 - P-selectin
   Mac-1 - endothelial ligands
4. Formation of atherosclerotic plaque

furin. The NH$_2$ termini can then also become linked together by disulfide bonds. Such linking may proceed until multimers exceeding 10,000 kDa are formed. Electron microscopy shows the largest such multimers to be up to 1300 nm in length and 200 to 300 nm in cross section. vWF multimers are synthesized intracellularly and stored in Weibel–Palade bodies in endothelial cells and α-granules in platelets and megakaryocytes (large cells that give rise to platelets). Some vWF is **secreted constitutively** by endothelial cells, giving a residual plasma concentration of this protein. **Regulated secretion** by endothelial cells and platelets occurs in response to vascular injury. The vWF monomer has binding sites for collagen and the platelet adhesion receptors GPIb (part of the GPIb-IX-V complex) and GPIIbβ$_3$, as well as for collagen and the blood-clotting protein factor VIII. Thus, the multimeric complexes are literally strings bristling with binding sites. The larger the multimers, the more effective they are at promoting thrombus formation.

Initial platelet adhesion is mediated by the binding of GPIb on the platelet to vWF, which is, in turn, bound to collagen. This attachment is easily broken and does not result in firm adhesion. Instead, platelets exhibiting this type of adhesion roll along the surface under the force of blood flow. It is possible that binding between selectin molecules on the platelet surface and the carbohydrate chains on vWF also contribute to initial adhesion.

GPIb-vWF interaction generates an intracellular signal within the platelets that involve changes in intracellular Ca^{2+} concentration and the signaling enzyme **protein kinase C**. The function of these signals is to bring about activation of the platelet integrin GPIIbβ$_3$ that then mediates firm adhesion to vWF.

Because it has multiple binding sites for platelet adhesion molecules, vWF can also mediate platelet aggregation by bridging between them. Platelets also adhere to fibrinogen and fibrin, other important participants in the clotting process. It appears that vWF and fibrin have complementary roles in thrombus formation. Thus, vWF mediates rapid thrombus formation at high shear rates in the absence of fibrinogen, but the thrombi are unstable. In the presence of fibrinogen thrombus development is slower but more stable. Thus, patients having congenital defects in either vWF or fibrinogen have clotting disorders.

As seen with fibrin, the process of thrombus formation by vWF requires regulation. This is done extracel-lularly by a plasma enzyme called **ADAMTS13**, which cleaves vWF multimers to restrict their size and possibly to prevent excessive thrombus formation. The normal function of ADAMTS13 is clearly important because mutations in the gene for this enzyme cause a disease called **chronic relapsing thrombocytopenic purpura**. It is important to understand the mechanisms involved in thrombus because it is likely that their abnormal function cause thrombocytic diseases such as **stroke, coronary thrombosis, phlebitis,** and **phlebothrombosis.**

SUMMARY

Cell adhesion and the ECM are fundamental to the normal structure and function of human tissues. Adhesion is mediated by molecules that mostly belong to one of four families of adhesion receptors, the cadherins, the Ig family, the selectins, and the integrins. Adhesion receptors are commonly clustered in cell junctions; desmosomes and adherens junctions mediate cell-cell adhesion, whereas hemidesmosomes and focal contacts mediate cell-matrix adhesion. Intercellular junctions also include tight junctions that regulate the permeability of paracellular channels and cell polarity, and gap junctions that facilitate intercellular communication. In addition to participating in cell adhesion, adhesion receptors transduce signals that regulate many aspects of cell behavior including movement, proliferation, differentiation, and survival. Cell adhesion is a dynamic process that is particularly evident where nonadhesive cells rapidly become adhesive, for example, leukocytes in inflammation and platelets in blood clotting.

The ECM has many components of which the most abundant are fibrillar collagens. These provide the strength of tendons and the dermis, and form the basis of bone and cartilage. Much of the bulk of tissues resides in GAGs and proteoglycans, negatively charged polymers that absorb water and resist compression especially in cartilage. Tissue elasticity is dependent on elastic fibers composed of elastin and fibrillin. The cells of many tissues, such as epithelia, adhere to basement membranes, the principal components of which are laminin and type IV collagen. Basement membranes of different tissues have specific properties dependent on differences in molecular composition. Most extracellular matrices are semipermanent in nature, but blood clots form rapidly in response to injury. Key components of the clot matrix are vWF and fibrin, both of which provide a substratum for platelet adhesion.

Suggested Readings

Cell Adhesion Molecules and Signaling

Juliano RL. Signal transduction by cell adhesion receptors and the cytoskeleton: functions of integrins, cadherins, selectins and immunoglobulin family members. *Annu Rev Pharmacol Toxicol* 2002; 42:283–323.

Intercellular Junctions

Aijaz S, Balda MS, Matter K. Tight junctions: molecular architecture and function. *Int Rev Cytol* 2006;248:261–298.
Garrod DR, Merritt AJ, Zhuxiang N. Desmosomal adhesion: structural basis, molecular mechanism and regulation. *Mol Membr Biol* 2002;19:81–94.

Wei C-J, Xu X, Lo C. Connexins and cell signalling in development and disease. *Annu Rev Cell Dev Biol* 2004;20:811–838.

Cell-Matrix Junctions

Frame MC, Carragher NO. Focal adhesion and actin dynamics: a place where kinases and proteins meet to promote invasion. *Trends Cell Biol* 2004;14:241–249.

Functional Aspect of Cell Adhesion

Ramasay I. Inherited bleeding disorders: disorders of platelet adhesion and aggregation. *Crit Rev Oncol Hematol* 2004;49:1–35.

Extracellular Matrix

Gelse K, Pöschl E, Aigner T. Collagens: structure, function, and biosynthesis. *Adv Drug Deliv Rev* 2003;55:1531–1546.

Kalluri R. Basement membranes; structure, assembly and role in tumour angiogenesis. *Nat Rev Cancer* 2003;3:422–433.

Mendolicchio GL, Ruggeri ZM. New perspectives on von Willebrand factor functions in hemostasis and thrombosis. *Semin Hematol* 2005;42:5–14.

Mosesson MW. Fibrinogen and fibrin structure and function. *J Thromb Haemost* 2005;3:1894–1904.

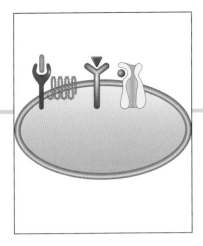

Intercellular Signaling

GENERAL MODES OF INTERCELLULAR SIGNALING

Intercellular Signaling Molecules Act as Ligands

Regardless of chemical structure, all cell-to-cell signaling molecules act as **ligands** and initiate biological responses in target cells by binding to specific **receptors** (Fig. 7–1). Molecules that are too large or too hydrophilic to cross the plasma membrane use receptors at the plasma membrane to relay the signal to the inside of the target cell. Plasma membrane receptors contain an extracellular ligand-binding domain, a transmembrane domain, and an intracellular domain that initiates the intracellular events that lead to a biological response. These plasma membrane receptors can be ion channels, enzymes, or linked to enzymes. Smaller, hydrophobic ligands diffuse across the plasma membrane and activate receptors located inside the target cell. These intracellular receptors are most often transcription factors that initiate changes in gene transcription. Intercellular signaling ligands and receptors are expressed at low levels and are not amenable to isolation and characterization using traditional biochemical approaches. Recent advances in molecular biology and pharmacology have circumvented these issues and significantly increased the rate at which intercellular ligands and receptors are identified/characterized. This information provided new insights regarding the complex cellular and molecular events that cells use to respond to their ever-changing environment.

Cells Exhibit Differential Responses to Signaling Molecules

Cells must be able to selectively respond to some signals and at the same time disregard other signals. Differential responses can be attributed to variations in the combination of ligands, receptors, and/or intracellular signaling pathways involved. If the ligand is absent, or present in reduced quantities, there will be no response. Even if the ligand is present in high concentrations, there will be no response if the target cell(s) does not express the appropriate receptor. Furthermore, the same ligand does not necessarily produce the same biological effect in all cells. For example, acetylcholine regulates contraction in the heart and skeletal muscle, but regulates secretion in the salivary gland (Fig. 7–2).

Ligand interaction with a different receptor subtype or coupling of the same receptor to different intracellular signaling pathways can generate differential responses. Finally, it is the sum of the cell's response to all of the individual ligands it is exposed to that determines the alteration in cell behavior that occurs. Highly divergent responses can be observed with only subtle changes in the combination of ligands or in the physiological state of the target cell, or in both.

A CELL-SURFACE RECEPTOR

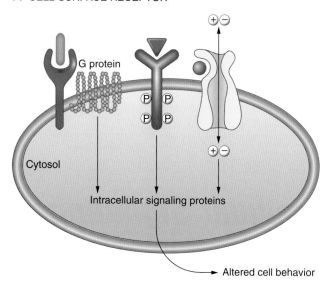

B INTRACELLULAR RECEPTOR

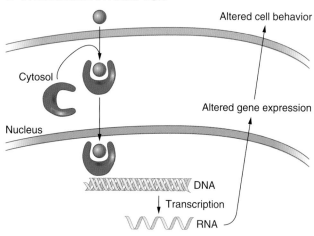

Figure 7–1. Intracellular signaling molecules are ligands and exert their effects via interaction with specific target cell receptors. Signaling molecules that are large and hydrophilic cannot enter the target cell by diffusion and exert their effects via interaction with cell-surface receptors (**A**). These receptors can be ion channels, G-protein coupled, or enzyme coupled. Receptor-mediated changes in downstream second messenger pathways alter cell behavior. Small/hydrophobic signaling molecules readily diffuse into the target cell and interact with receptors located inside the cell (**B**). These ligand-receptor complexes then bind to regulatory regions in DNA and promote the transcription of new gene products that alter cell behavior.

Intercellular Signaling Molecules Act via Multiple Mechanisms

As our understanding of the events involved in intercellular signaling has increased, the distinctions between classes of intercellular signaling molecules have become blurred. Traditionally, signaling molecules have been classified as endocrine, paracrine, autocrine, or juxtacrine, depending on the distance over which they act and the source of the ligand (Fig. 7–3).

However, individual signaling molecules can act via multiple mechanisms. For example, epidermal growth factor (EGF) is a transmembrane protein that can bind to and signal a neighboring cell via direct contact (juxtacrine signaling). However, it can also be cleaved by a protease, released into the circulation, and act as a hormone (endocrine signaling). Epinephrine functions as a neurotransmitter (paracrine signaling) and a systemic hormone (endocrine signaling). This broad diversity in ligand structure/biochemistry and ligand synthesis/distribution/metabolism, as well as target cell receptors, allows only a hundred intercellular signaling molecules to generate an unlimited number of signals. In the following sections, examples of clinically relevant intercellular signaling pathways are used to present the unique cell biological properties of the different classes of intercellular signaling molecules.

HORMONES

Hormones allow organisms to coordinate the diverse activities of cells over long distances. Specialized endocrine glands (pituitary, thyroid, parathyroid, pancreas, adrenal glands, gonads) and other organs secrete/release a long list of chemically diverse hormones (Table 7–1). These hormones enter the circulation and act on target cells located throughout the body. Because hormones distribute throughout the body, they can induce simultaneous changes in many different organs. However, the long distances that they travel to reach their target cells limits their actions to the regulation of physiologic processes—reproduction, growth and development, and metabolism—with a timescale of minutes to years. Hormones can be grouped into two distinct classes: small lipophilic molecules that interact with **intracellular receptors,** and hydrophilic molecules that interact with **cell-surface receptors.** The distinctive features of each class are described in the following section. Finally, the hypothalamic-pituitary axis is used to illustrate the interplay between these two classes of hormones, as well as the complex positive and negative feedback mechanisms that fine-tune hormone function.

Lipophilic Hormones Activate Cytosolic Receptors

The small, lipophilic hormones control a diverse list of biological processes. Sex hormones such as progesterone, estradiol, and testosterone are produced by the gonads and regulate sexual differentiation and function. Corticosteroids are synthesized by the adrenal gland and divided into two groups based on function: glucocorticoids increase glucose production in many different cell types, and mineral corticoids regulate salt and water balance in the kidney. Thyroxine is synthesized in

A SALIVARY GLAND   B HEART MUSCLE   C SKELETAL MUSCLE

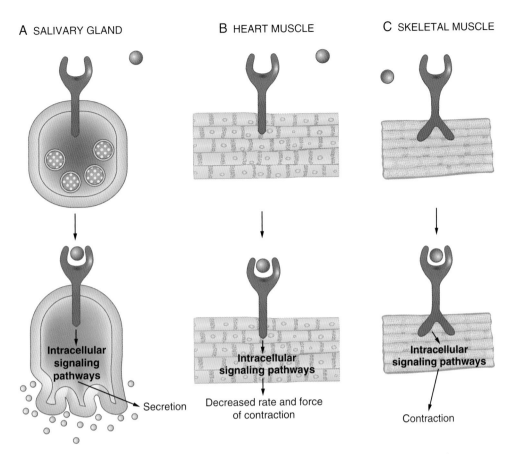

Figure 7–2. *Signaling molecules are versatile and induce differential responses.* In the salivary gland (**A**), acetylcholine activates a muscarinic receptor subtype, resulting in secretion. In the heart (**B**), acetylcholine activation of the same muscarinic receptor subtype has a different biological effect, decreased rate and force of contraction. The differential effects on cell behavior are due to the coupling of the muscarinic receptor to different intracellular signaling pathways in the two cell types. In skeletal muscle (**C**), acetylcholine activates a different receptor subtype, the nicotinic receptor, resulting in depolarization of the muscle cell and contraction.

the thyroid gland and regulates metabolism in virtually every organ. Vitamin D_3 regulates Ca^{2+} metabolism and bone growth. Retinoic acids and other retinoids play important roles in development. All lipophilic hormones freely diffuse across the lipid bilayer, interact with cytosolic receptors, and alter the expression of specific genes. It is the particular genes that are activated/inactivated that determine the ultimate effect of the hormone on the target cell/tissue.

The time frame via which lipophilic hormones act is determined by their synthesis and metabolism. All steroids are synthesized from cholesterol and have similar chemical backbones. Steroid-producing cells store only a small supply of hormone precursor; they do not store the mature active hormone. When stimulated, cells convert the precursor to the active hormone, which diffuses across the plasma membrane and enters the circulation. This process can take several hours to several days. Because their solubility in the aqueous environment is poor, steroids are tightly bound to carrier proteins in the bloodstream. This dramatically slows their

degradation and allows steroid hormones to remain in the circulation for hours to days. Thus, once a steroid–hormone response develops, it can persist for a prolonged period.

Although structurally related to the steroid hormones, retinoids are synthesized from retinol (vitamin A), not cholesterol. Retinol is found in high concentrations in the liver and in the bloodstream, where it is complexed with serum-binding proteins. Retinol diffuses across the plasma membrane and forms a complex with a cytosolic retinol-binding protein. Cytosolic retinol is converted to retinoic acid by a series of dehydrogenases. The newly synthesized retinoic acid exits the cell by diffusion and acts on neighboring cells. Retinoic acid is unique in that it can remain in the cytosol and signal within the synthesizing cell.

Thyroid hormone synthesis is unusual and complex. The thyroid gland stores large amounts of thyroid hormone as amino acid residues on a large, multimeric, extracellular glycoprotein called **thyroglobulin.** Thyroglobulin is endocytosized at the apical membrane and

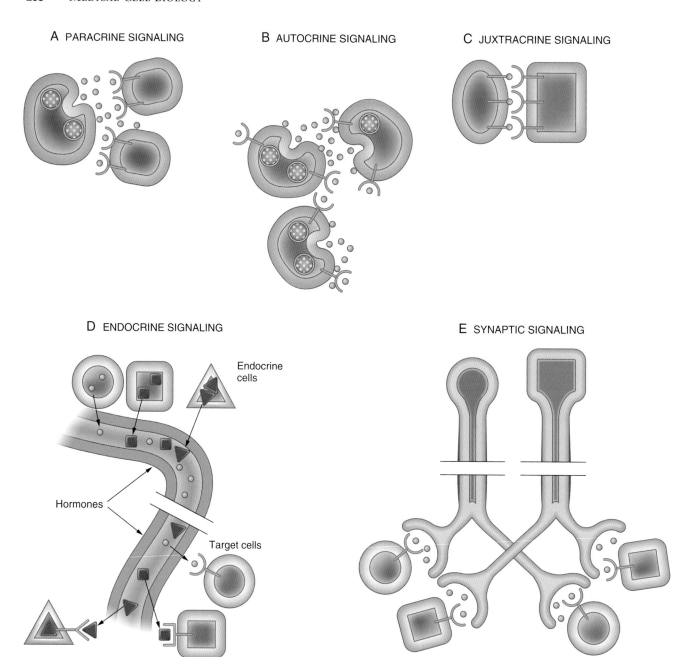

A PARACRINE SIGNALING

B AUTOCRINE SIGNALING

C JUXTRACRINE SIGNALING

D ENDOCRINE SIGNALING

Endocrine cells

Hormones

Target cells

E SYNAPTIC SIGNALING

Figure 7-3. **General schemes of intercellular signaling.** Cell-to-cell signaling can occur over short (A–C) or long distances (D, E). In paracrine signaling (**A**), chemicals are released into the extracellular environment and exert their effects on neighboring target cells that express the appropriate receptor. In autocrine signaling (**B**), the cell that synthesizes/releases the signaling molecule is also the target cell. In juxtacrine signaling (**C**), the signaling molecule remains attached to the plasma membrane and interacts with receptors on adjacent target cells. In endocrine signaling (**D**), hormones are released into the circulation and distribute throughout the body but alter only the behavior of cells that express the appropriate receptor. Synaptic signaling (**E**) is a specialized form of paracrine signaling that occurs over long distances because the signal is transmitted along neuronal cell processes that can span the entire length of the organism. Specificity in synaptic signaling is generated by the formation of synaptic contacts and not the signaling molecule/neurotransmitter.

appears as colloidal droplets within the cell. These droplets fuse with lysosomes where the thyroglobulin is degraded and thyroxine (T_4) and 3,5,3′-tri-iodothyronine (T_3) are liberated. This process is extremely inefficient and can take hours to days. The liberation of 2 to 5 thyroxine molecules requires the complete breakdown of 1 thyroglobulin molecule—approximately 5500

amino acid residues and 300 carbohydrate residues. The T_3 and T_4 then diffuse across the basement membrane and enter the circulation. Because their solubility in the aqueous environment of the blood is low, T_3 and T_4 are tightly bound to carrier proteins. This inhibits their degradation and allows responses to persist for hours to days.

TABLE 7–1. Major Biological Activities of Representative Hormones

Hormone	Site of Origin	Major Biological Activity
Proteins/polypeptides		
Insulin	Pancreatic β cells	Carbohydrate utilization
Growth hormone releasing hormone (GHRH)	Hypothalamus	Stimulate growth hormone secretion
Growth hormone	Anterior pituitary	General stimulation of growth
Leuteinizing hormone (LH)	Anterior pituitary	Stimulate LH secretion
Parahormone	Parathyroid	Increased bone resorption
Follicle-stimulating hormone (FSH)	Anterior pituitary	Stimulate ovarian follicle growth and spermatogenesis
Thyroid-stimulating hormone (TSH)	Anterior pituitary	Stimulate thyroid hormone secretion
Erythropoietin	Kidney	Increase red blood cell production
Prolactin	Anterior pituitary	Stimulate milk production
Glucagon	Pancreas	Stimulate glucose synthesis
Insulin-like growth factor-1	Liver	Stimulate bone and muscle growth
Small peptides		
Somatostatin	Hypothalamus	Inhibit growth hormone release from anterior pituitary
TSH-releasing hormone (TRH)	Hypothalamus	Stimulate TSH release from anterior pituitary
LH-releasing hormone (LRH)	Hypothalamus	Stimulate LH secretion from anterior pituitary
Vasopressin/antidiuretic hormone (ADH)	Posterior pituitary	Elevate blood pressure; increase kidney water resorption
Oxytocin	Posterior pituitary	Stimulate smooth muscle contraction
Amino acids		
Norepinephrine	Adrenal medulla	Increase blood pressure and heart rate
Dopamine	Hypothalamus	Inhibit prolactin secretion
Lipophilic hormones		
Estradiol	Ovary, placenta	Develop/maintain secondary male sex characteristics
Cortisol	Adrenal cortex	Metabolism; suppress inflammatory reactions
Progesterone	Ovary, placenta	Prepare uterus for pregnancy; maintain pregnancy
Testosterone	Testis	Develop/maintain secondary male sex characteristics
Thyroxine	Thyroid	Increase metabolic activity in many cells
Retinoic acid	Diet	Epithelial cell differentiation

Receptors for Lipophilic Hormones Are Members of the Nuclear Receptor Superfamily

Although the receptors for lipophilic hormones are not identical, they are evolutionarily related and belong to a large superfamily called the **nuclear receptor superfamily**. This superfamily also includes receptors that are activated by intracellular metabolites. Family members that have been identified by only DNA sequencing and for which ligands have not been identified are referred to as **orphan receptors**. Members of the nuclear receptor family contain related domains for ligand binding, DNA binding, and transcriptional activation.

Lipophilic hormone receptors are located in the cytosol, nucleus, or both, where are they are complexed with other proteins (Fig. 7–4). After ligand-binding the complex dissociates, the receptor dimerizes, undergoes phosphorylation, binds DNA, and induces the transcription of specific target genes. Inactive thyroid hormone receptors are located in the nucleus and complexed with DNA in a conformation that inhibits transcription (see Fig. 7–4). The ligand receptor complex remains associated with the DNA and undergoes a conformational change that results in activation of transcription. When the ligand dissociates, the receptor is dephosphorylated and returns to its inactive state/

location. Initially, these receptors directly activate the transcription of a small number of specific genes. These primary phase gene products activate other genes to produce a delayed, secondary response. Although the activity of these receptors has become synonymous with transcription, increasing evidence exists that some effects of small hydrophobic hormones may be due to direct regulation of cellular processes without genomic effects.

Peptide Hormones Activate Membrane-Bound Receptors

Peptide hormones range in size from a simple tripeptide (thyrotropin-releasing hormone) to a 198-amino acid protein (prolactin), to a glycosylated multisubunit oligomer (human chorionic gonadotropin). Because these agents mediate rapid responses to the environment, they are stored in secretory vesicles adjacent to the plasma membrane and are available for immediate release. The synthesis and release of peptide hormones occurs via regulated exocytotic pathway (see Chapter 4). The environmental signals that trigger peptide hormone release also stimulate synthesis of peptide hormones to ensure that the released hormone is replaced. Released hormones are in the blood for only a few seconds or minutes before they are degraded by blood/tissue proteases or

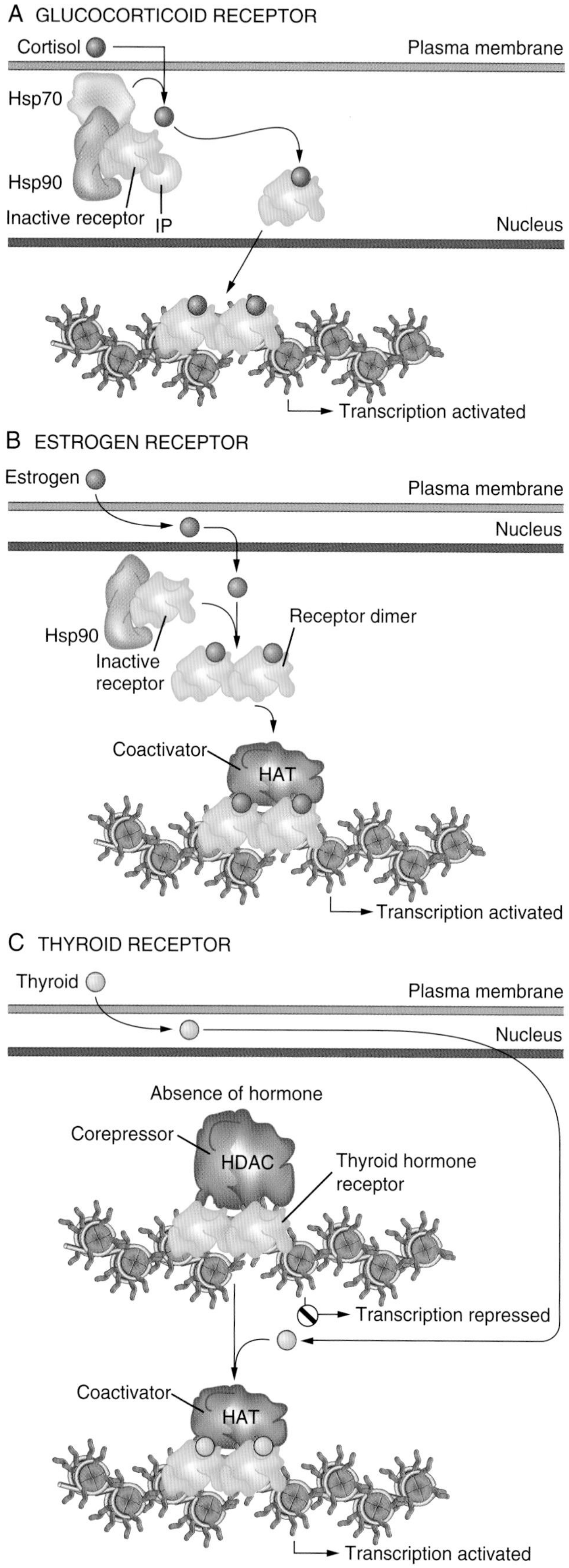

A GLUCOCORTICOID RECEPTOR

B ESTROGEN RECEPTOR

C THYROID RECEPTOR

Figure 7–4. Gene regulation by members of the nuclear receptor superfamily. **A:** The inactive glucocorticoid receptor is located in the cytosol in a complex that contains heat shock proteins 90 and 70 (Hsp90 and Hsp70) and immunophilin (IP). Cortisol binding to the receptor displaces the accessory proteins, and the activated ligand-receptor complex translocates to the nucleus where it activates target genes. **B:** The inactive estrogen receptor Hsp 90 complex is located in the nucleus. Estrogen binding displaces Hsp90, the activated receptors dimerize, bind to DNA, associate with the coactivator histone acetyltransferase (HAT), and activate target genes. **C:** The thyroid hormone receptor binds DNA in the presence and absence of ligand. In the absence of ligand, the receptor is complexed with a corepressor, histone deacetylase (HDAC), which prevents gene transcription. In the presence of hormone, the ligand-receptor complex binds the coactivator (HAT) and activates target gene expression.

taken up into cells. Water-soluble hormones cannot diffuse across the plasma membrane and exert their effects by binding to receptors on the surface of the target cell. The signal is carried to the inside of the cytoplasmic region of the receptor and involves the generation of a second messenger (see Chapter 8). Some receptors activate a cascade of phosphorylation events and others activate G proteins. In contrast with lipophilic hormones, the effects of peptide hormones are almost immediate and usually persist for only a short period. One exception is growth hormone, which can produce long-lasting and even irreversible changes as a result of downstream changes in gene transcription.

Clinical *Case 7-1*

On physical examination, Carolyn is a thin, but well-developed girl who appears her stated age. She is showing normal secondary sex characteristics. Her blood pressure and pulse are normal. Her HEENT examination (head, ear, eyes, nose, throat) is normal, including her eye grounds. Her heart and lungs are clear, and her abdominal examination is normal. However, in the middle of that examination, she excuses herself to urinate and get a drink of water. Her neurologic examination is normal. The doctor sends a blood sample to the blood laboratory next door and asks for a urine sample.

Her hemogram comes back with a hemoglobin level of 10.5, a reduced white cell count of 3400, and platelets of 110,000. The automated cell counter provides a mean cell volume of her red cells, even though it was not requested. It is 105. Her hemoglobin A1c is 8.5%, and her blood sugar is 360 mg/100 cc. Her urine has a low specific gravity and shows 4+ for sugar and a trace of acetone by dip-strip tests in the physician's office.

On questioning the mother, the physician determines that there is no family history of premature heart disease, cancer, or diabetes. Carolyn's maternal grandfather died of pernicious anemia.

Cell Biology, Diagnosis, and Treatment of Type I Diabetes Mellitus

Carolyn has Type 1 diabetes mellitus. She is symptomatic now from hyperglycemia, glucose-induced osmotic diuresis, and probable dehydration. This so-called juvenile form of diabetes mellitus usually occurs in younger patients. It is not unusual to start with the stress of puberty, and it is not associated with a family history of diabetes. Rather, it appears to be due to an "autoimmune" destruction of the insulin-secreting cells of the pancreas by a β-cell–specific antibody. It is also often associated with other autoimmune phenomena. The possibility that Carolyn's macrocytic anemia is an early case of pernicious anemia, as well as the previously suggested thyroiditis, both agree with this likely cause. Prompt treatment with insulin will control her hyperglycemia and polyuria. The possible pernicious anemia will be worked up with a bone marrow biopsy and, if positive, will be treated with regular intramuscular vitamin B12 injections.

Carolyn is an intelligent young lady with a supportive family. It is likely that she will understand the need for careful glycemic control to avoid the extremes of hypoglycemia and ketoacidosis. Her physician has already told her about self-monitoring of blood glucose levels and the potential utility of a portable insulin pump. Furthermore, initiation of instruction in dietary basics will start her on lifelong attention to her personal glucose metabolism.

As she gets used to the sudden discovery of her chronic disease, her physician will help her to understand some of its cardiovascular, renal, and ocular complications. He will also encourage her to recognize that with good glycemic control, which she can watch by following her HbA1c (glycosylated hemoglobin) levels, she may be able to delay or avert those complications.

The Hypothalamic-Pituitary Axis

The synthesis of classical pituitary hormones is controlled by a complex integrated feedback loop that involves hypothalamic neurons and peripheral endocrine glands (Fig. 7–5). These complex **positive and negative feedback** loops ensure that pituitary hormone secretion is in tune with all aspects of the organism's environment.

The classical pituitary contains five different cell types, each of which secretes a particular biologically

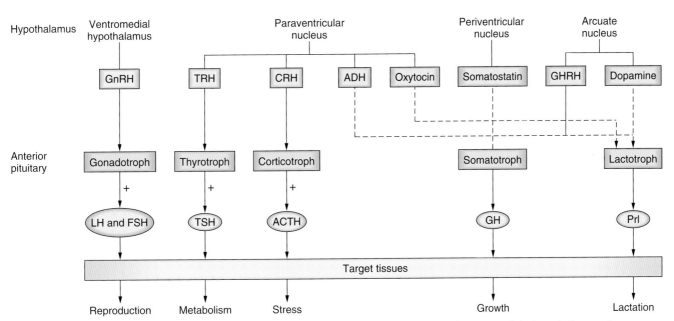

Figure 7–5. **Hypophysiotrophic control of pituitary hormones.** Neuronal cells in various nuclei/regions of the hypothalamus release signaling molecules into the bloodstream, rather than synapses. These signaling molecules (blue) travel through the bloodstream to the pituitary where they stimulate *(solid lines)* or inhibit *(dashed lines)* the release of one or more pituitary hormones (purple). ACTH, adrenocorticotropic hormone; ADH, antidiuretic hormone; CRH, corticotrophin-releasing hormone; FSH, follicle-stimulating hormone; GH, growth hormone; GHRH, growth hormone–releasing hormone; GnRH gonadotropin-releasing hormone; LH, leuteinizing hormone; TRH, TSH-releasing hormone; TSH, thyroid-stimulating hormone.

important hormone(s): somatotropes secrete growth hormone; lactotropes secrete prolactin; corticotrophs secrete adrenocorticotropic hormone (ACTH); thyrotropes secrete thyroid-stimulating hormone (TSH); and gonadotrophs secrete follicle-stimulating hormone (FSH) and leuteinizing hormone (LH). The pituitary also synthesizes many nonclassical hormones including growth factors, cytokines, and neurotransmitters. The secretion of the classical pituitary hormones is regulated by neurons in the hypothalamus that receive inputs from target tissues, hormonal feedback, and stimuli from other brain areas. These neurons fire at regular intervals, generating the pulsatile release of hypothalamic peptide hormones that is essential for proper endocrine system function. Because these agents are released into the bloodstream, specifically the superior or inferior hypophyseal artery, they are classified as hormones. Hypothalamic hormones travel through the portal circulation and activate receptors on specific pituitary cells to stimulate or inhibit the release of one or more pituitary hormones. The pituitary hormones then act on target tissues, such as the gonads, and stimulate the release of still other hormones, both peptides and steroids. Growth hormone secretion is an excellent example of the various factors that impact pituitary hormone secretion. Growth hormone levels must be coordinately regulated with metabolic fuel availability during prenatal and postnatal growth spurts and again during puberty (Fig. 7–6).

GROWTH FACTORS

Growth factors represent a large number of polypeptides that play an important role in cell growth and survival from conception to death. All cells synthesize one or more growth factors (Table 7–2). In most cases, the original names of growth factors do not reflect currently known biological activities. Although their usual physiologic effects are paracrine, some growth factors also act over longer distances. Thus, it is not uncommon for some growth factors to be classified as hormones and for hormones that regulate cell growth to be classified as growth factors. Growth factors act in one of three ways: **mitogens** stimulate cell proliferation, **trophic factors** promote growth, and **survival factors** inhibit apoptosis. Many growth factors are **pleiotropic**—that is, they have multiple effects within the same cell or elicit different responses in different cell types, or both. Depending on the environment of the cell, growth factors can promote cell growth at one time and inhibit it at another time. Abnormalities in growth factor signaling are the basis for many types of cancers. Therapeutic agents for the treatment of neurodegenerative disorders, the side effects from chemotherapy and viral infections, as well as the harvesting of stem cells for bone marrow transplantation target growth factor signaling pathways.

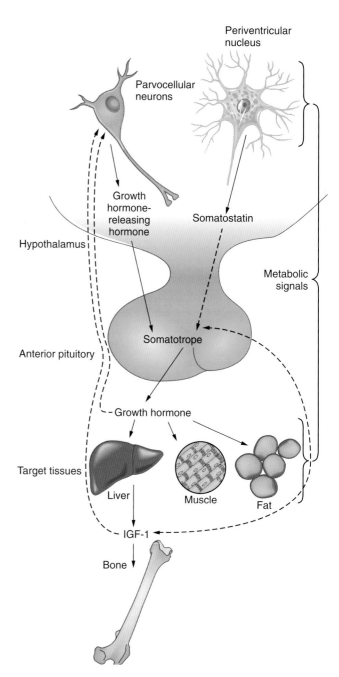

Figure 7–6. Regulation of growth hormone (GH) release. GH release from the pituitary is stimulated by growth hormone–releasing hormone (GHRH), which peaks at night, and inhibited by somatostatin, which reaches high levels during the day. GH then stimulates the growth of muscle and adipocytes, as well as the release of insulin-like growth factor-1 (IGF-1) from the liver. IGF-1 then stimulates bone growth. If there are adequate metabolic fuels, metabolic signals from the peripheral target tissues will act on the hypothalamus and stimulate GH release. GH and IGF-1 provide negative feedback and inhibit the release of GHRH from the hypothalamus. IGF-1 also inhibits GH release from somatotropes in the pituitary.

TABLE 7–2. Major Growth Factor Families

Signaling Molecule	Source	Major Biological Activity
Neurotrophins		
Nerve growth factor (NGF)	Brain, heart, spleen	Neuronal differentiation and survival
Brain-derived neurotrophic factor (BDNF)	Brain and heart	Neuronal differentiation and survival
Neurotrophin 3 (NT-3)	Brain, heart, kidney, liver, thymus	Neuronal differentiation and survival
Epidermal growth factor (EGF) family		
EGF	Salivary gland	Cell proliferation
Transforming growth factor-α (TGF-α)	Many cells/tissues	Cell proliferation
Fibroblast growth factor (FGF) family		
Fibroblast growth factors (22 total)	Many cells/tissues	Mitogenic
Transforming growth factor (TGF)-β family		
TGF-β	Ubiquitous	Inhibits proliferation
Inhibins/activins	Gonads and hypothalamus	Inhibits follicle-stimulating hormone (FSH) secretion
Bone morphogenetic proteins (>30 total)	Many cells and tissues	Osteogenesis
		Establishment of embryonic axis
Platelet-derived growth factor (PDGF) family		
PDGF	Platelets	Tissue repair
Vascular endothelial growth factor (VEGF)	Neural tissue	Endothelial cell proliferation
	Vascular smooth muscle	↑ Vascular permeability
Hematopoietic growth factors		
Erythropoietin	Kidney	↑ Red blood cell production
Colony-stimulating factors (CSFs)	Endothelium, T cells, fibroblasts, macrophages	↑ Red blood cell production
Thrombopoietin	Liver	↑ Platelet production
Insulin-like growth factor (IGF) family		
IGF-1	Many cells/tissues	Mitogenic, trophic, € survival
IGF-2	Many cells/tissues	Fetal growth
Tumor necrosis factor (TNF) family		
TNF-αα and -β	Macrophages, natural killer cells, T lymphocytes	Tumor regression
Interferons (types I and II)	**Helper T cells**	**Antiviral activity**
Interleukins (29 total)	**T, B, and mast cells predominantly**	**Proliferation and differentiation of T and B lymphocytes**

Nerve Growth Factor

Nerve growth factor (NGF) was discovered in the 1950s and was the first growth factor to be characterized. NGF, as well as other members of the neurotrophin family (see Table 7–2), regulate the development and survival of neurons. The main distinction among neurotrophin family members is their site(s) of synthesis and the target cells on which they act. During nervous system development, 50% or more of the neurons in the brain, spinal cord, and peripheral nervous system routinely die. This excess of neurons ensures that all postsynaptic cells will be innervated. NGF secreted from the future postsynaptic cells binds to NGF receptors on the growth cones of the closest approaching axons. NGF receptor activation alters gene expression; in particular, it down-regulates genes that promote programmed cell death (see Chapter 10) and up-regulates genes that promote survival and neurite extension. Neurons that do not receive NGF will eventually die. Only those neurons that receive NGF will survive and innervate the postsynaptic cell. There has been great interest in using NGF, and possibly other neurotrophins, to minimize neuronal cell death in neurodegenerative disorders such as Alzheimer's disease, Parkinson's disease, Huntington's disease, multiple sclerosis, encephalomyelitis, diabetic neuropathies, and spinal cord injury.

Growth Factor Families

All growth factors are classified into families based on their amino acid sequences and the receptors that they activate (see Table 7–2). Considerable variation in the size and biological activities of the growth factor families exists. The two main members of the EGF family are EGF and transforming growth factor-α (TGF-α). EGF and TGF-α interact with the same receptor and act as mitogens in a large number of tissues. The two members of the platelet-derived growth factor (PDGF) family, PDGF and vascular endothelial growth factor

(VEGF), are important in tissue repair after injury. The fibroblast growth factor (FGF) family is one of the largest and contains 22 mitogenic growth factors. Members of the FGF family play a central role in angiogenesis, the formation of new blood vessels. The TGF-β family is also large, and the biological activities of its members are quite diverse. At low concentrations, TGF-β stimulates growth; at high concentrations, it inhibits growth. The bone morphogenetic proteins promote osteogenesis and establishment of the early embryonic axis. The inhibins/activins act as classical hormones to inhibit FSH secretion and as growth factors to regulate formation of the embryonic notochord, somites, and neural tube. Another family that overlaps with hormones is the insulin-like growth factor (IGF) family. Although insulin and the IGFs (IGF-1 and IGF-2) have similar structures, their biological activities are quite different: Insulin promotes anabolic activity and has no mitogenic activity, whereas the IGFs are mitogenic, trophic, and survival factors. Two other clinically important growth factor families are the interferon family, whose members exhibit antiviral activity and are used to treat hepatitis C, and the tumor necrosis factor (TNF) family, whose members act as tumor regressors.

Growth Factor Synthesis and Release

Virtually all cells synthesize polypeptide growth factors and secrete them by a "classical" regulated exocytosis (see Chapter 4). The major exception is members of the hematopoietic growth factor family that are not stored, but rather rapidly synthesized when needed. The mechanisms responsible for the release/processing of growth factors are diverse. Most members of the FGF family have a classical leader sequence that ensures efficient secretion from cells via classical regulated exocytotic pathway. However, FGF-1 is released by a nonclassical release mechanism under stress conditions. Some FGF members accumulate in the cytosol and nucleus, where they are complexed with other proteins. The two members of the EGF family, EGF and TGF-α, are synthesized as membrane-bound precursors that are cleaved to yield smaller soluble peptides. Both the precursor and the soluble peptide have biological activity. In the mammary gland, preferential expression of the precursor form occurs. PDGF can also be retained on the plasma membrane via electrostatic interactions—it does not have a membrane-spanning domain. IGF-1 and IGF-2 are found in the serum associated with IGF binding proteins. This sequestration serves several functions: it prolongs the half-life of growth factor, forms a reservoir of growth factor, and inhibits growth factor activity by preventing interaction with its receptor. The major mechanism for termination of growth factor signaling is receptor-mediated endocytosis and lysosomal degradation (see Chapter 4).

Growth Factor Receptors Are Enzyme-Linked Receptors

All growth factor receptors are membrane-bound enzyme-linked receptors. Like other membrane receptors, they contain three domains: an extracellular ligand (growth factor) binding domain, a transmembrane domain, and a cytoplasmic domain that acts as an enzyme or forms a complex with another protein that acts as an enzyme. The majority of growth factor receptors are receptor tyrosine kinases. Growth factor binding leads to phosphorylation of tyrosine residues on a number of intracellular signaling molecules, and these molecules transmit the signal to the inside of the cell. The activation of FGF receptor (FGFR) tyrosine kinases is detailed later in this chapter. However, not all growth factor receptors are tyrosine kinase receptors. TGF-β activates receptor serine-threonine kinases that phosphorylate the SMAD protein transcription factor, resulting in downstream changes in gene transcription. Erythropoietin and the cytokines signal through the Janus kinase pathway. Growth hormone, prolactin, and colony-stimulating factors (CSFs) signal through the JAK-STAT pathway. These intracellular signaling pathways are detailed in Chapter 8.

Growth Factors Are Paracrine and Autocrine Signalers

Growth factors use a variety of different signaling modes to exert their biological effects. Most growth factors signal in a paracrine fashion; that is, the growth factor is synthesized in one cell and exerts its effects on a neighboring cell. However, growth factors can also signal in an autocrine fashion. One example is the proliferation of helper T cells during an immune response. Antigen-presenting macrophages secrete interleukin-1 (IL-1), which binds to IL-1 receptors on resting helper T cells, converting them to activated helper T cells. These activated cells synthesize and secrete IL-2 and express IL-2 receptors. The secreted IL-2 binds to an IL-2 receptor on the synthesizing cell and induces proliferation. If T cells do not proliferate, then an immune response will not be mounted. Another example is PDGF, which is best known for its paracrine actions on smooth muscle cells and fibroblasts during wound healing. Formation of the placenta requires the rapid clonal proliferation of cytotrophoblasts. This proliferation occurs as a result of secretion of PDGF by the cytotrophoblasts. Because these cells also express a PDGF receptor, they respond to the PDGF and proliferate.

Some Growth Factors Can Act over Long Distances

The ability of growth factors to signal over long distances has permitted their use as therapeutic agents. In fact, the administration of CSFs to cancer patients undergoing chemotherapy has become standard practice. Although the source is different, the exogenous CSFs act just like the endogenous CSF and mobilize the stem-cell populations needed to replenish the rapidly dividing hematopoietic cell populations destroyed by chemotherapy. This same approach is used to ensure that the harvested marrow used in bone marrow transplantation surgery contains high numbers of stem cells. More recently, granulocyte CSF has been shown to reduce infarct size in animal models of stroke by crossing the blood–brain barrier and interacting with its receptors in the brain.

Some Growth Factors Interact with Extracellular Matrix Components

A variety of growth factors and cytokines have been reported to bind specific carbohydrate moieties. The interaction of FGF with carbohydrates has been studied extensively (Fig. 7–7). All four FGFRs are tyrosine kinase receptors. In the presence of the extracellular matrix proteins heparan sulfate (HS) and glycosaminoglycans (see Chapter 6), a rigid FGF/FGFR/HS (2:2:2) dimer is formed. This dimer then activates the cytoplasmic domains and transphosphorylation of the receptors occurs. Once phosphorylated, the receptor activates downstream signal transduction cascades. In addition to activating the receptor, interactions between growth factors and extracellular matrix components may provide a mechanism for storing and concentrating these particular growth factors. In addition to FGF, TNF-α and interleukin-2 also recognize carbohydrate sequences, and this recognition modulates their biological activity. However, this is not a general feature of all growth factor receptors—EFGR dimerization requires the binding of EGF only.

HISTAMINE

Histamine mediates a diverse list of processes including immediate hypersensitivity reactions, allergic responses, gastric acid secretion, bronchoconstriction, and modulation of neurotransmitter release. Although all tissues contain histamine, the greatest histamine concentrations are found in the skin, bronchial mucosa, and intestinal mucosa. Endogenous histamine is released into the local environment and acts in a paracrine fashion to influence the activity of neighboring cells. Venoms, bacteria, and plants are sources of clinically important exogenous histamine. Many of the popular over-the-counter drugs that are used to treat allergic reactions (Benadryl and Dramamine) and ulcers/heartburn (Tagamet) antagonize histamine signaling.

Endogenous histamine is synthesized locally and in most organ systems is stored in secretory granules. Cytosolic histidine is converted to histamine by the enzyme histidine decarboxylase. Histamine is transported into secretory granules, where it is complexed with heparin or chondroitin sulfate proteoglycans and stored until the cell is activated. In peripheral tissues, mast cells and basophils are the primary sources of histamine. A variety of endogenous and exogenous compounds stimulate the exocytotic release of histamine. Some drugs, venoms, high-molecular-weight proteins, and basic compounds (X-ray radiocontrast media) alter the mast cell membrane and induce the nonexocytotic release of histamine. In addition, thermal and mechanical stresses (scratching) displace histamine. Once a cell releases its histamine stores, it may take weeks for the supply to reach normal levels again. Alternatively, some cells in the immune system (platelets, monocytes/macrophages, neutrophils, and T and B cells) synthesize large amounts of histamine that are released immediately. Histamine signaling is terminated by its metabolism to inactive forms and/or uptake into cells by specific transporters.

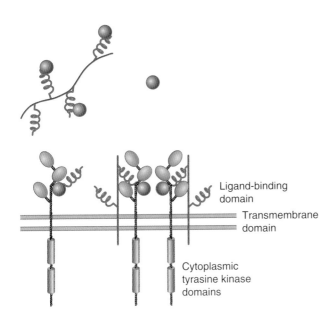

Ligand-binding domain

Transmembrane domain

Cytoplasmic tyrasine kinase domains

Figure 7–7. Extracellular matrix components modulate fibroblast growth factor (FGF) activity. FGF interacts with the heparin sulfate side chains found on proteoglycans in the extracellular matrix and bound to the plasma membrane. These proteoglycans act as low-affinity receptors for FGF, sequester the FGF, and prevent its degradation. FGF interaction with heparan sulfate is required for high-affinity interaction with its receptor (FGFR). The formation of an FGF/FGFR/heparan sulfate (HS) (2:2:2) dimer induces the transphosphorylation of the cytoplasmic receptor tyrosine kinase domains that activates FGF signaling and alters cell behavior.

Histamine Receptor Subtypes

Four different receptors, designated H_1, H_2, H_3, and H_4, mediate the actions of histamine (Table 7–3). These G-protein–coupled receptors use Ca^{2+} or cAMP second messenger systems, or both (see Chapter 8). The H_1 and H_2 receptors are the main targets for antihistamine therapies and were first described in the mid-1960s. The H_3 and H_4 receptors have only recently been characterized. H_1 receptors are expressed on smooth muscle and endothelial cells and are responsible for many of the symptoms of allergic disease and anaphylaxis. H_2 receptors are found in the gastrointestinal tract and are the main mediators of gastric acid secretion. H_3 receptors are expressed on histamine-containing neurons and act as presynaptic autoreceptors that mediate feedback inhibition of histamine release and synthesis. H_4 receptors are preferentially expressed on hematopoietic and immunocompetent cells and are responsible for mast/eosinophil chemotaxis and recruitment. Recent studies have demonstrated that the ineffectiveness of H_1 receptor antagonists in treating asthma is attributable to their inability to block H_4 receptor–mediated recruitment of mast cells, basophils, and eosinophils to lung. Additional characterization of histamine receptor subtypes will be needed to delineate the cellular events that mediate histamine signaling.

Mast Cell Histamine Release and the Allergic Reaction

One the best characterized actions of histamine is its role in allergic reactions (Fig. 7–8). Initial exposure to an antigen, such as bee venom protein or a food product, activates B lymphocytes to produce antibodies that recognize various regions of the antigen. Some of these antibodies will be IgE molecules. Mast cells express Fc receptors that bind these IgE antibodies, and the resulting IgE-Fc receptor complex forms a receptor for the antigen. When a second antigen exposure occurs, the new antigen will bind to and dimerize these immobilized IgE-Fc receptor complexes. Dimerization activates a cascade of intracellular signaling events that increase intracellular Ca^{2+} levels and trigger **degranulation,** the exocytosis of histamine-containing secretory granules. The released histamine then binds to H_1 receptors on the endothelial cells, resulting in increased capillary permeability. White blood cells and eosinophils migrate into the area to neutralize the antigen and repair tissue damage. Activation of endothelial cell H_1 receptors also stimulates production of local vasodilator substances, including nitric oxide (NO). Histamine activation of smooth muscle cell H_1 receptors also contributes to vasodilation. In tissues where local mast-cell degranulation has occurred, the histamine concentration will be sufficient to activate H_4 receptors on nearby eosinophils, and these cells will migrate into the site of the allergic reaction. In the event that these responses are exaggerated, a potentially fatal anaphylactic reaction can occur. Prior sensitization is not required for histamine release from mast cells. In fact, the "allergic reaction" associated with some drugs can be attributed to their ability to activate mast-cell degranulation without prior desensitization.

GASES: NITRIC OXIDE AND CARBON MONOXIDE

Carbon monoxide (CO) and NO were once considered to be only toxic pollutants, but they are now recognized as important intercellular signaling molecules. NO is a major paracrine signaling factor in the nervous, immune, and circulatory systems. In the circulatory system, both CO and NO mediate blood vessel dilation. In fact, NO is the final common mediator for many endogenous and exogenous smooth muscle relaxants. When nitroglycerin is administered to patients with angina, it is rapidly converted to NO in the bloodstream; the resulting NO enters the coronary vasculature, causing vessel dilation and increased blood flow. In septic shock, cell wall

TABLE 7–3. Histamine Receptor Subtypes			
Receptor	G Protein Second Messengers	Distribution	Major Biological Activity
H_1	$G_{q/11}$ $\uparrow Ca^{2+}$; $\uparrow$ cyclic adenosine monophosphate (cAMP)	Smooth muscle Endothelial cells Nerve cells	Bronchoconstriction and vasodilation
H_2	G_s $\uparrow$cAMP	Gastric parietal cells Cardiac muscle Mast cells Nerve cells	Gastric acid secretion
H_3	$G_{i/o}$ $\downarrow$cAMP	Nerve cells	Neurotransmitter release
H_4	$G_{i/o}$ $\downarrow$cAMP; $\downarrow Ca^{2+}$	Mast cells Hematopoietic cells	Chemotaxis of mast cells and eosinophils

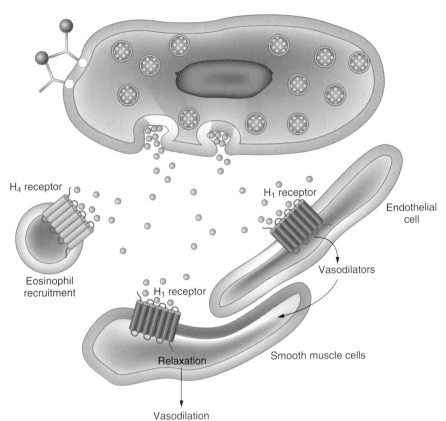

Figure 7–8. Mast-cell degranulation/histamine signaling. Mast cells and eosinophils in the blood contain large numbers of secretory granules that are filled with histamine. In the case of an allergic reaction, binding of the allergen to an IgE antibody immobilized on the mast cell triggers the release of histamine-containing granules. The released histamine binds endothelial and smooth muscle H_1 receptors to induce vasodilation. If concentrations are high enough, the released histamine acts on eosinophil H_4 receptors to induce their migration into the affected area. These eosinophils can then exacerbate the reaction by releasing additional histamine.

H_4 receptor

Eosinophil recruitment

H_1 receptor

Endothelial cell

Vasodilators

H_1 receptor

Relaxation

Smooth muscle cells

Vasodilation

material released from gram-negative bacteria triggers the release of NO from macrophages, resulting in widespread dilation of blood vessels and a dramatic, and sometimes fatal, decrease in blood pressure. Drugs for the treatment of erectile dysfunction inhibit the downstream second messenger for NO, guanylyl cyclase.

The vasodilator effects of NO have been well characterized (Fig. 7–9). The enzyme nitric oxide synthase (NOS) converts L-arginine to citrulline and liberates NO. Cells and tissues express three NOS isozymes: neuronal NOS (nNOS; NOS-1) was discovered in the brain, inducible NOS (iNOS; NOS-2) was originally purified from macrophages, and endothelial NOS (eNOS; NOS-3) was discovered in endothelial cells. These enzymes are the major site for regulating NO-mediated signaling. eNOS is constitutively expressed in endothelial cells and is responsible for the basal production of NO that is needed to maintain normal vascular tone. NO readily diffuses out of the cell. Because it has a half-life of only a few seconds, NO is able to act only on neighboring smooth muscle cells. NO binding to a heme group (iron) in the active site of the enzyme guanylyl cyclase activates the enzyme and increases intracellular cyclic guanosine monophosphate (cGMP) levels. This, in turn, activates the cGMP-dependent protein kinases responsible for the phosphorylation/activation of proteins that cause

smooth muscle relaxation. Under stress conditions, a new form of NOS, iNOS, is expressed. iNOS is essentially unregulated and once induced can maintain high NO levels for prolonged periods, resulting in clinically significant hypotension.

The intercellular mediator functions of CO have only recently been described. Significant amounts of cellular CO are produced as an elimination product during microsomal heme oxygenase catalyzed heme degradation. Heme oxygenases are found in all tissues, and CO is generated in all cells. CO is relatively inert and reacts only with iron-containing compounds. CO activates the same soluble guanylate cyclase enzyme that NO activates, but with much lower potency than NO. In some instances, CO may antagonize NO signaling. CO has also been implicated in oxygen sensing, oxygen-dependent changes in gene expression, and neuronal signaling.

EICOSANOIDS

The eicosanoids are a family of oxygenated derivatives of 20-carbon polyunsaturated fatty acids that includes **prostaglandins, thromboxanes, leukotrienes, endocannabinoids,** and **isoecosanoids.** Because they are rapidly broken down, eicosanoids are limited to autocrine and

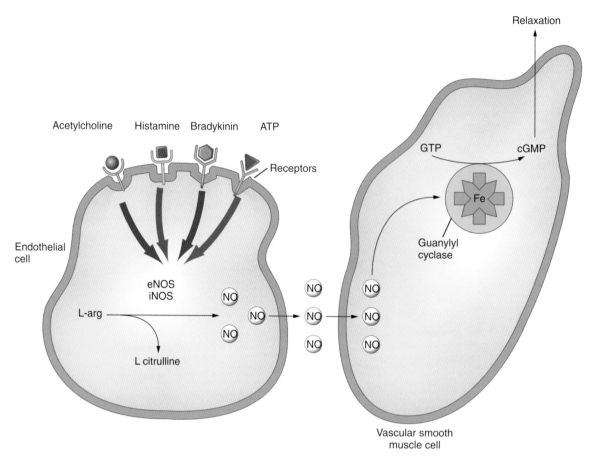

Figure 7–9. Nitric oxide (NO)–mediated endothelial cell relaxation. Endothelial nitric oxide synthase (eNOS) produces NO in endothelial cells in response to a variety of intercellular signaling molecules including acetylcholine, histamine, bradykinin, and ATP. NO diffuses out of the endothelial cell and into the vascular smooth muscle cell, where it binds to the heme (iron) moiety in guanylate cyclase. This increases cyclic guanosine monophosphate (cGMP) levels and activates the cGMP-dependent kinases that ultimately produce smooth relaxation/vasodilation. During stress, the inducible NOS (iNOS) will be activated and NO levels will increase.

paracrine signaling. Alterations in eicosanoid signaling are associated with failure of the ductus arteriosus to close at birth, platelet aggregation, inflammatory/immune responses, bronchoconstriction, and spontaneous abortion. The popular over-the-counter nonsteroidal anti-inflammatory drugs (NSAIDs) aspirin, acetaminophen (Tylenol), and ibuprofen (Advil) target eicosanoid-mediated signaling pathways.

Cells synthesize eicosanoids on demand and in response to physical, chemical, and hormonal stimuli. All eicosanoids are synthesized from a common precursor, arachidonic acid (Fig. 7–10). Phospholipase A hydrolysis of phospholipids located on the cytoplasmic surface of the plasma membrane releases arachidonic acid. In unstimulated cells, arachidonic acid is incorporated back into the membrane. In stimulated cells, several different pathways convert arachidonic acid to eicosanoids. There are two **cyclooxygenases,** COX-1 and COX-2, that convert arachidonic acid to prostaglandins. Both of these enzymes contribute to unstimulated and stimulated prostaglandin synthesis and are targets for NSAIDs. There are five human **lipooxygenases** that convert arachidonic acid leukotrienes. Endo-

cannabinoids, the newest members of the family, are synthesized primarily in the nervous system and are immediately released into the synaptic cleft—they are not stored in synaptic vesicles. The enzymes that convert arachidonic acid to endocannabinoids have not been characterized. Epoxyeicosatrienic acids, important regulators of renal and cardiovascular function, are synthesized in endothelial cells by the cytochrome P450 enzymes. Isoecosanoids are unique in that they are produced by a nonenzymatic free radical–based attack on arachidonic acid. Once synthesized, eicosanoids are transported out of the cell. Specific transporters have been identified for prostacyclin and thromboxane. Although the molecular mechanisms involved in terminating endogenous eicosanoid signaling have not been delineated, some signaling is terminated by uptake into target cells followed by enzymatic degradation.

Eicosanoids exert their effects by binding to a diverse list of G-protein–coupled cell-surface receptors (see Fig. 7–10). In the case of prostaglandins, eight different subclasses of receptors have been identified: EP_1, EP_2, EP_{3A-D}, and EP_4. The EP_3 receptor participates in the fever response, and the EP_4 receptor participates in closure of

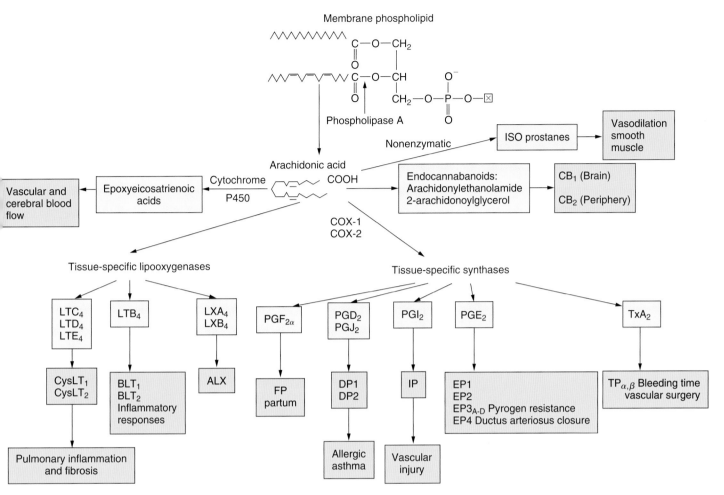

Figure 7–10. Eicosanoid-mediated intercellular signaling. Arachidonic acid, liberated from phospholipase A hydrolysis of plasma membrane phospholipids, is the precursor for all members of the eicosanoid family. The major signaling metabolites (blue), their receptors (yellow), and biological responses (yellow) are listed.

the ductus arteriosus. The endocannabinoids are endogenous ligands for the cannabinoid receptors CB_1 and CB_2. These are the same receptors for the psychoactive ingredients in leaves of the *Cannabis sativa,* marijuana and hashish. CB_1 is preferentially expressed in the nervous system and is responsible for the "high" produced by marijuana. Endocannabinoid signaling in the brain is associated with control of movement, learning and memory, pain perception, appetite, body temperature, and emesis. CB_2 is expressed primarily in the immune system. Despite their recent identification, a number of CB_1 and CB_2 agonists are already in clinical trials for the treatment of pain and multiple sclerosis. The wide variety of differential responses to only small changes in the concentration/complement of prostaglandins is due to that cells express multiple eicosanoids receptors. In addition, some members of the eicosanoids family can interact with the same receptor. For example, isoecosanoids alter vascular smooth muscle behavior via interaction with one of the thromboxane receptors.

NEUROTRANSMITTERS

A **neurotransmitter** is defined as a chemical that is released from a stimulated presynaptic neuron, binds to the membrane of a postsynaptic target cell, and induces a response (inhibitory or excitatory) in the target cell. Some neurotransmitters can act as both neurotransmitters and hormones. For example, the neurotransmitter epinephrine is also produced by the adrenal gland and signals glycogen breakdown in the cell. Defects in neurotransmission are associated with neurologic and psychiatric disorders. A basic understanding of synaptic transmission is essential for understanding the pharmacologic basis of therapeutic and recreational psychoactive drugs.

Electrical and Chemical Synapses

In the nervous system, information travels one cell to the next via specialized contact sites known as **synapses** (Fig. 7–11). Synapses that use chemicals to transmit

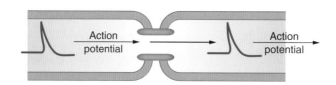

A ELECTRICAL SYNAPSE

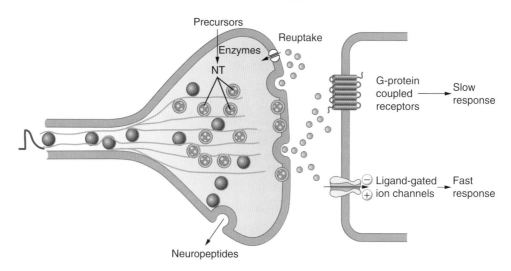

B CHEMICAL SYNAPSE

Figure 7–11. **Electrical and chemical synapses.** In an electrical synapse (**A**), gap junctions allow current to flow directly from the presynaptic cell to the postsynaptic cell. In a chemical synapse (**B**), the presynaptic cell converts the current into a chemical signal. Neurotransmitters (yellow), located in synaptic vesicles, and larger neuropeptides, located in dense core granules, form the chemical signal. The neurotransmitters are released into the synaptic cleft and interact with ligand-mediated G-protein–coupled receptors or ligand-mediated ion channel receptors on the postsynaptic cell. Neuropeptide (red) release occurs adjacent to the synaptic cleft. An action potential will be generated in the postsynaptic cell only if these chemicals produce a sufficient depolarization.

information are called **chemical synapses.** Chemical synapses are common and are restricted to the nervous system. **Electrical synapses** are relatively rare and are found in neuronal and nonneuronal cells. In an electrical synapse, membrane depolarization in presynaptic cells passes directly to the postsynaptic cells via **gap junctions.** Gap junctions are bidirectional and allow ionic current to pass in either direction. Signaling across electrical synapses occurs almost instantaneously and is fail-safe: an action potential in the presynaptic neuron will always produce an action potential in the postsynaptic cell. In invertebrates, electrical synapses mediate escape reflexes and allow animals to quickly retreat from danger. In humans, electrical synapses permit all of the cells within an organ to function as one large unit or **syncytium.** Electrical synapses make the synchronized beating of the heart, peristaltic movement in the gut, coordinated growth/maturation of the developing nervous system, and synchronized activity of neighbor-

ing neurons in specialized regions of the adult brain possible.

Chemical synapses are a specialized form of paracrine signaling that allows for fast and precise delivery of signals across large distances. In the central nervous system (CNS), chemical synapses carry signals between neurons. In the peripheral nervous system, chemical synapses carry information from nerves to myocytes and gland cells. In a chemical synapse, the presynaptic cell converts the electrical signal into a chemical signal. This chemical signal is transmitted to the target cell where it is converted back into an electrical signal. Unlike growth factors and hormones, it takes only a millisecond or less for chemicals to transverse the 20- to 50-nm distance across the synaptic cleft. Unlike the all-or-none electrical synapse, chemical synapses are capable of receiving and summing inputs from multiple sources; a process that is essential for learning and memory and other higher brain functions.

A Prototypical Chemical Synapse: The Neuromuscular Junction

The neuromuscular junction is formed between axon terminals of cranial or spinal motor neurons and skeletal muscle cells (Fig. 7–12). Before reaching the muscle, the axon divides into multiple synaptic varicosities or synaptic boutons. These synaptic boutons are ensheathed by the Schwann cell plasma membrane, but unmyelinated, and allow a single axon to innervate hundreds of muscle fibers within a muscle. Every synaptic bouton, or terminal axon end, is in close apposition with a junctional fold, or invagination, of the muscle membrane. Both the presynaptic and postsynaptic sides of the neuromuscular junction are highly organized. The specialized region of the muscle plasma membrane where innervation occurs is called the **motor end plate**.

The electrical action potential is converted into a chemical signal by the presynaptic cell. The action potential depolarizes the presynaptic membrane and opens voltage-sensitive Ca^{2+} channels. This influx of Ca^{2+} mobilizes synaptic vesicles, each of which contains approximately 1000 to 50,000 molecules of the neurotransmitter acetylcholine. Approximately 10% of the vesicles are docked at the plasma membrane in the active zone and immediately exocytosed. The remaining vesicles are located adjacent to the active zone, where they are reversibly tethered to the actin cytoskeleton via synapsins, a family of phosphoproteins. These vesicles move into the active zone and are primed for exocytosis as needed. The maintenance of two synaptic vesicle pools ensures that acetylcholine will be available for release if another action potential arrives in quick succession. The exocytosis of neurotransmitters involves many of the same vesicle targeting and fusion events that occur during regulated exocytosis in nonneuronal cells (see Chapter 4). However, secretion of neurotransmitters differs in that it is tightly coupled to the arrival of an action potential at the axon terminus, and synaptic vesicles are recycled locally. The entire process of retrieving and refilling synaptic vesicles takes less than 1 minute.

The exocytosed neurotransmitter then transmits the signal to the postsynaptic target cell. Acetylcholine and all neurotransmitters diffuse across the synaptic cleft and bind to specific receptors on the postsynaptic cell membrane. In the case of the neuromuscular junction, acetylcholine receptors are concentrated at the beginning of the junctional infolds of the muscle plasma membrane. Binding of acetylcholine to its receptor opens an ion channel for approximately 1 millisecond, during which time approximately 50,000 Na^+ molecules enter the muscle cell. These Na^+ molecules depolarize the end-plate potential, and the response/depolarization is proportional to the amount of transmitter released. If the depolarization exceeds the threshold, then voltage-dependent Na^+ channels open, and the action potential spreads throughout the muscle membrane and initiates the intracellular events needed for muscle contraction (see Chapters 2 and 3). Myasthenia gravis is an autoimmune disorder in which autoantibodies are made against the acetylcholine receptor subtype expressed on muscle cells. These antibodies block the receptor and promote its removal from the motor end plate. In the absence of the receptor the signal is not transmitted even though the chemical signal is still generated.

Clinical Case 7-2

Jane Crawford is a 37-year-old married mother who came to her internist reporting fatigue. She reports that she is overwhelmed by even minor physical effort that she could easily perform when she was raising her children. She also reports that she frequently has "heavy eyelids" and falls asleep while watching television in the morning. Her husband says that Jane's voice has changed and has recently become much more nasal.

Mrs. Crawford has been comparatively well physically previously. When she was in her teens, she had sore and swollen fingers bilaterally, which her college physician thought might be early rheumatoid arthritis. However, after a 2-month trial of high-dose aspirin, which bothered her stomach, those symptoms remitted and have not reoccurred.

On physical examination, Mrs. Crawford's blood pressure was 130/85 mmHg, her pulse was normal and regular, her heart normal sized without murmurs, and her thyroid was normal to palpation. She is not obese, and her abdominal examination was normal. She has no obvious signs of anemia, and her gross neurologic examination, including deep tender reflexes, was normal. She had difficulty, however, following the examiner's finger up and down during the eye motion tests. This was exaggerated with repetition. Furthermore, when asked to extend her arms out in front of her with her palms up, she could hold the position for only about 2.5 minutes. After receiving her permission for a brief test involving an intravenous injection and preparing a small amount of atropine if needed, her physician gave her an injection of 2 mg edrophonium in normal saline. Five minutes later, he asked her to repeat the arm-extension maneuver. To her pleasure, Jane found that she could now hold her arms out for at least 6 minutes. Furthermore, she reported that her eyelids no longer felt as heavy.

Cell Biology, Diagnosis, and Treatment of Myasthenia Gravis

Mrs. Crawford has early myasthenia gravis. This is an acquired disorder caused by the loss of the normal

Continued

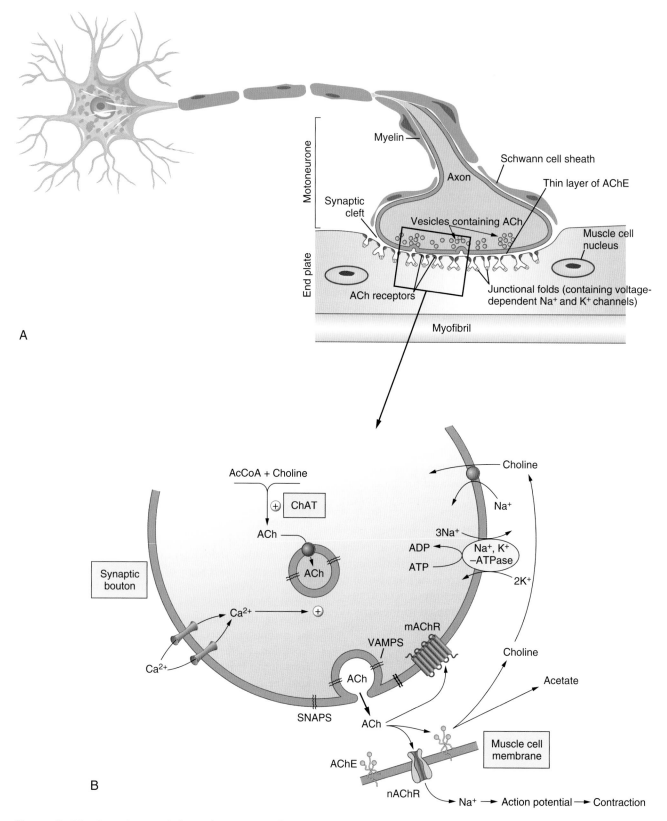

Figure 7–12. **Synaptic transmission at the neuromuscular junction. A:** The neuromuscular junction is composed of synaptic boutons from the innervating α-motor neuron and the specialized motor end plates on the muscle fiber. Acetylcholine is synthesized in the motor neuron terminal and taken up into synaptic vesicles (yellow). **B:** On release, the acetylcholine diffuses across the synaptic cleft and binds to a ligand-gated ion channel (red). The number of Na^+ molecules that enter the muscle cell is dependent on the number of acetylcholine molecules released from the neuron. If there is a sufficient depolarization, then action potential will be generated in the muscle cell and contraction will occur. Acetylcholinesterase, located in the synaptic cleft, hydrolyzes acetylcholine to acetate and choline. The choline is taken back up into the presynaptic cleft and used to refill synaptic vesicles.

number of acetylcholine receptors on her muscle membranes adjacent to the end plates of the innervating nerves. An autoimmune origin of this defect in Mrs. Crawford was suggested by her previous history of rheumatoid symptoms, and confirmed by the finding of a high-titer IgG anticholinesterase receptor antibody in her serum.

Before any effort at treatment, her physician ordered some blood tests for lupus erythematosus, rheumatoid arthritis, and thyroid diseases to rule out those disorders that may coexist with myasthenia. Also, he ordered a mediastinal computed tomography scan to rule out a possible thymoma. When all of those test results were negative, he started Mrs. Crawford on low-dose, oral pyridostigmine, an acetylcholinesterase mimetic drug, three times a day.

Mrs. Crawford has had an excellent response, with a marked return of muscular strength and reduction of fatigue. For the present, her physician does not think she will need any anti-immune therapy. He warns her, however, that she will need comparatively frequent follow-up for the rest of her life, and that a change in dose or other therapies may eventually be needed.

It is imperative that the chemical signal/neurotransmitter not influence neighboring cells and be removed before the next pulse of neurotransmitter is released. In the case of acetylcholine, an enzyme in the synaptic cleft,

acetylcholinesterase, hydrolyzes the neurotransmitter into acetate and choline. The time required to hydrolyze acetylcholine is less than a millisecond. A large number of nerve gases and neurotoxins inhibit acetylcholinesterase and prolong the action of acetylcholine. If the concentration of these agents is high enough, they will prevent the relaxation of muscles necessary for respiration, resulting in lethality. A Na^+/choline symporter on the nerve cell membrane transports the choline back into the nerve cell, where it will be used to synthesize more neurotransmitter.

Characteristics, Synthesis, and Metabolism of Neurotransmitters

Neurotransmitters can be divided into three categories: amino acids, amines, and peptides (Table 7–4). The amino acids and amines mediate fast synaptic transmission (submillisecond to millisecond) in the CNS. Acetylcholine mediates fast synaptic transmission at all neuromuscular junctions. The amino acid and amine transmitters are synthesized in the axonal terminal and stored in synaptic vesicles. The amino acids glutamate and glycine are found in proteins and are abundant in the cytosol of all cells. γ-Aminobutyric acid (GABA) and the other amines are found only in neurons that contain the specific enzymes needed to synthesize them from precursor molecules. Choline acetyltransferase uses acetyl coenzyme A from the mitochondria as a donor and catalyzes the acetylation of cytosolic choline. Once

TABLE 7–4. Major Neurotransmitters and Their Receptors

Transmitter	Receptor Subtype(s)	Signaling Mechanism
Amino acids		
Glutamate (Glu)	NMDAR; AMPAR; AMPAR	Ion channel
	$MGluR_1$, $MGluR_2$, $MGluR_3$, $MGluR_4$, $MGluR_5$, $MGluR_6$, $MGluR_7$	G-protein coupled
Glycine (Gly)	GlyR	Ion channel
γ-Aminobutyric acid (GABA)	$GABA_A$	Ion channel/↑ Cl^- conductance
	$GABA_B$	G-protein–coupled ↑ K+ conductance
	$GABA_C$	Ion channel ↑ Cl^- conductance
Nucleotides		
ATP	P2X	Ion channel
	P2Y	G-protein coupled
Amines		
Acetylcholine	Nicotinic (N_m and N_n)	Ion channels
	M1, M2, M3, M4, M5	G_q- and G_i-protein coupled
Norepinephrine	α1A, α1B, α1D	G_q coupled
Epinephrine	α2A, α2B, α2C	G_i- and ion-channel coupled
	$β_1$, $β_2$, $β_3$	G_s-protein coupled
Dopamine	D1, D2, D3, D4, D5	G-protein coupled
Serotonin	$5-HT_3$	Ion channel
	$5-HT_1$, $5-HT_2$, $5-HT_4$ $5-HT_5$	G-protein coupled £€cAMP; €PLC
Histamine	H_1, H_2, H_3, H_4	G-protein coupled
Neuropeptides		
Cholecystokinin (CCK)	CCK_1, CCK_2	G-protein coupled
Neuropeptide Y	Y_1, Y_2, Y_3, Y_4, Y_5	G-protein coupled
EnkephalinsDynorphin	μ, δ, κ	G-protein coupled

AMPAR, α-amino-3-hydroxy-5methylisoxazole-4-propionic acid receptor; cAMP, cyclic adenosine monophosphate; NMDAR, N-methyl-D-aspartate receptor; PLC, phospholipase C.

synthesized, these neurotransmitters are concentrated in **synaptic vesicles** by transporters embedded in the vesicle membrane. With the exception of acetylcholine, neurotransmitters are removed from the cleft by reuptake into the presynaptic cell via Na^+/neurotransmitter symporters. These transporters are a major site of action for therapeutic and recreational drugs: cocaine inhibits reuptake of norepinephrine, serotonin, and dopamine; older antidepressants inhibit norepinephrine and serotonin reuptake; and newer antidepressants (fluoxetine hydrochloride [Prozac]) specifically block serotonin reuptake.

Neuropeptide transmitters are found in the gut and throughout the nervous system. These peptides/polypeptides are synthesized by the endoplasmic reticulum in the cell body, packaged into secretory vesicles, and transported to the axonal terminal via fast axonal transport. The opioid neuropeptides are grouped into three distinct families: enkephalins, endorphins, and dynorphins. All three are synthesized as large precursor molecules that are subjected to complex cleavage and posttranslational modifications to generate the biologically active molecule. Individual secretory vesicles can contain multiple neuropeptides. **Secretory vesicles** are larger and more randomly distributed in the axonal terminal than synaptic vesicles (see Fig. 7–11). These vesicles are also referred to as large, dense-core vesicles because of their dark appearance in electron micrographs. Like synaptic vesicles, secretory vesicle exocytosis is triggered by an increase in Ca^{2+}. However, secretory vesicle fusion with the plasma membrane occurs at random sites on the periphery of the axon, not within the active zone/synaptic cleft. Because the secretory vesicles are located some distance from the synaptic cleft, several high-frequency action potentials may be needed for the Ca^{2+} levels in these areas to reach the level required to trigger secretory vesicle release. Thus, the time frame for neuropeptide release is much longer, 50 milliseconds or more, than the time frame for amino acid/amine release. Neuropeptides are degraded by extracellular proteases, and this terminates signaling.

Neurotransmitter Receptors

Two classes of neurotransmitter receptors exist: **transmitter-gated ion channels** and **G-protein–coupled receptors**. **Fast chemical synaptic transmission** is mediated by transmitter-gated ion channel receptors. These receptors are multimeric proteins that form a transmembrane pore. When a neurotransmitter binds to the receptor, the pore opens and ions flow into the cytoplasm. Receptors for excitatory neurotransmitters (acetylcholine and glutamate) are cation channels that permit an influx of Na^+, or Na^+ and K^+, that depolarizes the postsynaptic membrane toward the threshold potential for firing an action potential. Receptors for inhibitory neurotransmitters (GABA and glycine) are K^+ or Cl^- channels that hyperpolarize the postsynaptic membrane and suppress

action potentials. Barbiturates and tranquilizers, such as diazepam (Valium), bind to GABA receptors and potentiate the inhibitory actions of GABA by reducing the concentration of GABA needed to activate the ion-gated channel.

Slow synaptic transmission is mediated by G-protein–coupled receptors. These receptors are often referred to as metabotropic receptors because of the widespread metabolic effects they trigger. Neurotransmitters from all three categories interact with G-protein–coupled receptors and mediate slow synaptic transmission in the CNS. G-protein–coupled receptors are a single polypeptide containing seven transmembrane domains. Ligand binding to the receptor activates a small G protein located on the intracellular side of the postsynaptic membrane. This activated G protein then directly opens an ion channel, or it can initiate a cascade of intracellular events that indirectly results in the opening of ion channels (see Chapter 8). For example, the three classical opioid receptors μ, Δ, and κ inhibit adenylate cyclase, which reduces cAMP levels, resulting in activation of K^+ currents and inhibition of Ca^{2+} currents. Morphine produces its analgesic effects via interaction with these opioid receptors.

Divergence and Convergence of Neurotransmitter Function

Neurotransmitters exhibit both divergence and convergence. **Divergence** allows a single neurotransmitter to activate multiple receptors. Acetylcholine is an excellent example of the different effects that can be attributed to a single neurotransmitter—depending on the receptor that is activated. Nicotinic acetylcholine receptors are ligand-gated channels and result in excitatory responses that last only milliseconds. An example is the neuromuscular junction detailed in the previous section. In contrast, muscarinic acetylcholine receptors are G-protein–coupled receptors and result in a variety of different responses: the M2 subtype found in heart opens a K^+ channel and produces a hyperpolarization that lasts for a few seconds; the M1, M3, and M5 subtypes activate a second messenger phospholipase; and the M4 inhibits another second messenger adenylate cyclase (see Fig. 7–2). **Convergence** allows multiple transmitters acting through their individual receptors to activate the same downstream signaling pathway.

Spatiotemporal Summation

The CNS must integrate information from a variety of sources as it balances the needs of the organism and the demands from the environment. In the CNS, a single postsynaptic neuron will receive inputs from thousands of synaptic inputs. The process of integrating all of these inputs over time into a single neuronal output is called **spatiotemporal summation.** At each synapse, neurotransmitter release causes a local change in the membrane potential of the postsynaptic target cell. Some of these synapses are excitatory and produce small depolarizations called **excitatory postsynaptic potentials (EPSPs).** Other synapses are inhibitory and produce small hyperpolarizations referred to as **inhibitory postsynaptic potentials (IPSPs).** Both IPSPs and EPSPs vary in magnitude and duration. These signals travel to the axon hillock where they are spatially and temporally integrated. If excitatory inputs dominate, then the cell body will depolarize and voltage-sensitive Na$^+$ channels will open. If a sufficient number of channels are opened, an action potential will be generated. If inhibitory signals dominate, then there will be a hyperpolarization and inhibition of an action potential (Fig. 7–13).

SUMMARY

The coordination of day-to-day physiology and behavior in multicellular organisms requires that individual cells sense and respond to changes in their environment. To accomplish this task, cells use hundreds of different intercellular signaling molecules to generate an almost unlimited number of spatially and temporally coordinated signals. A basic understanding of the cellular/molecular events involved in intercellular signaling is essential for the diagnosis and treatment of human diseases. Deficiencies in intercellular communication are the precipitating event in many diseases, including endocrine disorders, neurodegenerative disorders, and all types of cancers. In addition, agents that mimic or antagonize the actions of cell–cell signaling molecules are common drug targets. These agents include the widely used over-the-counter analgesics such as aspirin, acetaminophen, and ibuprofen; prescription medications for hormone replacement therapy such as insulin and estrogen; abused medications for improved athletic performance such as growth hormone and erythropoietin; and recreational drugs such as marijuana and cocaine.

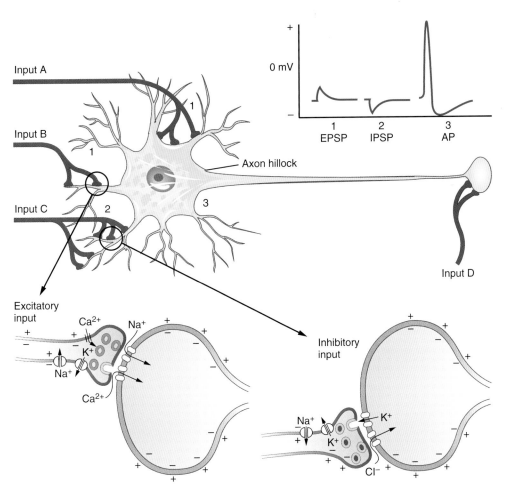

Figure 7–13. **Synaptic transmission in the central nervous system.** Neurons in the central nervous system have many synaptic contacts distributed on the cell body *(Input A)*, dendritic tree *(Inputs B and C)*, and terminal boutons *(Input D)*. Some of these contacts are excitatory, resulting in excitatory postsynaptic potentials (EPSPs; *Inputs A and B*), and others are inhibitory, resulting in inhibitory postsynaptic potentials (IPSPs; *Input C*). The ultimate decision of whether to fire an action potential (AP) is determined by the weighted integral of all of these inputs over time.

Suggested Readings

Beniot G, Malewicz, Perlmann T. Digging deep into the pockets of orphan nuclear receptors: insights from structural studies. *Trends Cell Biol* 2004;14:369–376.

Bishop A, Anderson JE. NO signaling in the CNS: from the physiological to the pathological. *Toxicology* 2005;208:193–205.

Bolander FF. *Molecular Endocrinology,* 3rd ed. San Diego: Elsevier Academic Press, 2004.

Branton LL, Lazo JS, Parker KL. *Goodman & Gillman's the Pharmacological Basis of Therapeutics.* New York: McGraw-Hill Medical Publishing Division, 2006.

Comoglio PM, Boccaccio C, Trusolino L. Interactions between growth factor receptors and adhesion molecules: breaking the rules. *Curr Opin Cell Biol* 2004;15:565–571.

Curtis-Prior P, ed. *The Eicosanoids.* West Sussex, United Kingdom: John Wiley & Sons, Ltd, 2004.

Davies RW, Morris BJ, eds. *Molecular Biology of the Neuron,* 2nd ed. New York: Oxford University Press, 2004.

Jutel M, Blaser K, Akdis CA. Histamine in allergic inflammation and immune modulation. *Int Arch Allergy Immunol* 2005;137: 82–92.

Ryter SW, Otterbein LE. Carbon monoxide in biology and medicine. *BioEssays* 2004;26:270–280.

Unsicker K, Krieglstein K, eds. Cell signaling and growth factors in development: from molecules to organogenesis. Weinheim, Federal Republic of German: Wiley-VCH Verlag GmbH and Company, 2006.

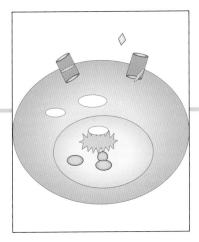

Chapter 8

Cell Signaling Events

The evolution of multicellular organisms into complex vertebrates would not be possible without a way for cells to communicate with each other. From inception through development and into postnatal life, vertebrates and invertebrates alike must produce and arrange cells into higher order structures such as tissues and sculpt tissues into functioning organs. At the root of these immensely complex processes is the seemingly simple process of one cell interacting and communicating with another cell. Without such interactions, the formation and function of tissues and organs would not be possible. Virtually all forms of communication between cells requires two essential components: (1) a signaling cell that produces an instructive signal, and (2) a responding cell that receives, acts on, and is changed by the instructive signal. The mechanism by which a cell receives, acts on, and alters its behavior in response to a signal from another cell is called *signal transduction*. The intracellular processes that allow that signal to be transduced from the outer cell membrane of the responding cell to the nucleus where it effects a new program of gene expression and cell behavior is called the *signal transduction pathway*. The overall cellular complexity of vertebrate tissues and organs is mirrored in the complexity of these signal transduction pathways, which are multiple, complex, and regulated in a myriad number of ways to elicit a precise and correct behavior in the responding cell. In general, however, virtually all signal transduction pathways essentially work the same way: a signal in the form of a diffusible extracellular molecule

binds to and activates receptors at the cell surface of the responding cell that, in turn, activates a cascade of intracellular signal transducers leading to the nucleus where the penultimate transducer promotes the expression of responsive genes. This chapter provides examples of different types of signal transduction pathways and how they convert extracellular signal into genetic response.

SIGNALING IS OFTEN MEDIATED BY CELL-SURFACE RECEPTORS

Many of the events involved in cell-to-cell communication invariably involve signaling molecules produced and secreted from the signaling cell that bind to and activate cell-surface receptors on the recipient or target cell. The binding of these molecules or ligands to cell-surface receptors activate signal transduction pathways that lead to the nucleus and to the activation of distinct genetic programs. Cell membrane receptors of the catalytic type usually have kinase activity, that is, the ability to transfer a phosphate molecule to a substrate protein by covalently linking the phosphate group to an amino acid on that protein, usually a serine, threonine, or tyrosine residue. In a number of signal transduction pathways and, in particular, those using receptor tyrosine kinases, the first molecule to be phosphorylated is another molecule of the same receptor type. In most instances, not one but two receptors of the same type bind the signaling ligand, with the binding of ligand providing a means for receptors to dimerize into a complex referred to as a *homodimer*. Within this complex, receptor tyrosine kinases can activate each other via cross-phosphorylation. Once activated in this way, receptor kinase activity can then be directed to proteins of the signal transduction cascade that are proximate to the cell membrane or in some way recruited to the cell membrane receptors. As with the receptors, these intracellular signal transducers are often kinases that are activated on phosphorylation to phosphorylate yet a second protein or kinase, thus setting up a chain of phosphorylation events that eventually activates transcription factors in the nucleus. The signal is thus carried along from membrane to nucleus by these phosphorylation events.

Signal transduction via catalytic receptor tyrosine kinases and serine-threonine kinases is a widespread and major form of communication between cells in most organisms, particularly in higher vertebrates. This type of signaling is central to many cellular and physiologic processes in the adult animal, as well as in the developing organism where the establishment (induction) and organization (morphogenesis) of tissues requires cell-to-cell communication. In fact, much of what is known about catalytic receptor kinases comes from studies of developing tissues in which a variety of extracellular signaling molecules have been identified and their signal transduction pathways in target cells elucidated. Recent advances in our understanding of the nature of these signals and how they are produced, received, and acted on has provided insights into the molecular basis of cell type induction, tissue morphogenesis, and organ function in normal and pathologic states. The heart is an exemplary organ with which to illustrate the role of various signal transduction mechanisms used in the development and response of the heart to environmental stress.

RECEPTOR TYROSINE KINASES AND RAS-DEPENDENT SIGNAL TRANSDUCTION

Cell-surface receptors that bind signaling molecules and are activated to phosphorylate tyrosine residues on substrate proteins are referred to as receptor tyrosine kinases. These receptors play an important role in numerous cellular interactions. For the most part, the signaling molecules they bind are growth factors with a wide range of biological activities. Some of these growth factors are vascular endothelial growth factor (VEGF), platelet-derived growth factor (PDGF), epidermal growth factor (EGF), insulin, nerve growth factor (NGF), and fibroblast growth factor (FGF). The receptors for these growth factors can transduce the growth factor signal down any one of three different signal transduction pathways, which include the phospholi-

pase C (PLC), the Grb2/Sos, and the phosphatidylinositol 3-kinase (PI3K) pathways (Fig. 8–1). Because all these receptors use the same array of downstream signal transduction pathways, the specificity of a cell's response to a given growth factor is determined largely by whether it expresses the appropriate receptor tyrosine kinase on its cell surface.

Fibroblast Growth Factors

The family of FGFs together play important roles in a wide range of biological events including the axial patterning of the embryo and embryonic tissues, the specification and differentiation of cell types, the morphogenesis of organs and organ systems such as the vascular system, and the regulation of cell proliferation and migration. The FGF signal transduction pathway likewise plays important and diverse roles in development and cell physiology and provides an excellent example of signaling through receptor tyrosine kinases. Activation of the FGF signal transduction pathway begins when the FGF ligand binds to FGF receptors (FGFRs) on the cell surface (see Fig. 8–1). Binding of FGF leads to dimerization of FGFRs and autophosphorylation of tyrosine residues in their intracellular domain, a process that serves as a mechanism for the assembly and recruitment of downstream signaling complexes. The FGF signal can be transmitted via three main pathways: the Retrovirus-Associated DNA Sequences (RAS)/Mitogen-Activated Protein Kinase (MAPK) pathway, the PLCγ/Ca²⁺ pathway, and the PI3K/Akt pathway (see Fig. 8–1). The major intracellular signaling pathway for FGF is the RAS/MAPK pathway. Activation of this pathway occurs when an activated FGFR binds to and phosphorylates tyrosine residues in a membrane-anchored docking protein called FGFR substrate 2α (FRS2α). Phosphorylation of FRS2α promotes binding of Grb2, a small adaptor molecule that is complexed with the nucleotide exchange factor Sos. Sos plays a pivotal role in activating the RAS pathway. In cells, RAS is active when it is bound to guanosine triphosphate (GTP) and inactive when it is bound to guanosine diphosphate (GDP). RAS signaling is initiated when the GDP bound to RAS is replaced by GTP, a reaction that is catalyzed by guanine nucleotide exchange factors (GEFs) such as Sos. RAS activation instigates a phosphorylation/activation cascade involving Raf, Mitogen-activated protein kinase/Extracellular-signal-related Kinase (MEK), and the MAPKs extracellular signal–regulated kinase 1 (ERK1) and ERK2. The ERKs enter the nucleus where they complete the transduction of the FGF signal by phosphorylating and activating transcription factors that then transcribe FGF-responsive genes.

Neuregulin

The RAS/MAPK signal transduction pathway is a major signal transduction pathway in vertebrate cell-to-cell

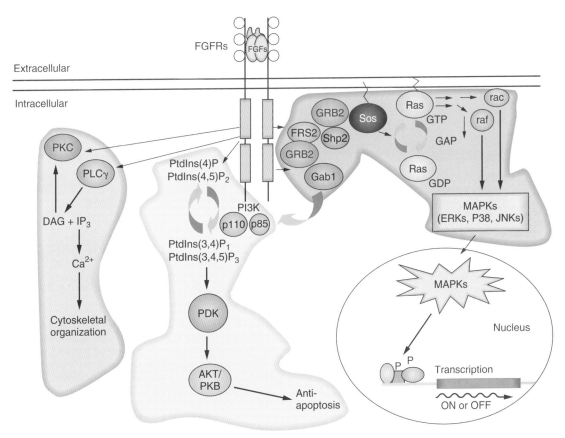

Figure 8–1. Fibroblast growth factor (FGF) signal transduction pathways. Activated FGF receptors (FGFRs; *red rectangles*) stimulate phospholipase Cγ (PLCγ) pathway *(blue highlight)*, the phosphatidylinositol 3-kinase (PI3K)-AKT/protein kinase B (PKB) pathway *(yellow highlight)*, and the FRS2-Ras-mitogen-activated protein kinase (MAPK) pathway *(green highlight)*. The activated MAPKs (extracellular signal–regulated kinases [ERKs], p38, or c-Jun N-terminal kinases [JNKs]) are translocated to the nucleus where they phosphorylate (P) transcription factors, thereby regulating target genes. (Modified from Dailey L, et al. *Cytokine Growth Factor Rev* 2005;16:233, by permission.)

communication and has been co-opted by numerous signaling molecules and their receptors to effect their signal transmission (Fig. 8–2). As with the catalytic tyrosine kinase receptors and their signal transduction pathways, the RAS/MAPK pathway can be directed to serve different biological ends by hooking up with different receptor-ligand pairs. An example of this can be seen in heart development. Cardiac cells are derived from progenitor cells within the mesoderm of the developing blastula. The induction of cardiac cells from mesodermal precursors is dependent on their receiving signals from the neighboring endodermal germ cell layer. Numerous studies have shown that a critical signal for this process is FGF. In those cells of the mesoderm expressing FGFRs, the endoderm-derived FGF signal activates their RAS/MAPK pathway as part of the process of inducing the cardiac cell lineage. Later in development, more differentiated heart cells express a new array of cell-surface receptors that direct the RAS/MAPK pathway toward different biological ends. A prime example of this is seen in the cell-to-cell signaling underlying the process of trabeculation. Trabeculation

is the extension and involuting of the compact myocardial layer of the nascent ventricular chamber into the chamber space, a process that helps the heart maintain blood flow before the heart myocardium is able to contract on its own. In this case, a signaling molecule called neuregulin-1 (NRG1) binds to Erb receptor tyrosine kinases on the surface of myocardial cells and activates the RAS/MAPK pathway (see Fig. 8–2). Unlike the earlier developmental signaling events between cells of relatively undifferentiated germ cell layers, NRG signaling provides an interesting example of how one differentiated cell type can direct the activity of another differentiated cell type within a nearly functional organ, such as the late embryonic heart. In this case, the NRG1 signal originates in the endocardial cell layer of the heart and signals to cells in the myocardial layer of the heart to form the trabeculated myocardium (Fig. 8–3).

In the heart, NRG1 signals to responsive cells via ErbB receptors that are heterodimers of ErbB2 and ErbB4, two of the possible four ErbB receptors encoded in vertebrate genomes (Fig. 8–4). ErbB2 is a

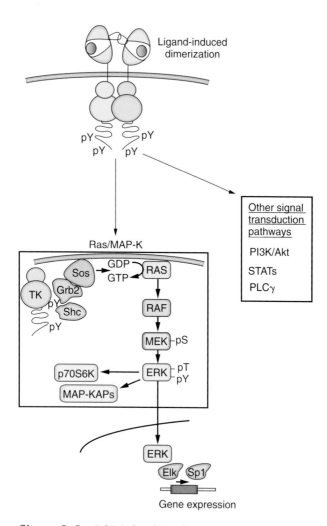

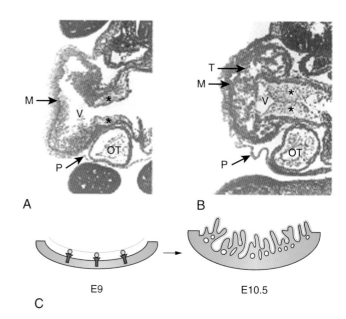

Figure 8–2. ErbB-induced signaling pathways. Ligand binding induces receptor homodimerization or heterodimerization, which leads to activation of tyrosine kinase activity and the phosphorylation of specific tyrosine residues (pY) within the carboxyl-terminal tail of ErbBs. Signaling effectors that contain either the SH2 or phosphotyrosine binding (PTB) domains for binding pY-containing peptides are recruited to activated receptors and direct the ErbB signal down specific signaling pathways. These pathways include the RAS-mitogen–activated protein kinase (MAPK), phosphatidylinositol 3-kinase (PI3K)-Akt, phospholipase C-protein kinase C (PLC-PKC), and the Jak/Stat signaling pathways. Virtually all ErbB receptors signal through the RAS/MAPK signal transduction pathway, which is shown. (Modified from Marmor MD, et al. *Int J Radiat Oncol Biol Phys* 2004;58:903, by permission.)

Figure 8–3. Loss of neuregulin-1 signaling through the ErbB2/ErbB4 heterodimer leads to defective trabeculation of the ventricles during heart development in embryonic day 10.5 mice. **A:** Heart section from a neuregulin knockout mouse, embryonic day 10.5. **B:** Heart section from a wild-type mouse, embryonic day 10.5. *M,* myocardium; *OT,* outflow tract; *P,* pericardium; *T,* trabeculated myocardium; *V,* ventricle. *Asterisks* denote cardiac cushions. **C:** Neuregulin-1 signaling from endocardium *(blue)* to ErbB2/ErbB4 heterodimers in the myocardium *(red).* (From Garratt AN, et al. *Trends Cardiovasc Med* 2003;13:80, by permission.)

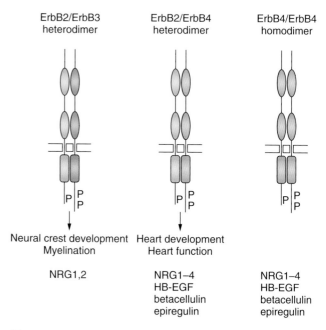

Figure 8–4. Schematic structure of the ErbB2, ErbB3, and ErbB4 receptors. The receptors contain two cysteine-rich domains that are located extracellularly *(ovals):* a transmembrane span and a cytoplasmically located tyrosine kinase *(rounded oblongs).* Both heterodimerization and homodimerization can occur, and it appears that the various ligands listed have specificity for a given receptor type. HB-EGF, heparin-binding epidermal growth factor; NRG, neuregulin. (From Garratt AN, et al. *Trends Cardiovasc Med* 2003;13:80, by permission.)

"ligandless" receptor that is a heterodimerizing partner with ErbB receptors that can bind ligand. In this case, specificity for NRG1 is likely to be in the ErbB4 receptor. ErbB receptors contain an extracellular ligand-binding domain and a single hydrophobic transmembrane domain. The intracellular portion of ErbB receptors consists of a highly conserved tyrosine kinase domain. Ligand binding by ErbB receptors induces either formation of a homodimer of the same receptor type or a heterodimer of different receptor types. In either case, the dimerization leads to phosphorylation of tyrosine residues on the partner receptor, resulting in an increase in its kinase activity. Additional tyrosine phosphorylation on residues within the carboxyl-terminal tail of the receptors enables the recruitment and activation of adaptor proteins containing Src homology 2 (SH2) and phosphotyrosine binding (PTB) domains (see Fig. 8–2). Proteins that contain these structural features are recruited to activated receptors where they assemble into multiprotein complexes that initiate ligand signaling to downstream signal transduction pathways. Binding of the activated receptor by these adaptor proteins comprises the first step in the intracellular signaling cascade and, as such, directly determines which of the various downstream signal transduction pathways will be used by the activated receptor to transduce the signal. In the case of ErbB signaling, this may be any of four different signal transduction pathways: the RAS/MAPK pathway, the PI3K-protein kinase B (PKB)/Akt pathway, the phospholipase C-protein kinase C (PLC-PKC) pathway, or the Jak/STAT pathway (see Fig. 8–2). For formation of trabeculae, the signal transduction pathway used by the ErbB2/4 receptor has not yet been formally demonstrated. What is known is that all ErbB ligands and receptors can activate the RAS/MAPK pathway, so in all likelihood this pathway plays a role in NRG1 signaling to myocardial cells. Signaling by this pathway begins with the binding of NRG1 to the ErbB2/4 receptor and the cross-phosphorylation of tyrosine residues in the intracellular domain of the receptors by the kinase activity of the receptors themselves. This has a twofold purpose, the first being to increase kinase activity of the receptors, and the second to create a conformational domain for the recruitment and binding of the RAS/MAPK effector complex of proteins. This complex links ErbB receptors to the RAS signal transduction pathway by bringing activators of RAS into close proximity with RAS to effect its activation. The major components of this complex are Grb2, an adaptor protein, and Sos, a guanine nucleotide exchange factor that is the actual activator of RAS. Grb2 and Sos form a complex that is recruited to phosphorylated ErbB receptors through binding of the SH2 domain of Grb2 to specific phosphotyrosine sites of the receptor. This brings Sos into close proximity with RAS at the plasma membrane where it activates RAS by exchanging a GDP nucleotide (RAS bound to GDP is inactive) for GTP, the nucleotide that binds and activates RAS. Once activated, RAS binds to and activates the Raf kinase. Raf then phosphorylates and stimulates the kinase activity of MAPK kinase (MAPKK, MEK) on a key serine residue. MAPKK then phosphorylates and activates MAPK (ERK), which can phosphorylate a variety of cytoplasmic and membrane-bound substrates or, more importantly, translocate into the nucleus where it phosphorylates and activates specific transcription factors to effect novel gene expression.

SIGNALING BY CATALYTIC RECEPTORS/SERINE-THREONINE KINASES

Cell-surface receptors that are activated to phosphorylate serine or threonine residues on substrate proteins comprise a large group of receptors that bind the transforming growth factor-β (TGF-β) family of signaling molecules. This family contains subfamilies of signaling molecules related by structural and functional homologies that include bone morphogenetic protein (BMP; organogenesis, cell induction events), activin (mesoderm induction), and TGF-β itself (hematopoietic, epithelial, mesenchymal cell proliferation, and differentiation) (Fig. 8–5).

Formation and activation of TGF-β receptors differs from that of receptor tyrosine kinases. TGF-β receptors are composed of two distinct receptor subtypes, types I (TβR-I) and II (TβR-II). Neither of these receptors on

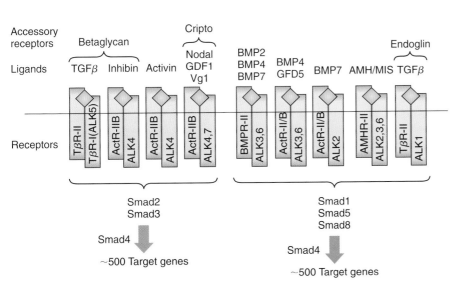

Figure 8–5. Relations between transforming growth factor-β (TGF-β) and TGF-β–like ligands and their type I and II receptors in vertebrates. The *Nodal* ligand binds to the ActR-IIB-Alk4 heterodimer. The activated receptor transmits the *Nodal* signal via Smads 2 and 3, which heterodimerize with Smad4 to activate target genes. The bone morphogenetic protein (BMP) signaling pathways are shown on the right for comparison. (Modified from Shi Y, Massague J. *Cell* 2003;113:685, by permission.)

their own can bind and transduce the TGF-β signal. The functional receptor is formed when ligand binding induces a complex of both receptors (Fig. 8–6).

In general, it appears that the type II receptor is the binding receptor and the type I receptor is the signaling receptor. TGF-β first binds to the type II receptor, causing a conformational change in the TGF-β molecule such that it can now be recognized by the type I receptor. Binding of the type I receptor to TGF-β brings the type I receptor into proximity with the type II receptor, which then phosphorylates the type I receptor on serine and threonine residues and activates its kinase activity. The type I receptor then transduces the TGF-β signal by phosphorylating downstream signal transducer molecules.

Bone Morphogenetic Proteins

Signaling by BMPs through receptors of the TGF-β receptor family represents a major signal transduction pathway both in early developmental events and in the differentiation of osteoblasts for later bone formation. BMPs are multifunctional proteins that play important roles in a wide range of biological activities on various cell types and provide a useful example of signaling through members of the TGF-β superfamily of receptors. Activation of the BMP signal transduction pathway begins when BMP binds two different BMP receptor subtypes, types I and II, acting as a bridge to bring these two receptors into juxtaposition and allowing phosphorylation of the type I receptor by type II receptor kinase. Once the type I receptor is phosphorylated, it acts as a kinase to transduce the BMP signal to downstream effectors. At this point, two signal transduction pathways can be activated: the TAK1-MKK3/

6-p38/JNK (c-Jun N-terminal kinase; see Fig. 8–6) and the Smad pathways (Fig. 8–7). TAK1 is a member of the mitogen-activated protein kinase kinase kinase (MAPKKK) superfamily. Phosphorylation of TAK1 triggers a cascade of phosphorylation reactions leading to activation of the nuclear transcription factor ATF-2 and up-regulation of subordinate genes.

Alternatively, the BMP signal can be transmitted via the Smad signal transduction pathway (see Figs. 8–6 and 8–7). On ligand stimulation by BMPs, Smad1 proteins are recruited to the type I receptor where they are phosphorylated and released. This BMP ligand-specific smad then associates with Smad4 (which does not bind receptors) and this Smad1/4 complex then translocates from the cytoplasm into the nucleus. In the nucleus, the Smad1/4 heterodimer binds to and activates the ATF-2 transcription factor that then transcribes BMP-responsive genes.

BMPs are a requisite signaling molecule used by endodermal cells to induce the cardiogenic cell lineage in mesodermal cells expressing BMP receptors on their surface. BMP2 and BMP4 appear to be the only BMP isoforms capable of inducing the formation of cardiogenic cells. These BMPs are secreted out of endodermal cells to bind and activate BMP receptors on the surface of precardiogenic mesodermal cells and activate the appropriate signal transduction pathways. BMP2/4 signaling has been shown not only to play an important role in these early steps of cardiogenic induction, but also in maintaining their cardiogenic potential until other later-acting signals complete their differentiation into cardiomyocytes. The early and continued requirement for the BMP signal transduction pathway illustrates that distinct signal transduction pathways can be used in a variety of different biological contexts, for

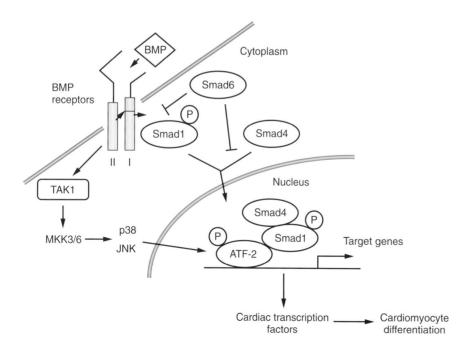

Figure 8–6. Schematic representation of bone morphogenetic protein (BMP) signal transduction pathways involved in cardiogenic induction. The BMP signal can be transmitted via the TAK1 signaling pathway or via Smad proteins, in particular Smad1 and 4. The Smad1/4 heterodimer can bind the ATF-2 transcription factor activating it to transcribe BMP-responsive genes. The same can be achieved by the alternate TAK1 pathway via the mitogen-activated protein kinases MKK3/6, which phosphorylate and activate the stress-activated protein kinases p38 and c-Jun N-terminal kinase (JNK) to go on and activate ATF-2. (Modified from Monzen K, Nagai R, Komuro I. *Trends Cardiovasc Med* 2002;12:263, by permission.)

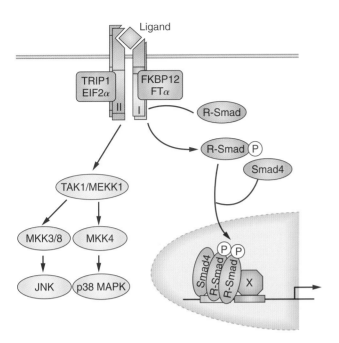

Figure 8–7. Transforming growth factor-β (TGF-β) receptor signaling through Smad-independent pathways. The TGF-β signal can be directed to different signaling pathways such as the TAK1/ MEKK1 or smad pathways. This will activate presumably different gene programs through the activation of different transcriptional effectors such as c-Jun N-terminal kinase (JNK), p38, mitogen-activated protein kinase (MAPK), or smad. (Modified from Derynck R, Zhang YE. *Nature* 2003;425:577, by permission.)

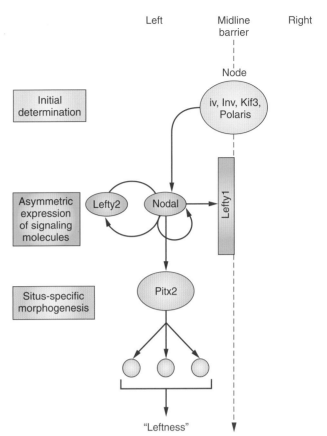

Figure 8–8. Signal transduction pathway for the determination of left-right asymmetry. A subtle but important process that initiates left-right asymmetry is the "sweeping" from right to left of potential effector molecules by cilia situated on or about the Node. Proteins of the kinesin superfamily such as Kif3, Polaris, Iv, and Inv are involved in this process. This initial break with embryonic symmetry leads to left-sided expression of *Nodal*, the transforming growth factor (TGF)-β–like signaling molecule. *Nodal*'s action is restricted by *Lefty*2, another TGF-β–like signaling molecule that acts as a feedback inhibitor of *Nodal*. *Nodal* signal transduction eventually leads to activation of the Pitx2 transcription factor and its subordinate genes that act to impart "leftness." *Nodal* also induces expression of *Lefty*1, another TGF-β–like signaling molecule that functions as a midline barrier to signals that determine "leftness." (From Hamada H, et al. *Nature Reviews* 2002;3:102, by permission.)

example, at different times during development, and that to effect the proper biologic response, multiple signals acting through a combination of different signaling pathways may be required.

Nodal

The catalytic receptor serine-threonine kinase signal transduction pathway is used throughout development and in a variety of different biologic contexts. What changes to direct this pathway toward different ends is primarily the nature of the ligand or signaling molecule inducer, where and when the ligand is produced, and how signaling is controlled. For example, whereas BMP signaling through TGF-β serine/threonine kinase receptors promotes the early developmental event of cardiac induction, another ligand called *Nodal* is also acting through serine-threonine kinase receptors to induce a left-right asymmetry in cardiac precursors that eventually results in the correct positioning of the heart in the left side of the chest cavity and in the formation of left and right ventricular chambers. The first step in establishing left-right asymmetry is the asymmetric distribution of the Nodal signaling molecule itself (Fig. 8–8). This is achieved by the use of the Hedgehog signaling pathway (see later) that establishes a Nodal expression domain in the left side of the developing embryo. All

Nodal signaling subsequent to this acts to maintain and delimit Nodal expression in a discrete area (i.e., the left side) of the embryo. As discussed later, this in itself provides an interesting example of how signal transducers and the activity of their downstream signaling pathways can be maintained through the use of a positive feedback type of regulation and how they can be delimited by the induction of negative inhibitors.

The *Nodal* ligand is a member of the TGF-β family of signaling molecules and differs from the BMP family of ligands by using type I and II receptors differently (Nodal has a greater affinity for the type II vs. type I receptors, a property that influences how its signaling receptor is formed), and it signals through a set of smad signal transducers that differs from that used by BMPs. *Nodal* signaling begins when *Nodal* ligands dimerize and bind to the type II ActR-IIb receptor (Fig. 8–9). This facilitates recruitment and complexing with the type I receptors, in this case referred to as ALK4 receptors, that are already bound to a coreceptor molecule called Crypto (or *Cryptic*). This coreceptor is a member of the epidermal growth factor (EGF)-CFC family of extracellular, glycosylphosphatidylinositol (GPI)-linked proteins that are essential for *Nodal* signaling. Once this trimeric receptor complex is formed, the type II receptor kinase phosphorylates the type I receptor, activating its kinase activity. Activated type I receptor then directly phosphorylates Smad2 and Smad3, which then translocate into the nucleus. Once in the nucleus, these smads complex with cell type–specific, DNA-binding, transcriptional activators to turn on *Nodal*-dependent genes.

Nodal establishes the left-right asymmetry in the embryo required for normal heart development and placement. As a diffusible signaling molecule, this asymmetry would be lost if Nodal were to diffuse uniformly throughout the embryo activating all cells. Thus, its function is critically dependent on its activity being maintained and restricted to the left-hand side of the embryo during the critical period of heart development. To maintain its expression, Nodal autoregulates the expression of its own gene; it achieves this by up-regulating Nodal gene transcription using a slightly different signal transduction pathway composed of a different type I receptor and a different array of smad proteins. To delimit its expression spatially to the left-hand side of the embryo, *Nodal* induces expression of two other genes, *Lefty*1 and *Lefty*2, two TGF-β–like

signaling molecules that are functional antagonists of *Nodal* signaling. Despite their name, they act to inhibit *Nodal*'s imposition of left-handedness on areas of the lateral plate mesoderm other than the left hand side of the embryo. *Lefty*1 is expressed in midline cells where it acts as a barrier preventing *Nodal* signaling from transgressing into the right side of the embryo, whereas *Lefty*2 is expressed within the left lateral plate mesoderm and acts to prevent further spread of the *Nodal* signal in this region and to limit the duration of *Nodal* activity. Because *Lefty*2 is an antagonist of *Nodal* signaling, it is likely that *Nodal* can signal uninhibited until levels of *Lefty*2 increase to a point where they can then antagonize *Nodal* signaling. It appears that the way in which *Lefty*2 antagonizes *Nodal* signaling is by binding to the EGF-CFC coreceptor *Cryptic* and preventing *Nodal* from assembling an active type I-II *Nodal* receptor complex (see Fig. 8–9). Despite being induced by *Nodal*, *Lefty*2 appears able to diffuse more readily into neighboring cells than can *Nodal*, thereby setting up a perimeter of cells in which *Nodal* can be completely inhibited. Together, these properties suggest that *Lefty*2 can delimit *Nodal* signaling temporally and spatially by "outracing" *Nodal* to cells with unoccupied EGF-CFC receptors, binding to them and precluding *Nodal*-dependent assembly of active receptors.

SIGNALING BY NONKINASE RECEPTORS

Ligand-based communication between cells is not limited to the use of tyrosine kinase receptors and their associated kinase signal transduction pathways. Three widespread and developmentally important signal trans-

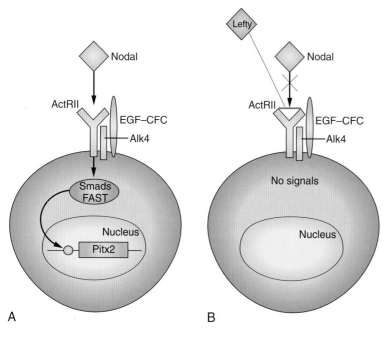

Figure 8–9. Actions of *Nodal* and the *Lefty* proteins. *Lefty* proteins act as antagonists of *Nodal* by competing with *Nodal* for binding to the ActR11 receptor. Epidermal growth factor (EGF)-CFC proteins, coreceptors for *Nodal* and *Lefty*, interact with Alk4 and not ActRII. **A:** The *Nodal* signal is transduced through the ActRII receptor, causing Smad and Fast transcription factors to translocate to the nucleus where they activate downstream genes such as *Pitx2*. **B:** When *Lefty* proteins are bound to ActRII, no signals are transduced. (From Hamada H, et al. *Nat Rev* 2002;3:102, by permission.)

A

B

duction pathways that fall into this category are the Wnt, Notch, and Hedgehog signal transduction pathways. These pathways do not use kinases at all for transducing signals. Instead, they act by blocking either the degradation of a transcriptional activator, as in the case of Wnt signal transduction, or the conversion of a transcriptional activator into a transcriptional inhibitor, as in the case of Hedgehog signal transduction. A third example, Notch signaling, transmits cell-surface signals to the nucleus via proteolytic cleavage of an intracellular signal transducer to yield a peptide that activates Notch-dependent gene transcription by displacing transcriptional repressors.

Wnts

Wnts are a family of secreted glycoproteins that have been implicated in a variety of developmental processes such as cell fate determination and differentiation, cell polarity, cell migration, and cell proliferation. Once secreted, Wnt proteins associate with glycosaminoglycans in the extracellular matrix and are bound tightly to the cell surface, where they can act on the producing cell or closely apposed neighboring cells. Despite this tethering to the Wnt-producing cell, Wnts can also act as long-range morphogens, eliciting different responses from responding cells at various distances from the Wnt-producing cell. This is likely achieved by the ability of Wnt to induce changes in responding cells that lead to production of other secreted signaling molecules in a sort of cell-to-cell relay mechanism of long-range signaling.

As far as signaling pathways go, Wnt signaling is complex. The presence of different Wnt isoforms (the human genome contains 19 Wnt genes, two distinct families of Wnt receptors, the Frizzled *(Fzd)* gene family and the low-density lipoprotein receptor–related protein (LRP) family, combined with the ability of Wnts to signal through either activating or inhibiting signal transduction pathways requires coordinated regulation between all factors to achieve a given biological effect. Wnt signaling begins when Wnt proteins bind to cell-surface receptors to activate intracellular signaling pathways. Wnt proteins exhibit a wide and diverse array of biological effects indicating that different Wnts might act through different signaling pathways. Pathway selection is determined in large part by which Frizzled receptor is activated by which Wnt ligand. Among the different signal transduction pathways that Wnts can activate are the canonical Wnt/β-catenin, the Wnt/Ca²⁺, and the Wnt/polarity pathways (Fig. 8–10).

Signaling through the canonical Wnt/β-catenin pathway is achieved with the activation of a latent group of transcription factors (the LEF/TCF family) by a molecule called β-*catenin*. In the absence of Wnt, a multiprotein complex binds to and degrades β-catenin,

preventing it from activating LEF/TCF transcription factors. In the presence of Wnt, this degradation process is blocked, resulting in an increase in free β-catenin that binds and activates LEF/TCF transcription factors and targets gene expression. The Wnt/Ca²⁺ pathway operates independently of β-catenin and is activated by a different group of Wnts and *Fzd* receptors. Wnt binding to *Fzd* receptors activates a heterotrimeric G protein that leads to an increase in intracellular Ca²⁺ levels and activation of calcium/calmodulin-regulated kinase II (CamKII) and protein kinase C (PKC; G-protein signaling pathways are discussed later). The choice between these two signaling pathways can also be influenced by secreted modulators of Wnt signaling that antagonize or block Wnt binding to *Fzd* receptors. The selective activation or inhibition, or both, of all three Wnt signal transduction pathways can together determine a given biologic effect. An example of such coordinated signal transduction can be found in the induction of cardiac precursors from mesodermal cells. The successful induction of cardiac mesodermal progenitors requires the spatially controlled expression of two opposing activities: inhibition of the cardiac-inhibiting Wnt/β-catenin signaling pathway and activation of the Wnt/Ca²⁺ and Wnt/polarity pathways. Among those inhibitors that appear to play a role in cardiogenesis are Dickkopf (Dkk-1) and Crescent. Dkk-1 blocks activation of Wnt signaling by interacting with the extracellular domain of LRPs, whereas Crescent appears to bind directly to Wnts and modulate their activity in a context-dependent manner.

Hedgehogs

Hedgehog (HH) proteins (the name originates from the "hedgehog-like" appearance of *Drosophila* embryos bearing a mutated Hedgehog gene) are secreted proteins that function in short-range signaling to neighboring cells, that is, on the order of a few dozen or so cell diameters. This property results from a unique type of posttranslational modification in which HH undergoes an internal autoproteolytic cleavage into separate N- and C-terminal peptides followed by the covalent addition of lipid molecules to the N-terminal peptide (Fig. 8–11).

The N-terminal peptide stays tightly associated with receptors on the surface of the cell in which it was synthesized, whereas the C-terminal peptide freely diffuses. The N-terminal peptide contains all the signaling function of the cleaved HH molecule. It remains tightly associated with receptors on the surface of the cell in which it was synthesized and cleaved and eventually undergoes limited diffusion to give a steep high-to-low HH concentration gradient that leads away from the HH-producing cell. Neighboring cells sense this concentration gradient and interpret it accordingly: high concentrations of HH lead to the production of a

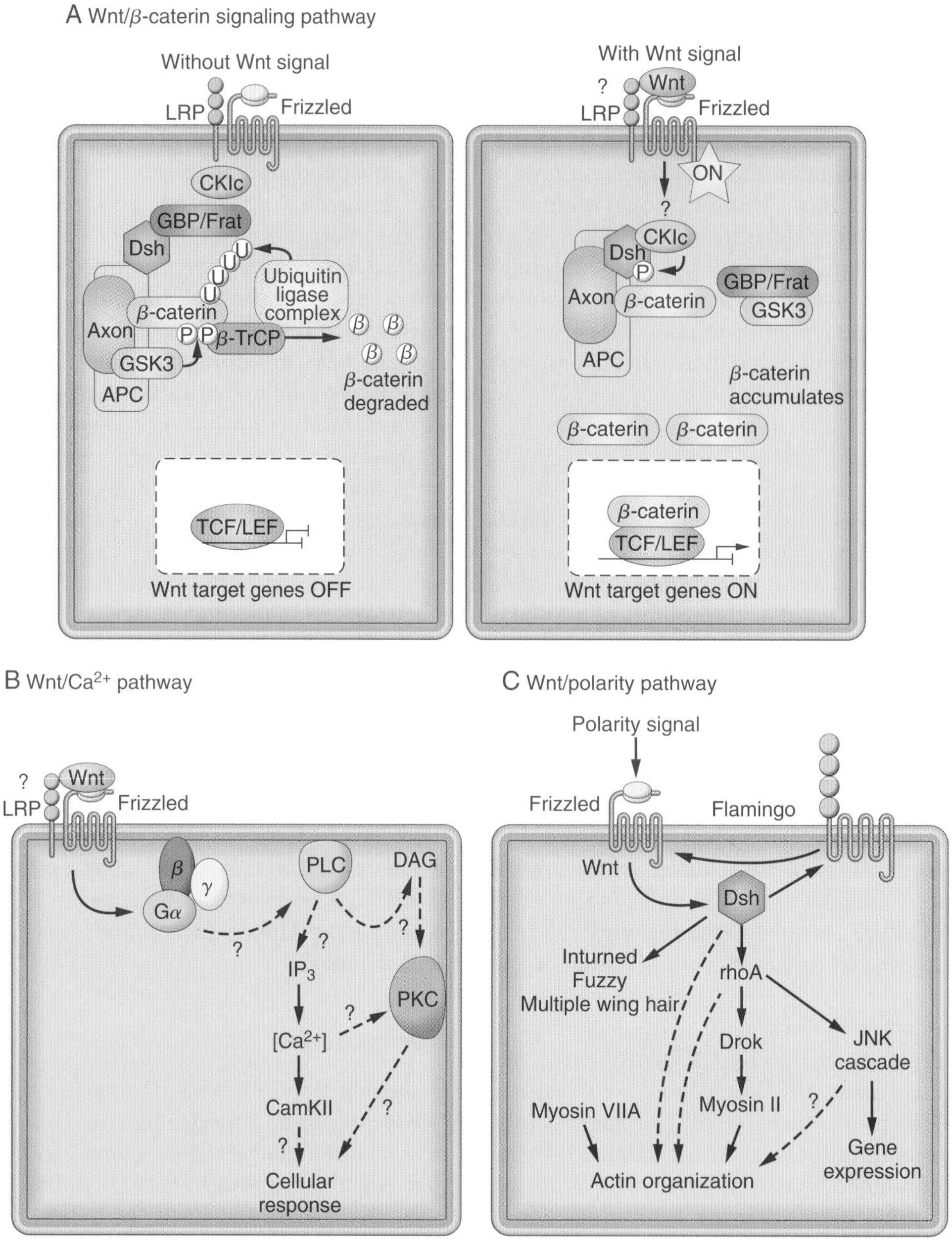

Figure 8–10. The known Wnt signaling pathways. **A:** The Wnt/β-catenin pathway. Wnt signaling depends on the steady-state levels of the multifunctional protein β-catenin. In the absence of Wnt signal, β-catenin is phosphorylated, leading to enhanced binding of β-catenin to β-TrCP. β-catenin becomes ubiquitinated and destroyed by the proteasome. Exposure of cells to Wnt leads to phosphorylation of disheveled (Dsh), a cytoplasmic scaffold protein that prevents the phosphorylation, ubiquitination, and degradation of β-catenin. β-catenin accumulates in the cell, enters the nucleus and interacts with members of the TCF/LEF transcription factor family, and activates target genes. **B:** Signaling through the Wnt/Ca^{2+} pathway appears to involve G-protein activation that leads to an increase in intracellular Ca^{2+} and activation of Ca^{2+}/calmodulin-dependent kinase II and protein kinase C, which both effect a Wnt-dependent response. **C:** The Wnt/polarity pathway (as shown for *Drosophila*). In vertebrates, Wnt/polarity signaling is thought to control polarized cell movements during gastrulation and neurulation. Wnt11 may use this pathway to regulate cell movement during gastrulation. (From Miller JR. The Wnts. *Genome Biol* 2001;3:3001.1, by permission.)

transcriptional activator in responding cells that up-regulates target genes, whereas lower concentrations attenuate production of a transcriptional repressor, allowing for activation by other non-HH pathways. The concentration-dependent activation of genes, many of which are transcription factors important in specifying cell type, is a critical feature of the ability of HH to pattern complex tissues. This property is put to evident use in the patterning of specific cell types in the nervous system and the patterning of digits in the vertebrate limb. In addition, HH signaling is involved in heart development as indicated by analysis of cardiogenesis in mice mutant for the HH ligand or for its receptor and downstream signal transducers. These studies showed that HH signaling plays an important role in specifying

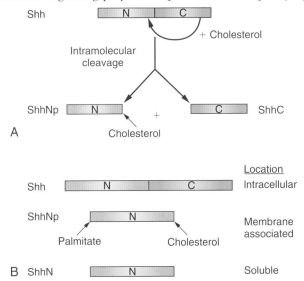

A

B

Figure 8–11. Cleavage forms of Shh. A: Schematic model illustrating the intramolecular cleavage reaction that Shh undergoes. Although both domains are secreted, the majority of ShhNp is membrane associated whereas ShhC is not. B: Shh forms and their known locations. The role of cholesterol on ShhNp is to tether Shh to the cell membrane, limiting its range of action. (Modified from Goetz JA, et al. *Bioessays* 2002;24:157, by permission.)

mesodermal cells to enter the cardiac muscle cell lineage (induction) while also playing an essential role in the correct looping of the heart tube (morphogenesis). Again, this illustrates the idea that signal transduction pathways can be used in a number of different developmental contexts to subserve different aspects of the developmental program.

HH signaling can be viewed as the transition between two states; one a basal state of transcriptional repression that occurs in the absence of HH, and the other a state of transcriptional derepression and gene activation that is initiated by HH binding to its cell surface receptor called Patched *(Ptc)*. HH signaling is essentially the transition between these two transcriptional states, a process that is mediated by a molecular "gatekeeper" of sorts called Smoothened *(Smo)* that controls the accessibility of the *Gli* family of transcription factors to HH-responsive gene promoters (Fig. 8–12).

In the basal, repressed transcription state (i.e., absence of HH ligand), the unoccupied HH receptor, Patched, inhibits the activity of Smoothened, another transmembrane protein. This allows formation of a multiprotein complex that prevents access of *Gli* transcription factors to HH-responsive genes. This complex consists of at least three proteins: the kinase Fused protein (Fu); the kinesin motor protein Costal2 (Cos2); and the Suppressor of Fused (SUFU), an antagonist of HH signaling. The complex binds tightly to cytosolic microtubules and to *Gli* transcription factors, sequestering them and

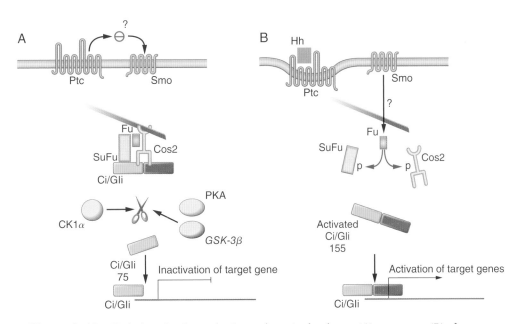

Figure 8–12. Hedgehog signal transduction pathway in the absence (A) or presence (B) of Hedgehog. A: In the absence of Hedgehog, the Patched receptor *(Ptc)* inhibits the membrane protein Smoothened *(Smo)*, allowing the complex of Fused (Fu), Costal2 (Cos2), and Suppressor of Fused (SUFU) to sequester the zinc finger transcription factor *Gli*. Sequestered *Gli* is cleaved to release a 75-kDa fragment that suppresses Hedgehog target genes; no signal is transmitted. B: Binding of Hedgehog to *Ptc* nullifies its inhibitory activity on *Smo*. Active *Smo* signals through unknown mechanisms to the Fu-Cos2-SUFU complex where phosphorylation of Fu and Cos2 disrupts the microarchitecture of the complex releasing *Gli*, which then transits to the nucleus, where it activates transcription of Hedgehog target genes. (From Bijlsma MF, et al. *Bioessays* 2004;26:387, by permission.)

cleaving them into two fragments, one of which contains the zinc-finger DNA-binding domain minus any transcriptional activation domains. This fragment translocates to the nucleus and binds to sites within the promoter of HH-responsive genes, effectively preventing functional full-length *Gli* proteins from binding and activating transcription. This state of transcriptional repression is relieved with the binding of HH ligand to the Patched receptor, a 12-transmembrane cell-surface receptor. The HH-Patched ligand-receptor complex is brought into the cell by endocytosis and degraded by the lysosomal pathway. Destruction of Patched relieves the inhibition of Smoothened, which then activates the HH pathway. Active Smoothen in some way promotes release of the Fu-Cos2-SUFU-*Gli* complex from microtubules, freeing *Gli* proteins from proteolytic cleavage. Free, full-length *Gli* translocates to the nucleus, where it binds promoter elements in HH-responsive genes, activating their transcription and completing the HH signaling pathway.

Notch

Unlike the signaling pathways discussed so far, Notch denotes the receptor of the pathway and not the signaling molecule. (The term *Notch* is derived from the gene mutation in *Drosophila* in which parts of the wing are missing, as if notches of tissue were cut out of the wing.) Another difference is that the ligands that interact with the Notch receptor, called Jagged and Delta, and the Notch receptor itself are type I transmembrane proteins. With both the signaling molecule and its receptor membrane bound, Notch signaling can only occur when the signaling cell and target cell are closely apposed. This may be a way of controlling and delimiting the extent of cells activated by Notch signaling. Notch signaling

may thus be considered a form of paracrine signaling in which the signaling cell and the target cell are in the same location or tissue. One way in which this type of signaling has been put to use is in controlling the delamination of cells from compact cell layers to allow for their migration. A prime example of such signaling and cell behavior is in the early steps of valve formation in the developing heart. The incipient events in heart valve formation require cells of the endocardial layer to delaminate and migrate into the cardiac jelly separating the myocardial and endocardial cell layers (Fig. 8–13). To achieve this, they undergo an epithelial-to-mesenchyme transition that allows cells to migrate into the cardiac jelly, where they proliferate and cellularize the jelly to form "cardiac cushions," the precursors of heart valves. Notch activity within the endocardial cell layer triggers this process (Fig. 8–14).

A prerequisite of Notch signaling is the establishment of cells competent to receive the Delta or Jagged signaling ligand. This is achieved by the proteolytic cleavage of Notch precursor proteins in target cells at a site within the molecule (called S1) to form a heterodimerized receptor of sorts on the cell surface (Figs. 8–15 and 8–16). One proteolytic product consists of an ectodomain called *Notch extracellular domain* (or NECD) and a membrane-tethered intracellular domain called NTM. Notch signaling from neighboring cells begins when the ligands Delta or Jagged bind the NECD causing the NTM to be cleaved by extracellular proteases at a site called *S2*. S2 cleavage releases the ectodomain of Notch and generates an activated membrane-bound form of Notch called *Notch extracellular truncation* (NEXT). NEXT is further cleaved at two sites, S3 and S4, to release a peptide called *Notch intracellular domain* (NICD) into the interior of the cell. NICD translocates into the nucleus and

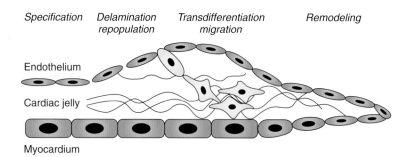

Figure 8–13. Anatomic overview of heart valve development. The developing heart tube contains an outer layer of myocardium and an inner lining of endothelial cells separated by an extracellular matrix referred to as the cardiac jelly. During heart valve formation, a subset of endothelial cells overlying the future valve site are specified to delaminate, differentiate, and migrate into the cardiac jelly, a process referred to as *endothelial-to-mesenchymal transformation* or *transdifferentiation*. Locally expanded swellings of cardiac jelly and mesenchymal cells are referred to as *cardiac cushions*. In a poorly understood process, cardiac cushions undergo extensive remodeling from bulbous swellings to thinly tapered heart valves. (From Armstrong EJ, Bischoff J. *Circ Res* 2004;95:459, by permission.)

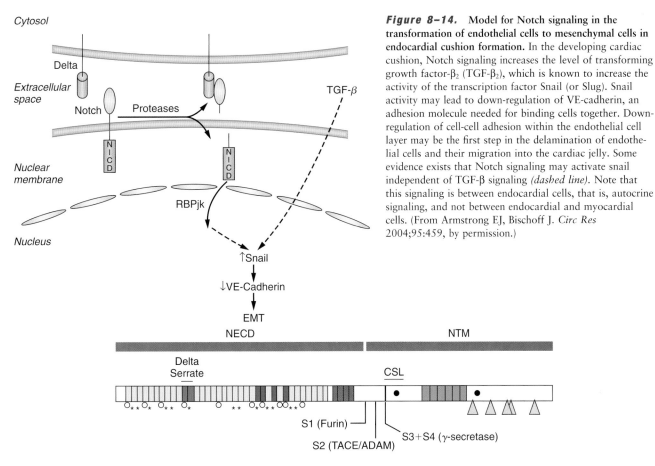

Figure 8-14. Model for Notch signaling in the transformation of endothelial cells to mesenchymal cells in endocardial cushion formation. In the developing cardiac cushion, Notch signaling increases the level of transforming growth factor-β_2 (TGF-β_2), which is known to increase the activity of the transcription factor Snail (or Slug). Snail activity may lead to down-regulation of VE-cadherin, an adhesion molecule needed for binding cells together. Down-regulation of cell-cell adhesion within the endothelial cell layer may be the first step in the delamination of endothelial cells and their migration into the cardiac jelly. Some evidence exists that Notch signaling may activate snail independent of TGF-β signaling *(dashed line)*. Note that this signaling is between endocardial cells, that is, autocrine signaling, and not between endocardial and myocardial cells. (From Armstrong EJ, Bischoff J. *Circ Res* 2004;95:459, by permission.)

Figure 8-15. Domain composition of Notch. The extracellular domain of Notch consists of epidermal growth factor (EGF) repeats *(light blue)* that vary in number between species and between Notch subtypes. It also includes three Lin12/Notch repeats (LNRs; *dark gray*). The EGF repeats 11 and 12 *(dark blue)* are necessary and sufficient for binding the Delta and Serrate ligands. The intracellular domain includes six ankyrin repeats *(green)* and two nuclear localization signals *(black dots)*. The binding sites for the Delta/Serrate ligand and for the CSL DNA-binding protein are shown. The positions of the S1-S4 cleavage sites are depicted *(arrows)*. Cleavage at the S1 site releases the Notch receptor into polypeptides, an ectodomain called *Notch extracellular domain* (NECD) and a membrane-tethered intracellular domain called *NTM*. (Modified from Schweisguth F. *Curr Biol* 2004;14:R129, by permission.)

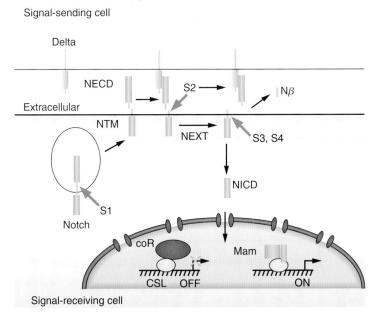

Figure 8-16. A model for Delta-dependent Notch signaling to the nuclear transcription factor CSL. Delta at the surface of the signaling cell binds S1-cleaved Notch at the surface of the responding cell. Ligand-dependent S2 cleavage of Notch generates an activated membrane-bound form of Notch called *Notch extracellular truncation* (NEXT), which is further processed at the S3 and S4 sites. This releases the Notch intracellular domain (NICD), which translocates into the nucleus where it derepresses CSL by displacing the corepressor coR. (From Schweisguth F. *Curr Biol* 2004;14: R129, by permission.)

assembles into a ternary complex with a DNA-binding protein called *CSL*. In the absence of Notch signaling and nuclear NICD, CSL recruits transcriptional repressors to Notch target genes to repress their expression. Binding by NICD converts CSL to a transcriptional activator by replacing these transcriptional repressors. In this way, activated Notch receptors can be considered transcriptional coactivators.

In considering signal transduction pathways individually, we risk losing sight of the fact that, in multicellular organisms, cells are continually subjected to a multiplicity of instructional signals that they must receive and integrate into an orchestrated behavior. These instructional signals direct the expression of specific genes that determine a cell's differentiation state and how it functions and interacts with other cells to form higher order structures such as organs. The failure of cells to receive, transduce, or integrate these instructive signals into appropriate behavior can lead to aberrant organ formation or function. With the recent advances in genetic analysis of human disease, many diseases and syndromes are now being linked to the aberrant expression or function of a few critical genes. In newborns, one of the most prevalent defects is atrial septal defect (ASD), in which the septum separating the right and left atria is aberrantly formed. The left and right atria are formed from a common atria by the intercession of a dividing septum that prevents direct communications between the atria. This septum results from the growth and fusion of two septa, the septum primum and the septum secundum, that originate in the dorsal wall of the single atrium and grow toward the endocardial cushions with which it fuses to seal off left from right atria (Fig. 8–17).

The proper formation of an intact septa requires the controlled growth and reabsorption of myocardium, that is, the coordinated proliferative, migratory, and apoptotic behavior of myocardial and nonmyocardial cells. This is most likely achieved by signaling between cells, either myocardial to myocardial or endocardial to myocardial signaling similar to what is seen in the formation of septa and valves in heart vessels. This latter process has been well studied and is known to be controlled by a complex network of signaling events between endocardium and myocardium (Fig. 8–18).

It thus may be that similar complex signaling occurs in the formation of the atrial septa. It is known that mutations in two transcription factors, Tbx5 and Nkx2.5, can lead to ASD. When taken in the context of valve signaling, it suggests that the high incidence of ASD in newborns could be attributable to defects in other genes, perhaps those in an equally complex network of signal transduction pathways that control septa formation via their activation of specific transcription factors and gene programs.

SIGNALING BY STEROID HORMONE RECEPTORS REQUIRES LIGAND INTERACTION WITHIN THE CYTOPLASM OR NUCLEUS

A third category of cell-to-cell communication that does not rely on kinases or proteolytic processing is signaling via peptide and nonpeptide hormones. Peptide hormones are secreted from one cell to affect the behavior of a target cell bearing specific receptors for that hormone. This form of communication results in biochemical changes in the target cell that occur independent of changes in the gene expression of that cell. This distinguishes hormonal communication between cells from the signal transduction mechanisms considered in this chapter in which the main mode of action is to turn on specific target genes in the responding cell. There is, however, a class of hormones whose main mode of action is to regulate expression of specific target genes.

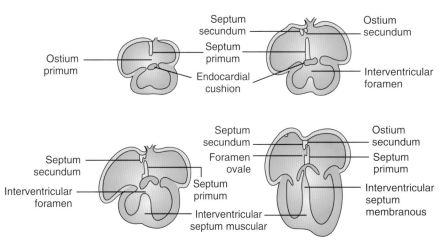

Figure 8–17. Schematic representation of the formation of the interatrial and interventricular septum. In humans, the transition from a tubular heart into a four-chambered structure is completed between 4 and 8 weeks gestation. This involves four events: septation of the common atrial chamber, septation of the ventricular chamber, proliferation of the endocardial cushions, and the growth of the bulbar ridges that divide the bulbus cordis or outflow tract into the future aorta and pulmonary tract. (From *Congenital Heart Disease: A Deductive Approach to Its Diagnosis.* Chicago: Burton W. Fink Year Book Medical Publishers, 1985, by permission.)

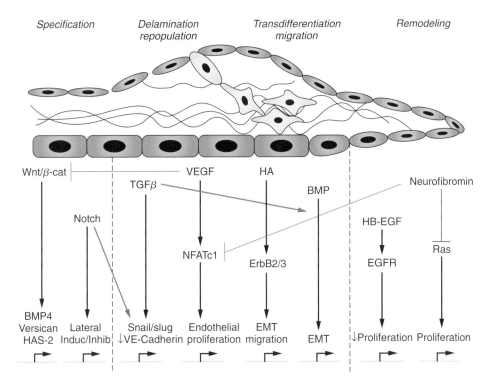

Specification *Delamination repopulation* *Transdifferentiation migration* *Remodeling*

Figure 8–18. Signaling network model for heart valve development and remodeling. In the signaling network model for cardiac cushion development, numerous signaling pathways and transcriptional regulators act to coordinately regulate the process of heart valve formation. Remodeling refers to the formation of valve leaflets, the eventual results of cushion formation, which is presently not completely defined at the molecular level. Thin red cells are endothelial cells, yellow cells are delaminated endothelial cells that have undergone the transformation to mesenchymal cells, and the thicker red cells are myocardial cells. (From Armstrong EJ, Bischoff J. *Circ Res* 2004;95:459, by permission.)

This class of hormones is composed of steroid hormones and their receptors. These hormones are nonpeptide molecules structurally derived from cholesterol and include the hormones progesterone, 17β-estradiol, testosterone, cortisol, aldosterone, and 1,25-dihydroxyvitamin D_3. An important subclass of hormone-like molecules includes retinoic acid (vitamin A) and 3,5,3′-triiodothyronine (thyroid hormone). Despite being structurally different from steroid hormones, they are considered steroid hormone–like by virtue of the structural and functional similarity of their receptors to those of steroid hormones.

Although steroid hormone signaling is similar to cytokine or growth factor signaling in that changes in gene expression are the ultimate goal of their action, steroid hormones achieve this in a completely different way. This is due largely to the nature of the steroid hormone receptor itself and how it works. Rather than being an integral membrane receptor protein, steroid hormone receptors exist in a free state within the cytoplasm or nucleus. Thus, unlike nonhormonal signal transduction in which signals from ligand-bound receptors at the cell surface are transmitted to the nucleus via a cascade of intracellular protein kinase signal transducers, steroid hormones must enter the cell to bind to their intracellular receptors. This is achieved, in part, by the lipophilic nature of steroid hormones that allow them to diffuse through the plasma cell membrane to encounter their receptors in the cytoplasm or nucleus. A second feature of steroid hormone signaling that distinguishes it from nonhormonal signal transduction is that steroid hormone receptors can affect gene expression directly

without the need for an intervening cascade of protein kinase signal transducers; in a sense, steroid hormones receptors can be considered ligand-activated transcription factors.

Many of the functional aspects of signal transduction by steroid hormones can be gleaned from the structure of the steroid hormone receptors themselves. Steroid hormone receptors have three major functional domains: a steroid binding domain located at the C terminus, a DNA-binding domain in the middle of the molecule, and a transcriptional activation domain located at the N terminus of the protein (Fig. 8–19).

Signaling by steroid hormones begins with the synthesis and secretion of the hormone from the signaling cell into the extracellular space. In fetuses and postnatal vertebrates, the hormone is then taken up into the bloodstream and disseminated throughout the body by the circulatory system. Diffusion of the hormone through the cellular plasma membrane brings it in contact with its intracellular receptors that can be either in the cytoplasm, as in the case of the glucocorticoid receptor, or in the nucleus, as in the case of receptors for estradiol, progesterone, androgen, and vitamin D_3 (Fig. 8–20).

There is some indication that unoccupied nuclear steroid hormone receptors may actually be bound to specific DNA sequences in the promoter of the genes they regulate and in the unoccupied state act as transcriptional inhibitors. With binding of hormone, cytoplasmic receptors transit to the nucleus and nuclear receptors undergo conformational changes to an active form. In both cases, the hormone-bound receptors then bind to sequence-specific response elements in the pro-

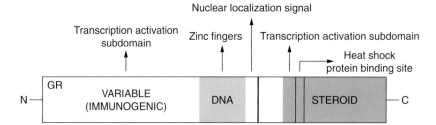

Figure 8-19. Model of a typical steroid hormone receptor. The glucocorticoid steroid hormone receptor provides a model for steroid hormone receptor structure. The structural features leading to function are: (1) STEROID: the steroid-binding domain in the C terminus, and (2) DNA: the DNA-binding domain that binds the receptor to specific response elements in the promoters of steroid hormone–responsive genes. Other functional domains include transcription activation subdomain, which recruits molecules of the transcriptional apparatus to the responsive gene's promoter; nuclear localization signal, which is used in translocating hormone-bound receptor to the nucleus; heat shock protein binding site, which binds heat shock protein 90 (Hsp90) in the unbound state to prevent the unoccupied receptor from binding DNA; and zinc fingers, which are protein structural motifs that intercalate into DNA-helical grooves to provide physically tight binding of receptor to DNA. (From Devlin TM, ed. *Textbook of Biochemistry with Clinical Correlations*, 5th ed. New York: Wiley-Liss, 2002.)

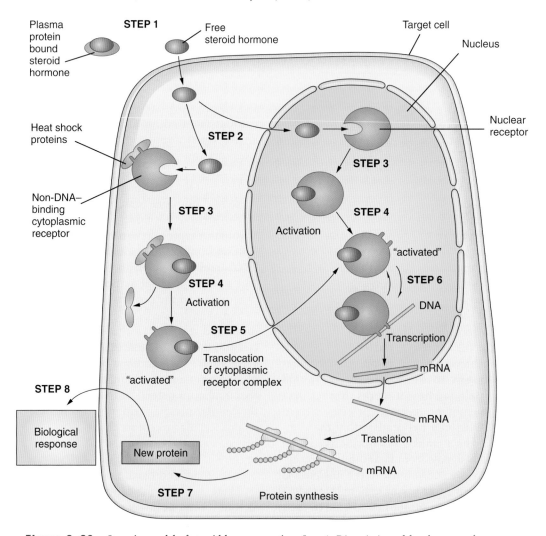

Figure 8-20. Stepwise model of steroid hormone action. Step 1: Dissociation of free hormone from circulating transport protein. Step 2: Diffusion of free ligand into cytosol or nucleus. Step 3: binding of ligand to unactivated cytoplasmic or nuclear receptor. Step 4: activation of cytosolic or nuclear hormone-receptor complex to activated, DNA-binding form. Step 5: translocation of activated cytosolic hormone-receptor complex into nucleus. Step 6: binding of activated hormone-receptor complexes to specific response elements within the DNA. Step 7: synthesis of new proteins encoded by hormone-responsive genes. Step 8: alteration in phenotype or metabolic activity of target cell mediated by specifically induced proteins. (From Devlin TM, ed. *Textbook of Biochemistry with Clinical Correlations*, 5th ed. New York: Wiley-Liss, 2002.)

moters of hormone-responsive genes. The transcriptional activation domain of the receptors then recruits and assembles components of the transcriptional apparatus at the gene's promoter that then transcribe the gene into messenger RNA. The genes so activated encode proteins that alter the behavior of the cell in response to the steroid hormone.

An example of steroid hormone action is the effect of aldosterone on heart function. Aldosterone regulates the transport of Na^+, K^+ ions in cells and also promotes collagen deposition in blood vessels influencing vascular remodeling, stimulates cardiac fibrosis and hypertrophy, and causes an increase in the inflammatory response, highlighting the role of aldosterone as a key hormone affecting the cardiovascular system. Aldosterone synthesis is regulated primarily by the renin-angiotensin-aldosterone system (or RAAS; see later) and also by the hormone ACTH (adrenocorticotropic hormone). Increased levels of extracellular K^+ stimulate aldosterone secretion. In the extracellular space, aldosterone binds to mineralocorticoid receptors (MRs) on epithelial cells, and the aldosterone-MR complex translocates to the nucleus, where it binds steroid-responsive elements in target genes and modulates their expression.

SIGNALING BY G-PROTEIN– COUPLED RECEPTORS INVOLVES CLEAVAGE OF GUANOSINE TRIPHOSPHATE TO GUANOSINE DIPHOSPHATE

A major form of signal transduction in cells is via G-protein–coupled receptors. Once the receptor interacts with the ligand, G protein dissociates into G_α and $G_{\beta\gamma}$ subunits, where G_α exchanges GTP for bound GDP, generating active GTP-bound G_α complex. G_α-GTP interaction regulates the extent of the signaling that is terminated on hydrolysis of GTP to GDP by the G-protein signaling (RGS) proteins that carry an intrinsic GTPase activity (Fig. 8–21).

Both G_α and $G_{\beta\gamma}$ subunits play a significant role in mediating the activation of defined second messengers, such as cyclic adenosine monophosphate (cAMP), inositol 1,4,5-triphosphate (IP_3), and diacylglycerol (DAG). Two major signaling systems associated with G-protein–coupled receptors are the RAAS (see later) and the sympathetic β-adrenergic pathway. A common target of both the RAAS and the β-adrenergic response is the seven-transmembrane G-protein–coupled receptor (GPCR), which also includes the adrenergic receptors (ARs). One example of signaling via G-protein–coupled receptors is in the heart. The RAAS and the sympathetic β-adrenergic pathway are the most studied signaling receptor/hormone interactions that mediate the noncompensatory phase of cardiac hypertrophy. This is the phase of cardiac hypertrophy in which the heart myo-

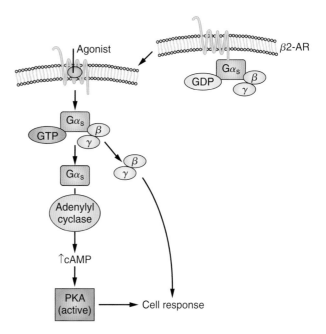

Figure 8–21. Seven-transmembrane receptor signaling. In response to agonists binding, a transient high-affinity complex of agonist-activated receptor and G protein is formed. GTP is released from the G protein and is replaced by GDP. This leads to dissociation of the G-protein complexes into α subunit and β/γ dimers, which activate several effectors. The increase in cyclic adenosine monophosphate (cAMP) activates protein kinase A (PKA), which phosphorylates many different substrates including the seven transmembrane receptors and transcription factors. (From Pierce, et al. *Nat Rev Mol Cell Biol* 2002;3:639–650, by permission.)

cardium, thickened by the hypertrophic growth of myocardial cells in the heart's attempt to normalize ventricular wall pressure, fails to compensate for this increased workload and undergoes apoptosis and deterioration. Signaling through β_1 receptors triggers a response to hypertrophic stimuli that allows rapid enhancement of cardiac contractility by influencing the firing rate of heart pacemaker cells. There are two ARs types in heart, β_1 and β_2, with four times as much β_1 than β_2. However, during heart failure, this ratio is reversed because of decreased expression or desensitization, or both, of β_1 AR. The β-adrenergic response is signaled through $G_{\alpha s}$ (β_1 receptor) or $G_{\alpha i}$ (β_2 receptor). $G_{\alpha s}$ activates the adenyl cyclase activity, triggering the activation of PKA, which, in turn, phosphorylates the L-type Ca^{2+} channel, causing an increase in intracellular Ca^{2+} (initial phase of systole). PKA also phosphorylates the ryanodine receptors (RyR2) that increase the intracellular Ca^{2+} by the release of Ca^{2+} by the sarcoplasmic reticulum (SR). Chronic β_1 receptor signaling triggers activation of β-adrenoceptor protein kinase, which down-regulates and desensitizes the β_1 receptor. During heart failure, the desensitization of β_1 receptors decreases the adenyl cyclase activity and alters the balance of intracellular Ca^{2+}.

Clinical Case 8-1

Leonard Chambers was a 6'3", 165 lb., 17-year-old black high school sophomore. Because of his size and natural athletic ability, he tried out for the basketball team and won a coveted starting position with the defending regional champions. During the fourth game of the season, while on a fast break, he blacked out, collapsed in midstride, and fell to the floor. After less than a minute he regained consciousness, nothing was broken, and he said he felt fine. Although he wanted to continue playing, his coach insisted he sit out the remainder of the game. (They were ahead and eventually won by 12 points.)

The next morning, they sent him to the school physician who found that he was a strapping young black male who appeared completely normal. His blood pressure was 115/75 mmHg, with a forceful pulse. His neurologic examination was entirely normal, as was his HEENT (head, eyes, ears, nose, and throat) examination. His lungs were clear, and his heart was normal except for a midsystolic murmur along the lower left sternal border. His abdominal examination was negative, except that when palpating for the spleen, the doctor felt a double impulse that coincided with Leonard's pulse. His extremity examination was normal except for right knee and elbow abrasions from the fall. His fingers were not elongated.

He had two brothers and one sister who were all in good health. One older brother, an accountant, had had "some fainting spells" when trying out for the wrestling team years ago. His mother was in good health; he knew nothing about his father, who had died suddenly when Leonard was 2 years old. The school physician made appointments for him to have an electrocardiogram and an echocardiogram. Despite the doctor's advice to "take it easy" until he had his appointments, Leonard practiced with the team the following day.

Again, after a particularly fast run and layup, Leonard suddenly collapsed—and died.

Cell Biology, Diagnosis, and Treatment of Idiopathic Subaortic Cardiac Hypertrophy (Braunwald's Disease)

Leonard had idiopathic hypertrophic subaortic stenosis, or Braunwald's disease. The scheduled electrocardiogram would have shown left ventricular hypertrophy, probably with Q-waves in the left precordial leads, and an electrically normal right ventricle. The echocardiogram would have shown a hypertrophied left ventricular outflow tract and an especially disproportionate increase in the thickness of the ventricular septum. Dynamic echo studies would likely have shown that the anterior leaflet of the mitral valve intermittently blocked the narrowed and hypertrophied ventricular outflow tract.

Biochemical-genetic studies would have disclosed abnormalities in the gene for β-myosin on chromosome 14. Histologic examination of Leonard's heart at autopsy showed disorganized and fractured myofibrils and hypertrophied intramuscular arteries and arterioles.

Unfortunately, sudden death, especially after intense exercise, is often the first sign of this disease. The mechanism is not clear, although paroxysmal arrhythmias induced by sudden outflow obstruction are likely to be involved. For others, symptoms include fatigue, syncope, chest pain on exertion, cardiac arrhythmias, and most frequently, dyspnea. Dyspnea occurs because of decreased filling and elevated diastolic pressures in the noncompliant hypertrophied heart.

Many therapeutic maneuvers including β-blockers, calcium channel blockers, amiodarone-type antiarrhythmic agents, electrical pacing if atrial fibrillation develops, and even ethanol infarction or myomectomy of the septum sometimes relieve symptoms. Unfortunately, neither the incidence nor the management of symptoms appears to be related to the risk of sudden death, which can occur at any time.

SIGNALING BY THE RENIN-ANGIOTENSIN-ALDOSTERONE SYSTEM

The most active component of the RAAS is the octapeptide angiotensin II (Ang II), which is obtained by the cleavage of angiotensinogen by angiotensin-converting enzyme (ACE). Two receptors, AT1 and AT2, mediate Ang II signaling and belong to the superfamily of G-protein–coupled receptors. Most of the functions attributed to Ang II are via signaling through the AT1 receptors. Less is known about AT2 receptor function, but it appears to exert a counterbalancing effect on AT1. Ang II triggers a strong intracellular Ca^{2+} transient that is significant not only for the excitation-contraction coupling, but also for initiating the signal transduction pathways that are Ca^{2+} dependent. One of the better characterized signaling pathways through AT1 receptors involves $G_{\alpha q}$ that activates PLC, triggering formation of inositol 1,4,5-phosphate, which causes the release of Ca^{2+} from the SR by activating the IP_3 receptor (IP_3R) and formation of DAG. RAAS activates several signal transduction pathways associated with cardiac cell hypertrophy (the increase in cardiac cell size due to workload); prominent among them are the MAPK, the Janus kinase/signal transducers and activators of transcription (Jak/Stat), and Ca^{2+}/calmodulin (CaM)-dependent calcineurin (see later). RAAS has also been implicated in the proinflammatory response eliciting the cellular events that include production of reactive oxygen species and release of cytokines. ACE not only produces Ang II but causes the breakdown of bradykinin into inactive peptides. Bradykinin is angiogenic and a vasodilator and stimulates the production of nitric oxide. The RAAS is the most potent vasopressor mechanism that has been studied in relation to cardiovascular disease.

SIGNALING BY THE JAK/STAT PATHWAY

The tyrosine Janus kinases (Jak1, Jak2, Jak3, and Tyk2) and the signal transduction and activators of transcription (STAT) proteins (Stat1, Stat2, Stat3, Stat5, and Stat6) are the main components of the Jak/Stat signal transduction pathway. The C-terminal cytoplasmic domain of the AT1 receptor directly interacts with the tyrosine kinase Jak2, and its phosphorylation mediates activation via phosphorylation of STATs and their translocation into the nucleus. STATs bind to the conserved GAS (γ-interferon–activating sequence) DNA site in the promoter of target genes (Fig. 8–22).

Mobilization of specific STAT proteins is triggered by ligand-receptor interactions, such as Ang II/G-receptor or IL-6/gp130 interactions. Activated STAT proteins are translocated into the nucleus and promote alteration of the gene expression program intrinsic to cell survival, hypertrophy, and/or cell death. STAT3 plays a prominent role in the onset of cardiac hypertrophy by promoting gene expression patterns characteristic of the hypertrophic phenotype. Surprisingly, however, STAT3 plays a role in a cell survival (antiapoptosis) gene expression program as well. The precise mechanism of the antithetical roles played by STAT3 in growth and hypertrophy is not clearly understood. It is known, however, that cytokine-induced STAT3 activation

mediates the transcriptional induction of the prohormone angiotensinogen and, consequently, the activation of the RAAS. Jak2 activation is coupled to the transcriptional activation of the prohormone angiotensinogen. The regulatory region of the angiotensinogen promoter contains the GAS cis-element that serves as the binding sites for STAT proteins. STAT1, in contrast, is known to target the promoter of the proapoptotic Bax gene, and as such, it is associated with apoptosis in the heart. The mechanism(s) that balances the proapoptotic and cell survival programs mediated by activated STATs is not known. The complexity of RAAS/Jak2 signaling is evident from the finding that removal of that part of the AT1 receptor that binds G proteins results in lack of receptor coupling to G proteins and leads to activation of other signaling pathways, such as the Src and ERK pathways, and produces a notable cardiac hypertrophy. When mice lacking the $G_{\alpha q}$ and $G_{\alpha i}$ receptors were subjected to pressure overload, the onset of left ventricular hypertrophy occurs, albeit with less apoptosis and extracellular deposition compared with the wild-type AT1 transgenic mice.

CALCIUM/CALMODULIN SIGNAL TRANSDUCTION

In many organisms, the extracellular concentration of Ca^{+2} is almost 10^4 times the concentration within cells. Ca^{+2}-based signaling takes advantage of this concentration differential, placing it under the control of a primary signal transducer, such as a neurotransmitter or a hormone, that triggers the influx of Ca^{+2} into the cell in a signal-dependent way. Because it is used to transduce the primary extracellular signal, the signal-dependent influx of Ca^{+2} is often referred to as a second messenger.

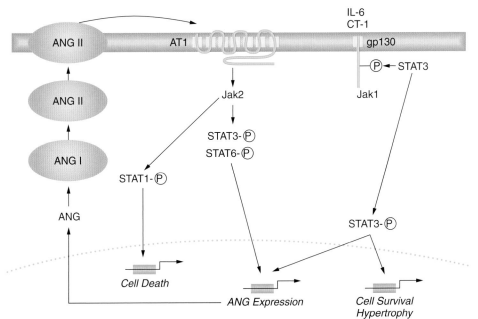

Figure 8–22. Model for agonists induced Jak/STAT signal transduction pathway. Agonist binding to the respective receptors triggers the tyrosine phosphorylation and activation of tyrosine kinase Jak2, which associates with the AT1 receptor and activates downstream signaling components such as STATs. Activated STATs are heterodimerized and translocated into the nucleus, where they activate transcription of target genes.

The signal-dependent influx of Ca^{+2} elevates the intracellular Ca^{+2} concentration to levels that can activate numerous Ca^{+2}-regulated enzymes, such as protein kinases and phosphatases, phospholipases, proteases, and endonucleases. Because of these wide-ranging effects, the influx of Ca^{+2} into cells can, if left unchecked, lead to Ca^{+2} overload, which results in the deleterious hyperactivation of enzymes and cell death. To avoid this, a number of Ca^{+2}-buffering and Ca^{+2}-sensing proteins have evolved that modulate, buffer, and direct the Ca^{+2} signal to make it more specific. These proteins, along with Ca^{+2} channels, impart specificity on the Ca^{+2} signal by controlling its route of entry and its intracellular location. An additional means of imparting specificity on the Ca^{+2} signal involves the ways Ca^{+2} enters and remains in cells. Recent studies have shown that Ca^{+2} enters cells in "waves" or puffs, causing intracellular levels to fluctuate at a defined frequency and amplitude. This frequency and amplitude apparently can encode specific signaling information that can be differentially acted upon (e.g., by differential activation of certain genes).

One of the most important functions of calmodulin is chaperoning Ca^{+2} to specifically activate three Ca^{+2}/calmodulin-dependent enzymes, each of which has important roles in a variety of cell types: Ca^{+2}/calmodulin-dependent protein kinase (CaM kinase or CaMK), protein phosphatase 2B (calcineurin), and myosin light chain kinase (MLCK, found only in cardiac cells). CaMKs consist of CaMKI, -II, and -IV and are ubiquitous mediators of Ca^{+2} signaling (Fig. 8–23). CaMKs can phosphorylate multiple substrates and regulate numerous cellular functions. One major cellular function regulated by CaMKs is Ca^{+2} signal-dependent transcription. Both Ca^{+2} and CaM can translocate to the nucleus, where they activate a nuclear isoform of CaMKII, CaMKIIδ_B, to phosphorylate and activate certain widely used transcription factors. As with the activation of specific enzymes, the specificity of gene activation also appears to be a function of the amplitude, frequency, source, and subcellular localization of Ca^{+2} signals.

Another Ca^{+2}/calmodulin-regulated effector is protein phosphatase 2B, or calcineurin, which is a serine/threonine protein phosphatase activated by sustained elevations in intracellular Ca^{+2}. Unlike kinases, phosphatases remove phosphate groups from serine and threonine residues. One major function of calcineurin evident in Ca^{+2}-regulated cell types such as cardiac cells and neurons is the regulation of the inositol-3-phosphate receptor (IP_3R). IP_3Rs are intracellular Ca^{+2} release channels expressed on the endoplasmic reticulum. They respond to the IP_3 generated from activation of G protein-coupled receptors on the plasma membrane by releasing Ca^{+2} from internal stores and facilitating the opening of plasma membrane-bound Ca^{+2} channels and the inward rush of Ca^{+2}. As with CaMKII, calcineurin

also appears to mediate Ca^{+2}-dependent gene expression, but unlike CaMKII, calcineurin achieves this by remaining in the cytoplasm, where it controls the cytoplasmic to nuclear transit of NFAT, a widely used transcription factor, via its phosphorylation state (see the following). From these examples, the importance of calmodulin in buffering, modulating, and directing the Ca^{+2} signal in cells cannot be underestimated. Indeed, perturbation of the Ca^{+2}-sensing/Ca^{+2}-buffering system, in general, has been shown to lead to the enlargement of cardiomyocytes, as seen in cardiac hypertrophy, and the cytotoxic death of neurons, as occurs in neurodegenerative diseases and instances of neuronal loss (e.g., stroke).

SIGNALING BY THE CALCINEURIN/NFAT PATHWAY

Another signaling pathway activated during maladaptive cardiac hypertrophy (see later) via G-protein–coupled receptor in a Ca^{2+}-dependent manner is the calcineurin/NFAT pathway. On stress, the $G_{\alpha q}$ and $G_{\alpha 11}$ subunits of G receptors recruit PLC to the membrane, mediating the hydrolysis of phosphatidylinositol 4,5 biphosphate and releasing IP_3 and DAG. IP_3 triggers the release of Ca^{2+} from the SR with a consequent increase in cytosolic Ca^{2+}. Calmodulin activates the protein phosphatase calcineurin, which dephosphorylates (activates) the cytoplasmic transcription factor NFAT, facilitating its translocation to the nucleus and promoting transcriptional activation of hypertrophic genes. NFAT also interacts with other transcription factors such as MEF2 and GATA4 to promote the transcriptional response. DAG activates PKC, which targets the corepressor histone deacetylases (HDAC 5 and 9) for phosphorylation, resulting in inactivation of HDACs and their export out of the nucleus.

SIGNALING BY ION CHANNEL RECEPTORS

The transition from left ventricular hypertrophy to heart failure is characterized by impaired removal of cytosolic calcium, reduced loading of the cardiac SR, and defective calcium release, culminating in impairment of cardiac diastolic and systolic functions. The precise regulation of Ca^{2+} channel receptors (RyR2) by the RAAS and the sympathetic response is key to the mechanism by which extracellular and intracellular Ca^{2+} levels are modulated. A maladaptive cardiac cell response to hypertrophic stimuli that interferes with the increase of intracellular Ca^{2+} during systole or its decrease during diastole is observed during the transition from left ventricular hypertrophy to heart failure. The increase of cAMP, as a consequence of the initial activation mediated by the sympathetic response via the β_1 receptor,

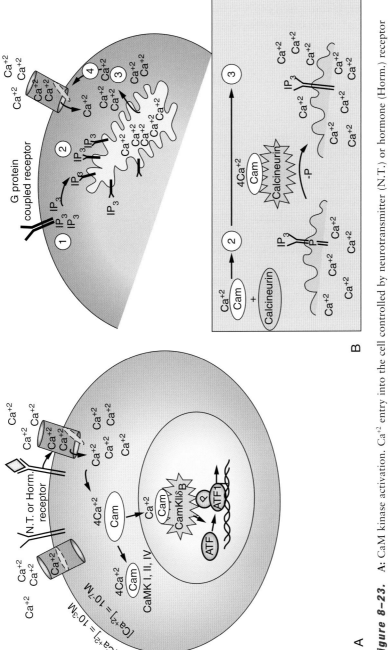

Figure 8–23. **A:** CaM kinase activation. Ca^{+2} entry into the cell controlled by neurotransmitter (N.T.) or hormone (Horm.) receptor activation increases the intracellular Ca^{+2} concentration. Ca^{+2} ions are bound by calmodulin (Cam), and the Ca^{+2}/calmodulin complex activates kinases such as CamKI, -II, and -IV. Ca^{+2}/calmodulin can also translocate to the nucleus and activate a nuclear isoform of CaMKII, CaMKIIδ$_B$, to phosphorylate and activate certain widely used transcription factors (e.g., ATF1). **B:** Calcineurin activation. Ca^{+2}-dependent activation of calmodulin (as in Panel A) leads to activation of the phosphatase calcineurin. Activation of G-protein coupled receptors causes formation of IP$_3$ (1) that binds to and activates IP$_3$ receptors (2) on the endoplasmic reticulum (ER). Calcineurin dephosphorylates the IP$_3$ receptor, facilitating release of Ca^{+2} stored in the ER into the cytosol (3), and subsequent activation of plasma membrane-bound Ca^{+2} channels (4).

triggers activation of PKA. PKA phosphorylates and triggers the release of calstabin 2 from the ryanodine receptor RyR, facilitating the opening of the RyR channels and increase in intracellular Ca^{2+} during systole. Signaling via the AT1 receptor activates PLC, triggering the formation of inositol 1,4,5-phosphate, which causes release of Ca^{2+} from the SR. DAG produced via the signaling mediated by AT1 receptor activates PKC, which phosphorylates the Na^+-H^+ exchanger, leading to a Na^+ uptake and stimulation of the Na^+/Ca^{2+} exchanger that further increases intracellular Ca^{2+}.

SIGNALING IN MYOCARDIAL HYPERTROPHY

Left ventricular hypertrophy is the single most important contributor to cardiovascular morbidity and mortality. Several factors contribute to the development of left ventricular hypertrophy; among them are hypertension, coronary artery disease, valvular pathologies, obesity, or a combination of these factors. One hypothesis is that hypertension leads to left ventricular hypertrophy followed by ventricular dilation and contractile dysfunction. The initial response to physiologic workload is a compensatory one that triggers changes in contractility and ventricular remodeling. However, when the heart is subjected to sustained mechanical overload, it leads to a maladaptive response characterized by an increase in oxygen and energy consumption, chamber dilation, reduced contractility, and eventually to heart failure (Fig. 8-24).

It is believed that at the cellular level both the compensatory and maladaptive responses involve activation of multiple second messengers, altered ion balance, activation of the signaling pathways described earlier, and specific changes in gene expression. The signaling pathways involved in the maintenance of the excitation-contraction coupling events represent the main area of study in several laboratories investigating new strategies for therapeutic approaches to combat the deleterious cardiovascular effects during the transition from adaptive to maladaptive left ventricular hypertrophy.

SUMMARY

Signaling between cells is essential for the development and function of multicellular organisms, and its disruption is a leading cause of many disease states. How a cell receives, acts on, and alters its behavior in response to a signal from another cell is called *signal transduction*. Processing of that signal into a genetic response is conducted via intracellular signal transduction pathways of which there are many. Among these are the catalytic receptor class of signal transducers that include receptor tyrosine kinases and serine-threonine kinase receptors that transduce extracellular signals via a kinase-dependent cascade of protein phosphorylation events. Other pathways transduce signals by controlling the activity of transcriptional repressors and activators via proteolytic cleavage of key regulatory proteins. A third category of signaling is via nonpeptide hormones and their receptors that act as ligand-activatable transcription factors. In addition to these signal transduc-

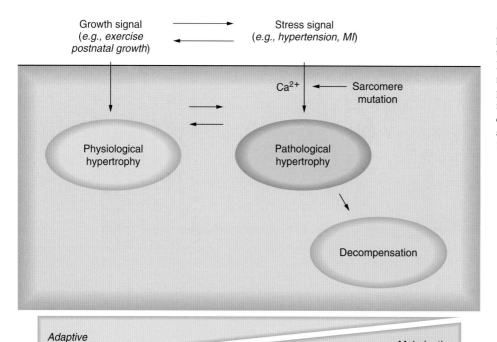

Figure 8–24. Physiologic and pathologic hypertrophy in response to growth and stress signals. Physiologic response of the myocardium (adaptive) and pathologic response (maladaptive) involve activation of signaling cascades. (Adapted from Frey, et al. *Circulation* 2004;109:1580–1589, by permission.)

tion pathways, there are other signaling pathways that play important roles in the adult and are often involved in the cause of disease states, particularly that of heart disease. These include signaling by G-protein–coupled receptors, by the RAAS, the Jak/STAT pathway, the calcineurin/NFAT pathway, and ion-channel receptors. Examples of how all these signaling pathways contribute to the development, function, and dysfunction of organ systems can be seen in the heart. Understanding how these pathways mediate cell communication will shed some light on the underlying molecular cause of heart and other disease states and provide a platform for devising future therapies.

Clinical *Case 8-2*

Susan Peters is a 27-year-old professional tennis player who was in excellent health until she had a serious fall during a match, tearing multiple ligaments in her left knee. She immediately underwent extensive reconstructive surgery at an athletic orthopedic center. The surgery went well technically, and her surgeon believed that after a week's bed rest she would be weight bearing and, with adequate physical therapy, be back on the courts within 3 months. However, the surgical trauma was more extensive and painful than expected, and she was not able to get out of bed for 5 days. Then, when she tried to walk, she told the nurse assisting her that her right leg was painful, and she had to return to bed. The nurse noted she was flushed and found that she had a fever of 39°C.

When her orthopedist saw her on his evening rounds, he found that her right calf was hot, swollen, and very tender, and that the tenderness extended to her inner thigh. She had a positive Homan's sign. His examination of her chest showed normal breath sounds and a soft midsystolic murmur most prominent at the base of the heart. He immediately diagnosed a deep vein thrombosis and ordered intravenous (IV) heparin from the pharmacy. As he was leaving, he elicited the further history that Susan had been taking birth control pills for the past 3 years.

A few minutes later, before the IV heparin could be administered, Susan coughed, cried out with sudden pain in her chest, and developed acute dyspnea and an increased pulse of 128. The nurses were alarmed and re-called the orthopedist. When he arrived, 20 minutes later, the patient's dyspnea had subsided somewhat, and he was about to administer morphine for her chest pain. However, while she was complaining of the pain, he recognized that her speech was slurred, and she started to complain that her right arm was weak and numb. The orthopedist canceled the heparin order and called for an emergency hematology consult.

Cell Biology, Diagnosis, and Treatment of Paradoxical Embolus after Deep Vein Thrombosis

This young woman did indeed have a deep vein thrombosis subsequent to her orthopedic surgery and bed rest, perhaps influenced by her estrogen-containing birth control medicine. Furthermore, she had acute pulmonary embolization from that thrombosis, which produced the dyspnea and chest pain when emboli lodged in the terminal circulation of her lungs.

However, the most unusual part of her illness was that she also had paradoxical embolization of a small fragment of her leg clot that was not trapped in the lung, but rather passed through a congenital developmental defect in the septum between her right and left atria, referred to as ASD. It then entered the arterial circulation and lodged in a branch of her left midcerebral artery. It is likely that the pain and coughing from her initial embolization induced a momentary right-to-left shunt between her atria to select that small embolic fragment for its paradoxical trip through the circulation.

Fortunately, the resulting stroke was immediately recognized, and the hematologist began prompt thrombolytic therapy with recombinant tissue plasminogen activator (tPA). Her dysarthria abated the next day, and sensation and strength returned to normal in her right arm 3 days later. The tPA was discontinued and the patient was switched to heparin. Her deep vein thrombosis resolved in about 10 days. Two weeks later, she was discharged on long-term coumadin anticoagulation pending surgical correction of her *fossa ovalis* midseptal atrial defect.

One year and several months of intense physical therapy later, she was again playing tennis.

Suggested Reading

Tyrosine Kinases

Marmor MD, Kochupurakkal BS, Yarden Y. Signal transduction and oncogenesis by ErbB/HER receptors. *Int J Radiat Oncol Biol Phys* 2004;58:903–913.

Thisse B, Thisse C. Functions and regulations of fibroblast growth factor signaling during embryonic development. *Dev Biol* 2005;287:390–402.

Serine/Threonine Kinases

Shi Y, Massague J. Mechanisms of TGF-β signaling from cell membrane to the nucleus. *Cell* 2003;113:685–700.

Nonkinase Signal Transduction

Bijisma MF, Spek A, Peppelenbosch MP. Hedgehog: an unusual signal transducer. *BioEssays* 2004;26:387–394.

Schweisguth F. Regulation of Notch signaling activity. *Curr Biol* 2004;14:R129–R138.

Wodarz A, Nusse R. Mechanisms of Wnt signaling in development. *Annu Rev Cell Dev Biol* 1998;14:59–88.

Steroid Hormone Signaling

Nemere I, Pietras RJ, Blackmore PF. Membrane receptors for steroid hormones: signal transduction and physiological significance. *J Cell Biochem* 2003;88:438–445.

Signaling in Cardiac Development

Zaffran S, Frasch M. Early signals in cardiac development. *Circ Res* 2002;91:457–469.

G-Protein–Linked Receptors

Rockman HA, Koch WJ, Lefkowitz RJ. Seven-transmembrane-spanning receptors and heart function. *Nature* 2002;415:206–212.

Jak STAT Signal Transduction

Rawlings JS, Rosler KM, Harrison DA. The JAK/STAT signaling pathway. *J Cell Sci* 2004;117:1281–1283.

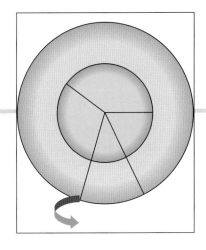

Chapter 9

The Cell Cycle and Cancer

CELL CYCLE: HISTORY

The adult human body is composed of approximately 200,000 billion (2×10^{14}) cells, all of which derive from a single cell, the fertilized egg. In adults, there are sub-populations of cells that continuously divide or retain the capacity to divide to replace cells that die or are otherwise lost. The process of cell multiplication and division requires multiple intricately regulated steps. The cells must increase in size, precisely duplicate their DNA and chromosomes, and segregate their chromosomes between the two daughter cells so that each daughter cell acquires an exact copy of the parental chromosome complement. Each of these processes is coordinately regulated as the cells proceed through their cell cycle. In 2001, the Nobel Prize in Physiology or Medicine was awarded to Leland Hartwell, Paul Nurse, and Tim Hunt, three pioneers of cell biology, all of whom made seminal discoveries using a combination of genetic and molecular biology approaches to uncover the mechanisms that regulate the cell cycle. With contributions from many other deserving investigators, they found that proteins, designated cyclins, and cyclin-dependent kinases (CDKs) drive cells from one cell cycle phase to the next in a regulated fashion. One might compare the CDKs to an engine and the cyclins to the gearbox that determines whether the engine will idle so that the cells remain stationary or whether the engine will be in gear and drive the cells forward through cell cycle.

The mechanisms that control the progress of cells through the cell cycle are highly conserved throughout evolution. Hence, lessons regarding how the cell cycle works have been drawn from a wide spectrum of single-cell and multicellular organisms ranging from yeast to plants to sea urchins to clams to frogs to mammalian cells. Our fundamental understanding of the compartmentalization of the cell cycle is based on a seminal article by Alma Howard and Stephen Pelc published in 1953. Before the 1950s, cell biologists and pathologists recognized only two phases of the cell cycle that were visible by microscopy: interphase and mitosis. Using the broad bean, *Vicia faba*, Howard and Pelc first deduced that the interphase could be divided into three compartments or phases. Using a labeling approach with ^{32}P and autoradiography, they demonstrated that DNA synthesis occurred during a discreet interval of interphase that they designated the S phase (synthesis). They also demonstrated that there was a gap between the end of mitosis and the start of the next S phase, and also between the end of the S phase and the start of the ensuing mitotic division. These gaps became known as G1 and G2, respectively (Fig. 9–1).

By the end of the 1960s, Hartwell had recognized the power of genetics for dissecting the cell cycle. As a model system, he used baker's yeast, *Saccharomyces cerevisiae*, which proved highly amenable for cell-cycle analysis. In an elegant series of experiments, he isolated yeast cells in which genes controlling the cell cycle were conditionally mutant, allowing cell proliferation at a

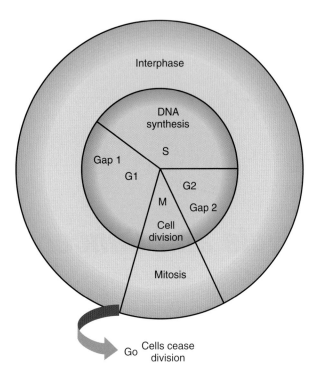

Figure 9–1. Schematic depicting the different phases of the cell cycle.

permissive temperature, but interfering with proliferation at an increased, nonpermissive temperature. By this approach, he successfully identified more than 100 genes directly involved in cell-cycle control, which were designated CDC genes (cell division cycle genes). One of these genes, *cdc28* in *S. cerevisiae,* is a CDK that controls the first step in the progression through the G1 phase of the cell cycle, and was therefore also called *start.*

Another important concept in cell-cycle regulation, also introduced by Hartwell, is that of cell-cycle checkpoints, which are activated after stress or environmental exposure. In addition to dissecting the genetics of cell-cycle progression in unperturbed cells, Hartwell studied the sensitivity of yeast to DNA damage by ionizing radiation. Based on his observations that cells are transiently arrested when their DNA is damaged, he coined the term *checkpoint* to underscore the notion that the cells arrest, repair the damage, and then proceed to the next phase of the cell cycle.

Paul Nurse used an approach similar to that of Hartwell's but took advantage of a different type of yeast, *Schizosaccharomyces pombe,* as a model organism. This yeast is distantly related to baker's yeast, having separated early during evolution. In the mid-1970s, Nurse identified the *cdc2* gene in *S. pombe* and showed that it played a key role in the transition from G2 to mitosis (M). He later discovered that it participates in additional cell-cycle transitions, that it is identical to the *cdc28* ("start") gene identified earlier by

Hartwell in baker's yeast, and that it controls the transition from G1 to S. Subsequently, Nurse isolated the corresponding gene from humans, which encodes a CDK and was later designated CDK1 (cyclin-dependent kinase 1). Nurse then showed that CDK1 activation is dependent on reversible phosphorylation (i.e., its reversible modification by the addition or removal of phosphate groups). Since this initial finding, several different human CDKs have been described.

The CDKs are activated by forming complexes with one of several cyclins. The cyclins were discovered serendipitously by Tim Hunt as part of the physiology course at the Marine Biological Laboratories (MBL) at Woods Hole in the early 1980s. He noticed that during synchronous cleavage divisions of the sea urchin *Arbacia punctulata* embryo, a specific protein was destroyed at each cell division and resynthesized in the next cycle. He coined the term *cyclin* to reflect the recurring sequential synthesis and degradation of this protein. His finding was subsequently confirmed by Joan Ruderman in the cleavage embryos of the clam and later in mammalian cells, where there are multiple cyclins that interact with different CDK molecules at different times of the cell cycle. These interactions confer specificity to cell-cycle regulation by defining and selecting the regulatory proteins that are activated or deactivated by phosphorylation. Notably, most of the important principles underlying cell-cycle regulation derive from studies on plants, two evolutionarily diverged yeast, sea urchin embryos, clam embryos, and frogs, reinforcing the importance of nonmammalian systems for discovering and analyzing fundamental biologic processes.

One of the hallmarks of cancer is uncontrolled cell proliferation. In most, if not all, human tumors, one or more of the cell-cycle checkpoints is compromised such that cells proliferate in an unregulated manner. A detailed but still incomplete understanding of the mechanisms that coordinate events that govern cell proliferation has come from molecular analyses of human tumors, normal tissues, and animal models. In addition to the cyclins and the CDKs, there are numerous proteins and pathways that regulate cell cycle kinetics, many of which are categorized as oncogenes and tumor suppressors. Oncogenes are mutant versions of normal genes (protooncogenes) whose products are important in facilitating normal cell-cycle progression. When they are mutant, the oncogenes promote uncontrolled growth. Tumor suppressors serve to restrain uncontrolled proliferation, so that when they are mutant or lost, the cells will proliferate abnormally. Thus, oncogenes can be viewed as accelerators of cell proliferation, and tumor suppressors as the brakes. Oncogenes are generally activated by activating mutations, such as those found in the *Ras* genes. In contrast, tumor suppressor gene function is lost by mutation, by epigenetic modification (e.g., DNA methylation), and/or by loss of heterozygosity (LOH) as a consequence of mitotic recombination (recombination between chromosome homologs), deletion, or chromosomal loss.

THE CELL CYCLE IS REGULATED BY CYCLIN AND RELATED PROTEINS

Cyclins

The cyclins are a family of proteins that are centrally involved in cell-cycle regulation and are structurally identified by conserved "cyclin box" regions. The cyclin boxes are composed of about 150 amino acid residues, which are organized into 5 helical regions and are important in binding partner proteins, including the CDKs. More than 20 cyclins or cyclin-like proteins have been identified, many of which have no known function. Those whose functions have been defined are about 56 kDa in size and play critical roles in allowing the progression of cells through all phases of the cell cycle, including mitosis (Fig. 9–2).

Cyclins are the regulatory subunits of holoenzyme CDK complexes that control progression through cell-cycle checkpoints by phosphorylating and inactivating target substrates. The cyclins associate with different CDKs to provide specificity of function at different times during the cell cycle (see Fig. 9–2). The involvement of aberrant cyclin expression with cancer was first realized when a chromosome breakpoint common in B-cell lymphomas was cloned and shown to encode cyclin D1. At about the same time, cyclin D1 was independently cloned at a chromosome 11 inversion junction, common in parathyroid adenomas, that juxtaposed parathyroid hormone gene regulatory elements with the body of the cyclin D1 gene. Subsequently, overexpression of several cyclins, particularly cyclin E, has been associated with cell-cycle deregulation in tumors. Cyclin E overexpression can shorten the time spent in G1 and allow premature entry of cells into the S phase. Cyclin E is also important in facilitating centrosome replication, the precise regulation of which is important in maintaining chromosome stability. Figure 9–2 shows the principle cyclins involved in regulated cell-cycle progression and the CDKs with which they partner at dif-ferent times of the cell cycle, including cyclins A, B1, D1, D3, and E. Consistent with the notion that disruption of regulatory mechanisms that govern cell proliferation leads to tumorigenesis, several of the cyclins are frequently overexpressed in tumors.

Cyclin-Dependent Kinases

The CDKs are the catalytic subunits of the cyclin/CDK complexes. About 20 CDKs or CDK-like proteins have been identified; the functions and substrates for some CDKs remain unknown. The CDKs have a common cyclin-binding domain characterized by a consensus sequence of seven amino acids (PSTAIRE). They are serine/threonine kinases that have the capacity to phosphorylate multiple substrates, including the retinoblastoma (Rb) proteins and the E2F transcription factor. Progression through G1 is regulated by sequential involvement of at least three CDKs: CDK4, CDK6, and CDK2 (see Fig. 9–2). Among other activities, CDK4 in complex with cyclin D appears to mediate phosphorylation of the Rb family of proteins. Further phosphorylation of the retinoblastoma protein (pRb) is mediated by cyclin D/CDK6, and finally by cylinE/CDK2. These sequential phosphorylations, coupled with other cellular activities, allow the ordered progression of cells through the G1 phase. The activity of cyclinE/CDK2 is essential for the transit of cells from G1 into S phase and for the duplication of centrosomes whose function is required for proper chromosome segregation at mitosis. Progression through S phase is mediated principally by cyclinA/CDK2. During S phase, cyclinA/CDK2 phosphorylates numerous proteins involved in transcription and DNA replication and repair, as well as proteins thought to be necessary for the completion of S phase and the entry of cells into G2. Among the proteins phosphorylated by this complex are the transcription factors E2F1 and B-Myb, proteins such as the DNA single-strand binding protein RPA that is involved in DNA replication, and proteins that participate in DNA repair such as BRCA1, BRCA2, and Ku 70. As cells complete the S phase and begin to transit into G2, the A-type cyclins associate with CDK1 (also designated Cdc2 for historical reasons). As cells proceed through G2, cyclin A is degraded by a ubiquitin-mediated proteolytic mechanism as cyclin B is actively synthesized. During mid-to-late G2, cyclinB/CDK1 complexes form and play multiple essential roles during the G2/M transition and during the progression of cells through mitosis (see Fig. 9–2). At least 70 proteins have been identified as substrates for the cyclin B/CDK1 complexes. In the cytoplasm, cyclin B/CDK1 complexes associate with centrosomes during prophase (see later) and facilitate centrosome separation by phosphorylating a centrosome-associated motor protein. These complexes also participate in chromosome condensation by phosphorylating a subset of histones, in the fragmentation of the

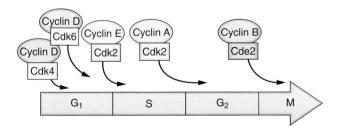

Figure 9–2. The association of cyclin-dependent kinases (CDKs) with their respective partner cyclins during different phases of the cell cycle.

Golgi apparatus during prophase, and in the breakdown of the nuclear lamina by phosphorylating the lamin B receptor. Inactivation of this same complex is also important for the exit of cells from mitosis into G1. This inactivation is achieved by ubiquitination of cyclin B and its proteolytic degradation mediated by the anaphase-promoting complex (APC), a ubiquitin ligase.

Cyclin-Dependent Kinase Inhibitors

If kinases remained constitutively active and if kinase substrates remained permanently phosphorylated, the cell-cycle regulatory circuit would be compromised. It is not surprising, therefore, that there are naturally occurring inhibitors of the CDKs. There are at least two distinct families of CDK inhibitors (CKIs). The prototype of the CIP/KIP family is p21, which was identified almost simultaneously by three different groups and consequently had multiple separate designations. It is now most commonly referred to as p21, or p21$^{waf1/cip1}$, which is reflective of its molecular weight. In addition to p21$^{waf1/cip1}$, this family of kinase inhibitors includes related proteins p27 and p57, which are all capable of binding to and forming complexes with CDK1/cyclin B, CDK2/cyclin A, CDK2/cylin E, CDK4/cyclin D, and CDK6/cyclin D and inhibiting the kinase activity of these complexes. The prototypical member of the second CKI class is p16, one of the INK4 class of proteins, which is dependent on Rb or one of the related proteins, p107 or p130. Other members of this class of CKIs, based on structure and function, include proteins designated p15, p18, and p19. The designation of this family of kinase inhibitors commonly contains a superscript "INK4" (e.g., p16^{INK4}). Unlike the p21 class of kinase inhibitors, the INK4 proteins bind exclusively to CDK4 and CDK6, prevent the association of these CDKs with their regulatory cyclins, and thereby inhibit their kinase activities.

Cdc25 Phosphatases

A second strategy to ensure that the phosphorylation state of proteins is reversible involves the activity of phosphatases with specificity for defined substrates. An example of phosphatases that participate in cell-cycle regulation is the Cdc25 family of phosphatases. Their primary function in cell-cycle control is to dephosphorylate the various phosphorylated CDKs to allow sequential passage through the phases of the cell cycle. For example, CDK2 is phosphorylated on threonine 14 and tyrosine 15, and when so phosphorylated, the cells do not transit from G1 into S phase. When Cdc25A dephosphorylates CDK2 at phosphotyrosine 15, the cells will enter S phase. Similarly, CDK1 is phosphorylated at the same two residues by the specific kinases, wee1 and CDK-activating kinase. Unless Cdk1 is dephosphorylated at tyrosine 15, the cells arrest at the G2/M bound-

ary and do not progress through mitosis. The Cdc25C phosphatase is instrumental in dephosphorylating CDK1, thereby allowing the cells to progress through mitosis. Notably, in the absence of Cdc25C, its close relative Cdc25A can assume this activity.

p53

One of the tumor suppressors most commonly mutated or lost in human tumors is p53, which is mutant in more than 50% of all human tumors. The p53 protein is a key regulator of the G1/S and G2/M checkpoints. Its functions are so important that it has been called the "guardian of the genome" and "the cellular gatekeeper for growth and division." The amino-terminal sequences of p53 serve as a transcriptional activation domain, and the carboxy-terminal sequences appear to be required for p53 to form homodimers and homotetramers. The p53 protein activates transcription of numerous genes that participate in the control of the cell cycle, including those encoding *p21$^{waf1/cip1}$*, *GADD45* (a growth arrest, DNA damage-inducible gene), and MDM2 (a protein that is a known negative regulator of p53). One of the essential roles performed by p53 is to participate in the arrest of cells in G1 after genotoxic damage. Presumably, this arrest provides opportunity for DNA repair to occur before DNA replication and cell division. If the damage is sufficiently great, p53 triggers cell death through an apoptotic pathway instead of allowing severely damaged cells to progress through the cell cycle. Tumor cells that lack normal p53 function do not arrest efficiently in G1 and are more likely to progress into S or G2/M. Occasionally, some of these damaged cells will escape cell death and contribute to tumor progression.

The p53 protein, when activated by appropriate phosphorylation, can function as a transcription factor that in turn activates many target genes. In the context of cell-cycle checkpoints, one of its most important targets is the *p21$^{Waf1/Cip1}$* gene. After damage to DNA, p53 is activated by phosphorylation on specific amino acids, which allows it to induce transcription of the *p21$^{Waf1/Cip1}$* gene and results in increased levels of the p21 messenger RNA (mRNA) and protein. The p21 protein, in turn, binds to the cyclin E/CDK2 and cyclin D/CDK4 or cyclin D/CDK6 complexes and inhibits the kinase activity of the CDKs. One target of CDK2 is the pRb protein, another prominent tumor suppressor whose phosphorylation by CDK is required for progression from G1 to the S phase. Thus, by binding to the CDK/cyclin complex, *p21$^{Waf1/Cip1}$* inhibits phosphorylation of pRb and produces a G1/S-phase arrest.

pRb

As indicated earlier, the retinoblastoma (pRb) protein is another major tumor suppressor that is mutant in almost half of all human tumors. Its name derives from the childhood retinal tumor, retinoblastoma. In 40% of the cases, the disease is hereditary, and most patients have bilateral tumors. The remaining cases are mostly sporadic, and the patients usually have tumors only in one eye. Cytogenetic studies had shown that deletions within chromosome 13, particularly in the long arm, are common in patients with hereditary retinoblastoma who inherit one defective chromosome 13. In the early 1970s, this observation led Alfred Knudson to propose the "two-hit" theory of cancer, for which he subsequently received the Lasker Award. This theory argues that when an individual inherits a mutant tumor suppressor allele, only one additional (or second) "hit" is required to facilitate tumorigenesis and tumor progression. The second hit can be in the form of an independent mutation in the functional allele or as loss of the functional allele because of mitotic recombination (recombination between chromosome homologues) and resultant LOH. This hypothesis has been validated for retinoblastoma and for several other hereditary and nonhereditary cancers. In nonhereditary cases, the first "hit" occurs in a somatic cell and is followed by a second "hit" in the same cell at a later time.

Clinical Case 9–1

Priscilla Waxman is an 18-month-old white baby girl. She was in excellent health and had achieved all expected milestones, when, while playing with a new puppy, she scratched her eyelid. Examining this scratch, her mother noted that her left eye seemed to have changed its iris color. When she took the baby for her 18-month checkup 2 weeks later, she mentioned this to the pediatrician. He looked into Priscilla's eyes with an ophthalmoscope and immediately noted that the usual red reflection from her retina was not present, and there was a small white mass behind the lens. He promptly sent Priscilla to an ophthalmologist, who performed several evaluative tests. These included ultrasound and computed tomographic (CT) examinations of both of her orbits, a lumbar puncture for cytology, and an immunologic study for *Toxocara canis* infestation. Priscilla had two brothers, aged 6 and 8, who were in excellent health, as were both her parents.

Cell Biology, Diagnosis, and Treatment of Retinoblastoma

Priscilla had retinoblastoma. The echo and CT studies suggested that there was no extension beyond the eye, and the lumbar puncture showed no abnormal cells. Anticipating radical surgery, the appropriately compulsive ophthalmologist went one step further and ordered magnetic resonance imaging (MRI) of the orbit, which ruled out extension to the optic nerve or orbital soft tissues.

With the report of the negative serology for *Toxocara*, the most important nonmalignant differential alternative was eliminated. Because of the possibility of seeding of malignant cells, the ophthalmologist chose not to perform any biopsy studies and arranged to have her left eye and optic nerve carefully enucleated the following week. At surgery, she also received a plastic orbital implant for cosmesis.

It is likely that Priscilla had the more common, sporadic, nonfamilial form of retinoblastoma. The disease presented unifocally, and only in one eye; there was no family history. Retinoblastoma is the most common ocular tumor in children. It usually presents before the age of 4, and if no extension has occurred, enucleation is likely to be curative. The sporadic form is probably due to a stepwise loss of both alleles of the RB1 tumor suppressor gene found on chromosome 13 in the 14q region. In contrast, the familial form is inherited as an autosomally dominant defect in the same region. There is limited penetrance, but there is still a substantial likelihood of a family history with a risk for nearly 45% for children of a parent with the disease.

The RB1 nuclear protein appears to have a fundamental role as a transcription factor in the regulation and control of cell division and differentiation. It has served as the archetype of a tumor suppressor gene in tumor biology.

The pRb protein has multiple and diverse functions in the cell cycle. One prominent role is as a regulator of progression from G1 to S phase. The pRb protein in its hypophosphorylated state (low level of phosphorylation) binds the cyclin D/CDK4 and cyclin E/CDK2 complexes. The pRb protein, in its hypophosphorylated state, also binds and inactivates the transcription factor E2F, which controls the transcription of a subset of genes whose products are important for the transition from the G1 to S phase. Therefore, in its hypophosphorylated state, pRb inhibits the progression of cells from the G1 into S phase. The pRb is phosphorylated at a subset of sites by cyclin D/CDK4 in early to mid G1 and by cyclin E/CDK2 near the end of G1. Phosphorylation of pRb at its carboxy-terminal region in late G1 by cyclin D/CDK4 releases E2F from its complex, making it available to act as a transcription factor. Hypophosphorylation of pRb allows it to bind promoter regions of E2F target genes; subsequently, its phosphorylation causes it to dissociate from these DNA sequences. The consequence of pRb hyperphosphorylation is transcription of E2F responsive genes whose products are required for the transition from the G1 into S phase (Fig. 9–3).

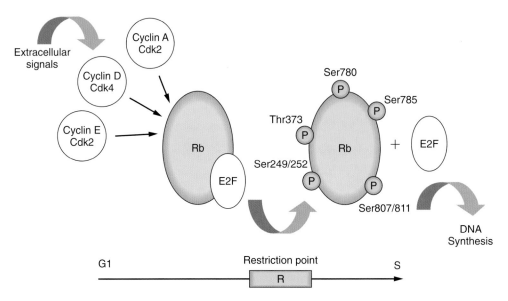

Figure 9–3. Phosphorylation of pRb by cyclin/cyclin-dependent kinase (CDK) complexes, inactivating pRb and releasing the E2F transcription factor.

MITOSIS

What Is Mitosis?

Mitosis is the tightly regulated process of cell division that includes both nuclear division (karyokinesis) and the division of cytoplasm to two daughter cells (cytokinesis). This process can be divided into distinct phases including prophase, prometaphase, metaphase, anaphase, telophase, and finally, cytokinesis. The products of mitosis are two daughter cells that have identical DNA content that is also identical to the DNA content of the original parental cell. Depending on the nature of the parental cell, the daughters may be essentially identical in phenotype, or they may differ. In the case of undifferentiated adult stem cells or progenitor cells, one daughter may remain undifferentiated (self-renewing), whereas the other becomes committed to a differentiated lineage.

Interphase and Mitosis

Until the early 1950s, histologists and cytologists believed that interphase was essentially a period of cell quiescence and that most of the "action" in cells occurred during mitosis. This concept was based on visual images of cells in histological sections. In most cases, more than 95% of cells were in "interphase" and a minority of cells were in various phases of physical cell division (i.e., mitosis). It was not until the seminal experiments of Alma Howard and Stephen Pelc (see earlier) that scientists appreciated the extent of cellular activity that occurred during interphase. This activity involves general cellular metabolism, including replication, tran-

scription, and translation, much of which is preparatory for the mechanics of mitosis and cell division in proliferating cells (see later and also Fig. 9–4).

Mitotic Stages

Prophase

In prophase, the nuclear envelope begins to disaggregate and chromatin in the nucleus begins to condense and becomes visible by light microscopy as elongated, spindly chromosomes. The nucleolus disappears. Centrosomes begin migrating to opposite poles of the cell, and microtubule fibers extend from centrosome to centrosome and from centrosome to the kinetochore of the centromere of each chromosome to form the mitotic spindle.

Prometaphase

In prometaphase, the nuclear membrane disaggregates, marking the beginning of prometaphase. Multiple proteins associate with the kinetochore within the centromeres, and microtubules attached at the kinetochores begin to move the chromosomes to the center of the cell.

Metaphase

As the chromosomes continue to condense and become more compact, they align themselves along a central axis of the cell, which is sometimes referred to as the *metaphase plate.* At this stage, each chromosome is composed of paired chromatids that are the ultimate

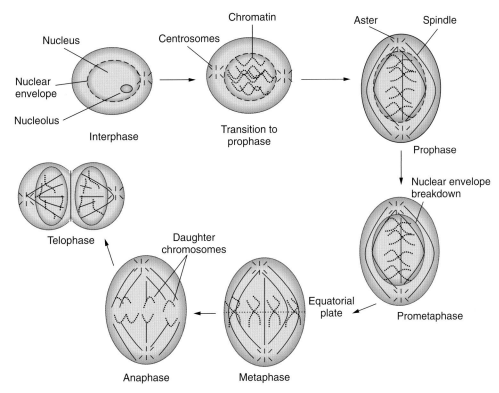

Figure 9–4. Progression of cells through the various phases of mitosis.

product of DNA replication. This alignment and organization of chromosomes helps to ensure that in anaphase, the next phase of mitosis at which chromatids segregate, each daughter nucleus will receive a complete copy of each chromosome.

Anaphase

In anaphase, the paired chromatids separate at the kinetochores by which they were previously tethered and randomly segregate and move to opposite poles of the cell. The migration of the chromatids (now designated chromosomes) results from a combination of kinetochore movement along the spindle microtubules away from the center of the cell and toward the cell poles and the physical interaction of polar microtubules.

Telophase

At telophase, the chromosomes (formerly chromatids) arrive at opposite poles of the cell, and new membranes form around the daughter nuclei. The chromosomes disperse and are no longer visible as discrete chromosomes by light microscopy. The spindle fibers disappear and cytokinesis is initiated.

Cytokinesis

In animal cells, cytokinesis results when a ring of actin fibers encircles the cells roughly equidistant from the

two poles and participates in the cytokinesis process by contracting and constricting the cell into its two daughter cells, each with one nucleus. The newly formed daughter cells are now ready to resume a new cell cycle or to initiate a program of cell differentiation and exit the cell cycle in a G0 state.

Cell-Cycle Checkpoints

The concept of cell-cycle checkpoints was introduced by Leland Hartwell in the 1980s. The principle of a checkpoint is that after damage to a cell, the cell will arrest in a nonrandom manner to allow for repair of the damage before proceeding through the cycle. The various cell-cycle checkpoints are the end result of inhibitory signaling pathways that produce a delay or arrest of cell-cycle progression at specific points in the cell cycle in response to DNA damage. Historically, checkpoints were defined as specific junctures in the cell cycle during which the integrity of DNA is assessed before cells progress to the next phase of the cell cycle. The term *checkpoint* has recently become more ambiguous because it has been applied to the entire ensemble of cellular responses to DNA damage, including the arrest of cell-cycle progression, induction of DNA repair genes, and apoptosis. This expanded definition is not unreasonable because activation of proteins involved in cell-cycle arrest also leads to expression of genes that participate in DNA repair and apoptosis. It is important, however, to appreciate that DNA repair pathways are functional in the absence of damage-induced cell-cycle arrest, and that apoptosis can occur independently of the cell-cycle arrest machinery. Accordingly, the term *DNA damage checkpoint* should be reserved for events that specifically retard or arrest cell-cycle progression in response to DNA damage.

As discussed earlier, the G1/S and G2/M transition points, as well as S-phase progression, are tightly controlled in unperturbed cells. Many of the same proteins involved in regulating the orderly progression through the cell cycle are also involved in checkpoint responses. Thus, the DNA damage checkpoints are not unique pathways activated by DNA damage, but rather are biochemical pathways that operate under normal growth conditions whose effects are amplified after an increase in DNA damage. The principal checkpoints are at the G1/S-phase boundary, an intra-S-phase checkpoint, at the G2/M boundary, and within mitosis.

Molecular Components of the DNA Damage Checkpoints

The DNA damage checkpoints are the end result of complex signaling pathways composed of three principle components: sensors of damage, signal transducers, and effectors. Genetic and biochemical studies in yeast, and more recently in human cells, have identified several proteins involved in sensing DNA damage, in transducing the signals from the sensors, and in carrying out the effector steps of the DNA damage checkpoints (Fig. 9–5). These functional classifications, however, are not absolute because there is considerable overlap in function between each of these components. The damage sensor ATM, for example, also functions as a signal

transducer. Furthermore, a fourth class of checkpoint proteins, sometimes referred to as *mediators*, has been identified that includes BRCA1, Claspin, 53BP1, and MDC1. Conceptually, this class of proteins has been positioned between the sensors and signal transducers. As in the case of ATM, which functions both as a sensor and transducer, these mediator proteins also appear to participate in more than one step of the checkpoint response.

Although the G1/S, intra-S, and the G2/M checkpoints represent distinct checkpoints within the cell cycle, the damage sensor molecules that activate the various checkpoints appear to either be shared by each of the checkpoints or to play a primary sensor role in one or more cases and a backup role in the others. Similarly, the protein kinases and phosphatases that participate as signal-transducing molecules are shared to varying degrees by the different DNA damage response pathways. Although the effector components (proteins that directly inhibit phase transition) of the checkpoints are what define checkpoint specificity, the various sensors, mediators, and signal transducers may play more prominent roles in one checkpoint pathway than in others.

MEIOSIS

Mitosis differs from meiosis in both function and mechanics. The purpose of mitosis is to ensure that both daughter cells receive identical complements of their genomes that are also identical to that of their parent cell. Thus, the DNA complement of a dividing somatic cell goes from 2N (where N is the haploid complement

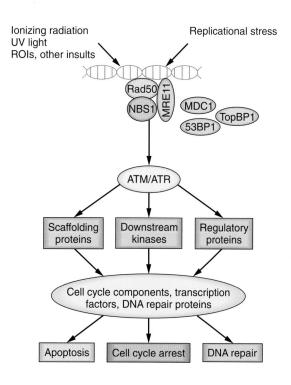

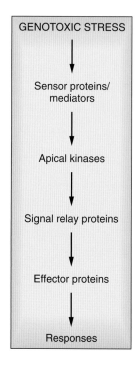

Figure 9–5. **Signaling cascades initiated by damage to DNA.** *(right)* The class of signaling molecule at each level of the cascade. ROI, region of interest; UV, ultraviolet.

of DNA) to 4N after DNA replication and back to 2N after mitosis. In meiosis, spermatogonia and oogonia enter meiotic division as cells with a 4N complement and undergo two successive reduction divisions, with no intervening S phase, to produce four gametes with a 1N (haploid) complement of DNA. The products of spermatogenesis are four functional spermatocytes, whereas those of oogenesis are one functional oocyte and three polar bodies. Besides reducing the DNA complement from 4N to 1N, a primary function of meiotic division is to ensure genetic diversity so that none of the gametes produced are genetically identical to any other gamete produced by that series of meiotic divisions. This is accomplished by recombination between chromosome homologues that appears to be required for subsequent proper chromosomal segregation.

Meiotic divisions begin with proliferating germ cells that have completed replication and that have a 4N DNA complement. In males, they are designated spermatogonia, and in females, they are oogonia. In the first meiotic division, the nuclear membrane disaggregates and the two chromatids of each chromosome remain attached to each other by a common kinetochore (meiotic prophase I). The homologous chromosomes pair and form a "bivalent" composed of two homologous chromosomes, each with two paired chromatids, and align themselves at the center of the cell. Thus, there are four chromatids in a bivalent. The paired chromosomes, one of paternal and the other of maternal origin, associate with one another along their entire length in what is know as the *synaptonemal complex*. This stage of meiotic prophase I is known as pachytene because the paired chromosomes are condensed and appear fat. During the next stage of the first meiotic prophase (diplotene), the chromosome pairs begin to separate except at a limited number of discreet points called *chiasmata*, which are the sites of reciprocal recombination between

the paired chromosomes. These recombination events are required for proper chromosome segregation in meiosis I and form the basis for evolutionary diversity.

The synapsed homologous chromosome pairs, now aligned at the cell center (metaphase I), begin to separate randomly to the opposite poles of the cell during meiotic anaphase I. Telophase I follows, resulting in two cells whose paired chromatids are now chromosomes that are homozygous throughout except in the region distal to the site of crossover. Nuclear membranes may reform or the cells may rapidly enter meiosis II. The second meiotic division proceeds in the absence of DNA replication and follows the morphologically recognizable steps that one observes for a mitotic division. Because there is no intervening S phase, the resulting gametes are haploid (1N complement of DNA). A generic depiction of the meiotic reduction divisions and the outcome of meiotic recombination is shown in Figure 9–6.

Although the mechanics of meiosis are essentially the same for males and females, significant differences do exist. In both cases, meiotic prophase I is the most protracted of each of the phases. In the human female body, meiotic divisions begin during embryogenesis, and by the fifth month of gestation, the oocytes become arrested in the diplotene stage of meiotic prophase I and remain arrested at that stage for decades. The meiotic divisions of individual oocytes are not completed until just before ovulation, at which time the oocyte undergoes two successive meiotic divisions to cause a mature ovum and three polar bodies. The division of cytoplasm is unequal so that only the future egg receives the majority of the cytoplasm whereas the polar bodies receive sufficiently little cytoplasm that ultimately they do not survive. The sequence of events during oogenesis involves proliferating oogonia, some of which arrest in G2, with a 4N

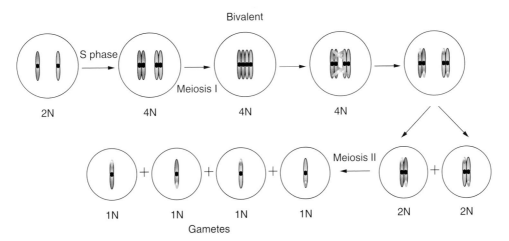

Figure 9–6. Schematic depiction of reduction divisions during meiosis, illustrating the pairing of chromosome homologues during meiosis I and recombination between maternally and paternally derived chromosomes to produce genetically different gametes.

complement of DNA and begin to differentiate. The cells arrest at the diplotene stage of meiotic prophase I and are designated a primary oocyte. Before ovulation, meiotic division is triggered, producing a secondary oocyte (2N) and a 2N polar body that eventually will be discarded. Meiosis II creates a haploid (1N) egg and another polar body. The 2N polar body from the first meiotic division also divides, ultimately yielding an egg and three polar bodies, all of which are 1N. As indicated earlier, meiosis in the male is mechanistically similar, but its timing and outcome are different. Meiosis in male germ cells does not begin until puberty and continues for many decades. Once the meiotic process is initiated, it proceeds with no periods of interruption except for a somewhat protracted first meiotic prophase. The proliferating spermatogonia produce a primary spermatocyte that undergoes the first meiotic division to produce two secondary spermatogonia. Each of the secondary spermatogonia undergoes the second meiotic division to produce a total of four functional haploid spermatocytes. The progression of the meiotic process and the fundamental differences between oogenesis and spermatogenesis are outlined in Figure 9–7.

Some of the players that regulate the mitotic cell cycle also regulate meiosis. In the 1970s, Yoshio Masui discovered that a factor that became known as MPF (maturation-promoting factor), was critical for frog oocyte maturation. Later, it was discovered that MPF is a protein complex that contains Cdk1 and cyclin B, the same factors that control entry of somatic cells into mitosis. Notably, the mos protooncogene, first isolated as an oncogene from the Maloney murine sarcoma

virus, is highly expressed in mouse and human germ cells and appears to be required for proper meiotic progression. When its activity is inhibited, meiotic maturation does not occur.

SENSORS RECOGNIZE SITES OF DNA DAMAGE

ATM and ATR

Arrest first requires the sensing and recognition of sites of damaged DNA sites by sensors such as ATM (*ataxia telangiectasia mutated*) to initiate a checkpoint. Mutations in ATM are responsible for a rare genetic syndrome, ataxia-telangiectasia (A-T), which is characterized by cerebellar degeneration, immunodeficiency, genome instability, clinical radiosensitivity, and cancer predisposition. ATM is a 350-kDa oligomeric protein that exhibits significant sequence homology to the phosphatidylinositol 3-kinases (PI3Ks), but lacks lipid kinase activity. It does, however, have protein kinase activity that is stimulated by agents that induce double-strand DNA breaks. ATM preferentially binds free DNA termini at the site of DNA breaks, apparently in monomeric form. After exposure of cells to ionizing radiation, ATM is activated by autophosphorylation and phosphorylates and activates a large number of target proteins, including Chk2, p53, NBS1, and BRCA1, at serines and threonines that precede a glutamine in the sequence motif of SQ and TQ. Neighboring sequences must also confer some specificity because $S_{15}Q$ of p53 is phosphorylated by ATM, whereas $S_{37}Q$ is not.

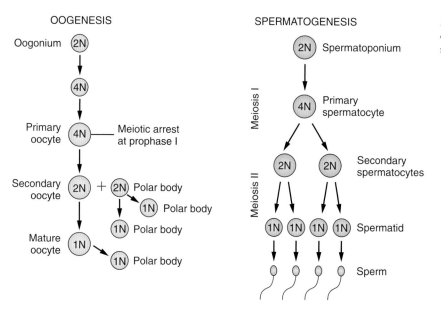

Figure 9–7. Comparison between oogenesis and spermatogenesis and their meiotic outcomes.

Clinical *Case 9-2*

Jennie Marlowe is a 16-year-old white girl who comes to the ophthalmologist for increasing problems with her eyes, including intermittent double vision while watching television.

Examination of her eyes shows strabismus and substantial nystagmus on lateral gaze with image displacement between her two eyes on single-eye testing. While examining her eyes, the ophthalmologist noticed that she had dilated blood vessels in her conjunctiva and also a "capillary rash" in a butterfly distribution over the bridge of her nose and the middle of her face.

Jennie had always had a peculiar unstable gait while she was growing up, and it was almost impossible for her to walk in the dark without falling after a few steps. She also had a history of frequent upper respiratory infections, at least four episodes of bacterial pneumonia requiring antibiotics, and "too many sinus infections to remember." As yet, Jennie has not had any menstrual periods and only shows minimal secondary sexual characteristics.

Her father is a food chemist. Her mother was an X-ray technologist but is no longer working because she has metastatic breast cancer. Jennie has a sister and two brothers who are all in excellent health.

The ophthalmologist prescribed corrective lenses with substantial prismatic components in an effort to relieve her nystagmus, and sent her to an internist for further evaluation. The internist ordered several tests. On sinus films, he found that Jennie indeed had the ground-glass appearance of long-standing chronic sinusitis. More alarmingly, on chest film, he found a mediastinal mass strongly suggestive of lymphoma. Blood tests also disclosed a high level of α-fetal protein, and a quantitative serum protein electrophoresis showed a marked reduction in IgA and IgE levels.

Cell Biology, Diagnosis, and Treatment of Ataxia Telangiectasia

Jennie has ataxia telangiectasia. This is an autosomal recessive disorder due to a mutant ATU gene that appears to influence the phosphatidyl inositol kinase signal transduction pathway. It is likely that this pathway is involved in the regulation of the development of cerebellar function because the cardinal sign of the disorder is cerebellar ataxia. In addition, reproductive development may be affected, because ovarian dysgenesis or agenesis is not infrequent in female patients. Importantly, the mutant gene is also associated with defects in DNA repair, which make homozygous carriers particularly susceptible to radiation damage and other forms of DNA-mediated

carcinogenesis. Because this is an autosomally recessive disorder, both of Jennie's parents are requisite heterozygous carriers, and they, too, are susceptible to radiation-induced cancers.

There is also associated humeral immunodeficiency. Though the mechanism of this is not well understood, resultant chronic sinusitis and other bacterial infections are common.

In addition to her new glasses, Jennie is also going to receive intravenous IgG therapy in an attempt to forestall further infections. Most importantly, she will have a CT-directed mediastinal lymph node biopsy to evaluate the possibility of lymphoma.

A related PI3K, ATR, was discovered in the human genome database as a gene with sequence homology to ATM and to SpRad3 (from *S. pombe*), hence the name ATR (*ATM* and *R*ad3 related). The gene encodes a large protein of 303 kDa with a C-terminal kinase domain and regions of homology to other PI3K family members. Absence of ATR in mice results in embryonic lethality. In humans, mutations causing partial loss of ATR activity are associated with Seckel syndrome, an autosomal recessive disorder that shares features with A-T. Like ATM, ATR is a protein kinase that has specificity for serine and threonine residues in SQ/TQ motifs, and that can phosphorylate essentially all of the proteins that are phosphorylated by ATM. In contrast with ATM, ATR is activated by ultraviolet (UV) light rather than by ionizing radiation, and it is the main PI3K family member that initiates signal transduction after UV irradiation.

The relative specificity of ATR for UV irradiation has raised the possibility that the enzyme might directly recognize specific types of DNA damage produced by UV. ATR binds directly to DNA, with preference for sites of UV-damaged DNA containing a (6-4) photoproduct. Electron microscopy of ATR-DNA interactions has shown that extent of DNA binding by ATR increases with increasing UV irradiation, and that unlike ATM, ATR rarely binds termini of linear DNA, indicating that ATR does not recognize DNA double-strand breaks. ATR phosphorylation of p53 *in vitro* is stimulated by the addition of undamaged DNA to such a kinase assay, and is even further stimulated by the addition of UV-damaged DNA in a dose-dependent manner. Thus, ATM is both a sensor and transducer molecule that responds to double-strand breaks, and ATR serves an analogous role for a cellular response to base damage incurred by UV irradiation.

Mediators Simultaneously Associate with Sensors and Signal Transducers

The mediator proteins simultaneously associate with DNA damage sensors proteins and with signal transducers at defined phases of the cell cycle. Consequently, they help provide specificity to the various signal transduction pathways that are activated. The prototype mediator protein is the yeast scRad9 protein, which functions in the signal transduction pathway from yeast scMec1 (ATR in mammals) to yeast scRad53 (CHEK2 or Chk2 in mammals). Another mediator, Mrc1 (*mediator of replication checkpoint*) that is found in both types of yeast, *S. cerevisiae* and *S. pombe,* is expressed only during S phase and is necessary for S-phase checkpoint signaling from scMec1/spRad3 (ATR) to scRad53/spCds1 (CHEK2). Humans have at least three proteins that contain the conserved BRCT motif involved in protein–protein interaction that fit into the mediator class of proteins. These include the p53 binding protein, 53BP1, the topoisomerase binding protein, TopBP1, and the *mediator of DNA damage checkpoint 1,* MDC1. These proteins interact with DNA damage sensors such as ATM, with DNA repair proteins such as BRCA1 and the Mre11/Rad50/NBS1 (M/R/N) complex, with signal transducers such as CHEK2, and with effector molecules such as p53. The DNA damage checkpoint response is compromised in cells that have decreased levels of, or lack of, these proteins.

Signal Transducers CHEK1 and CHEK2 Are Kinases Involved in Cell-Cycle Regulation

Humans have at least two kinases, CHEK1 and CHEK2, whose function is primarily that of a signal transducer in cell-cycle regulation and checkpoint responses. These kinases were identified based on homology with yeast scChk1 and scRad53/spCds1, respectively. Both CHEK1 (Chk1 in mice) and CHEK2 (Chk2 in mice) are serine and threonine kinases that have limited homology to each other and that display moderate substrate specificities. In mammalian cells, the double-strand break signal sensed by ATM is transduced predominantly by CHEK2, and the UV damage signal sensed by ATR is mainly transduced by CHEK1. There is some overlap, however, between the functions of the two proteins. It is notable, however, that mice lacking Chk1 (Chk1[−/−]) exhibit embryonic lethality, whereas mice deficient in Chk2 (Chk2[−/−]) are viable and display near-normal checkpoint responses. Mutations in human CHEK2 increase risk for breast cancer, as well as for a multitumor Li–Fraumeni (LFS)–like cancer syndrome.

Effectors p53 and Cdc25 Phosphatases Are Important Effector Proteins in Cell-Cycle Regulation

Two of the principal effector proteins are p53 and the Cdc25 phosphatases (see earlier descriptions). The Cdc25 family of proteins dephosphorylates the various CDKs to allow cell-cycle progression. The Cdc25 proteins are themselves targets for phosphorylation, which causes their sequestration by binding to 14-3-3 proteins, causes their exclusion from the nucleus, or causes their ubiquitination and proteosome-mediated degradation. In each case, after DNA damage, the Cdc25 phosphatase is rendered unavailable to dephosphorylate its target CDK protein resulting in a cell-cycle arrest. As examples, the hypophosphorylated form of Cdc25A promotes the G1/S transition by dephosphorylating CDK2, and the hypophosphorylated form of Cdc25C promotes the G2/M transition by dephosphorylating CDK1 (Fig. 9–8).

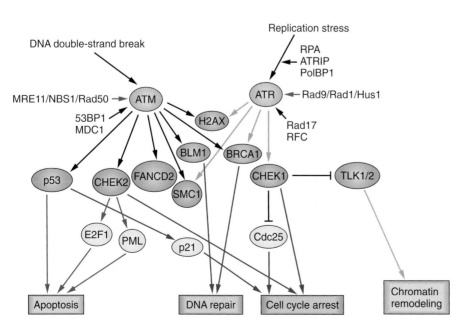

Figure 9–8. Signaling cascades originating from DNA double-strand breaks and from replication stress, as well as the various cellular responses and outcomes.

G1/S Checkpoint

After DNA damage, cells arrest at the G1/S checkpoint, which prevents their entry into the S phase by inhibiting the initiation of replication. When cells are unperturbed, those in G1 become committed to entering the S phase at the "restriction point" (mammalian cells) or "start" (budding yeast). In proliferating human cells, the restriction point precedes the actual start of DNA synthesis by about 2 hours. When there is DNA damage, however, entry into S phase is prevented regardless of whether the cells have passed the restriction point. If the DNA damage involves DNA double-strand breaks caused by ionizing radiation or by radiomimetic agents such as some chemotherapeutics, ATM becomes activated and phosphorylates and activates numerous target molecules, notably p53 and CHEK2. One consequence of these phosphorylation events is that of a G1/S arrest. In one case, ATM phosphorylates and activates CHEK2, which in turn phosphorylates Cdc25A phosphatase, causing its inactivation by nuclear exclusion and ubiquitin-mediated proteolytic degradation. Loss of active Cdc25A results in the accumulation of the phosphorylated (inactive) form of CDK2, which is incapable of phosphorylating Cdc45, a protein whose phosphorylation is required to initiate replication at sites of preformed replication origin complexes. In the second slower but more sustained case, ATM phosphorylates p53, as does the newly activated CHEK2. The former phosphorylates p53 at serine 15 and the latter at serine 20. These phosphorylation events inhibit nuclear export and degradation of p53 and allow it to serve as a transcriptional activator that induces transcription of p21$^{Waf-1/Cip1}$, the inhibitor of CDK2/cyclin E and of CDK4/cyclin D. Thus, cells accumulate at the G1/S phase boundary at the G1/S checkpoint.

If the incurred DNA damage is caused by UV light or by UV-mimetic agents, rather than by ionizing radiation, the types of DNA damage are different and the cellular response to this damage is similar to, but not identical to, that induced by double-strand DNA breaks. Damage to DNA that is caused by UV irradiation is not sensed by ATM, but rather by ATR, Rad17-RFC, and the 9-1-1 complex (a complex containing Rad9, Rad1, and Hus1). Once the damage is recognized, ATR phosphorylates the transducing kinase CHEK1, which, in turn, phosphorylates and inactivates Cdc25A, leading to G1 arrest (see Fig. 9–8).

Intra-S-Phase Checkpoint

The intra-S-phase checkpoint is activated by damage that occurs while cells are replicating their DNA or by unrepaired DNA damage in cells that escape the G1/S checkpoint. In both cases, cells manifest a block in replication. Because the original definition of a checkpoint implied the involvement of a biochemical pathway that ensured the completion of a reaction before progression to the next cell-cycle phase, the intra-S-phase checkpoint does not strictly meet that definition. Nevertheless, as our understanding of molecular events that underlie checkpoints has become clearer, the intra-S checkpoint does meet a definition that is modified to include this checkpoint. A cell-cycle checkpoint is the end product of a molecular regulatory pathway or signaling cascade that ensures an ordered succession of cell-cycle events, which when perturbed, results in cell-cycle arrest (Fig. 9–9).

The damage sensors for the intra-S checkpoint encompass a large set of checkpoint and DNA repair proteins. When DNA damage is a frank double-strand break or a double-strand break resulting from replication of a nicked or gapped DNA, ATM, the M/R/N complex, and BRCA1 contribute to activation of the checkpoint. These repair proteins presumably function as sensors because they bind to either double-strand DNA breaks (ATM) or to special branched DNA structures (the M/R/N complex, BRCA1, BRCA2). They participate in the intra-S checkpoint by initiating a kinase signaling cascade that proceeds through pathways involving CHEK1 and CHEK2 (see Fig. 9–8).

A striking characteristic of the intra-S-phase checkpoint is radioresistant DNA synthesis (RDS), which illustrates the intimate relation between this checkpoint and double-strand DNA break repair. In wild-type cells, ionizing radiation causes immediate cessation of active DNA synthesis. In contrast, cells from A-T patients challenged in like manner continue to proceed through S phase without delay. This RDS also occurs in cells with mutations in subunits of the M/R/N complex and

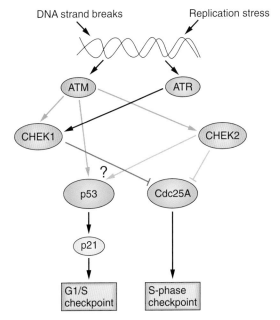

Figure 9–9. Signaling pathways leading to checkpoint arrest in S phase and G1/S boundary.

in FANCD2 (BRCA2) and represents a compromised intra-S checkpoint.

When DNA is damaged by UV radiation that produces pyrimidine dimers or by chemicals that produce bulky DNA adducts, it is the ATR protein, or more precisely the ATR-ATRIP heterodimer, that serves as the primary sensor of these types of DNA damage. ATR binds to chromatin, preferentially to the UV-induced lesions, and is activated. Activated ATR phosphorylates CHEK1, which in turn phosphorylates and down-regulates Cdc25A by causing its degradation and inhibiting the firing of replication origins (see Figs. 9–6 and 9–7).

G2/M Checkpoints

The G2/M checkpoints actually represent multiple checkpoints that regulate cellular entry into mitosis, transit through mitosis, and exit from mitosis. When DNA is damaged during the late S phase, after cells are committed to divide, the cells will arrest in late G2. Again, the arrest is mediated by ATM after introduction of DNA double-strand breaks or by ATR if the lesions are a consequence of chemical challenge or UV exposure. Signaling occurs through CHEK 2 and CHEK 1, respectively, resulting in the phosphorylation and activation of p53 and the induction of the CDK inhibitor p21. These pathways also inhibit Cdc25A and Cdc25C, which prevent CDK1/cyclin B activation and consequent progression of cells from G2 into mitosis (Fig. 9–10). Although these are the inhibitory pathways that we understand best, other regulatory proteins, such as the aurora and the polo kinases, that are less well understood play significant roles in the transition from late G2 into mitosis. Another key protein that controls the onset of prophase is designated Chfr. This protein

appears to monitor microtubule and centrosome integrity. When cells are challenged with drugs that disrupt microtubules or their dynamics, chromosome condensation and centrosome separation are delayed. After such drug treatment or other stress, Chfr may function by interfering with the import of CDK1/cyclin B into the nucleus, which is necessary for progression of cells into and through prophase.

Once mitosis is initiated, cell progression is subject to yet another checkpoint, the "spindle assembly checkpoint." This checkpoint is complex and is mediated by inhibition of a large macromolecular complex known as the APC that has ubiquitin ligase activity that is necessary for the degradation of several key proteins such as cyclin B and securin. Degradation of cyclin B is required for exit of mitosis, and destruction of securin is needed for chromatid separation. Additional proteins that associate with or regulate APC activity in selective protein degradation include MAD, BUB1, BUBR1, and Cdc20, among others. Loss of any of these factors compromises the ability of cells to arrest properly when spindle microtubule assembly is perturbed. One consequence of the failure to arrest is "mitotic catastrophe," whereby aberrant mitotic figures are formed, and ultimately cells with aberrant chromosome numbers emerge and undergo cell death.

CELL-CYCLE ALTERATIONS AND CANCER

The preceding discussion describes many of the signaling molecules and pathways that are required for regulated cell-cycle progression in normal proliferating cells. When mutations or epigenetic modification perturb any one or combinations of these events or pathways, the risk for tumorigenesis is increased. Almost every human tumor is accompanied by one or more mutations in genes that affect control of cell cycle, leading to unregulated cell proliferation. In principle, the cell-cycle lesions can occur at any cell-cycle checkpoint or in any phase of the cell cycle, and they represent an important step in the progression to full-blown cancer. Thus, human tumors can be characterized in part by the type of cell-cycle control that the tumor cells have lost.

Perturbations in Regulating the G1-to-S Transition in Cancer

Cells in which cyclin D1 forms aberrant complexes with CDK4 or CDK6, caused by overexpression or by other mechanisms that have rendered these complexes hyperactive, behave like cells in which pRb has been lost or inactivated and, therefore, are insensitive to mitogenic signals. When such cells are subjected to physiological insults that normally result in pRb-mediated inhibition of progression into S phase, they continue to proliferate.

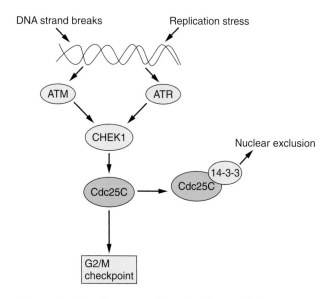

Figure 9-10.   Signaling pathways leading to a G2/M checkpoint.

Such aberrant CDK activation or loss of pRb has obvious implications for cancer, and indeed, pRb loss or hyperactivation of CDK4, CDK6, or both is found in a large fraction of human tumors. Hyperactivation of CDK4 and CDK6 can occur through deregulated expression of the D-type cyclins, loss or epigenetic silencing of p16^{INK4a}, or mutations that compromise the inhibitory functions of p16^{INK4a}. Hence, every element of the core pRb pathway (p16^{INK4a}, D-type cyclins, CDK4, CDK6, and pRb itself) is a potential oncogene or a tumor suppressor.

Molecular analyses of human cancers fully support this notion. Amplification or rearrangement of the *cyclin D1* gene, for example, as well overexpression of the cyclin D1 protein, have been described in a wide spectrum of human cancers such as squamous cell carcinomas of the head and neck, carcinomas of the uterine cervix, astrocytomas, non-small cell lung cancers, and soft-tissue sarcomas, among others. A well-documented example is the frequent involvement of *cyclin D1* in human breast cancer. The *cyclin D1* gene is amplified in about 15% to 20% of human mammary carcinomas and is overexpressed in more than 50% of such tumors. Overexpression of cyclin D1 is common at the earliest stages of breast cancer, such as ductal carcinoma *in situ* (DCIS), but not in premalignant lesions, such as atypical ductal hyperplasia. Thus, overexpression of cyclin D1 can serve as a marker of malignant transformation of mammary epithelial cells. When cyclin D1 is overexpressed in tumor cells, this overexpression is maintained at the same level throughout breast cancer progression, from DCIS to invasive carcinoma, and is preserved even in metastatic lesions.

Other members of the cyclin D gene family, such as those encoding cyclins D2 and D3, are also often amplified and the encoded proteins overexpressed in many human cancers. Cyclin D2 overexpression is common in B-cell lymphocytic leukemias, lymphoplasmacytic lymphomas, chronic lymphocytic leukemias, as well as in testicular and ovarian germ cell tumors. Cyclin D3 overexpression is seen in glioblastomas, renal cell carcinomas, pancreatic adenocarcinomas, and several B-cell malignancies such as diffuse large B-cell lymphomas or multiple myelomas.

Overexpression of CDK4 occurs in breast cancers, in gliomas, glioblastoma multiforme, sarcomas, and meningiomas, often as a consequence of gene amplification. Moreover, because of the loss, mutation or silencing of the gene encoding the CDK4 inhibitor, p16^{INK4a}, a different subset of tumors, including retinoblastoma, osteosarcoma, small-cell lung carcinoma, and bladder carcinoma, is associated with the loss of the pRb protein.

pRb Pathway in Cancer

Notably, the majority of human tumors show an alteration of only one member or step of the pRb pathway in any given tumor. For example, if cyclin D1 is overexpressed in a tumor, pRb and p16^{INK4a} are usually unaffected and expressed normally. This is presumably due to that excess cyclin D1 in complex with CDK4 or CDK6 is sufficient to evade the inhibitory effects of p16^{INK4a} and neutralize the function of pRb. This observation suggests that the pRb pathway is linear, and that CDK4 and CDK6 converge on pRb as their primary critical substrate, at least as far as cancer cell proliferation is concerned. In this view, inactivation of pRb or any component in this pathway, with subsequent loss of regulation of the E2F transcription factor, is critical to cell-cycle control and is sufficient to produce unregulated cell cycle and cell proliferation kinetics in tumors. Loss of this negative control mechanism contrasts with the mechanism by which cyclin E is oncogenic and contributes, for example, to the pathobiology of breast tumors. The frequency with which cyclin E is overexpressed in tumors suggests that it may accompany other mutations in the core pRb pathway. This apparent selection for cyclin E overexpression, even in the presence of pRb pathway mutations, is likely due to "downstream" roles of cyclin E1 in replication control and serves to emphasize the more pRb-centric roles of D-type cyclins and CDK4 and CDK6, whose functions are generally dispensable in cells that lack pRb.

ATM in Cancer

As discussed earlier, different types of genotoxic agents evoke distinct and overlapping cellular responses that correspond to the particular type of DNA damage. The different responses are mediated by specific molecules involved in the detection of distinct lesions, such as bulky DNA adducts or DNA strand breaks. Double-strand DNA strand breaks are generated by a variety of genotoxic agents, including ionizing radiation, radiomimetic agents, and inhibitors of topoisomerase, all of which activate the ATM kinase. Cells from patients with ataxia-telangiectasia (A-T) are extremely sensitive to these genotoxic agents and display several impaired signal transduction pathways that mediate cellular responses to this type of genotoxic stress. A-T is characterized by multiple phenotypic aberrations including lymphoreticular malignancies and extreme sensitivity to ionizing radiation. The mutant *ATM* gene that is responsible for this pleiotropic phenotype is inactivated in the majority of A-T patients by mutations that cause frameshifts that truncate its protein product and impair its activity. Disruption of the corresponding gene in the mouse creates a phenotype that reflects most of the features of A-T in humans.

The possible role of the mutant *ATM* gene in cancer predisposition in heterozygous individuals is of concern. The specific issue is whether female A-T carriers are at increased risk for cancer after low-dose radiation, and therefore should avoid diagnostic mammography. LOH at *ATM* induced by the low-dose X-irradiation may be sufficient to increase risk for cancer if a woman is otherwise predisposed. The association of mutant ATM with at least one type of malignancy, T-prolymphocytic leukemia (T-PLL), is apparent and has implicated *ATM* as a tumor suppressor gene in somatic cells. Tumor tissue from T-PLL patients shows somatic inactivation of both *ATM* alleles caused by rearrangements or point mutations in greater than 50% of cases, supporting the role of *ATM* as a tumor suppressor and linking the biology of T-PLL to ATM function.

Mice genetically engineered to be ATM deficient (*Atm*$^{-/-}$) develop aggressive malignant thymic lymphomas and usually die by 4 months of age. In a similar mouse model but in a different genetic strain, the average time of tumor onset was considerably longer, with more than 50% of the animals surviving to 10 months. The difference in cancer susceptibility between the two mouse models is probably due to differences in genetic background between the two mouse strains and reflects the variability that one sees in A-T patients.

p53 in Cancer

The p53 protein, which is mutant in more than half of all cancers, functions as a tetrameric transcription factor that is present at low levels in normal, unperturbed cells. After cells are stressed, p53 undergoes posttranslational modification and is stabilized, resulting in its accumulation. After cells are stressed, the p53 protein acts in its capacity as a transcription factor and facilitates the transcription of multiple genes involved in cell cycle arrest or apoptosis, depending on the cellular context, the extent of damage, or other unknown parameters. In about 90% of cases in which cells or tumors have lost p53 function, the protein is mutant. In the remaining 10% of cases, the protein is completely absent. Most tumors with mutant p53 have lost heterozygosity at that locus, consistent with the behavior of a classic tumor suppressor. The model, which has been substantiated for several tumor types, is that mutation occurs at one allele of p53, and that the remaining wild-type allele is lost as a consequence of LOH caused by mitotic recombination (recombination between chromosome homologues). This finding is consistent with the observation that tumors with LOH at p53 also have lost heterozygosity at multiple linked loci on the short arm of chromosome 17, which houses the *p53* gene.

The p53 protein is an integral component of several signaling pathways. Thus, it is not surprising that mutations or changes in expression of other genes within these pathways can affect p53 function and, consequently, the signaling cascades regulated by p53. Predominant among these is MDM2, a protein that regulates the stability of p53 by physically binding p53 and causing its ubiquitination and transport to the pro-

teasome for degradation. The level of the MDM2 protein, which is elevated in many tumors, especially sarcomas, is the result of amplification of the *MDM2* gene, enhanced transcription of the gene, or enhanced translation of its mRNA. When p53 is phosphorylated by ATM in normal cells, its affinity for MDM2 is reduced, which allows p53 to accumulate and to act in its capacity as a transcription factor. When MDM is overexpressed, it presumably depletes the cell of available p53 and mimics a phenotype consistent with loss of p53.

Several virally encoded proteins have a mode of action that mimics that of MDM2 and renders the cell effectively p53 deficient. Some human papillomavirus (HPV) subtypes, particularly HPV16 and HPV18, are associated with cervical and laryngeal cancer and produce two oncogenic proteins, E6 and E7. The E6 protein specifically binds p53 and causes its degradation, which explains why p53 mutations in cervical cancers are rare. Notably, the viral E7 protein binds cellular pRb.

THE CHECKPOINT KINASES AND CANCER

CHEK1 and Cancer

Although CHEK1 mutations in tumors are extremely rare, they do occur. They appear to be restricted to carcinomas of the colon, stomach, and endometrium. Some colon and endometrial carcinomas with microsatellite instability have been reported to have frameshift mutations in *CHEK1* caused by insertion or deletion of single adenine in a polyadenine tract. The resulting truncated CHEK1 proteins are defective because of the lack of the C-terminal end of the catalytic domain and the complete loss of the SQ-rich regulatory domain. Also, a subset of small-cell lung carcinomas produces a CHEK1 mRNA that encodes a shorter CHEK1 isoform lacking a conserved sequence in the catalytic domain. The deleted region appears to be involved in substrate selectivity, but the significance of the predominant expression of this alternative CHEK1 isoform in fetal lung and in small-cell carcinomas of the lung, but not in normal adult lung tissue or other types of lung tumors, remains to be established. Notably, the complete absence of Chk1 in genetically engineered mice results in early embryonic lethality. Deletion of *CHEK1* in a p53-deficient chicken tumor cell line is tolerated by these cells, as is the earlier mentioned heterozygous truncation mutation of CHEK1 in some human tumors. Given the role of CHEK1 in the DNA damage response, it is likely that either hypomorphic mutations in CHEK1 or loss of function mutations that may occur during the progression of cancer contribute to enhanced genetic instability in some tumors.

CHEK2 and Cancer

The first evidence that genetic alteration in *CHEK2* may predispose to cancer was the detection of rare *CHEK2* germline mutations in families with a form of LFS. LFS is a familial cancer syndrome first associated with germline mutations in p53. The syndrome is characterized by multiple types of tumors, predominantly breast cancer and sarcomas, that arise during childhood. The *CHEK2* polymorphic variant, *CHEK2*1100delC*, has a single C deletion that produces a truncated protein, and that was first identified in a small subset of families in Finland that manifested an LFS-like syndrome. The fact that *p53* was wild-type in all of these cases suggested that germline mutations in *CHEK2* represented an alternative genetic defect predisposing to LFS.

The observation that the *CHEK2*1100delC* variant occurred in LFS stimulated large epidemiologic studies asking whether *CHEK2*1100delC* was overrepresented in hereditary breast cancer. The major finding of these studies is that *CHEK2*1100delC* is a low-penetrance breast cancer susceptibility allele with an allele frequency of about 1.5% in the general human population. Notably, *CHEK2*1100delC* confers no increased cancer risk in breast cancer families that already have mutations in the two previously identified breast cancer susceptibility genes, *BRCA1* and *BRCA2*. This observation is consistent with the concept that *CHEK2*, *BRCA1*, and *BRCA2* all participate in the same DNA damage response network, whose function can be perturbed by mutations in any one of these genes. At the molecular level, the 1100delC truncation eliminates the kinase domain and therefore the CHEK2 kinase activity, and the truncated protein is unstable. The remaining wild-type allele is often lost in the tumors. Thus, CHEK2 function is often lost accompanied by a marked reduction in or absence of CHEK2 protein. Interestingly, the *CHEK2*1100delC* variant is particularly common in those rare families predisposed to combined breast and colon cancer.

Unlike the *CHEK2*1100delC* variant whose allele frequency is similar in Western European, North American, and Finnish populations, the missense variant CHEK2 I157T, also originally detected in rare LFS families, is much more common in normal Finnish population (at 5–6%) than elsewhere. This variant allele is significantly overrepresented in an unselected cohort of breast cancer patients, suggesting that it contributes to the cause of breast cancer in many patients within the Finnish population. Mechanistically, the I157T exerts its effect differently than the *CHEK2*1100delC* variant. The CHEK2*1100delC variant protein is unstable, and when homozygous behaves much like a null mutant. The I157T variant, in contrast, is stable and behaves like a dominant-negative mutant, exerting an effect even when the remaining allele is wild-type. The I157T mutant protein interferes with the ability of the wild-type CHEK2 to associate with many of its physiological substrates after exposure to ionizing radiation.

SUMMARY

During the last decade, our understanding of the cell cycle and its regulation has grown exponentially. We are beginning to put together the pathways and circuits that govern normal and compromised cell-cycle progression and cell division. Despite its enormous complexity, various components of the cell-cycle circuitry are beginning to prove clinically valuable as diagnostic or prognostic markers, and others are emerging as promising therapeutic anticancer targets. The pathways that lead to programmed cell death, apoptosis, are mediated by many of the same molecular players that participate in cell-cycle control and are also providing therapeutic targets for cancer. The basic strategies are to find compounds that arrest tumor cells so that they will not contribute to growth of the tumor, or to push the tumor cells into apoptosis to rid the lesion of tumor cells. Most of the proteins that have been discussed in this chapter are currently considered to be potential therapeutic targets and are under active investigation in many academic laboratories and pharmaceutical companies. Tumor cells, however, have evolved to escape most antitumor therapies, indicating that most new therapeutic approaches will have to be combination therapies rather than one individual therapeutic strategy. Nevertheless, as the cell cycle and apoptotic pathways are more thoroughly dissected, there is hope that in the near future many cancers for which we currently have no effective treatments will become manageable with concomitant improvement in quality of life.

Suggested Readings

Howard A, Pelc S. Synthesis of deoxyribonucleic acid in normal and irradiated cells and its relation to chromosome breakage. *Heredity* 1953;6(Suppl.):261–273.

Lukas J, Lukas C, Bartek J. Mammalian cell cycle checkpoints: signalling pathways and their organization in space and time. *DNA Repair* 2004;3:997–1007.

Murray A, Hunt T. *The Cell Cycle, an Introduction.* New York: W. H. Freeman and Company, 1993.

Sánchez I, Dynlacht BD. New insights into cyclins, CDKs, and cell cycle control. *Sem Cell Dev Biol* 2005;16:311–321.

Santamaria D, Ortega S. Cyclins and CDKs in development and cancer: lessons from genetically modified mice. *Front Biosci* 2006;11:1164–1188.

Chapter 10

Programmed Cell Death

Normal animal development depends on the co-ordinated regulation of cellular processes to build and maintain diverse cell populations present in different tissue types that together compose complex organisms. An obvious example of such coordinated regulation is the proliferation of precursor cells, or progenitor cells, and their subsequent specification or differentiation into distinct cell types. Less obvious is the role of programmed cell death, another cellular fate critical to the formation and maintenance of most tissue types. The term **programmed cell death** initially was used to emphasize that these cell deaths occur in predictable body locations, at predictable developmental periods, and as part of a predetermined developmental plan of the organism. During development, programmed cell death sculpts body parts and eliminates transitory structures. Examples of this include the elimination of cells within the tail of a tadpole as it changes into a frog, the dele-tion of interdigital tissue in the vertebrate limb bud leading to the formation of digits, and the hollowing out of tissue to create lumina in vertebrates (Fig. 10–1).

In addition, programmed cell death rids the develop-ing animal of superfluous or unwanted cells. For example, large numbers of self-reactive lymphocytes or lymphocytes that fail to produce useful antigen-specific receptors are deleted by programmed cell death. In the developing nervous system, about half of the neurons produced undergo programmed cell death soon after they mature in a process that ensures that optimal con-nectivity between newly generated neurons is estab-lished. Programmed cell death continues to play a critical role into adulthood where it maintains a balance with cell division, thus controlling tissue, organ, and body size. Cells dangerous to the organism, such as trans-formed cells, damaged cells, or those infected with pathogens such as viruses, are also eliminated by the activation of programmed cell death.

A clear morphologic distinction exists between cells undergoing programmed cell death and the pathologic cell death that occurs in acute lesions, such as trauma. In the latter case, the cells and their organelles swell and rupture, spilling their contents in a process called **necrosis.** Because the release of intracellular material includes lysosomal enzymes, necrotic cell death damages neighboring cells and provokes an inflammatory reaction in the surrounding tissue. The activities and secretions of macrophages and other cells of the immune system contribute to the damage of tissues in the vicinity. In programmed cell death, leakage of cellular contents does not occur. In fact, the extent of naturally occurring programmed cell death was grossly underes-timated until relatively recently because the death process and the subsequent processing of material from the dead cells proceeds in such an orderly fashion and without inflammation. It is now estimated that in the average adult human, billions of cells die every day so that the survival of the organism as a whole benefits. In fact, the health of all multicellular organisms is critically dependent on the ability of individual cells to self-destruct when they become unwanted.

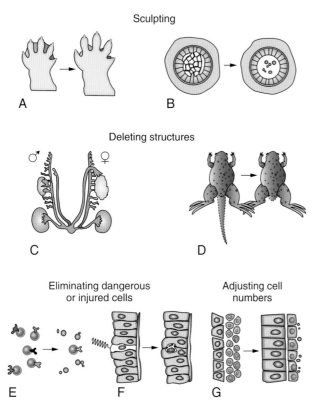

Sculpting

A   B

Deleting structures

C   D

Eliminating dangerous   Adjusting cell
or injured cells   numbers

E   F   G

Figure 10–1. Programmed cell death during animal development. **A:** Elimination of cells between developing digits and (**B**) hollowing out of solid structures to form lumina. **C:** Elimination of the Mullerian duct (which forms the uterus and oviducts in females) in males. Conversely, the Wolffian duct (which forms the epididymis, vas deferens, and seminal vesicle in males) is eliminated in females. **D:** Disappearance of the tail in the developing frog and (**E**) disappearance of self-reactive lymphocytes. **F:** Cells that are damaged by radiation or chemicals or that are transformed and (**G**) overproduced or superfluous cells are eliminated by programmed cell death. (*Adapted from Jacobson et al.* Cell, *1997;88:347–354*).

DISTINCT FORMS OF PROGRAMMED CELL DEATH

More recent ultrastructural studies of vertebrate tissue have shown that programmed cell death encompasses at least three distinct forms of cell death distinguishable by morphologic criteria using light and electron microscopy. The first to be described, and the best-studied form of programmed cell death, is called **apoptosis** (derived from Greek, meaning "the seasonal falling of leaves from plants or trees"), or more infrequently, **type 1 cell death.** Cells undergoing apoptosis shrink and display nuclear and chromatin condensation. Externally, the cell appears to boil as the membrane becomes convoluted, an event referred to as membrane "blebbing." The cell then fragments into membrane-bound bodies called *apoptotic bodies* that contain intact cellular organelles and the occasional chunk of condensed chromatin. Apoptotic bodies are then engulfed by phagocytosis.

Autophagic cell death, or **Type 2 cell death,** another, less-studied form of programmed cell death, is charac-

terized by the appearance of large numbers of cytoplasmic vacuoles and is associated with increased lysosomal activity. In autophagic death, double-membrane sheets, derived largely from the endoplasmic reticulum (ER), form cytoplasmic vacuoles that engulf intracellular organelles and cytoplasmic materials. These vesicles fuse with lysosomes to become structures called *autolysosomes,* where the sequestered cellular components are digested. This process is independent of phagocytosis (Fig. 10–2). It may be noted that **autophagy** (derived from Greek, meaning "to eat oneself") is an evolutionarily conserved process that occurs at a low level even in healthy cells and is, in fact, the cell's major mechanism for degrading organelles and long-lived proteins. In autophagic cell death, this process that is normally regulated and used to the benefit of the cell occurs at a highly increased level, causing the cell to die.

Type 3 cell death, the least-studied type of programmed cell death, is characterized by the swelling of intracellular organelles and lysosome-independent formation of "empty spaces" in the cytoplasm; it has some similarities to necrosis. No chromatin condensation occurs, although chromatin clustering is sometimes seen. In contrast with apoptosis and autophagic cell death, which are observed in many types of developing animals, Type 3 cell death (or nonlysosomal cell death) does not appear to be as common in nonpathologic conditions. More recently, other forms of cell death have been described such as anoikis, excitotoxicity, and caspase-independent cell death, although it remains unclear whether these are actually distinct modes or simply variants of the major forms of cell death described earlier.

Because apoptosis is the principal mechanism for cell elimination in metazoan organisms, most research on the mechanisms underlying programmed cell death has centered on death by apoptosis. Relatively little is known about the mechanisms regulating autophagic cell death, and virtually nothing is known regarding the regulation of Type 3 cell death or other less characterized forms of programmed cell death. This chapter focuses primarily on death by apoptosis.

Naturally Occurring Neuronal Death Is Regulated by Factors Provided by Other Cells

Why do some cells die while others around it live? What mechanisms regulate the survival of a cell? All cell types possess an intrinsic "suicide" program that has to be continuously inactivated for the cell to survive. Cell survival depends on signals generally provided to the cell by other cells that serve to neutralize the suicide program. That is, all cells are programmed to die unless they receive extracellular death-suppressing signals. The most common type of such signals are secreted polypep-

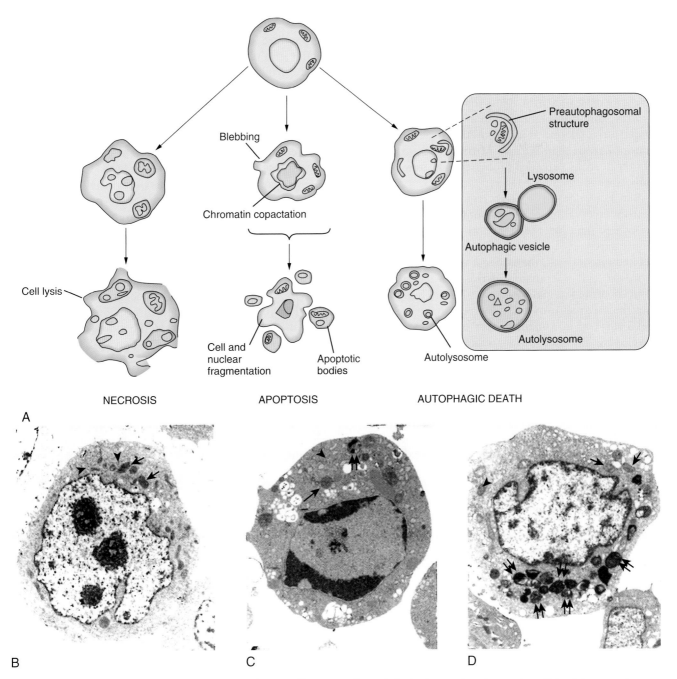

Figure 10-2. Different forms of cell death. **A:** Schematic illustration of morphologic changes that characterize cell death by necrosis, apoptosis, and autophagic cell death. Whereas necrotic cell death culminates in cell lysis and provokes inflammation, apoptotic cells are packaged into apoptotic bodies that are then engulfed by adjacent cells without an inflammatory response. Autophagic cell death is characterized by the appearance of cytoplasmic vesicles engulfing bulk cytoplasm and organelles. The contents of the vesicles are digested by the lysosomal system of the same cell after the fusion of the autophagic vesicles with lysosomes. Inset: Formation of autolysosome by fusion of autophagic vesicle and lysosome within the cell undergoing autophagy. **B–D:** Ultrastructural features of cells undergoing apoptotic and autophagic cell death: (**B**) a normal cell, (**C**) an apoptotic cell, and (**D**) a cell undergoing autophagic cell death are shown. Polyribosomes *(arrowhead)*, mitochondria *(arrow)*, and autophagic vacuoles *(double arrows)* are indicated. Although autophagic vacuoles can be seen in healthy cells and apoptotic cells, they are much more abundant during autophagic cell death. *(From Bursch, et al. J Cell Sci 2000;113:1189–1198, by permission.)*

tide factors, referred to as trophic factors, which bind to specific receptors located on the cell surface. Depending on the specific types of receptors expressed, different cell types need distinct sets of trophic factors for their survival. Failure to obtain the appropriate trophic factors in adequate quantities activates the suicide program within the cell, leading to its demise.

In the 1940s, Rita Levi-Montalcini and Viktor Hamburger first described the requirement for trophic factors for cell survival in the developing chick nervous system. These scientists observed that the number of spinal cord motor neurons and sensory neurons that survived during development depended on the size of the target to which they connected. Thus, surgical removal of the wing bud or limb bud in the developing chick resulted in an increase in the amount of cell death in sensory and motor neuron populations that normally connect to these appendages (Fig. 10–3).

Partial removal of the limb bud resulted in a less severe increase of neuronal loss than that seen with total removal, and transplantation of an extra limb bud next to the original one resulted in the survival of more motor neurons than normal. Moreover, these supernumerary neurons extended processes that grew preferentially toward the added target.

But how does the target regulate the survival of neurons innervating it? Levi-Montalcini and Hamburger discovered that, like the extra limb bud, when a specific mouse **sarcoma** (a connective tissue tumor) was transplanted into a chick, the sarcoma was invaded by nerve processes and a dramatic increase in the survival of some neuronal populations occurred. This suggested that the sarcoma provided a factor that promoted neuronal survival in a manner akin to that seen with the developing limb bud. A key observation made in these experiments was that the survival of certain neuronal populations with no apparent physical connection with the tumor was also greatly increased by the tumor, indicating that the survival-promoting signal was a diffusible factor. An *in vitro* assay was developed to isolate the survival-promoting factor in which ganglia of sympathetic neurons were placed in a tissue culture dish, either alone or next to the sarcoma tumor. When grown next to the tumor, the neurons grew a dense halo of processes and survived in the culture dish. It was also observed that the tumor itself was not necessary; the neurons survived even if they were provided with culture medium lacking the tumor but in which the tumor had

been maintained. This indicated that the neurons were responding to a soluble factor that was secreted by the tumor into the medium. The idea that cells could affect each other through secreted or diffusible factors was a novel one at the time these experiments were performed (the 1940s). As a step toward characterizing the trophic factor secreted by the tumor, Levi-Montalcini proceeded to test whether it was a protein or a nucleic acid. In a stroke of serendipity, she used snake venom. Snake venom was known to contain high levels of phosphodiesterase, an enzyme that breaks down nucleic acids. If the survival-promoting activity was provided by a nucleic acid, the addition of snake venom to the culture dish would abolish the ability of the tumor to promote neuronal survival. Not only did it not inhibit the trophic activity of the tumor-secreted factor, but the snake venom itself was found to have potent survival-promoting activity. This suggested that the survival-promoting activity would also be found in a mammalian equivalent, the salivary gland. In fact, the mouse submaxillary gland was found to be a rich source of the survival factor, enabling Levi-Montalcini and Steve Cohen, a biochemist working with her, to purify it to homogeneity and to sequence it. The factor was called **nerve growth factor** (NGF). The addition of purified NGF to explanted sympathetic neurons or sensory ganglia elicited a massive outgrowth of processes from the ganglia. Direct injection of NGF purified from snake venom or mouse submaxillary gland into neonatal rodents produced a dramatic increase in the number of sensory and sympathetic neurons that survived, whereas injection of an antibody directed against the NGF protein led to the death of neurons that normally survive (Fig. 10–4).

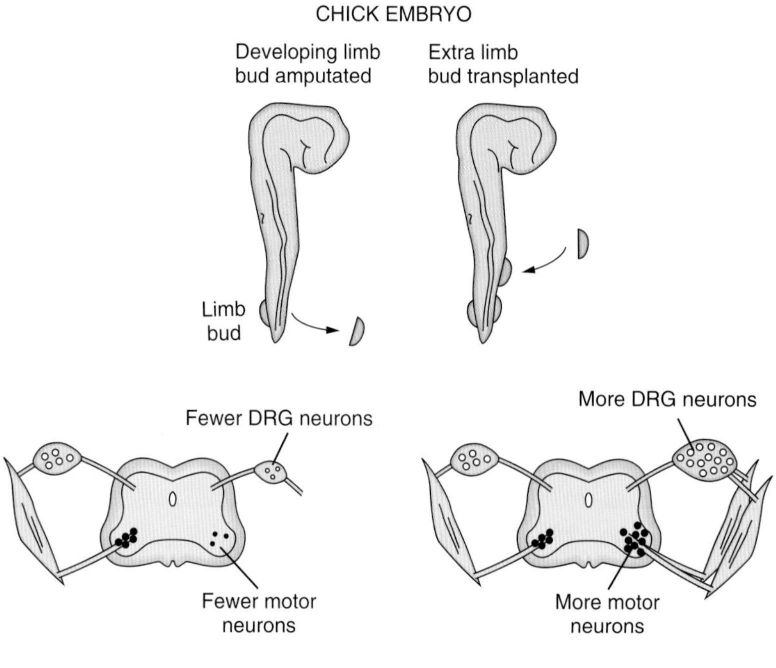

CHICK EMBRYO

Developing limb bud amputated

Extra limb bud transplanted

Limb bud

Fewer DRG neurons

More DRG neurons

Fewer motor neurons

More motor neurons

Figure 10–3. **The survival of motor neurons is regulated by the amount of target tissue.** The chick embryo has been used to test the influence of target tissue on the survival of neurons. The limb bud (which is the target of motor neurons) is surgically removed, or an extra limb bud is transplanted on one side of the embryo. An examination of the embryo several days later shows that there is a substantially reduced number of motor neurons on the side of the spinal cord missing the limb bud (left). In contrast, transplantation of an extra limb bud results in the survival of a greater number of motor neurons than normal (right). Many more sensory neurons (dorsal root ganglion or DRG neurons) are also observed on the side with the extra limb bud. *(Adapted from Hamburger V, 1943.)*

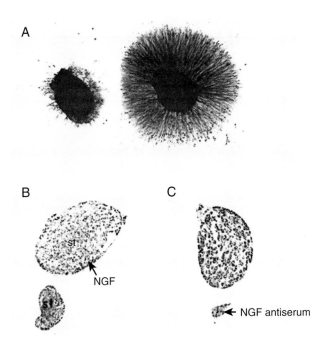

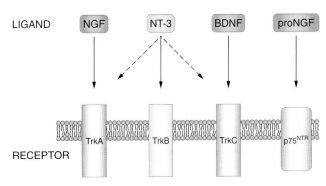

Figure 10–5. **Neurotrophins and their receptors.** Each neurotrophin binds with high affinity to a member of the tyrosine receptor kinase (Trk) family indicated by the *solid arrow* from the ligand to the receptor. Neurotrophin-3 (NT-3) can also bind with lower affinity to both TrkA and TrkB as indicated by the *dashed arrow*. In addition, neurotrophins also bind with lower affinity to a distinct receptor called p75NTR, either alone or in combination with Trks. The unprocessed proform of nerve growth factor (proNGF) has recently been shown to serve as a high-affinity ligand for p75NTR. Such an interaction between proNGF and p75NTR activates a signaling pathway that leads to neuronal death. BDNF, brain-derived neurotrophic factor.

Figure 10–4. **Levi-Montalcini's experiments identifying nerve growth factor (NGF) as a neurotrophic factor. A:** NGF purified from the salivary gland of mice was added to sensory ganglia explanted from chick embryos. The appearance of the ganglia in medium (left) and in medium supplemented with salivary NGF (right) is shown. Addition of NGF results in a robust outgrowth of nerve processes. **B:** Appearance of stellate ganglia (St) from 19-day-old control mice and mice injected with salivary NGF. The ganglion from NGF-administered mice is much larger than normal. **C:** Appearance of superior cervical ganglia from 9-month-old control mice and mice injected for a week after birth with antiserum to salivary NGF. The ganglion from mice administered with NGF antiserum is reduced in size.

With the advent of genetic engineering technology, it was possible to reproduce the effects of antibody treatment by generating mice in which the NGF gene is deleted. The analyses of these mutant mice revealed that not all peripheral and central nervous system neurons are dependent on NGF for their survival during development, raising the possibility that the survival of neurons not responsive to NGF is dependent on other trophic factors. In fact, NGF belongs to a family of structurally and functionally related trophic factors collectively referred to as **neurotrophins.** The other members of this family are brain-derived neurotrophic factor (BDNF), neurotrophin-3 (NT-3), and NT-4/5. Two other neurotrophins called NT-6 and NT-7 are synthesized only in fish. All neurotrophins are synthesized as precursor proteins that are then processed posttranslationally to generate peptides of about 120 amino acids. The biologically active neurotrophin is a homodimer of two identical peptides.

The neurotrophins are not the only trophic factors for neurons. Growth factors such as insulin-like growth factor (IGF-1) and fibroblast growth factor (FGF) can maintain the survival of some neuronal types. Several cytokines including ciliary neurotrophic factor (CNTF), glial-derived neurotrophic factor (GDNF), and hepatocyte growth factor/scatter factor (HGF/SF) also promote neuronal survival.

NEUROTROPHIN RECEPTORS

Like other growth factors, the neurotrophins exert their biologic effects by binding to receptors located on the surface of neurons. Binding studies using ^{125}I-labeled neurotrophins demonstrate two types of receptors: one with a high-affinity binding site and one with a low-affinity binding site. The high-affinity neurotrophin receptor belongs to a family of tyrosine kinases called Trks (for tyrosine receptor kinases). Three Trk proteins have been identified: TrkA, TrkB, and TrkC (Fig. 10–5).

Each Trk receptor has two immunoglobulin-like repeats in the extracellular domain and a tyrosine kinase domain in the cytoplasmic region. The extracellular domains are about 50% homologous, but each Trk displays a specific affinity for one or two neurotrophins. TrkA is activated by NGF, TrkB is activated by BDNF and by NT-4/5, and TrkC is activated by NT-3. Although TrkC is its preferred receptor, NT-3 can also activate TrkA and TrkB. In the absence of ligand, Trk is monomeric. Binding of dimeric neurotrophin to the Trk protein brings two receptor monomers together to form a dimeric receptor-ligand complex. Each subunit within the dimeric Trk receptor phosphorylates one or more tyrosine residues in the cytoplasmic domain of the other subunit, resulting in receptor activation. The low-affinity neurotrophin receptor is a 75-kDa glycoprotein known as p75^{NTR}. It binds the different members of the neurotrophin family with similar affinity, indicating that specificity in the action of each of the neurotrophins comes from the Trk receptor to which it binds. Although not strictly required for the survival-promoting activity of the neurotrophins, p75^{NTR} cooperates with Trk receptors, greatly enhancing ligand binding and phosphorylation, and thus helps neuronal survival. Somewhat unexpectedly, p75^{NTR} can also promote death of neurons when it is bound by NGF in the absence of TrkA. More recent studies have demonstrated that it is the unprocessed form of NGF (proNGF), rather than the mature NGF, that binds p75^{NTR} with high affinity, conferring it with death-promoting activity.

Apoptosis Is Regulated by a Cell-Intrinsic Genetic Program

In general, cell death by apoptosis is achieved by gene activation resulting in the synthesis of proteins that proceed to kill the cell. In fact, treatment with pharmacologic blockers of RNA or protein synthesis can often delay or prevent cell death. Therefore, in contrast with the passive nature of necrotic death, apoptosis is considered to be an active process. Genetic studies in the nematode worm *Caenorhabditis elegans* were critical in the initial characterization of the core components of the cell death program. During the development of this worm, exactly 1090 cells are generated, of which 131 die. Each of the 131 cell deaths occurs at a specific time and location that is essentially invariant from animal to animal. Two genes, *CED-3* and *CED-4*, were identified as genes required for cell death: Inactivating mutations in either of these two genes prevents all developmentally occurring cell deaths in *C. elegans*. This observation also provided the first direct evidence that cells die by an intrinsic suicide program. CED-3 is a protease, whereas CED-4 is a protein that activates CED-3 by interacting with it in dying cells. Another gene, *CED-9*, acts to protect cells from programmed cell death (Fig. 10–6).

CED-9 blocks death by interacting with CED-4, thus preventing it from activating CED-3. Inactivating mutations of CED-9 lead to the death of cells that would normally survive through development; thus, worms with CED-9 mutations die early. Complicating the issue, CED-9 is also expressed in many of the 131 cells that are fated to die during development. How then do these cells die? Death of these cells requires the expression of another gene, *EGL-1*. EGL-1 antagonizes CED-9 by physically interacting with it, thus freeing CED-4 to activate CED-3. Genetic studies have shown

that the overexpression of EGL-1 or mutations that hyperactivate EGL-1 (gain of function mutation) can induce ectopic cell death in *C. elegans,* which can be suppressed by gain-of-function mutation of CED-9 or loss of function mutations of CED-3 or CED-4. In contrast, a loss of function mutation of EGL-1 cannot suppress cell deaths caused by the loss of CED-9 activity (loss of function CED-9). These results are consistent with *EGL-1* acting upstream of the *CED-3, CED-4,* and *CED-9* genes in the genetic pathway regulating cell death.

The identification of key components of the genetic program in the nematode led to the search for their homologs in mammals. These studies have resulted in the identification of a family of proteins, collectively called *caspases*, as the mammalian homologs of CED-3. The homolog of CED-4 in mammals is a protein called Apaf-1, whereas proteins related in sequence to CED-9 comprise the Bcl-2 family. In contrast with the situation in *C. elegans,* however, where CED-9 is antiapoptotic, the Bcl-2 family of proteins is composed of both antiapoptotic and proapoptotic members. Some of these proapoptotic Bcl-2 proteins function as the mammalian counterparts of EGL-1. Consistent with the evolutionary conservation of the mechanisms underlying apoptosis, the overexpression of the nematode proteins CED-3 or CED-4 in mammalian cells results in apoptosis, whereas overexpression of Bcl-2 can compensate for the lack of CED-9 in the nematode. Given their importance to the regulation of apoptosis in mammals, the following section takes a closer look at the caspases and Bcl-2 family of proteins.

EGL-1 ———| CED-9 ———| CED-4 ———▶ CED-3

Figure 10–6. Basic genetic pathway of programmed cell death in *C. elegans.* Activation of *CED-3* and *CED-4* leads to cell death in the nematode. *CED-9* can inhibit cell death by preventing *CED-4* from activating *CED-3.* EGL-1 can promote cell death by binding to *CED-9*, thus displacing it from *CED-4.*

CASPASES

Caspases are a family of cysteine proteases that represent the mammalian homologs of CED-3. Currently, 12 caspases have been identified in humans (only 10 have been identified in mice).

Although members of this family differ in primary sequence and substrate specificity, they are all synthesized and exist normally in healthy cells as zymogens. The procaspases are cleaved to form heterodimers of two small (approximately 10–13 kDa) and two large subunits (approximately 17–21 kDa), which is the active form of the enzyme (Fig. 10–7).

Active caspases recognize a tetrapeptide sequence within substrates cleaving after an aspartic acid residue at the P1 position of the sequence. For example, the preferred recognition sequence for caspase-1 is Tyr-Val-Ala-Asp (Y-V-A-D), whereas the recognition sequence preferred by caspase-3 is Asp-Glu-Val-Asp (D-E-V-D).

Although most caspases function to regulate apoptosis, some members of the caspase family (e.g., caspase-1, -4, and -5) are involved in cytokine maturation. Caspases that regulate apoptosis have been classified into two groups referred to as **initiator caspases** and **effector caspases**. Initiator caspases (e.g., caspase-8, -9, and -10), characterized by long prodomains, have low-intrinsic–enzymatic activity even in their zymogen form. In this state, the initiator caspases are unable to harm the cell. When these procaspases aggregate or oligomerize, however, they cleave and activate themselves. Oligomerization requires the prodomain. Deletion of the prodomain of these initiator caspases by molecular biologic means results in the absence of oligomerization and abolishes their apoptotic activity. Once activated, initiator caspases cleave effector caspases (e.g., caspase-3, -6, and -7), leading to their activation. The effector caspases, also referred to as **executioner caspases,** cleave a variety of cellular substrates, thus destroying normal cellular functions and leading to the demise of the cell. More than 250 cellular substrates for caspases have been identified to date. Among these are proteins involved in scaffolding of the cytoplasm and nucleus, signal transduction and transcription-regulatory proteins, proteins involved in DNA replication and repair, and cell cycle components. The characteristic morphologic features of apoptosis are the direct result of caspase activation. For example, membrane blebbing, a characteristic feature of apoptotic cell death, is caused by caspase-mediated cleavage of the actin-binding protein gelsolin as well as kinases involved in cytoskeletal function, such as ROCK-1 and PAK-2.

Activation of the caspases represents a pivotal step in dooming the cell to death. Two major pathways of caspase-mediated cell death have been described in mammals: the **extrinsic pathway** (also called the death-receptor–mediated pathway), which plays an important role in the maintenance of tissue homeostasis, especially in the immune system, and the **intrinsic pathway** (or the mitochondrial pathway), which is used generally in response to external cues, such as trophic factor withdrawal, and internal insults such as DNA damage (Fig. 10–8).

The extrinsic pathway is initiated at the cell membrane by the binding of extracellular ligands to members of a small family of receptors called *death receptors.* The low-affinity NGF receptor, p75^NTR, is an example of a death receptor that can initiate the extrinsic pathway of cell death. Two other death receptors that have been studied extensively are the tumor necrosis factor receptor-1 (TNF-R1) and the Fas ligand receptor, Fas. These receptors all share a domain of approximately 80 amino acids in their cytoplasmic region, commonly referred to as the **death domain.** Binding of ligand (NGF in the case of p75^NTR or TNF in the case of TNF-R1) to these receptors leads to their activation. The activated receptor now interacts with an adaptor protein (e.g., Fas-associated protein with death domain [FADD]), and via this adaptor protein recruits procaspase-8 to the plasma membrane. The principal role of the adaptor protein is to bring multiple procaspase-8 molecules into close proximity, permitting autocatalytic cleavage to occur, liberating active caspase-8. The receptor-adaptor-procaspase-8 protein complex is called a **death-inducing signaling complex (DISC).**

The intrinsic pathway of caspase activation, in contrast, is initiated by activation of procaspases within the cytosol. Because the mitochondria play an essential role in this mechanism of cell death, this is also called the *mitochondrial pathway.* In response to many death stimuli, cytochrome *c* is released into the cytoplasm from the intramembrane space of mitochondria, where it is normally involved in electron transport. Once in the cytoplasm, cytochrome *c* binds to Apaf-1 (the mammalian homolog of the *C. elegans* CED-4 protein). This facilitates the binding of ATP with Apaf1, altering its conformation, which leads to its oligomerization. Several molecules of procaspase-9 bind to the oligomerized Apaf-1. Such a high-molecular-weight complex of cytochrome *c*, Apaf-1, caspase-9, and ATP is referred to as an **apoptosome.** Proximity-induced aggregation of caspase-9 within the

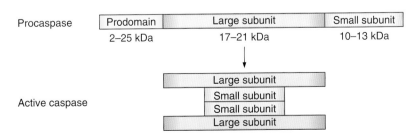

Procaspase

Active caspase

Figure 10–7. Activation of caspases. Caspases are synthesized in their zymogen form requiring proteolytic processing for activation. Processing of the proenzyme at specific aspartate residues removes the N-terminal prodomain. Cleavage between the large and short subunits and association of two heterodimers to form a tetramer results in the generation of active caspase.

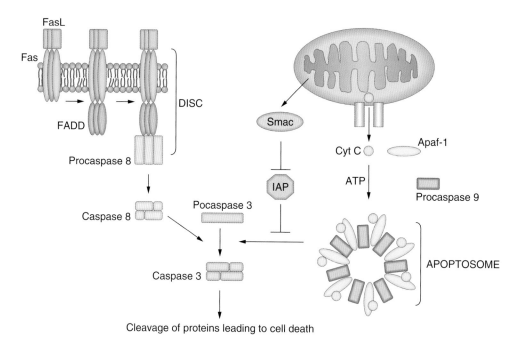

Figure 10–8. Caspase activation by the extrinsic and intrinsic pathways. The extrinsic or death receptor pathway (left) is activated by the binding of a ligand such as FasL to its receptor Fas, followed by the sequential binding of the adaptor protein Fas-associated protein with death domain (FADD) and procaspase-8. The formation of the death-inducing signaling complex (DISC) permits the activation of procaspase-8, which then proteolyzes and activates caspase-8. The mitochondrial pathway (right) also leads to the activation of caspase-3. This pathway involves the release of cytochrome *c* from the mitochondria. Once in the cytoplasm, cytochrome *c* promotes the assembly of Apaf-1 and procaspase-9 into a large complex containing seven molecules each of Apaf-1 and procaspase-9. Active caspase-9 within the apoptosome can cleave caspase-9. In addition to cytochrome *c*, other proapoptotic proteins such as Smac/Diablo are released. Smac/Diablo acts by sequestering inhibitors of apoptosis (IAPS) away from caspases, thus permitting cell death to occur.

apoptosome leads to its self-cleavage and activation. More recent studies have demonstrated that although caspase-9 is cleaved in cells undergoing apoptosis through the mitochondrial pathway, its cleavage is not important for its enzymatic activity and might instead regulate turnover of the active enzyme. In any case, activated caspase-9 cleaves and activates effector caspases such as caspase-3, -6, and -7, which proteolyze a number of cellular proteins leading to cell death.

Although sharing considerable resemblance to cell death in *C. elegans*, apoptosis in mammals shows some important differences. For example, in contrast with Apaf-1, which requires cytochrome *c* for its activation, CED-4 is constitutively active. Therefore, cell death in *C. elegans* depends not on cytochrome *c* release, but on the disruption of the interaction of CED-9 with CED-4. This striking difference is attributed to a motif called the *WD40 repeat*, which is present in Apaf-1 but not in CED-4 (Fig. 10–9). The WD40 repeats suppress the ability of Apaf-1 to interact with procaspase-9. Biochemical experiments have shown that the Apaf-1 protein in which the WD40 domain is deleted is active without cytochrome *c*, resembling CED-4.

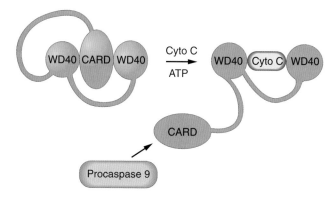

Figure 10–9. Molecular model for the activation of procaspase-9 by Apaf-1. Apaf-1 has 12 or 13 WD40 repeats at its carboxy terminus that act as a negative regulatory region. Under normal conditions, the caspase recruitment domain (CARD) within Apaf-1 is inaccessible to procaspase 9. The release of cytochrome *c* from the mitochondria, in the presence of ATP, enables cytochrome *c* to bind to the WD40 repeats of Apaf-1, altering its conformation to expose its CARD domain. *CED-4* lacks WD40 repeats and does not require cytochrome *c* and ATP for its activation.

Although exactly how cytochrome *c* and other mitochondrial intermembrane constituents are released in mammalian cells remains unclear, two major models have been proposed to explain the release. According to one of these models, exposure of the cell to certain apoptotic signals causes the opening of a pore in the inner membrane of the mitochondrial wall called the **permeability transition pore** (PTP). Opening of the PTP allows water to enter and ions to equilibrate, leading to a dissipation of mitochondrial potential. Mitochondrial swelling caused by entry of water leads to the rupturing of the outer membrane and permits release of cytochrome *c* and other proteins that are located in the intermembrane space. In the second mechanism that has been gaining wide aoceptance, specific proapoptotic members of the Bcl-2 family of proteins (see later), perhaps in association with other proteins, act directly to form pores in the outer mitochondrial membrane through which cytochrome *c* is released.

More recent work has demonstrated that besides the mitochondria, the endoplasmic reticulum (ER) may be a second compartment that participates in the intrinsic pathway of apoptosis. A major function of the ER is to ensure that only properly folded and modified proteins are passed along to their different destinations. However, prolonged stress to the ER, caused by chemical toxicity or by oxidative stress, leads to an accumulation of unfolded proteins and enhanced calcium release from the ER. The calcium released from the ER is taken up by mitochondria, resulting in a dramatic stimulation of cytochrome *c* release from the mitochondria, which then activates caspases. Another molecule that plays an important role in ER stress-mediated apoptosis is caspase-12, a member of the caspase family that is normally localized to the ER. Caspase-12 is activated within the ER after exposure to stress-causing stimuli and translocates to the cytosol, where it directly cleaves procaspase-9 to activate caspase-3.

In addition to the extrinsic and intrinsic pathways, other mechanisms of caspase activation have been described in mammalian cells. For example, cytotoxic T cells can trigger apoptosis of their target cells through a mechanism called the **perforin/granzyme B-dependent pathway**. This pathway, used mainly by the immune system in its surveillance of transformed or virally infected cells, involves a pore-forming protein, perforin, and a serine protease, granzyme B, which is injected into the cytoplasm of the target cell by the activated cytotoxic T cell. Granzyme B can directly cleave and activate procaspase-3, thus inducing apoptosis in the target cell.

Caspase Inhibition

The regulation of apoptosis is subject to a complex system of checks and balances. The activity of caspases can be negatively regulated within cells by endogenously produced proteins. Evidence of the existence of such caspase inhibitory proteins first came from viruses. One of the first viral inhibitors to be identified was CrmA, a protein produced by the cowpox virus. CrmA is structurally similar to serpins, a family of serine protease inhibitors. But rather than inhibiting serine proteases, CrmA targets caspases. Several forms of herpesviruses inhibit apoptosis by producing a protein called v-FLIP (viral-FLICE-inhibitory protein), which binds the adaptor protein FADD, preventing it from recruiting caspase-8 to the TNF death receptor. Baculoviruses synthesize a potent caspase inhibitor called p35, which prevents the death of insect cells that are infected by the virus. In addition to p35, baculoviruses express a second antiapoptotic protein called the **inhibitor of apoptosis (IAP)**. Although bearing no structural relationship to p35 or CrmA/serpins, IAPs are found in mammalian cells and are conserved across species including yeast, nematodes, and *Drosophila*. Mammals express eight different IAP proteins. Structurally, IAPs are characterized by the presence of one or more conserved amino acid stretches referred to as baculovirus inhibitory repeat domains. Although some IAP proteins have other functions unrelated to the regulation of apoptosis, most IAP family members do inhibit caspase activity. IAPs act by binding to the cleaved and active form of certain effector caspases, inhibiting their ability to cleave cellular proteins. IAPs may also bind the proform of caspase-9, blocking its interaction with Apaf-1 and the processing of procaspase-9 into an active enzyme. The activity of IAPs may itself be negatively regulated by other proteins that are released from the mitochondria together with cytochrome *c* during intrinsic apoptosis signaling. One of these proteins has been called Smac (or Diablo). Smac/Diablo binds to the IAPs, preventing their ability to inhibit caspases.

As described earlier, each caspase recognizes a tetrapeptide sequence within its substrates. Researchers have coupled these tetrapeptide sequences to chemical groups such as fluoro-methyl ketones (FMK) and modified the peptides to make them membrane permeable. Such modified but uncleavable tetrapeptides are recognized by caspases and are bound by the enzyme irreversibly. When administered to cells, these pseudo-substrate peptides function as potent and highly selective inhibitors of caspases and can inhibit cell death in tissue culture systems and in living animals in response to a variety of different apoptotic stimuli (Fig. 10–10).

Several caspase-deficient mice have been generated-that exhibit tissue- and cell-specific or stimulus-dependent defects in apoptosis. Mice lacking caspase-9 have a highly enlarged brain and die *in utero*. Caspase-3 is not active in these mice. The brain deformity indicates that caspase-9 is required for the developmentally regulated elimination of neurons that occurs during normal brain development. Not unexpectedly, mice lacking Apaf-1 also have a larger brain and die prenatally. Mice

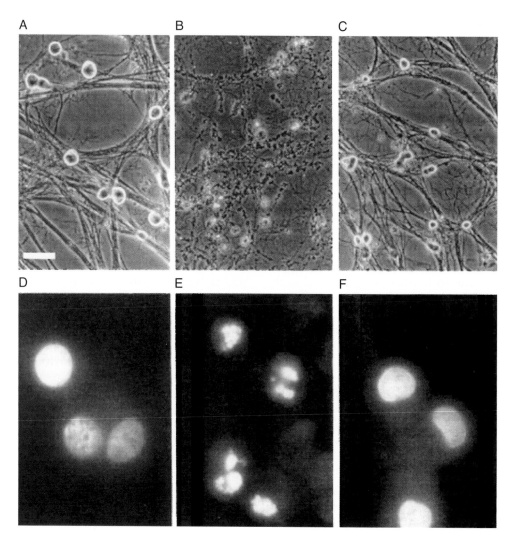

Figure 10–10. Inhibition of caspases protects neurons from death. Sympathetic neurons cultured from the superior cervical ganglion of rats or mice will survive in tissue culture medium supplemented with nerve growth factor (NGF). Removal of NGF from the medium leads to apoptosis. **A–C:** Images from phase-contrast microscopy. When maintained in the presence of NGF (**A**), the neurons appear bright and their processes are smooth and abundant. When deprived of NGF (**B**), the cell bodies and the processes of the neurons degenerate. Neuronal cell death induced by NGF deprivation is prevented if a pan-caspase peptide inhibitor, z-VAD-fmk, is added to the culture medium (**C**). **D–F:** Images on nuclei stained with a fluorescent DNA-binding dye. Whereas chromatin staining is diffuse within the nuclei of neurons receiving NGF (**D**), chromatin within the nuclei of cultures deprived of NGF is condensed and fragmented (**E**), which are hallmarks of apoptotic cell death. Inhibition of caspases by the addition of z-VAD-fmk (**F**) prevents the condensation and fragmentation of chromatin even in the absence of NGF. *(From Deshmukh, et al. J Cell Biol 1996;135:1341–1354, by permission.)*

lacking caspase-3 are smaller than normal, display ectopic masses containing excess neurons in several brain regions, and only survive to 3 weeks of age. The milder phenotype of the caspase-3 knockout mice as compared with the caspase-9 knockout mice is likely to be due to the functioning of other caspase-9–activated effector caspases, such as caspase-7. Caspase-8 and FADD-deficient mice have profound cardiac defects and die *in utero*, indicating that the extrinsic pathway of caspase activation is critical for embryonic development.

BCL-2 PROTEINS

The Bcl-2 protein family is named after Bcl-2, which was the first member of the family to be discovered. Members of the Bcl-2 protein family are critical apoptosis regulators that act by controlling the release of cytochrome *c* and other apoptotic factors such as Smac/Diablo from the mitochondria. Humans have more than 20 Bcl-2 family proteins. All Bcl-2 family proteins possess at least one of four conserved Bcl-2 homology (BH) domains, designated BH1-4. It has been found that

Bcl-2 proteins that possess all four BH domains protect cells from apoptosis (Fig. 10–11).

These antiapoptotic members generally also possess a transmembrane domain and are most often localized to the mitochondria membrane, where they serve to block cytochrome c release. The antiapoptotic Bcl-2 proteins are regarded as the homologs of *C. elegans* CED-9. This group of Bcl-2 proteins is generally referred to as the Group I Bcl-2 proteins. Two of the best studied members of the Group I family are Bcl-2 itself and Bcl-x_L. Proapoptotic Bcl-2 proteins, in contrast, come in two forms: the multidomain members that lack the BH4 domain (referred to as the Group II family), and those that have only the BH3 domain (referred to as the Group III family). In response to a death signal (e.g., the lack of NGF for certain types of neurons), proapoptotic Bcl-2 family proteins that reside in the cytoplasm translocate to the mitochondrial membrane, where they undergo a conformational change, oligomerize, and insert themselves into the membrane to form pores through which cytochrome c is released. Two multidomain proapoptotic Bcl-2 proteins called Bax and Bak are particularly crucial for cell death because cells lacking these two proteins fail to undergo apoptosis in response to a wide range of apoptotic stimuli. Antiapoptotic proteins such as Bcl-2 and Bcl-x_L heterodimerize with Bax and Bak, preventing the formation of functional pores. Although some BH3-only proteins promote apoptosis by heterodimerizing with the antiapoptotic Bcl-2 proteins, thus inhibiting them, others exert their proapoptotic action by enhancing pore formation activity of Bax and Bak. In general, therefore, the relative balance between the levels of antiapoptotic and proapoptotic Bcl-2 proteins may be an important determinant of whether a cell lives or dies.

As observed in other cell types, Bcl-2 proteins are critical regulators of neuronal apoptosis. Overexpression of Bcl-2, the founding member of the family and one that is antiapoptotic, protects against trophic factor-deprivation–induced death of cultured neurons. Mice engineered to overexpress Bcl-2 have more than normal numbers of neurons in many brain areas. In contrast, mice lacking Bcl-2 die within a few months of birth. Whereas apoptosis occurs to the normal extent during brain development, all neuronal types degenerate postnatally in these mice, suggesting that Bcl-2 might be important to maintain survival of neurons after birth. These mice also suffer extensive apoptosis in the thymus and spleen, indicating the importance of Bcl-2 in organs outside the nervous system.

The normal survival of neurons during development in Bcl-2–deficient mice points to the involvement of other members as being more important for the maintenance of neuronal survival before birth. One such member is Bcl-x_L. Mice lacking Bcl-x_L display massive cell death of immature neurons in several different areas of the developing central and peripheral nervous systems and die embryonically. In contrast, mice lacking the proapoptotic protein Bax display increased numbers of neurons in several brain regions. These findings suggest unique roles for different Bcl-2 proteins, with the opposing actions of Bcl-x_L and Bax controlling the generation of the proper numbers of neurons during development, and Bcl-2 being responsible for the survival of neurons after birth.

Many viruses encode functional Bcl-2 homologs. Among the best studied of these is the E1B 19-kDa protein encoded by adenovirus. Host cells infected with viruses are recognized by the immune system leading to the release of many inflammatory cytokines including

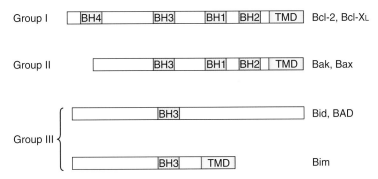

Figure 10–11. **Bcl-2 family of proteins.** From functional and structural criteria, the Bcl-2 family has been divided into three groups. Members of the Group I family are antiapoptotic and possess all four BH domains (BH1-4). Most Group I Bcl-2 proteins also have a C-terminal transmembrane domain (TMD) with which they attach to various intracellular membranes such as the mitochondrial membrane. Group II Bcl-2 proteins are strongly proapoptotic and lack the N-terminal BH4 domain. Group III consists of a large group of proapoptotic proteins that contain the BH3 domain only. Some members also sport a TMD. The figure also lists representative members of each subfamily.

TNF. Through activation of its receptor, TNF induces apoptosis in the infected cell. The ability of TNF to kill the host cell is blocked by the E1B 19-kDa protein, thus giving adenovirus a specific replicative advantage by evading one arm of the immune response.

Engulfment of the Apoptotic Cell

Perhaps the most important aspect of apoptosis is the rapid phagocytosis of the dying cell. The engulfment process is so closely coordinated with the events occurring within the cell primed to die that it is often regarded as part of the cell death program itself. The engulfing cell may use one or more of a range of possible recognition mechanisms, whereas apoptotic cells may display their "eat me" signal in different ways. For example, early in the apoptotic process and as a result of membrane blebbing, apoptotic cells display on their surfaces molecules such as phosphatidylserine, a phospholipid that is normally present on the cytoplasmic side of the cell membrane. Many different cell-surface receptors in the phagocytosing cell, such as certain scavenger receptors and integrins, can recognize and respond to the eat-me signals. Specific proteins in serum have been identified that can bind to phosphatidylserine and act as extracellular bridging molecules linking the apoptotic cell to the cell engulfing it.

In *C. elegans,* six genes have been identified that are required by the engulfing cell for phagocytosis of apoptotic cells. One of these genes encodes a protein called CED-7, which is localized to the plasma membrane and may be involved in the recognition of signals on the dying cell. Another protein, CED-5, is required to envelope apoptotic cells by extension of the cell surface of the engulfing cell, presumably through reorganization of the cytoskeleton. Indeed, mutation of proteins within the engulfing cell that blocks its ability to phagocytose sometimes allows cells that would normally die to survive instead. The more elaborate genetic mechanisms regulating the phagocytosis of apoptotic cells in mammalian systems are poorly understood but are actively being deciphered.

SIGNALING PATHWAYS THAT PROMOTE CELL SURVIVAL

Survival factors such as neurotrophins act by suppressing the activation of caspases and other proapoptotic molecules. In the case of neurotrophins and other related growth factors, binding to their receptors leads to receptor dimerization and autophosphorylation of the receptor on specific tyrosine residues. The cytoplasmic domain containing these phosphorylated residues serves as a docking site for a variety of proteins, including those that regulate the activity of the phosphatidylinositol 3-kinase (PI3K)/Akt and the Raf/mitogen-activated

protein kinase (MEK)/extracellular signal-regulated kinase (ERK) pathways. These two signaling pathways are responsible for the majority of survival signaling in most other cell types.

Phosphatidylinositol 3-Kinase/Akt Signaling Pathway

Although activated by different stimuli, the PI3K/Akt pathway has been best studied in the context of growth factor signaling. PI3K is a lipid kinase, which is composed of a regulatory subunit, p85, and a tightly associated catalytic subunit, p110. The binding of neurotrophic growth factors to their cognate receptor tyrosine kinases (RTKs) leads to the recruitment of the regulatory subunit together with its associated p110 catalytic subunit (Fig. 10–12). Recruitment can be direct or, more often, involve other proteins such as the small guanosine triphosphatase (GTPase) Ras and certain adaptor proteins such as Shc and Grb-2.

Active PI3K phosphorylates phosphatidylinositol 4,5-bisphosphate, PI(4,5)P2, to phosphatidylinositol 3,4,5-bisphosphate, PI(4,5)P3, within the membrane. PI(4,5)P3 binds to the pleckstrin homology (PH) domains of target proteins, thus recruiting them to the membrane. A key target of PI3K action in the promotion of cell survival is the serine-threonine kinase Akt (also called protein kinase B). Over the past few years, a number of proteins have been identified as substrates of Akt that are involved in the regulation of apoptosis. One class of Akt substrates is proapoptotic proteins that are inactivated when phosphorylated by Akt. Among these are the members of the FOXO family of Forkhead-related transcription factors. Phosphorylation by Akt sequesters FOXO in the cytosol. In the absence of survival factors, when Akt is inactive, FOXO translocates to the nucleus and activates the expression of proapoptotic genes. Another proapoptotic protein phosphorylated by Akt is the proapoptotic Bcl-2 family member BAD. When it is not phosphorylated, BAD will inhibit Bcl-x$_L$ and other antiapoptotic Bcl-2 family members at the mitochondria by direct binding. Once phosphorylated, however, BAD is bound with high affinity by 14-3-3 proteins, thus localizing BAD to the cytosol and effectively neutralizing its apoptotic activity. Although phosphorylation of FOXO and BAD increases their affinity for interaction with 14-3-3 proteins, phosphorylation of certain proapoptotic proteins, such as glycogen synthetase kinase (GSK)-3β, inhibits their enzymatic activity. Depriving cells of survival-promoting trophic factor leads to the inactivation of Akt and, as a consequence, the dephosphorylation of Gsk-3β, which results in its activation and the unleashing of its proapoptotic activity.

A second class of proteins that is phosphorylated by Akt has antiapoptotic functions. In this case, Akt-

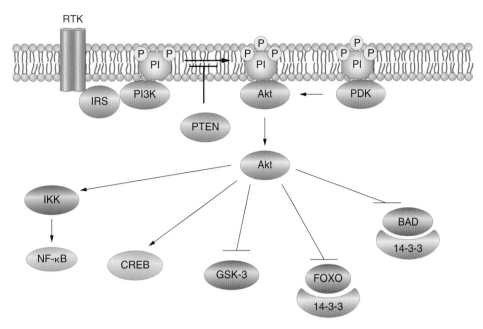

Figure 10-12. The phosphatidylinositol 3-kinase (PI3K)/Akt signaling pathway. Activation of receptor tyrosine kinases (RTKs) leads to the recruitment and activation of PI3K, either directly or via adaptor proteins (such as insulin receptor substrate-1 [IRS]). PI3K catalyzes the phosphorylation of PI(4,5)P2 to PI(3,4,5)P3, leading to the recruitment of Akt and 3-phosphoinositide-dependent protein kinase-1 [PDK-1] to the membrane, where Akt is activated. Once activated, Akt phosphorylates and inactivates several proapoptotic proteins such as GSK-3β, FOXO, and BAD. Phosphorylation of FOXO and BAD leads to their sequestration in the cytoplasm by 14-3-3, preventing their translocation to the nucleus and mitochondrial membrane, respectively. Akt phosphorylates IκB kinase kinase (IKK) and the antiapoptotic protein, CREB, leading to their activation. Phosphorylation of IKK results in activation of nuclear factor-kappa B (NF-κB), a protein with strong antiapoptotic effects. The action of PI3K is opposed by PTEN (pentaerythritol tetranitrate), a lipid phosphatase that converts PI(3,4,5)P3 back to PI(4,5)P2.

mediated phosphorylation leads to their activation. For example, Akt phosphorylates and activates the transcription factor CREB (cAMP response element binding) protein. Phosphorylated CREB enters the nucleus and increases expression of antiapoptotic proteins, such as Bcl-2. Akt also phosphorylates the IκB-kinase (IKK), a protein whose activation leads to stimulation of the activity of the nuclear factor-kappa B (NF-κB) transcription factor. NF-κB plays a pivotal role in promoting the survival of many different cell types.

Raf/MEK/ERK Signaling Pathway

The Raf/MEK/ERK pathway is a mitogen-activated protein kinase (MAPK) signaling cascade that is best known for its role in cell proliferation. This pathway is also important for the survival of many cell types. As described in Chapter 8, members of the different MAPK families (which also include the JNK and p38 MAPKs) are activated through a chain of sequentially activated serine-threonine kinases. Activation of the Raf-MEK-ERK signaling pathway actually begins with the interaction of the GTP-binding protein Ras with the activated

growth factor, cytokine, or G-protein–coupled receptor in the plasma membrane (Fig. 10–13).

Other adaptor proteins (such as Shc, Grb, and SOS) are also involved in this process. GTP-Ras then recruits Raf, a serine-threonine kinase, to the plasma membrane where it is activated. On activation, Raf phosphorylates MEK (for MAP kinase or ERK kinase), which, in turn, phosphorylates and activates extracellular signal-regulated kinases (ERKs). In cells in which the Raf/MEK/ERK pathway sustains survival, ERK activation can also lead to the activation of CREB. Activated ERK also promotes survival by phosphorylating BAD, as described earlier for Akt. Yet another target of ERK may be procaspase-9. Phosphorylation of procaspase-9 inhibits its processing and subsequent caspase-3 activation, thereby blocking the caspase cascade during apoptosis.

Depending on the cell type and the survival-promoting stimulus, either the PI3K/Akt or the Raf/MEK/ERK pathway may be activated. These signaling pathways promote cell survival by acting on common molecules such as BAD and CREB, as well as distinct ones such as IKK or GSK-3β, which are selective for the PI3K/Akt signaling pathway.

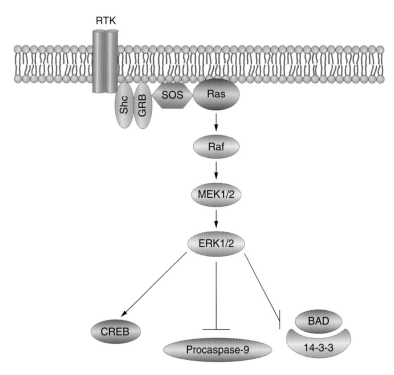

Figure 10–13. The Raf/Mitogen-Activated Protein Kinase (MEK)/ Extracellular Signal–Regulated Kinase (ERK) pathway. Activation of the GTPase Ras occurs after stimulation of receptor tyrosine kinases (RTKs), a process that also involves adaptor proteins such as Shc, Grb, and Sos. Once activated, Ras activates Raf, which turns on MEK, which, in turn, activates ERK. ERK activation leads to the phosphorylation of procaspase-9 inhibiting its activation, inactivating it. ERK also phosphorylates BAD, leading to its sequestration in the cytoplasm by 14-3-3. Finally, ERK activation leads to the activation of CREB, a protein with antiapoptotic effects.

Clinical Case 10-1

Peter Gregory is a 68-year-old obese Manhattan cab driver who had "never been ill a day in my life," until 3 months ago. At that time, his left big toe was sore, and he had trouble suddenly braking to avoid a skateboarder at 42nd Street. His "stomach" was bruised by the bottom of the steering wheel. He thought nothing of this until that evening, when his wife felt that "his stomach was too hard." Thinking about it, he noted that although he had lost about 10 pounds in the past 2 months, his pants were tighter than usual. The following day he saw his neighborhood general practitioner, who found his physical examination normal except for some pallor and a large, firm, nontender mass in his upper left quadrant descending on respiration more than a handbreadth below the ribcage. Believing that this was an enlarged spleen, and that Mr. Gregory "looked a little peaked," the physician drew a blood sample and put the tube aside in a cup while he discussed insurance forms with

Mr. Gregory. Fifteen minutes later, when he returned to pick up the blood, he made the diagnosis.

Cell Biology, Diagnosis, and Treatment of Chronic Myelogenous Leukemia

Mr. Gregory has chronic myelogenous leukemia (CML). The sample tube left standing in the cup had surprisingly separated into plasma and cellular fractions by gravity alone in just 15 minutes, almost as if it had been centrifuged. Moreover, the red cell fraction was small, and instead of a faint, thin "buffy coat" layer between the red cells and the plasma, a thick cloud of greenish-yellow white cells was filling most of the plasma space. The physician rapidly confirmed his diagnosis with a simple blood count, which showed a hematocrit of 17% and a white cell count of 190,000. Because he still had his medical school microscope, he went one step further and made a blood smear, which showed that most of the white cells were mature-looking polys, with a few large, round cells.

More sophisticated studies at the hospital showed a myeloblast count of 6%, a white cell leukocyte alkaline phosphatase level of less than 10, and 600,000 platelets. The uric acid concentration was 10.2. A bone marrow examination was hypercellular, with extensive myeloid hyperplasia; and a quick "squash" preparation of the fresh marrow specimen showed a tiny atypical chromosome fragment. Later, fluorescent *in situ* hybridization stains of the marrow sample specifically demonstrated the Philadelphia chromosome, and polymerase chain reaction analysis was positive for the BCR-ABL site of the reciprocal translocation of chromosomes 9 and 22.

Now, 3 months after treatment with the tyrosine kinase inhibitor imatinib, Mr. Gregory's pants fit, his toe does not bother him, and he is picking up fares in Times Square.

APOPTOSIS AND HUMAN DISEASE

Although critical for normal tissue homeostasis, the failure of cells to regulate apoptosis appropriately can have severe consequences to the health of the organism. Diseases involving deregulated apoptosis can be divided into two groups: those in which apoptosis fails to occur, resulting in abnormally prolonged cell survival, and those in which apoptosis occurs prematurely, leading to increased and undesirable cell death. Among the diseases resulting from insufficient apoptosis are a variety of cancers. Although initially believed to be a consequence of uncontrolled cell proliferation, it is now well established that many types of cancers result from the inability of cells to undergo apoptosis. Bcl-2 was originally identified as a protooncogene whose chromosomal translocation is responsible for a variety of lymphomas. Increased Bcl-2 expression has been observed in many types of cancers and is currently considered to be a predictive factor for worse prognosis in prostrate and colon cancer. Mice engineered to overexpress Bcl-2 specifically in the B-cell lineage develop follicular lymphoma at 12 months old, providing proof that cancers can arise solely from the inhibition of apoptosis. Akt was also initially identified as a protooncogene. Gene amplifications and increased activity of Akt are found in several human cancer diseases. Blockade of Akt signaling within the cell using molecular biologic approaches can reverse the transformed cellular phenotype in tissue culture models of oncogenesis.

A majority of chemotherapeutic drugs used to treat human cancers work by activating the apoptotic machinery in tumor cells. An example of this is imatinib (Gleevec), a drug that is used in the treatment of CML, a cancer of the blood caused by the chromosomal translocation and overexpression of the c-Abl oncogene. Gleevec acts by inhibiting survival-promoting pathways activated by c-Abl, thus inducing apoptosis in the cancerous cells. Disturbingly, the resistance of some cancers to conventional chemotherapeutic drugs and irradiation is also due to their enhanced resistance to apoptosis. For instance, in malignant melanoma, a deadly cancer that fails to respond to conventional chemotherapy, reduced expression of Apaf-1 has been observed. Restoration of Apaf-1 expression by treatment with the methylation inhibitor 5-aza-2'-deoxycytidine enhances the sensitivity of cultured malignant melanoma cells to standard chemotherapeutic drugs and rescues the apoptotic defects in them. The IAPs are also an attractive target for cancer therapy. Preclinical studies have shown that reducing the levels of specific IAP proteins by the expression of antisense oligonucleotides against their messenger RNA reduces tumor growth.

Clinical Case 10-2

Ruth Daly is a 75-year-old retired real estate agent who was well until 2 months ago, when, after spending several hours at the beach with her grandchildren, she became dizzy while driving home. She stopped the car, noticed that she was not exactly sure where she was, and became aware of her heart "beating strangely." She fell asleep for about 5 minutes, and when she awoke felt "pretty good, but all worn out." When she got home, a half hour later, she felt fine.

Today, Ruth was taking a walk with her dog in her own neighborhood, when she suddenly realized she did not know where she was. Her dog led her to the door of her next-door neighbor, a retired gynecologist. When he answered her knock, she vaguely recognized him and was able to give a garbled version of her confusion during the previous 5 minutes. He found it difficult to understand her slurred and incomplete words, and he immediately called for an ambulance to take her to the nearby university hospital. He then gave her a glass of water and an aspirin.

At the emergency ward 18 minutes later, Ruth was not speaking coherently and could not move her right arm. Her blood pressure was 190/140 mm Hg; her pulse was 90 and irregular. A quick panel of laboratory tests was normal except for a blood sugar level of 290.

Ruth's previous medical problems were limited to mild diabetes and hypertension. The diabetes had developed 15 years earlier and was well controlled with one dose of long-acting insulin daily. Her hypertension was controlled in the range of 130/85 mm Hg with a diuretic and low-dose atenolol. She also took a statin drug each evening to control her mild hyperlipidemia.

Physical examination showed that, except for her pulse, her heart and lungs were unremarkable. There

Continued

was no bruit over either of her carotid arteries. Her abdomen was moderately obese, but she had no organomegaly. Her legs showed no edema, and her calves were soft and nontender. Ruth's neurologic examination showed that though conscious, she was disoriented in time and place, and her right arm was completely paralyzed and insensitive. She could move her right leg, but only slowly, and pinprick sensitivity was markedly reduced.

The emergency department physicians diagnosed an acute stroke and ordered an immediate computed tomography (CT) scan of her brain. This showed no acute changes, and no evidence for old or new hemorrhage. Brain magnetic resonance imaging (MRI) was scheduled for the following day. Before any further studies, she was started on an intravenous (IV) infusion of recombinant tissue plasminogen activator (tPA). The total time elapsed from the neighbor's call for the ambulance was 2 hours and 10 minutes. She was then given IV insulin to reduce her blood sugar to normal, and put on a cooling blanket to reduce her temperature as much as possible.

Cell Biology, Diagnosis, and Treatment of Cerebral Thrombosis

Mrs. Daly had a stroke involving a small branch of her left midcerebral artery. It was probably embolic and presaged by a transient ischemic attack consequent to atrial fibrillation, which began on the way home from the beach. On her second hospital day, an initial diffusion-weighted MRI of her brain showed a small area of diffusion defect in this region. Because of this finding, a subsequent gadolinium-contrast perfusion study was performed. This showed an appreciably larger area of reduced perfusion surrounding the same site. This was interpreted as an "ischemic penumbra" by the radiologist.

The following day, Ruth's dysarthria and disorientation were markedly improved, and the tPA infusion was discontinued. Most of her right-sided weakness gradually abated over the next 3 weeks. Three months later, after extensive recuperative physical and occupational therapy, she was able to return to her own home.

Ruth had minimal weakness remaining in her right arm, but her leg was fine, her speech had returned to normal, and she was again able to play with her grandchildren.

Decreased apoptosis also contributes to certain autoimmune diseases. During the normal development of the immune system, self-reactive lymphocytes are eliminated by apoptosis. Failure of these self-reactive lymphocytes to die causes autoimmunity. Some autoimmune diseases such as lupus erythematosus are associated with mutations in proapoptotic molecules rendering the autoreactive lymphocytes resistant to apoptosis. Devel-

opmental defects such as cleft lip are also caused by the failure of cells to undergo apoptosis during embryogenesis.

Premature or excessive apoptosis, in contrast, occurs in neurodegenerative diseases such as Alzheimer's disease, Parkinson's disease, amyotrophic lateral sclerosis (ALS), and retinitis pigmentosa. Apoptosis also plays an important role in ischemic stroke and in degeneration of neurons caused by neurotoxic agents. Although the immediate death of neurons in strokes, such as cerebral thrombosis (a type of stroke caused by the development of a clot within a blood vessel of the brain), is necrotic in nature, much of the long-term neurologic damage suffered by stroke patients is due to apoptotic neurodegeneration that occurs around the necrotic lesion caused by the stroke. Prevention of this delayed apoptotic death would greatly reduce the neurologic deficits in such patients.

Human clinical trials are under way to examine whether infusion of neurotrophic factors into the brains of patients with neurodegenerative diseases can slow down the loss of neurons by activating the survival-promoting signaling pathways within vulnerable cells. Similarly, chemical drugs that block apoptosis in tissue culture and animal models of neurodegeneration are being tested in clinical trials involving patients with Parkinson's disease and other neurologic pathologies. Familial ALS, a progressive disorder associated with the death of motor neurons in the spinal cord and brain, is often caused by mutations in the gene encoding Cu/Zn superoxide dismutase-1 *(SOD1)*. Mice that are genetically engineered to express the mutated human *SOD1* gene acquire a motor neuron degenerative pathology resembling ALS. The mutant mice suffer paralysis similar to humans with ALS and have therefore served as useful models for understanding the mechanisms underlying disease pathogenesis and for developing potential treatment strategies. If Bcl-2 is overexpressed in these "ALS mice," disease symptoms are delayed. Likewise, the administration of a pharmacologic caspase inhibitor or the overexpression of IAP proteins in these mice attenuate disease progression and increase life span. These encouraging results raise the possibility that drugs and strategies that block apoptosis may also be useful in the treatment of ALS in humans, as well as in the treatment of other neurodegenerative pathologies.

Outside the nervous system, increased apoptosis has also been implicated in the pathologic depletion of $CD4^+$ cells in HIV-infected patients and in diseases that involve tissue damage such as myocardial infarction, congestive heart failure, renal damage, and cirrhosis of the liver.

SUMMARY

Apoptosis is a morphologically and biochemically distinct form of programmed cell death that plays an essential role during embryologic development, after birth, and during adulthood. However, deregulation of apoptosis is involved in the pathogenesis of a variety of human diseases. Since the late 1990s, the core components of the mammalian apoptotic machinery have been identified, and much information on how this complex machinery is regulated has been gathered. Current work is focused on completely unraveling the mechanisms regulating this process and using our knowledge in the development of therapies for a variety of human diseases.

Suggested Readings

Apoptosis and Disease

Fadeel B, Orrenius S. Apoptosis: a basic phenomenon with wide-ranging implications in human disease. *J Int Med* 2005; 258:479–517.

Caspases and Bcl-2 Proteins

Cory S, Adams JM. The Bcl2 family: regulators of the cellular life-or-death switch. *Nat Rev Cancer* 2002;2:647–656.

Jiang X, Wang X. Cytochrome C-mediated apoptosis. *Annu Rev Biochem* 2004;73:87–106.

Scorrano L, Korsmeyer SJ. Mechanisms of cytochrome c release by proapoptotic BCL-2 family members. *Biochem Biophys Res Comm* 2003;304:437–444.

Neurotrophic Factors and Signaling Pathways

Cantrell DA. Phosphoinositide 3-kinase signalling pathways. *J Cell Sci* 2001;114:1439–1445.

Ip NY, Yancopoulos GD. The neurotrophins and CNTF: two families of collaborative neurotrophic factors. *Annu Rev Neurosci* 1996;19:491–515.

Kaplan DR, Miller FD. Neurotrophin signal transduction in the nervous system. *Curr Opin Neurobiol* 2000;10:381–391.

Programmed Cell Death

Jacobson MD, McCarthy N, eds. *Apoptosis*. New York: Oxford University Press, 2002.

Kaufmann SH, Hengartner MO. Programmed cell death: alive and well in the new millennium. *Trends Cell Biol* 2001;11:526–534.

Metzstein MM, Stanfield GM, Horvitz HR. Genetics of programmed cell death in *C. elegans:* past, present and future. *Trends Genet* 1998;14:410–416.

Index

Exam Preparation and Review

CD-ROM
for
Windows®

- Simulates the actual board examination experience
- Draws from the required basic and clinical science medical subjects
- Provides detailed feedback on strengths and weaknesses
- Create your own custom exam and study sessions
- Thousands of questions and explanations

Prep for the
USMLE
UNITED STATES MEDICAL LICENSING EXAMINATION*

Fast
Efficient
Affordable

EXAM MASTER®

"If you can do EXAM MASTER,
you can do the boards!"

http://www.exammaster.com/usmle/index.html

*USMLE is a joint program of The Federation of State Medical Boards of the U.S., Inc. and the National Board of Medical Examiners.